PHYSICAL GEOLOGY

PHYSICAL

Allan Ludman and Nicholas K. Coch
Department of Earth and Environmental Sciences
Queens College of City University of New York

McGraw-Hill Book Company
New York St. Louis San Francisco Auckland Bogotá Hamburg
Johannesburg London Madrid Mexico Montreal New Delhi
Panama Paris São Paulo Singapore Sydney Tokyo Toronto

GEOLOGY

PHYSICAL GEOLOGY

1 2 3 4 5 6 7 8 9 0 R M R M 8 9 8 7 6 5 4 3 2 1

This book was set in Times Roman by Ruttle, Shaw & Wetherill, Inc.
The editors were Jay Ricci, Sibyl Golden, Janet Wagner, and Scott
Amerman; the designer was Merrill Haber; the production supervisor
was Dennis J. Conroy. The photo editor was Mira Schachne.
The drawings were done by J & R Services, Inc.
Rand McNally & Company was printer and binder.

Cover photo: "Slot Canyon," Dewitt Jones Productions, Inc.

Library of Congress Cataloging in Publication Data

Ludman, Allan.
    Physical geology.

    Includes bibliographies and index.
     1. Physical geology. I. Coch, Nicholas K.
II. Title.
QE28.2.L82   551   81-4552
ISBN 0-07-011510-9   AACR2

To Carol, Elaine, Jessica, and Kenneth

Contents in Brief

Contents

Preface

In the last 25 years, there has been a revolution in both the scope and the philosophy of geology that has changed forever the nature of the science. Geologists no longer are restricted to the 25 percent of planet earth that is dry land—indeed, we are not even confined to our own planet. Today we study rocks and sediments from the deepest parts of the oceans and from the surface of the moon; compare glaciers in Antarctica with those on Mars; and try to explain volcanic eruptions in the state of Washington and on Io, the second satellite of Jupiter.

With this increased scope has come a new understanding of the way in which our planet works. The once-ridiculed concept of moving continents is now seen as the most logical process by which earth's present geography has evolved. As we have learned more about the earth, we have become more aware of the environment in which we live. Geology began as a "relevant" science devoted to finding mineral resources and fuels. To this role we now add the responsibility for planning our interaction with natural processess in ways that will preserve our environment for future generations.

We have used a traditional outline of topics in our book, and have worked into it two threads of thought that help tie material together in a comprehensible and, we hope, interesting way. The first of these is the relevance of geology to everyday life and the second is the use of the scientific method to solve problems in geology. In our teaching, we have found that students appreciate geologic processes and materials more if they can understand their application to everyday life. Thus, we have tried to show, *in every chapter,* how such apparently academic topics as bonding in minerals and the angle of sediment repose affect human lives. In an attempt to show students *how* geologists think in addition to what we work with, we present five hypotheses for the origin of earth's mountains and oceans in the first chapter, and in subsequent chapters we examine the evidence favoring each of the models.

The first chapter introduces the scope of physical geology and begins application of the scientific method to geologic problems. Chapters 2 through 7 deal with minerals and rocks—the materials of which earth is made—and with the energy needed for earth processes. In Chapter 8 we introduce geologic time and discuss the ways in which the ages of earth materials are determined. In Chapters 9 through 16 we describe the processes by which gravity, water, wind, and ice move sediments at the earth's surface and in its oceans, and show how these processes form the earth's landscapes. Chapters 17 and 18 describe how rocks are deformed and raised into mountains. Chapters 19 and 20 deal with earthquakes, seismology, gravity, and magnetism, and show how geologists interpret the composition of, and the processes active in, the earth's interior. In Chapter 21 we evaluate the five hypotheses set out in Chapter 1 and present a model for earth's evolution that best fits the present evidence.

Two guest authors have written chapters at the end of the book. In Chapter 22, David Speidel of Queens College, City University of New York, discusses the origins of earth resources and their present consumption and future reserves.

In Chapter 23, Jeffrey Warner of NASA shows what the space program has told us about our neighboring planets and what they in turn tell us about the early evolution of the earth.

We have included a variety of features that will arouse students' interest and help them learn. Self-contained illustrated *perspective sections* in many of the chapters present fascinating related topics such as the Galapagos Rift and weathering on the moon in more detail than could be included in the main body of the text. Unique *diagrams* and numerous *photographs* illustrate the concepts discussed. *Color plates* have been chosen not only for their beauty but also for their ability to depict key geological points. *New terms* are indicated in boldface type and defined when they are first mentioned, and each is also included in a *glossary* at the end of the book. Chapter *summaries* and end-of-chapter *questions* are designed both for review and for further thought. To aid students who wish to learn more about the topics discussed in each chapter we have included an *annotated list of references* at the end of each chapter and have ranked them according to their difficulty. *Appendixes* on metric units, mineral identification, and soil classification provide detailed information that will be of use to many students. Finally, an *instructor's manual* prepared by the authors is available from the publisher.

We have shared equally in the preparation of this book and our names appear in the order given on the title page on the basis of a coin toss. Out task was made easier through the assistance of many people. Preliminary art sketches were prepared by Howard Craig, Susan Knapp, and Kathy Tonnies. Mary O'Shea typed the numerous drafts of the manuscript. These various revisions were reviewed in whole or in part by Robert W. Baker, University of Wisconsin at River Falls; Victor R. Baker, University of Texas at Austin; Robert Behling, West Virginia University; Michael Bikerman, University of Pittsburgh; William H. Blackburn, University of Kentucky; Robert E. Boyer, University of Texas at Austin; Charles W. Byers, University of Wisconsin; Richard Enright, Bridgewater State College; William R. Farrand, University of Michigan; Irving S. Fisher, University of Kentucky; Ralph Gram, San Jose State University and California State University at Hayward; Warren D. Huff, University of Cincinnati; Mead L. Jensen, University of Utah; Donald H. Lindsley, State University of New York at Stony Brook; Marshall E. Maddock, San Jose State University; Donald B. Moore, Contra Costa College; Donald F. Palmer, Kent State University; Ivan D. Sanderson, Purdue University; and V. L. Yeats, Texas Technological University. The following people provided specialist reviews of individual chapters: S. Bhattacharji, Brooklyn College, City University of New York; Laurie Brown, University of Massachusetts; G. Gordon Connally; Donald Doehring, Colorado State University; Paul Enos, State University of New York at Binghamton; Joseph Hartshorn, University of Massachusetts; Walter D. Keller, University of Missouri at Columbia; John C. Kraft, University of Delaware; Peter H. Mattson, Queens College, City University of New York; M. Morisawa, State University of New York at Binghamton; Albert J. Rudman, Indiana University; and S. K. Saxena, Brooklyn College, City University of New York. We are grateful for the guidance and editing assistance of Janet Wagner and Sibyl Golden of the McGraw-Hill College Division and for the photo and illustration research of Mira Schachne. Finally, we reserve special thanks for our families and friends for their encouragement and patience during the preparation of this book.

Allan Ludman and Nicholas K. Coch

CHAPTER 1

Introduction

WHAT IS GEOLOGY?

Geology is the study of the earth. You are holding this book and reading these words because you are interested in some aspect of your planet. Perhaps you are intrigued by earthquakes, volcanoes, dinosaurs, glaciers, or the origin of mountains. You may be interested in the applications of geology to environmental planning or to exploration for energy and mineral resources. These topics are covered in this book, as are many others. *Physical* geology examines many aspects of the earth (Figure 1.1): its internal structure and composition; its atmosphere; surface processes such as erosion by streams, wind, and glaciers; and internal processes such as the formation of lava, the folding of solid rock, and the movement of earth materials in earthquakes. Before studying any of these features, it is best to recognize the full scope of the science. In that way the ultimate goal of the geologist will always be kept in mind: to understand how the earth works. With a background in physical geology, scientists can study the evolution of life and the sequence of events that has affected our planet since its creation.

What is the earth made of? What atoms exist in our planet? How do they combine to form the minerals, rocks, soils, and oceans that we see every day? Finding answers to these questions is not just intellectually satisfying. Our civilization depends on the ability of geologists to find enough water, mineral, and energy resources to meet our growing needs.

We must also understand the processes that act on these earth materials. How did the Colorado River carve the Grand Canyon? (Figure 1.2). How can masses of ice more than 3 kilometers (km) (10,000 ft) thick flow across the land? How does wind pick up and move grains of sand? How can solid rock be folded? What makes rocks break? How do mountains form?

Some of these processes are easy to study. We can go to Arizona to observe the Colorado River in action or to Antarctica or Greenland to study the behavior of large masses of ice. We can travel to the Alps to see whether small glaciers behave differently from large ones. We can study arid environments in the Sahara or Mojave deserts, tropical environments in the Amazon jungle, and shoreline processes in Tahiti or Coney Island. The entire earth is our laboratory.

On the other hand, some processes are not so easily studied. We cannot travel into the earth to watch lava form or trace its rise toward the surface. We could not survive, much less observe or take notes, inside the earth, where rocks are being folded. We find it difficult to study mountain building because the rate at which most mountains form is so slow that in our lifetime, or in a thousand lifetimes, we can see only a small part of the total process.

Those processes which can be observed directly are understood more readily because their complexities can be unraveled in the field or recreated in the laboratory, where they can be studied under controlled conditions. Those processes which cannot be observed directly,

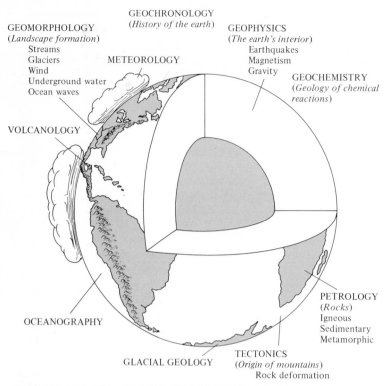

GEOCHRONOLOGY
(*History of the earth*)

GEOMORPHOLOGY
(*Landscape formation*)
Streams
Glaciers
Wind
Underground water
Ocean waves

METEOROLOGY

GEOPHYSICS
(*The earth's interior*)
Earthquakes
Magnetism
Gravity

GEOCHEMISTRY
(*Geology of chemical reactions*)

VOLCANOLOGY

PETROLOGY
(*Rocks*)
Igneous
Sedimentary
Metamorphic

OCEANOGRAPHY

GLACIAL GEOLOGY

TECTONICS
(*Origin of mountains*)
Rock deformation

ECONOMIC GEOLOGY and ENVIRONMENTAL GEOLOGY Combine all of the above
Mineral resources   Land use planning
Energy resources   Pollution control

Figure 1.1 The many fields of physical geology.

Figure 1.2 Grand Canyon of the Colorado River viewed from the South Rim. (*H. L. Mackey, Design Photographers International*)

such as the earth's inner workings, must be studied with instruments that enable us to probe deep within the earth, much as a doctor uses a stethoscope to study the internal workings of the human body. Just finding a way to begin research of this type is a difficult task. How, for instance, can we measure the temperature of lava erupting from a volcano or collect the gases that stream from its vent? Problems such as these make geology more challenging, and all the more rewarding once results are finally obtained.

THE EARTH: PAST, PRESENT, AND FUTURE

There is yet another dimension to geology, one that makes it unique among the sciences: time. Geologists not only must learn how the earth works today, but they must also determine whether it has operated the same way throughout its history. Earth processes acting while you read these words will make the rocks and carve the landscapes that geologists of the future will study. Ancient processes are responsible for the rocks we see today. How can we understand what these ancient events were if they took place long before there were human beings?

The Present Is the Key to the Past

The study of the past begins with the study of the earth today. Every earth process leaves some kind of record for geologists to examine (Figure 1.3). Melting glaciers leave behind the boulders, gravel, sand, and silt they had carried. The glaciers may be gone, but their deposits record their former existence. Winds deposit sand when they slow down. Molten lava cools and becomes solid rock. These processes leave behind materials as evidence of their operation.

In some cases the record is so clear that geologists can determine what process was involved, even if millions of years have elapsed since the process stopped. For example, sand can be deposited by both glaciers and wind. Can the two types of sand deposits be distinguished? Modern glacial deposits consist of sharp-cornered grains of many different sizes and types (including sand grains), while sand deposited by wind in modern deserts contains rounded grains of nearly uniform size. In this

(a) A pile of glacial debris. Note the wide range in particle size and the angularity of the boulders. (*I. J. Witkind, U. S. Geological Survey*)

(b) Photomicrograph of grains in a rock deposited by wind. Note the relatively homogeneous size of the sand grains and well rounded shapes. (*D. Carroll, U. S. Geological Survey*)

Figure 1.3 Comparison of particles deposited by glaciers and wind

example, grain shape and size permit identification of the agent of deposition.

Once we correlate modern processes with their results, we have a basis for studying rocks formed by ancient processes. There are very few features of ancient rocks that cannot be explained by modern processes. Mud cracks found in rocks millions of years old are interpreted as having been formed by drying of sediment, just as mud cracks form today (Figure 1.4). Fossil corals, sea fans, and clamlike creatures found in

(a) Modern mud cracks forming when sediment is dried. (K. Segerstrom, U. S. Geological Survey)

(b) Ancient mud cracks preserved in rock presumably formed the same way. (C. H. Dane, U. S. Geological Survey)

Figure 1.4   An application of uniformitarianism

rocks in Kansas, Illinois, Indiana, and Ohio are interpreted as evidence that these states were once part of a sea floor because similar creatures that live today are found only in the sea. Thus, the present is the key to the past.

This concept is known formally as the **Doctrine of Uniformitarianism.** It states that the processes operating on and in the earth today are basically the same as those which acted in the geologic past and that as a result, modern processes can be used to interpret ancient ones. The Doctrine of Uniformitarianism was first stated in 1785 by James Hutton, a natural scientist from Edinburgh who is considered by many scientists to be the father of modern geology.

Hutton's ideas were considered revolutionary because most scientists of his time had a very different view of the earth's past. They saw major mountain ranges but did not recognize any current processes that could have produced such features. Instead, they believed that geologic features such as mountains formed during cataclysmic events called **catastrophes.** Hutton showed that the same features could form over long periods of time through a series of slow-moving, everyday events. Hutton's ideas were accepted gradually by other geologists, and today they form the framework within which physical and historical geologists study the earth.

The Problems of Studying the Past

Uniformitarianism provides the key to unlock the door to the mysteries of the past, but there are some problems with its use. Geologists believe that the earth formed about 4.6 billion years ago (4.6×10^9 years). Some of the earliest processes—such as the initial formation of the planet, generation of an atmosphere, and formation of a solid outer crust—probably happened only once during earth history. Because they are not happening today, we cannot use uniformitarianism to interpret them. Uniformitarian principles probably apply only to the last two-thirds of the history of the earth.

Another problem is that the record of the past is incomplete. The earth is a dynamic planet—its surface and interior are always in a state of change. Mountains that existed millions of years ago are gone, eroded away along with all the evidence of the ancient processes that had been contained in their rocks. Other rocks are hidden from view, their evidence buried thousands of meters beneath the earth's surface. This problem is much like reading a detective novel and trying to figure out who the criminal is with several pages missing from the book. We are never certain that we have the whole story.

Do these problems make the study of geology impossible? Obviously not. There are many geologists who seem to have some success at interpreting earth processes, finding petroleum and natural gas, preventing landslides, and helping to design earthquake-proof buildings. The difficulties just make earth scientists work harder and in many ways make the science more interesting and enjoyable.

This book will focus mainly on the present in order to examine the processes active on and in the earth today; but if the present is the key to the past, the past and present are keys to the future. Once we have determined what controls the flow of water and deposition in streams, we can predict what areas are likely to suffer from flooding and what to expect when dams are built across rivers. If we can find out what causes earthquakes, we may be able to predict when and where they will take place and thus save thousands of lives.

HOW DO GEOLOGISTS WORK?

Geologists study the earth in many ways. Some work in the field, collecting rocks, soil, or water samples for later laboratory analysis. Others use delicate instruments to measure the earth's magnetism, gravity, or internal earthquake activity. Some geologists work in laboratories, trying to re-create the conditions that make rocks fold, break, or melt. Others build small-scale streams to learn about flooding and sediment movement, study radioactivity in rocks to learn their age, or bombard minerals with x-rays to learn their structure.

The Scientific Method

In the field or in the laboratory, using time-tested foot-and-hammer methods or sophisticated instruments, geologists all use the **scientific method.** This is an orderly, logical approach to whatever problem is being studied. In the scientific method, we ask three questions: What am I looking at? By what processes did it form? What caused these processes? The questions are asked in that sequence because in order to answer the last one, the first two must be answered.

In the scientific method, observations are made of an object or process. After extensive observations, the observer suggests a **hypothesis**—a tentative explanation for the phenomenon being studied. The hypothesis then is tested by further observations or by experiments, and it may be revised or discarded in light of the new data. The process of observation-hypothesis-testing is repeated until a final hypothesis is reached that explains satisfactorily all the observations. However, the final hypothesis may not be the "right" answer. Other scientists studying the same phenomenon might have made other observations or collected other data during their experiments that lead to equally plausible hypotheses. An example of the manner in which the scientific method is used in geology is given in the accompanying perspective, titled Do glaciers move?

Multiple Working Hypotheses

Geologists try to set up **multiple working hypotheses**—as many plausible explanations as can be conceived to fit the data. This is generally easy at the beginning of a study because many variables control the outcome of geologic processes. Field and laboratory data are then collected to evaluate each working hypothesis, and those which do not fit the new data are discarded. As more and more information is compiled, more restrictions are placed on the hypotheses, and the number of hypotheses dwindles rapidly.

This process is much like the steps followed by a detective in solving a murder. The investigation produces new leads to be checked (suggests new observations or experiments). These point to new suspects (suggest different hypotheses). Suspect after suspect is weeded out as new evidence is turned up (hypotheses are discarded) until there is a single one left.

How Are Mountains Formed?

In this book we hope to follow the scientific method. We will make many observations in the following chapters, but we do not expect you to memorize all of them. We will try to present the observations as stepping stones to the formulation of hypotheses about the large-scale work-

Do Glaciers Move?
An Early Application of the Scientific Method in Geology

Herders and farmers in the Alpine valleys of Switzerland and France saw that the angular boulders and cobbles strewn over the valley floors were very similar to the debris carried on and in the glaciers at the heads of the valleys. They (and some scientists) thought that the boulders had been deposited in the valleys by the glaciers, that at one time the ice sheets had advanced down the valleys, had melted, and had left the boulders as evidence of their former presence. Other scientists laughed at the idea that masses of solid ice hundreds of meters thick could move.

In the beginning of the eighteenth century, a simple experiment was designed and carried out by Swiss and French scientists to test the hypothesis that glaciers can move (see Figure 1). Stone markers were placed on the bedrock walls of a valley, and an accurately surveyed line of markers was set up on the surface of the ice between them. If the ice moved, the markers on its surface would change position when compared with the two stationary markers on the walls of the valley.

The results? The markers on the ice did indeed move down the valley, proving conclusively that glaciers can move. There was an added dividend, however, in the fact that the ice movement was shown to be uneven. Ice in the center of the glacier moved farther than ice at the edges. As is often the case in science, the very observations that confirm or refute one hypothesis present new problems to be solved. Why did the ice at the sides of a glacier move slower than the ice at the center? Scientists immediately suggested that friction between the ice and the valley walls was responsible. This then became a new hypothesis to be tested.

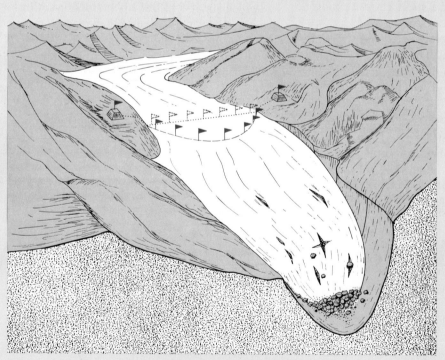

Figure 1   An early experiment to see whether glaciers move.

ings of the earth. As an example of the application of the scientific method, we will use observations to evaluate five hypotheses that have been proposed to explain the causes of mountain building.

From the first time human beings saw mountains, oceans, and plains, they have wondered how these large-scale features form. Why are there mountains in Alaska but not in Iowa? Why are the oceans and continents where they are? Several different hypotheses have been proposed in the past 200 years to answer these questions, but we shall consider only five. At one time or another, each of the five has received general scientific or public approval. As you read the following chapters, think about the plausibility of these hypotheses. We shall help by pointing out how different observations can be applied to these models.

Hypothesis 1 — The static earth

The hypothesis of a **static earth** dominated public discussion of mountain building for many years prior to the development of the Doctrine of Uniformitarianism. This hypothesis states that the earth that we see today is not greatly different from the way it was originally created. The mountains formed at creation are the same ones that exist now, as are the ocean basins, continents, lakes, and rivers. Erosion has perhaps lowered the elevation of the mountains slightly, and deposition of sand, gravel, and silt by wind, streams, and glaciers may have filled in some of the lowlands a bit; but in the overall scope of earth history, these alterations are insignificant. Some adherents of this hypothesis believed that the Bible describes the creation of the world accurately, and for this reason this hypothesis is also known as the **creationist hypothesis.**

Hypothesis 2 — The expanding earth

After uniformitarianism had been used to study earth history, a number of hypotheses developed postulating a slowly but continuously changing planet rather than a static one. Marine fossils were found in rocks at the tops of tall mountains in the Alps, indicating that a sea floor had somehow been destroyed and uplifted to become a mountaintop. Creationists claimed that the fossils were the result of Noah's flood, but uniformitarianists stated that even extreme changes like this must proceed at the same slow pace as most geologic processes.

One school of thought claimed that the earth has been getting hotter gradually since it first formed, and as a result has been expanding constantly (Figure 1.5). This is the hypothesis of an **expanding earth.** In this model, the earth's surface is compared to a thin, brittle coating on the surface of an expanding balloon. As the balloon grows larger, expansion cracks form in

Figure 1.5 The expanding earth. Expansion cracks are filled with lava from within the earth and form ocean basins.

Original earth

Expanded earth

the coating. On earth these cracks become new ocean basins or deep continental basins such as Death Valley. Upward arching at the margins of the cracks produces new mountain ranges. In this model, new ocean basins, mountains, and continents can appear at any time.

Hypothesis 3 — The shrinking earth

Other people hypothesized a **shrinking earth.** This model is diametrically opposed to the previous one, but it also holds that new continents and oceans can form at any time. It states that the earth has been getting cooler since it first formed, and as a result it has been shrinking constantly (Figure 1.6). In this model, the globe is compared with a slowly deflating balloon. As the balloon gets smaller, its originally smooth surface becomes wrinkled and creased. The creases become the oceans, and the wrinkles become the mountain ranges.

Hypothesis 4 — The pulsating earth

Some geologists thought they had detected evidence of both worldwide expansion and contraction. They developed the model of a **pulsating earth,** which combines the features of the expanding and shrinking earth hypotheses. According to this model, the earth experiences alternating periods of expansion and contraction like a beating heart. If expansion and contraction are related to worldwide heating and cooling, an extremely complex thermal history is required for the earth. Heat must be built up periodically and then released.

Hypothesis 5 — The plate tectonic earth

The twentieth century has seen a revolution in geology. Advanced technology has enabled geologists to study the ocean floors, the earth's gravity and magnetism, and rocks themselves in ways in which Hutton could not have dreamed. New and startling data have led to an equally new and startling hypothesis — the **plate tectonic hypothesis** — that has been fully developed only within the past 20 years.

According to plate tectonic models, the earth's surface is made up of a small number of relatively rigid blocks of material called **lithosphere plates** (Figure 1.7). Some of these plates are immense — our own North American plate extends from the center of the Atlantic Ocean to the Pacific — but some, such as the Cocos and Caribbean plates, are relatively small. Some plates lie beneath oceans, some lie beneath continents, and some, such as the North American plate, contain both ocean and continent. All plates can move freely and independently across the face of the earth. They may move apart, crash together, or grind past one another; in the

Figure 1.6 The shrinking earth. As the earth shrinks, wrinkles become continents and mountains; creases become ocean basins.

Original earth

Shrunken earth

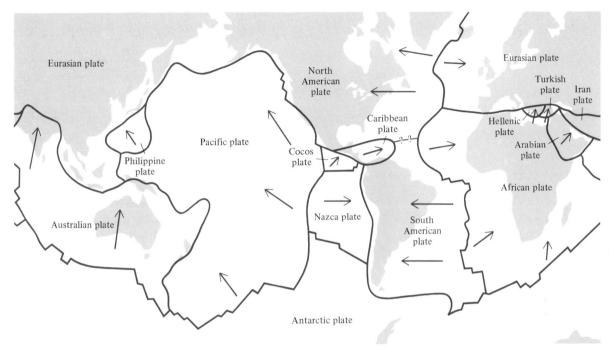

Figure 1.7 The earth's lithosphere plates. The color lines indicate the boundaries between the plates; the arrows show the directions in which the plates are believed to be moving.

process, they create the earth's mountains. The arrows on Figure 1.7 show the directions in which the plates are proposed to be moving today.

How can continents and oceans move? Figure 1.8 shows the postulated explanation. The upper 100 to 150 km of the earth consists of relatively rigid rock material called the **lithosphere,** and it is this material which forms the moving plates. Beneath the lithosphere is the **asthenosphere,** a region some 100 km thick in which rocks behave as if they had the plasticity of children's modeling clay and can flow freely. The lithosphere plates "float" on the relatively plastic asthenosphere, and the continents and oceans are carried as passengers.

In the plate tectonic model, the earth is expanding in some areas but simultaneously contracting in others. Expansion takes place in every ocean at elongate mountain chains called **ocean ridges** (see Figure 1-8). At each of these ridges a plate cracks and is wedged apart by lava that flows upward from deeper in the earth. As expansion continues, a conveyor belt effect de-

velops in which new material is added to a ridge from below and is then carried outward on both sides of the ridge's crest. The result is a process called **sea-floor spreading** that continuously enlarges oceans and drives plates apart from one another. The Mid-Atlantic Ridge is thought to be a spreading center that separates the westward-moving North American plate from the eastward-moving Eurasian plate.

Contraction takes place at the trenches that mark the deepest parts of the oceans. Plates driven away from spreading centers sometimes collide, and one is forced beneath the other at the trenches in what are called **subduction zones.** The collision of plates in subduction zones crushes and crumples rocks on so large a scale that mountains form. At the subduction zones, the material added to the plates at the ocean ridges is returned to the depths so that the size of the earth is kept relatively constant. The mountains, volcanoes, earthquakes, and folded rocks of Japan, Indonesia, Puerto Rico, and the Andes Mountains are the results of subduction.

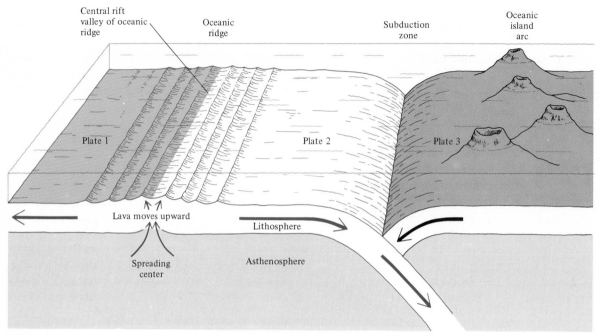

Figure 1.8   The plate tectonic model.   Expansion occurs at ocean ridges, where lava rises to the surface.   Contraction occurs in the subduction zones, where material is returned to the depths of the earth.

Instead of colliding head on in subduction zones, two plates may slide past each other along large earth fractures called **transform faults.** The San Andreas Fault in California may be a transform fault along which the North American and Pacific plates are in contact.

Testing the five hypotheses

Which of the five hypotheses is most correct?   Some, such as the expanding and contracting earth models, may seem attractive because of their simplicity, but a simple explanation is not always the best.   Others, such as the plate tectonic or pulsating earth hypotheses, are much more complex, but complexity alone is no guarantee of correctness.   Plate tectonics is the most recent hypothesis—does that mean that it must be correct?

Whether you instinctively like or dislike a model is irrelevant because each model must be tested objectively.   How can we test them? What kinds of data should geologists look for? What processes and rocks are involved?   Where are the best places to find the answers?   These

questions will not be answered here.   Keep them in mind as you read the following observations and hypotheses.   You may find that you are able to answer some of the questions yourself, and we will review them in Chapters 17, 18, 20, and 21.

THE RELEVANCE OF GEOLOGY

Geologic knowledge has helped the survival of human beings from the time when *Homo sapiens* first walked upright until the present day.   Primitive humans used flint for arrowheads and cutting tools, bright-colored cinnabar and malachite for pigments, and clay for pottery and brick.   Stone was quarried to build the pyramids of Egypt and the temples and palaces of many ancient cities. As we have become more "civilized," we have found more ways to use our knowledge of the earth.   Once humans learned to extract metals from rocks, the earth was scoured for ores of copper, tin, iron, and eventually aluminum. Today we also tap earth materials such as coal and petroleum for energy.

As our population increases, our demands for these resources grow greater, and easily found resources are no longer sufficient for our growing needs. Detailed knowledge of the earth and advanced technology are needed just to keep supplies of minerals and energy abreast of our requirements. Oil from the north slope of Alaska and from the North Sea would have been inconceivable just a few decades ago. Today we need it desperately and are exploring the sea floors actively for more.

As human population has increased, we have inhabited more and more of the surface of our planet. In so doing we have often chosen to live in areas that are rich in geologic resources: water, minerals, energy. In other instances we have elected to live in areas that are dangerous — areas where earth processes can cause catastrophic damage. We all know enough not to live on a volcano. Or do we? Look at Hawaii, Italy, Central America, Japan, and Iceland. We all know enough not to live where landslides, earthquakes, floods, and windstorms are common events. Or do we? If we look at newspaper accounts of natural disasters, we realize that appropriate use of geologic knowledge could save many lives and millions of dollars each year.

We take pride in being able to modify the environment to make it more hospitable. By using detailed geologic data on the surficial and bedrock features of an area, we can locate dams, reservoirs, pollution control facilities, power plants, and houses in the most advantageous sites. We can make long-range predictions about water supply to determine how large a population an area can support.

We often run into problems when we ignore or fail to seek geologic data. Reservoirs have been built in porous rocks through which all the water has leaked out. Houses built with inadequate foundations have sunk into the ground in the first major rainfall. We build dams to control flooding, jetties to prevent beach erosion, and highways to speed transportation, but all such construction interferes in some way with natural processes. The Aswan Dam of Egypt has helped generate electricity and control flooding, but it has also stopped the yearly replenishing of fertile soil downstream. Extensive paving in metropolitan areas has reduced the seepage of rainwater into the natural underground water supply, leading to water shortages. Careful attention to basic geologic principles can make our environmental management more effective. Ignoring these principles can lead to disaster.

Geology remains what it has always been — one of the most practical of the sciences. Nearly every aspect of geology has some economic or environmental application, and we will examine this potential in the following pages.

GEOLOGY TODAY AND TOMORROW

Geology today is in an exciting state of growth perhaps unmatched in the past. Ideas laughed at 30 years ago, such as plate tectonics, are considered seriously and argued today, and new methods of study are being developed that were science fiction a few years ago. Orbiting satellites for resource exploration, nuclear reactors, x-ray generators, and laser beams are as much a part of today's geology as field boots, hammers, shovels, and oil wells.

Geologists can now look for information at the North and South Poles, on the continental shelves, in the ocean deeps, and even on other planets. We have had earthquake (moonquake?) sensing stations on the moon, and a geologist (astronaut Harrison Schmitt) has run a scientific traverse there, much to the envy of his earthbound colleagues. We are becoming planetologists as well as geologists, and we are analyzing Voyager photographs of Jupiter's moon Io to see what the causes of its volcanic eruptions might be.

With all these advances there are still many basic questions. Some, such as why the earth has a magnetic field, have never been answered completely, but others arise as new methods provide new kinds of data. The state of constant change in the science parallels our concept of the ever-changing nature of the earth. We see landscapes change, feel climates fluctuate, and watch volcanoes erupt and then die out peacefully. The changing nature of the planet and of the science that studies it makes geology an exciting adventure. We hope you will enjoy this introduction to it.

CHAPTER 2

Matter and Energy

Mt. St. Helens, erupting in Washington, blasts tons of dust-sized particles into the air, while Mauna Loa, erupting in Hawaii, sends rivers of glowing lava down its flanks. What are the lava and dust made of? What made rock melt inside the earth and forced it to the surface as lava? In a very different kind of process, the Mississippi River flows into the Gulf of Mexico, carrying with it millions of tons of silt and sand and many other materials that it has dissolved. What are the sand grains made of? Are they the same material as the Hawaiian lavas or Alaskan volcanic dust? What makes the Mississippi River flow?

The events described above are examples of the many kinds of earth processes that geologists study. Volcanic eruptions and the flow of streams seem to be very different kinds of surface processes, but the questions asked about them are the same: What are earth materials made of? What makes earth processes happen? These two questions are fundamental to all of geology and must be answered before any single process or rock can be studied.

WHAT IS THE EARTH MADE OF?

All materials—gases in the atmosphere, water in the oceans, solid rocks in the mountains—are composed of tiny, submicroscopic particles called *atoms*. To understand the earth, we must understand what an atom is and how atoms can combine with one another to make different substances.

Atomic Structure

Atoms are composed of even smaller particles: electrons, protons, and neutrons (Table 2.1). **Electrons** are particles that have very little mass but contain a small unit of electric charge (-1). **Protons** have much more mass—1832 times the mass of an electron—and an electric charge ($+1$) that is equal but opposite to that of an electron. **Neutrons** have slightly more mass than protons—1833 times the mass of an electron—but have no electric charge. The protons and neutrons cluster in the center (**nucleus**) of every atom and are surrounded by a cloud of orbiting

TABLE 2.1 The major subatomic particles

Name	Electric charge	Mass
Electron	-1	1
Proton	$+1$	1832
Neutron	0	1833

electrons (Figure 2.1). Most of the mass and all of the positive charge of an atom are thus found in the nucleus, while the negative charge is concentrated in the electron cloud. Although orbiting electrons move rapidly, they cannot escape from the nucleus. They are held to the oppositely charged protons of the nucleus just as opposite poles of magnets attract one another. These forces of attraction are called **electrostatic forces**. The overall electric charge of an atom is 0 so that the number of positively charged protons in the nucleus must be equal to the number of negatively charged orbiting electrons.

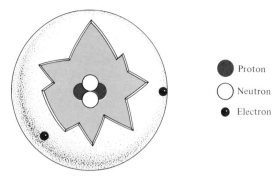

Figure 2.1 The structure of an atom.

Not all atoms are the same. They may differ in **atomic number** (the number of protons in the nucleus) or **atomic mass** (the sum of the protons and neutrons in the nucleus). The mass of an atom is measured in **atomic mass units** (amu), each of which is equal to the mass of a proton. The contribution of electrons to the mass of an atom is so small that generally it is ignored. Atoms may also differ in size. In general, the more electrons there are in orbit around the nucleus, the larger the atom is. Atoms with very large atomic numbers are thus much larger than atoms with very small atomic numbers.

Elements

All atoms with the same atomic number are said to be atoms of the same **element.** They have the same physical properties and behave the same way in natural processes. All atoms of the green, gaseous, and extremely poisonous element chlorine, for example, contain 17 protons, and all atoms of the heavy, solid, and unstable element uranium have 92 protons. Eighty-eight elements have been discovered in the earth and its atmosphere, and eighteen more have been made by nuclear physicists (Table 2.2). However, only a small number of these elements play important roles in earth processes, and these are shown in color in the table.

Isotopes

All atoms of an element must have the same atomic number but not necessarily the same atomic mass. Hydrogen atoms (atomic number

1) may have masses of 1, 2, or 3 amu depending on how many neutrons (zero, one, or two) are in the nucleus. Uranium atoms (atomic number 92) may have masses of 234, 235, or 238 amu. Atoms of an element that have different masses are called **isotopes** of that element. To identify an atom fully, we must know its atomic number and mass. For identification we will use a simple shorthand system:

Atomic mass $\longrightarrow$ $\nwarrow$ 1
$_1^1$H $\longleftarrow$ Chemical symbol (hydrogen)
Atomic number $\longrightarrow$ 1

The different isotopes of hydrogen and uranium thus may be designated as:

$$_1^1H, \,_1^2H, \,_1^3H \qquad _{92}^{234}U, \,_{92}^{235}U, \,_{92}^{238}U$$

The Structure of the Elements

Certain aspects of atomic structure are particularly important to an understanding of earth materials. A brief look at the atoms of some of the simpler elements will illustrate these features. Take a look at Figure 2.2. Notice that the simplest atom is hydrogen. Its nucleus consists of a single proton (with zero, one, or two neutrons), and it has a single orbiting electron. Helium, $_2^4$He, has two protons and two neutrons in its nucleus. Both of its electrons are the same distance from the nucleus as they orbit. However, the three electrons of lithium, $_3^7$Li, are not the same distance from the nucleus. Two circle the nucleus at an equal distance, but the third is farther out. Electrons that orbit a nucleus at equal distances are said to belong to the same **electron shell.** There is room for only two electrons in the innermost electron shell of lithium; the third electron must enter a different shell. The same behavior occurs in elements of atomic numbers 4 (beryllium) through 10 (neon). The fourth through tenth electrons are at approximately the same distance from the nucleus as the third, and they define a second electron shell outside the first.

Sodium, $_{11}^{23}$Na, is different. As shown in Figure 2.2, it has two electrons in its inner shell, as do atoms of elements with atomic numbers 2 through 10, and eight electrons in its second shell, as does neon, but the eleventh electron

TABLE 2.2 **Alphabetical list of the elements**

Elements in color are most important in earth materials.

Element	Symbol	Atomic number	Element	Symbol	Atomic number
Actinium	Ac	89	Mercury	Hg	80
Aluminum	Al	13	Molybdenum	Mo	42
Americium	Am	95	Neodymium	Nd	60
Antimony	Sb	51	Neon	Ne	10
Argon	Ar	18	Neptunium	Np	93
Arsenic	As	33	Nickel	Ni	28
Astatine	At	85	Niobium	Nb	41
Barium	Ba	56	Nitrogen	N	7
Berkelium	Bk	97	Nobelium	No	102
Beryllium	Be	4	Osmium	Os	76
Bismuth	Bi	83	Oxygen	O	8
Boron	B	5	Palladium	Pd	46
Bromine	Br	35	Phosphorous	P	15
Cadmium	Cd	48	Plantinum	Pt	78
Calcium	Ca	20	Plutonium	Pu	94
Californium	Cf	98	Polonium	Po	84
Carbon	C	6	Potassium	K	19
Cerium	Ce	58	Praseodymium	Pr	59
Cesium	Cs	55	Promethium	Pm	61
Chlorine	Cl	17	Protactinium	Pa	91
Chromium	Cr	24	Radium	Ra	88
Cobalt	Co	27	Radon	Rn	86
Copper	Cu	29	Rhenium	Re	75
Curium	Cm	96	Rhodium	Rh	45
Dysprosium	Dy	66	Rubidium	Rb	37
Einsteinium	Es	99	Ruthenium	Ru	44
Erbium	Er	68	Rutherfordium	Rf	104
Europium	Eu	63	Samarium	Sm	62
Fermium	Fm	100	Scandium	Sc	21
Fluorine	F	9	Selenium	Se	34
Francium	Fr	87	Silicon	Si	14
Gadolinium	Gd	64	Silver	Ag	47
Gallium	Ga	31	Sodium	Na	11
Germanium	Ge	32	Strontium	Sr	38
Gold	Au	79	Sulfur	S	16
Hafnium	Hf	72	Tantalum	Ta	73
Hahnium	Ha	105	Technetium	Tc	43
Helium	He	2	Tellurium	Te	52
Holmium	Ho	67	Terbium	Tb	65
Hydrogen	H	1	Thallium	Tl	81
Indium	In	49	Thorium	Th	90
Iodine	I	53	Thulium	Tm	69
Iridium	Ir	77	Tin	Sn	50
Iron	Fe	26	Titanium	Ti	22
Krypton	Kr	36	Tungsten	W	74
Lanthanum	La	57	Uranium	U	92
Lawrencium	Lr	103	Vanadium	V	23
Lead	Pb	82	Xenon	Xe	54
Lithium	Li	3	Ytterbium	Yb	70
Lutetium	Lu	71	Yttrium	Y	39
Magnesium	Mg	12	Zinc	Zn	30
Manganese	Mn	25	Zirconium	Zr	40
Mendelevium	Md	101	Unnamed	—	106

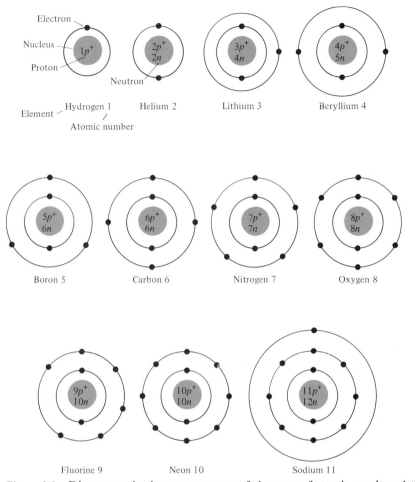

Figure 2.2 Diagrammatic electron structure of elements of atomic numbers 1 through 11.

begins a third shell. There are strict laws governing the number of electrons that can fit into a given shell—2 in the first, 8 in the second, 18 in the third—but the nature of these laws need not concern us.

Reactions Involving Atoms

Atoms undergo two kinds of transformation: **chemical reactions** that involve only the electrons orbiting the nucleus, and **nuclear reactions** that affect primarily the nucleus itself. During chemical reactions, atoms are attached (**bonded**) to one another without altering their nuclei in any way. During nuclear reactions, the nucleus changes by gaining or losing nuclear particles. In this process, one element is changed into another. Atoms of nearly all elements participate in chemical reactions, but only certain isotopes of specific elements take part in nuclear reactions. The three isotopes of potassium, $^{39}_{19}\text{K}$, $^{40}_{19}\text{K}$, and $^{41}_{19}\text{K}$, for example, all take part in chemical reactions, but only $^{40}_{19}\text{K}$ is involved in nuclear reactions. Both chemical and nuclear reactions are important in earth materials and geologic processes.

Nuclear reactions

In nuclear reactions, the nucleus of an atom of one element is converted into the nucleus of an atom of another element. As an example, consider the type of nuclear reaction involving one of the isotopes of uranium:

$$^{238}_{92}\text{U} \rightarrow\ ^{234}_{90}\text{Th} + ^{4}_{2}\alpha + \text{energy}$$

The uranium nucleus is changed into a thorium nucleus by losing a fragment of nuclear material composed of two protons and two neutrons. This particle is called an **alpha (α) particle,** and this type of reaction an **alpha decay.** In addition, some of the energy that held the uranium nucleus together is released also. Nuclear reactions such as this, in which a nucleus splits, are called **fission** reactions. Some nuclear reactions involve the combination of two atoms of an element to make a different element. Such reactions are called **fusion** reactions. Fusion reactions occur in the sun, where hydrogen nuclei join to make helium nuclei, but not in the earth.

When nuclear reactions take place spontaneously, without any human interference, the radiant energy released is called **radioactivity,** and the elements involved are said to be radioactive. The uranium reaction will be examined more closely along with the other kinds of radioactivity in Chapter 8, where the application of radioactivity to measuring the age of earth materials will be discussed.

Chemical reactions and compounds

Atoms of some elements occur in nature as isolated atoms, which are not combined with any other atoms. Examples are the inert gases helium, neon, and argon. Most atoms, however, are found combined with others in compounds. Some of these elements occur in what is called the **native state,** in which their atoms are combined only with one another. Common examples are gold, sulfur, carbon, and oxygen. In most compounds, however, atoms of one element combine with atoms of other elements. Typical examples include water, a combination of oxygen and hydrogen atoms, H_2O, and table salt, a combination of sodium and chlorine atoms, NaCl. Regardless of how the atoms occur, some forces must bond them together.

When atoms of two different elements are bonded together, they generally lose their distinguishing characteristics and are joined in a compound that exhibits properties of its own. Table salt, NaCl, for example, is a white solid vital to human survival; but if pure sodium and pure chlorine were taken internally, the result

probably would be fatal. Pure chlorine is a green poisonous gas, and sodium is a yellow solid that explodes and bursts into flame when placed in water.

What Controls Bonding?

A comparison of the **inert** elements — elements whose atoms are not bonded to those of other elements or even to each other — with the **reactive** elements indicates that bonding is controlled by an atom's outermost electrons. The inert gases do not enter into chemical reactions because their outermost electron shells are filled completely (as in helium and neon) (Figure 2.2) or are filled partially with a particularly stable number of electrons, as in argon. All other elements have unfilled outer shells or subshells and bond to other atoms.

The effect of the outer electrons is clarified when the elements are arranged in the periodic table (Table 2.3) so that those elements with the same number of electrons in their outer shells are aligned in columns. The inert gases are in the right-hand column, next to those elements which lack only one electron in their outermost shells or subshells. The left-hand column contains those elements which have only a single electron in their outermost shells. Elements in in the same column behave in very similar ways in chemical reactions, showing the importance of the number of outermost electrons in governing chemical behavior.

How does bonding occur?

Four types of bonding are of particular interest in geology: ionic, covalent, metallic, and van der Waals'. All earth materials are held together by one of these bond types or by a combination of several.

Ionic bonding The inert gases do not enter into chemical reactions because their filled electron shells are particularly stable. In **ionic bonding,** atoms add or lose electrons to fill their outermost shells. As an example, examine the bonding of lithium and fluorine in the compound lithium fluoride, LiF (Figure 2.3). An atom of lithium has a single electron in its outermost

TABLE 2.3   Periodic Table of the Elements*

1	2	3	4	5	6	7	8	9	10	11	12	13	14	15	16	17	18
1 H Hydrogen																	2 He Helium
3 Li Lithium	4 Be Beryllium											5 B Boron	6 C Carbon	7 N Nitrogen	8 O Oxygen	9 F Fluorine	10 Ne Neon
11 Na Sodium	12 Mg Magnesium											13 Al Aluminum	14 Si Silicon	15 P Phosphorus	16 S Sulfur	17 Cl Chlorine	18 Ar Argon
19 K Potassium	20 Ca Calcium	21 Sc Scandium	22 Ti Titanium	23 V Vanadium	24 Cr Chromium	25 Mn Manganese	26 Fe Iron	27 Co Cobalt	28 Ni Nickel	29 Cu Copper	30 Zn Zinc	31 Ga Gallium	32 Ge Germanium	33 As Arsenic	34 Se Selenium	35 Br Bromine	36 Kr Krypton
37 Rb Rubidium	38 Sr Strontium	39 Y Yttrium	40 Zr Zirconium	41 Nb Niobium	42 Mo Molybdenum	43* Tc Technetium	44 Ru Ruthenium	45 Rh Rhodium	46 Pd Palladium	47 Ag Silver	48 Cd Cadmium	49 In Indium	50 Sn Tin	51 Sb Antimony	52 Te Tellurium	53 I Iodine	54 Xe Xenon
55 Cs Cesium	56 Ba Barium	57 La Lanthanum †	72 Hf Hafnium	73 Ta Tantalum	74 W Tungsten	75 Re Rhenium	76 Os Osmium	77 Ir Iridium	78 Pt Platinum	79 Au Gold	80 Hg Mercury	81 Tl Thallium	82 Pb Lead	83 Bi Bismuth	84 Po Polonium	85* At Astatine	86 Rn Radon
87* Fr Francium	88 Ra Radium	89 Ac Actinium ‡	104* Rf Rutherfordium	105* Ha Hahnium	106* 106												

†Lanthanides	58 Ce Cerium	59 Pr Praseodymium	60 Nd Neodymium	61* Pm Promethium	62 Sm Samarium	63 Eu Europium	64 Gd Gadolinium	65 Tb Terbium	66 Dy Dysprosium	67 Ho Holmium	68 Er Erbium	69 Tm Thulium	70 Yb Ytterbium	71 Lu Lutetium
‡Actinides	90 Th Thorium	91 Pa Protactinium	92 U Uranium	93* Np Neptunium	94* Pu Plutonium	95* Am Americium	96* Cm Curium	97* Bk Berkelium	98* Cf Californium	99* Es Einsteinium	100* Fm Fermium	101* Md Mendelevium	102* No Nobelium	103* Lw Lawrencium

*Elements shown in color are most important in earth materials. Elements with an asterisk (*) are artificial.

shell. It can have the same stable outermost electron shell as the helium atom if it loses the single outer electron. Similarly, the fluorine atom can have the same stable electron structure as the inert neon atom if it can add one electron to complete its outer shell. The transfer of an electron from lithium to fluorine produces both stable electron structures simultaneously but also leaves lithium with a positive charge (three protons in the nucleus and only two orbiting electrons) and fluorine with a negative charge (nine protons in the nucleus but 10 electrons).

These charged particles are called **ions.** Positively charged ions such as lithium are called **cations,** and negatively charged ions such as fluorine are called **anions.** The oppositely charged lithium and fluorine ions attract one another and are bonded together electrostatically. Bonds between ions are called **ionic**

bonds and are very common in earth materials. Sodium and chlorine atoms, for example, are held together by ionic bonding in table salt, NaCl.

Covalent bonding Atoms of the same element, or of elements very close to each other in the periodic table, may be bonded to one another by **covalent bonding.** Consider the bonding that holds two oxygen atoms together in the compound O_2 (Figure 2.4). Each oxygen atom has two electrons filling its innermost electron shell and six of the possible eight in its outer shell. In O_2 the 12 outer-shell electrons of the two oxygen atoms form a special shell called a **molecular orbit** around the two nuclei (Figure 2.4). The electrostatic attraction between nuclei and molecular-orbit electrons holds the two nuclei together. The inner electrons are not

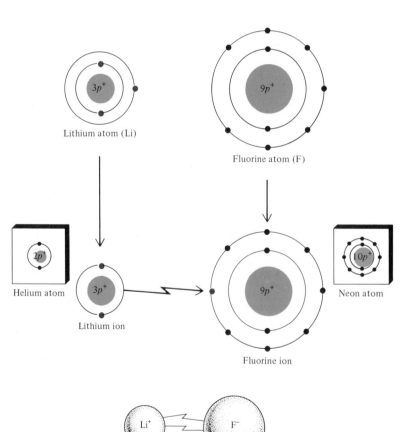

Figure 2.3 Ionic bonding in lithium fluoride, LiF.

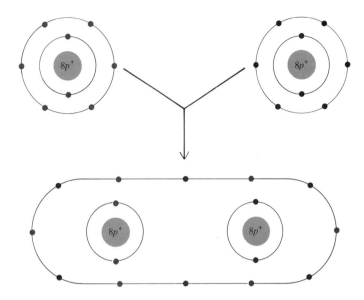

Figure 2.4 Covalent bonding in the oxygen molecule, O_2.

affected at all, and no ions are formed. The group of two covalently bonded oxygen atoms is called an oxygen **molecule,** the smallest unit of native oxygen that exists on earth. Covalent bonding occurs in solids also; it holds the carbon atoms together in the mineral diamond.

Metallic bonding Bonding between atoms of copper, gold, or silver occurs by a process similar to covalent bonding called **metallic bonding.** As in covalent bonding, the outer electrons are shared by different nuclei, but in metallic bonding the outer electrons are not shared by a small group of nuclei. Instead, all the outermost electrons of all the atoms present are shared by all the nuclei in the metal and are free to travel from one nucleus to another.

Van der Waals' bonding Some compounds possess additional forces, called **van der Waals' forces,** that assist covalent and ionic bonding in holding atoms together. These forces are different from the other types of bonding because they do not involve transfers or rearrangements of electrons between nuclei. Van der Waals' forces are present in all substances, even be-

tween atoms of inert gases such as helium (Figure 2.5).

As the two orbiting electrons circle the helium nucleus rapidly, they can be positioned instantaneously on the same side of the nucleus. The helium atom is electrically neutral, but with both electrons on one side of the atom, it has decidedly positive and negative sides. The atom then acts as a tiny bar magnet, affecting electrons of nearby atoms. Its positive side attracts the electrons of nearby helium atoms with a very weak electrostatic force. No earth materials are held together entirely by van der Waals' forces, but they are important forces of attraction in some substances, such as graphite and talc.

Figure 2.5 Van der Waals' bonding of helium atoms.

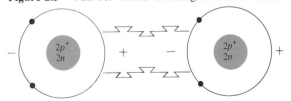

Both electrons are on one side of the atom.
Weak van der Waals' forces.

States of Matter

Native elements and compounds occur in three states: solid, liquid, and gaseous. Molecules in a gas are free to move about because they are held to each other very weakly. Gases fill completely whatever size container they are placed in because their molecules can move closer together or farther apart. Cohesive forces in liquids are stronger than those in gases. Molecules in liquids can also move freely, but they cannot move far apart or be compressed very much. Liquids are thus highly mobile, but unlike gases they retain their volume even when small amounts are placed in large containers. Cohesive forces in solids are far stronger than they are in liquids or gases. The positions of atoms and ions in solid earth materials are controlled rigidly by rules that will be discussed in Chapter 3.

Composition of the Earth

Our planet consists of three different physical/chemical regions, each with a unique combination of elements and a unique proportion of solids, liquids, and gases (Table 2.4, Figure 2.6).

The **atmosphere** is the gaseous envelope that surrounds the earth. It is composed mostly of nitrogen and oxygen but contains other gases and small amounts of solids (dust particles, hailstones, and snowflakes) and liquids (rain). The **hydrosphere** is the liquid outer covering of the earth that comprises the oceans, lakes, streams, and underground water supply. Most of the earth's surface is water —71 percent is covered by oceans alone. The hydrosphere is nearly all water, but a great deal of material is dissolved in both ocean and freshwater as ions. Solids are carried in suspension in the earth's surface waters, and gases are dissolved in these waters, but the hydrosphere is basically liquid.

The earth itself is nearly all solid and is divided into three regions—the **crust, mantle, and core** (Figure 2.6). Geologists have been able to sample materials only from the earth's crust. The composition and very existence of the mantle and core are inferred from the remote sensing methods described in Chapters 19 and 20. Oxygen and silicon are by far the most abundant elements in the crust and mantle, and most materials from these parts of the earth are compounds that contain both elements. The

TABLE 2.4 Composition of the earth in weight percent

	Solid earth		Hydrosphere	Atmosphere
	Crust	Entire earth	Hydrosphere	Atmosphere
Aluminum	8.2	1.09	—	—
Calcium	4.1	1.13	0.04	—
Carbon	0.2	Trace	0.003	Trace
Chlorine	0.01	Trace	1.8	—
Hydrogen	0.1	Trace	10.8	Trace
Iron	5.6	34.6	—	—
Magnesium	2.3	12.7	—	—
Nitrogen	0.002	Trace	—	75.50
Oxygen	46.3	29.5	86.3	23.15
Potassium	2.1	0.07	.04	—
Sodium	2.4	0.57	1.05	—
Silicon	28.2	15.2	—	—
Titanium	0.5	0.05	—	—
Nickel	Trace	2.39	—	—
Others	Trace	2.7	—	1.35

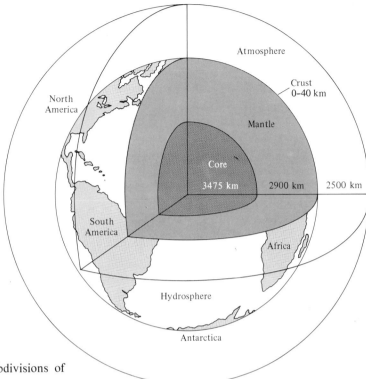

Figure 2.6 Chemical subdivisions of the earth.

core is thought to be composed of a combination of metallic nickel and iron because of its density, the properties of the earth's magnetic field, and comparisons with meteorites. In Chapter 1 the outer few hundred kilometers of the earth were divided into lithosphere and asthenosphere based on *physical* criteria — the different rigidities of the materials in the two regions. The division here into crust and mantle is based on *chemical* criteria, and the relationship between lithosphere, asthenosphere, crust, and mantle is shown in Figure 2.7. The lithosphere contains the entire crust and a portion of the upper mantle, while the asthenosphere is a region 100 to 150 km thick in the mantle.

The atmosphere is an extremely mobile group of gases but has a very low density. Movement of large masses of the atmosphere causes changes in the weather, and we can detect these movements as wind. The hydrosphere is considerably denser than the atmosphere, but it is also very mobile, as shown by oceanic currents such as the Gulf Stream and the daily

movement of the tides. The earth itself, being mostly solid, would seem to be immobile. We saw in Chapter 1, however, that even though ice is a solid, it can still move. We shall see several examples in later chapters of this apparent paradox: Solid earth materials are capable of movement.

Wherever solid, liquid, and gaseous parts of the earth interact, their differences bring about important physical and chemical processes. For example, shoreline erosion takes place where oceans and solid continents interact; rocks break down to produce soils where atmosphere and solid earth come into contact; and waves form and move across the ocean where atmosphere meets hydrosphere.

ENERGY

The earth's continents and oceans, lava, volcanic dust, and rivers are made of different combinations of the 88 naturally occurring elements.

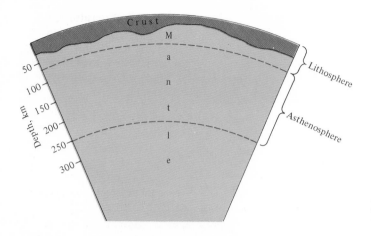

Figure 2.7 Physical and chemical divisions of the outermost part of the earth.

In Chapters 3, 4, 6, and 7, which describe the more common minerals and rocks, we shall take a close look at some of these combinations. But the first question asked at the beginning of this chapter has been answered—we know what earth materials can be made of. The second question remains: What makes earth processes happen?

The answer is a single word—energy. **Energy** is defined as the capacity to do work or to cause activity to take place. Without energy, the earth would be a dead, motionless planet; with it, the earth is a dynamic, active world. There are many kinds of energy: kinetic, including heat and electric; radiant, including light and x-ray; nuclear; and chemical. Most play some role in causing and sustaining earth processes. To understand the complex factors that govern earth processes, we must now look at the interplay of energy forms that makes the processes operate. Energy is, of course, also vital to human processes and work. The source and uses of energy for human needs are discussed in Chapter 22.

Types of Energy

Kinetic energy

The ability of a moving object to induce activity in other objects is called **kinetic energy.** In most cases, the composition of the moving object is unimportant. The kinetic energy comes from the movement, not from the composition or state of matter. A rolling billiard ball has

kinetic energy (Figure 2.8). When it strikes a stationary ball, it causes the second ball to move. Similarly, molecules of water flowing in the Mississippi River have kinetic energy and can therefore make sand grains in the riverbed move.

The amount of kinetic energy depends on the mass, m, and velocity, v, of the moving object:

$$\text{Kinetic energy} = 0.5mv^2$$

The faster an object moves, or the greater its mass, the more kinetic energy is involved. Large, fast-flowing streams can move more sand

Figure 2.8 Kinetic energy. A rolling ball has kinetic energy, and it can cause a stationary ball to move.

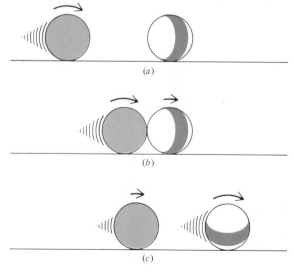

(a)

(b)

(c)

grains and do more geologic work than small, slow-moving streams because they have more kinetic energy. Imagine how much kinetic energy must be involved if lithosphere plates the size of North America and the western Atlantic move across the earth's surface as envisaged in the plate tectonic model.

Heat and electric energy are two special kinds of kinetic energy that involve movement of very small particles. Heat energy is one of the most important causes of earth processes. Electric energy is less important in nature but is one of the energy forms most commonly used in human society.

Heat The kinetic energy associated with the movement of atoms, ions, and molecules is called **heat energy.** The faster the atoms and molecules of gases and liquids move, the more heat energy they have. Even in solids, atoms and ions can move a little. They vibrate back and forth about their fixed positions; and the more they move, the greater the amount of heat

energy the solid possesses. Temperature is not the same as heat, although they are related. The **temperature** of a substance is the *average* kinetic energy of its atoms, whereas its heat energy is the *total* (sum) of the kinetic energy of all its constituent atoms. A match and a roaring bonfire may both be at the same temperature, but there can be little doubt about which generates more heat. Just try barbecuing a chicken over a match. In this book, temperatures will be given in degrees Celsius (°C). (Water freezes at 0°C and boils at 100°C; a room temperature of 68°F is 20° on the Celsius scale.)

Heat energy moves from place to place, always from hot objects to cooler ones, by the processes of conduction, convection, and radiation (Figure 2.9). **Conduction** transfers heat energy between materials that are in contact with one another because atoms and ions vibrating rapidly in the hotter material interact with those in the colder one, making them move faster. Some of the heat energy is thus transferred. For example, sunlight beating down on rocks in a

Figure 2.9 Types of heat energy transfer.

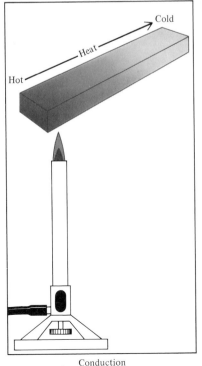

Conduction

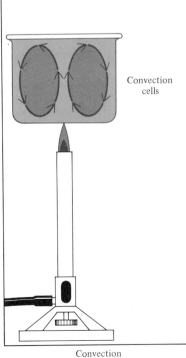

Convection

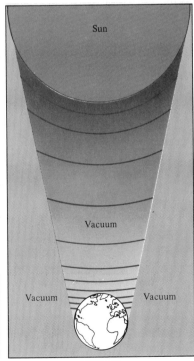

Radiation

desert heats their outer surfaces. The heat energy then is *conducted* inward to warm their interiors. In **convection,** heated particles do not just vibrate in place but move much greater distances, carrying the heat with them. As air is heated by conduction from the warm ground, it rises. Cooler air moves in to replace it, is also warmed, and rises. The rising air heats the surrounding atmosphere by conduction, is cooled, and sinks toward the ground to begin the cycle again. A circular air motion, called a **convection cell,** is established. In **radiation,** heat energy is changed into a different type of energy (radiant energy) and can be transmitted through a vacuum, where there are no particles to be moved. Heat from the sun is carried to the earth by radiation through the vacuum of space.

Electric energy   The special type of kinetic energy associated with the movement of electrons through a substance is called **electric energy.** In the form of lightning it plays a very small role in earth processes today, but experiments dealing with the origin of life suggest that lightning in the atmosphere of the primitive earth may have helped create organic molecules.

Radiant energy

Light, infrared and ultraviolet radiation, x-rays, and radio waves are all kinds of **radiant energy.** All are electromagnetic waves—which is to say they are energy forms that move in the wavelike motion shown in Figure 2.10. All forms of radiant energy can move through a vacuum, and they do so at the same velocity—the speed of light [3×10^9 meters per second (m/s)]. They differ from one another in **wave**length (the distance from one point to the next equivalent point on the wave) and **frequency** (the number of wavelengths that passes an observer in a given interval of time). The differences in the properties of the different radiant-energy forms are reflected by the vast differences in their wavelengths. Some radio waves have wavelengths on the order of hundreds of kilometers, whereas x-rays have wavelengths as small as 1 Angstrom unit (Å). (An **angstrom** is 10^{-8} cm.) Our eyes can detect radiant-energy wavelengths ranging from 3900 to 7700 Å (visible light), but we must use instruments to detect the others.

Nuclear (atomic) energy

We saw earlier that some of the energy that holds the nucleus of uranium 238 together is released when the atom is converted to a thorium atom. In fact, this type of nuclear-binding energy is released during all nuclear reactions and is called **nuclear,** or **atomic, energy.**

Chemical energy

During chemical reactions, some of the energy by which atoms and ions are bonded to one another may be released. This type of energy is called **chemical energy.** When petroleum products are used as fuel for automobiles, the chemical energy that holds the organic molecules together is released and is used to turn the wheels.

Energy Transformations

Most kinds of energy can be converted easily into other kinds, resulting in the complex net-

Figure 2.10   The form of electromagnetic waves.

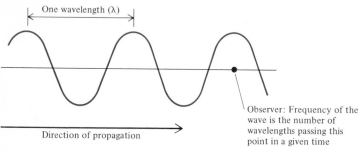

One wavelength (λ)

Direction of propagation

Observer: Frequency of the wave is the number of wavelengths passing this point in a given time

work of energy transfers that governs all earth processes. Energy is converted easily, but *it cannot be created or destroyed.* It may be transformed from one type to another, but it does not disappear. A rolling ball has kinetic energy, but where does that energy go when the ball stops rolling? Some is used to heat atoms in the material the ball is rolling on, and some is used to move air molecules out of the path of the ball. The transfer of kinetic energy to heat energy by this method is called **friction.**

Potential energy

When energy is not actually being used to produce activity or do work, it can be stored within materials in a passive state. Stored energy, regardless of its type, is called **potential energy.** A car battery contains potential electric energy that is released as the ignition switch is turned. An atom of uranium 238 is a reservoir of potential nuclear energy that is released when a nuclear reaction occurs. Organic molecules in petroleum are a source of potential chemical energy. A boulder poised at the top of a steep slope has potential kinetic energy that is released when it begins to move.

Energy conversions

Changes from one type of energy to another —from active to potential, or from potential to active—take place during all earth processes. A brief look at a geologic process shows how complex the earth's energy network is.

What makes a stream flow? The process begins in the sun. Nuclear fusion reactions in the sun release large amounts of nuclear energy. This is converted to radiant energy, and some of it reaches the earth, where it is converted to heat energy in the atmosphere and oceans. The heat energy counteracts the cohesive forces between water molecules in the oceans, causing a change of state from liquid water to water vapor (evaporation). The heat and released chemical energy are changed to kinetic energy, causing the vapor to rise into the atmosphere. Much of this kinetic energy is stored as the vapor moves laterally, but it is released when gravity pulls raindrops

down to earth. It is this kinetic energy that causes water to flow in streams.

Eventually, the water returns to the oceans, where it begins the cycle again. With all the energy conversions, the water is back in its original position and its original state. Its energy level is the same as when it entered the cycle—energy has been neither lost nor gained.

Sources of Energy for Earth Processes

Enormous amounts of energy are required each second to drive earth processes. All internal processes, such as earthquakes, melting of rock to make lava, and sliding of lithosphere plates, require energy. All surface processes, such as the flow of ocean currents and the carving of landscapes by glaciers, require energy. For the most part, internal processes are driven by sources of energy within the earth, and surface processes are driven by energy from the sun.

Internal energy sources

As miners descend into the earth, the temperature of the rock around them increases. In deep mines, such as the copper mines at Butte, Montana, and the diamond mines of South Africa, temperatures increase so rapidly that cool surface air must be circulated through the lower levels or they would be too hot for human beings to work in. Temperatures in the earth's crust and upper mantle increase at rates of 15 to 60°C per kilometer of depth. These vertical changes in the earth's temperature are called **geothermal gradients.** The fact that the temperature increases with depth indicates that there must be a source of heat energy inside the earth.

The most important internal source of heat energy is radioactivity. Part of the nuclear energy released during nuclear reactions is given off as radiant energy, and part is released as kinetic energy carried by the particles (such as alpha particles) ejected from the nucleus. Collisions with the ejected particles and interactions with the radiant energy cause atoms and ions in rocks to vibrate, thus producing heat energy. The radioactive elements uranium, thorium, potassium, and rubidium are concentrated in the earth's crust and upper mantle, where they supply the energy for a process such as the eruption of a volcano. Small amounts of heat energy are supplied by chemical reactions and by friction from movement of earth materials caused by the tides, but these are minor compared with the input from nuclear reactions.

External energy sources

Almost all the energy needed for surface processes comes from the sun by the mechanism described above. Some internal energy does reach the earth's surface, as in volcanic eruptions, but its contribution is minimal. Some kinetic energy is also added by meteorite impacts, but this too is of negligible importance.

Gravity

Although gravity, strictly speaking, is not an energy source like the sun, it does cause activity and is responsible for a considerable amount of geologic work. In fact, gravity may be thought of as a unifying theme in the study of earth sciences. In the following chapters we shall see how it controls a wide variety of processes including wind, stream, and glacial erosion of the land; the construction of different types of volcanic cones; and the types of collisions theoretically possible between different kinds of lithosphere plates.

Gravity is a force of attraction that objects have for one another due to their mass. Unlike objects that exhibit magnetic attraction, the objects need not be of a specific composition. Isaac Newton showed that the force of gravitational attraction between two bodies depends on the masses of the two objects involved, M_1, M_2, and the distance r between them:,

$$\text{Gravity} = \gamma \frac{M_1 M_2}{r^2}$$

where γ is a constant whose value depends on the units of mass and distance used. The larger the masses or the closer the objects are, the greater the force of attraction will be between them. The smaller the masses or the farther apart the objects are, the smaller the force of attraction will be.

The mass of the earth is almost incomprehensibly large (6.5×10^{21} metric tons) so that it exerts a strong attractive force on any object on its surface or in its atmosphere. A mass dropped from a cliff "falls" because of gravitational attraction. The farther it falls, the closer it gets to earth and the stronger the gravitational attraction becomes. Thus, a free-falling object does not fall at a constant velocity; it accelerates at 980 centimeters per second per second (cm/s/s). Gravity also plays a major part in processes in the earth's interior because its downward pull on rocks results in pressures that increase with depth in the planet. The pressure felt by any object at the earth's surface is caused by the effect of gravity on the molecules of gas in the atmosphere. At sea level, this pressure is 1033 grams (g/cm^2) [14.7 pounds per square inch (lb/in^2)], or 1 **atmosphere** (atm). Pressures in the earth are far greater, on the order of thousands of atmospheres. The standard unit of internal earth pressure is the **kilobar** (kb), equal to nearly 1000 atm (987 atm). Pressure increases continuously with increasing depth, and in the crust the rate of increase is approximately 1 kb for every 3.3 km.

Tides Gravity that affects earth processes does not all come from the earth. The gravitational pulls of the earth, moon, and sun combine to cause one of the major earth processes, the **tides**—the daily changes in sea level that sweep along the coastlines of the world. Twice a day the oceans lap up upon the shores of the east coast of North America at high tide, and twice a day they recede into the ocean basins at low tide. In other areas the tides are less regular, and in some areas only one high and one low tide are felt each day. The process is quiet and gradual at most shorelines, with the elevation differences of high and low water ranging only 1 or 2 m along the east and west coasts of North America and a little less in the Gulf Coast region. However, not all tidal changes are so slight or so regular. Few who have seen the tides in the Bay of Fundy can forget them. The tidal range

Figure 2.11   Tidal range at Dark Harbor, Grand Manan Island, Canada. *(a)* High tide; *(b)* low tide.   *(New Brunswick Department of Tourism)*

Figure 2.12   Causes of the tides.

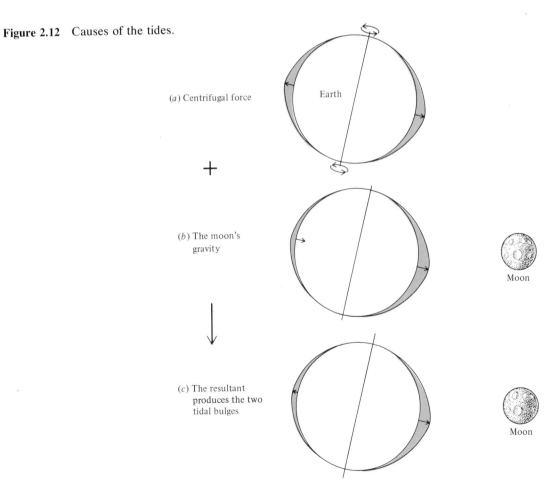

(a) Centrifugal force

Earth

+

(b) The moon's gravity

Moon

(c) The resultant produces the two tidal bulges

Moon

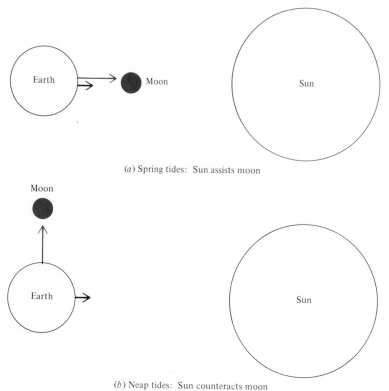

(a) Spring tides: Sun assists moon

(b) Neap tides: Sun counteracts moon

Figure 2.13 The role of the sun in tides.

there is up to 15 m (50 ft), and the difference in the shore between high and low tides is spectacular (Figure 2.11).

What causes the tides? As the earth rotates, centrifugal forces act on the oceans, causing them to bulge outward at the lower latitudes (Figure 2.12). At the same time, the moon exerts a considerable force of gravitational attraction on both the hydrosphere and the solid earth. The moon's gravitational pull is greater on the side of the earth closest to the moon than on the side farthest from it (Figure 2.12). On the side of the earth nearest the moon, centrifugal force and the moon's gravitational pull act in unison to pull the oceans toward the moon, resulting in a tidal bulge, which is what observers call high tide. On the opposite side of the earth, the centrifugal forces oppose the pull of the moon and actually overcome it so that the water there is also pulled away from the earth, and a second

tidal bulge forms. High tides are felt simultaneously on the side of the earth directly beneath the moon, and 180° away on the opposite side of the earth. Low tides are felt midway between the two high-tide areas. As the earth rotates on its axis, the geographic positions of high and low tide change; and during a 360° rotation (1 day), two high and two low tides are experienced.

Twice a month, when the moon and sun are aligned with the earth (Figure 2.13a), the gravitational pull of the sun acts to assist that of the moon. High tides are higher than usual then and low tides lower. These greater-than-usual tidal ranges are called **spring tides.** When the moon, earth, and sun are arranged as in Figure 2.13b, the sun's gravitational pull partially counteracts that of the moon. High tides are lower than usual and low tides higher, and these tides are called **neap tides.**

SUMMARY

All matter is composed of small particles called atoms, which are made up of protons, neutrons, and electrons. A single atom consists of a nucleus that contains positively charged protons and uncharged neutrons, and a cloud of negatively charged orbiting electrons. Atoms with the same number of protons are said to have the same atomic number and are atoms of the same element. Atoms with the same atomic number but different atomic masses are called isotopes.

An atom of one element may be transformed into an atom of another element by a nuclear reaction that causes an increase or decrease in the number of protons in its nucleus. Atoms may combine with one another in chemical reactions to produce compounds. Forces that bond the atoms together in compounds involve the outermost orbiting electrons and may be ionic, covalent, metallic, or van der Waals' types, depending on the nature of the interaction between the electron clouds. Compounds often exhibit physical and chemical properties that differ from those of their constituent elements.

The earth is divided into several distinct physical/chemical regions. The atmosphere consists of low-density gases, predominantly nitrogen and oxygen. The hydrosphere includes all the surface and underground waters, is liquid, and is composed mostly of H_2O. The solid earth can be divided into three parts: crust, mantle, and core. The crust is composed of eight major elements: oxygen, silicon, aluminum, iron, calcium, sodium, potassium, and magnesium. The compositions of the mantle and core are less well known. The mantle is thought to be richer in iron and magnesium and lower in sodium, potassium, and aluminum than the crust; and the core is believed to consist of metallic iron and nickel.

Energy is the capacity to do work or create activity. It is required for all earth processes and human activity. There are several kinds of energy. Kinetic energy is the energy associated with motion. Heat is a form of kinetic energy that results from the vibratory motion of atoms in solids, liquids, and gases. Heat may be transmitted through materials by conduction and convection, and through a vacuum by radiation. Radiant energy consists of electromagnetic waves defined by their specific wavelengths. Forms of radiant energy include visible light, infrared and ultraviolet radiation, x-rays, gamma rays, and radio waves.

All energy forms may be stored as potential energy until they are actually used. The energy forms are interconvertible so that in earth processes energy is neither created nor destroyed. Energy for internal earth processes comes mostly from nuclear reactions in the radioactive elements concentrated in the crust and mantle. Energy for surface processes comes mainly from the nuclear energy of the sun. Gravity is a major causal mechanism for earth processes. The combined gravitational pulls of the earth, moon, and sun produce the tides.

QUESTIONS FOR REVIEW AND FURTHER THOUGHT

1. How can atoms of two different elements have the same mass?

2. Describe the role of each of the three subatomic particles in chemical reactions.

3. Explain, with the aid of diagrams, the bonding between atoms of calcium and fluorine in the compound CaF_2 (calcium fluoride).

4. How are van der Waals' forces different from the other types of bonding?

5. What are the differences between nuclear and chemical reactions?

6. What are the different forms of kinetic energy? How do they differ?

7. What are the similarities and differences among the different forms of radiant energy?

8. What evidence is there that solar energy is not a major factor in causing internal earth processes?

9. Why is the sun's role in causing the tides less important than that of the moon?

ADDITIONAL READINGS

The following readings are freshman-level textbooks designed for introductory courses in physics and chemistry

Chemistry

These books are brief introductions to chemical principles, written for people with no previous background in the subject.

Hein, Morris, *Foundations of College Chemistry,* 4th ed., Dickenson Publishing Company, Belmont, Calif., 1976.

Roach, Don, and Edmund Leddy, *Basic College Chemistry,* McGraw-Hill, New York, 1979.

Physics

Bueche, Frederick, *Principles of Physics,* 3d ed., McGraw-Hill, New York, 1977.
(A somewhat more rigorous approach to classical physics, but still at the beginning level. Background in algebra and trigonometry is assumed, but no calculus is required)

Dittman, Richard, and Glenn Schmieg, *Physics in Everyday Life,* McGraw-Hill, New York, 1979.
(Principles of classic physics applied to modern human existence)

Romer, Robert H., *Energy: An Introduction to Physics,* W. H. Freeman, San Francisco, 1976.
(An introduction to principles of physics using the interlocking energy network as a vehicle)

CHAPTER 3

Minerals

The solid earth is composed of special types of compounds called minerals. The word *mineral* is used frequently, but it means different things to different people. To those concerned about health and diet, minerals are things to be eaten along with vitamins and proteins. To jewelers, minerals are stones to be cut, polished, and mounted in gold or silver settings. To some people, anything that is neither animal nor vegetable must be a mineral. In this chapter we will examine those features of minerals which make them special, discuss the processes by which they form, and give a brief introduction to the most common minerals in the earth's crust.

WHAT IS A MINERAL?

To a geologist, a **mineral** is a naturally occurring, inorganic solid with an ordered internal arrangement of atoms or ions, and a chemical composition that either is fixed or varies according to chemical laws. All minerals are solids so that neither water nor the gases of the atmosphere can be considered minerals. Some minerals are native elements, such as gold, sulfur, and diamond. Some are simple compounds, such as halite, $NaCl$, and hematite, Fe_2O_3. Others, however, are as complex as the mineral tourmaline, $Na(Mg,Fe,Mn,Li,Al)_3Al_6(Si_6O_{18})(BO_3)_3(OH,F)_4$. All minerals form by chemical reactions and are held together by ionic, covalent, metallic, or van der Waals' bonding.

When the definition of a mineral is applied to earth materials, several gems and "mineral" resources prove not to be minerals at all. Amber (the hardened sap of ancient trees), pearls (the secretions of shellfish), and ivory (the tusks of elephants and walruses) are not minerals because they are organic, the products of living creatures. Native mercury is not organic, but it too is not a mineral because it is a liquid. Oil and natural gas are also not solids and therefore cannot be minerals. The semiprecious gemstone opal is an inorganic solid, but it is not a mineral because it lacks the rigidly ordered internal structure required by the definition.

On the other hand, a snowflake *is* a mineral. It occurs naturally, is formed by inorganic processes, and has a constant chemical composition. Its hydrogen and oxygen ions are in fixed positions within a geometrically ordered internal structure, resulting in the familiar but always beautiful shapes of snowflakes (Figure 3.1). It is this ordered internal structure that sets minerals apart from other types of compounds, and it will be discussed at length below.

What's in a Name?

Minerals are given names, although, as compounds, they could be described by their chemical formulas. Sodium chloride, $NaCl$, is the mineral halite, silicon dioxide, SiO_2, the mineral quartz, and hydrous potassium aluminum silicate the transparent mica muscovite. In many cases, the name is easier to remember than the formula.

Mineral names come from several sources. Many minerals are named after the locality

Figure 3.1 The complex crystal shapes of snowflakes *(American Museum of Natural History)*

where they were first described or where particularly good specimens have been collected. For example, andalusite was named for Andalusia province, Spain; franklinite, for Franklin, New Jersey; and labradorite, for Labrador, Canada. Some names are derived from the Greek and Latin words that describe a mineral's physical properties. For example, the mineral orthoclase is named from the Greek words *orthos* (meaning *perpendicular*) and *klast* (meaning *to break*) because it breaks apart at right angles when struck with a hammer. Other minerals are named after people: sillimanite after Benjamin Silliman, an early American mineralogist and geologist; goethite after the German poet and philosopher Goethe; and armalcolite after the first Apollo astronaut crew to visit the moon (*Arm*strong, *Al*drin, and *Col*lins).

THE INTERNAL STRUCTURE OF MINERALS

Under favorable conditions of formation, minerals occur in regular geometric shapes called **crystals,** bounded by smooth, flat surfaces called **crystal faces** (Figure 3.2). Some are simple shapes with only a few faces, but others, such as snowflakes, are complex. It was crystals that first led scientists to conclude, as long ago as 1669, that minerals possess an ordered internal structure. However, it took nearly 250 years for scientists to prove that the structure of minerals involves the systematic arrangement of atoms or ions. In 1912, Max von Laue first used x-rays to study minerals, and within 2 years W. H. Bragg and W. L. Bragg had determined the ordered and geometrically perfect internal structure of the mineral halite (Figure 3.3).

Today, mineralogists using sophisticated instruments have determined the structures of the most abundant minerals. The ordered structure of minerals is called a **crystalline structure** because it was discovered first in crystals, but it is present in every sample of a mineral, even when conditions during formation prevented development of crystal faces. Each mineral's structure is a unique three-dimentional arrangement of its atoms or ions and can be used like a fingerprint to identify the mineral.

Figure 3.2 Unusually large crystals of Brazilian quartz with well-developed crystal faces. *(A. E. J. Engel, U. S. Geological Survey)*

Why Do Minerals Have Crystalline Structures?

Atoms in gases and liquids can move freely. For example, once two oxygen atoms are combined by covalent bonds to form the O_2 oxygen molecule, only very weak van der Waals' forces hold them to other molecules, and they are free to move. Each oxygen atom is bonded strongly to only one other atom. However, in the mineral kingdom, each atom or ion is bonded strongly to several others, and this locks them into place. In the mineral halite, NaCl, for example, each sodium ion is bonded to six chlorine ions and each chlorine is bonded to six sodiums. The reason for this arrangement lies in the sizes and electrostatic charges of the ions.

Ionic size

The role of ionic size in determining mineral structure is illustrated by halite. Ions form when an electron is transferred from a sodium atom to a chlorine atom, in the same manner as was shown for lithium and fluorine in Figure 2.3.

Figure 3.3 The internal atomic structure of halite, NaCl. This ordered cubic array of alternating Na^+ and Cl^- ions was interpreted from x-ray data by W. H. Bragg and W. L. Bragg in 1914.

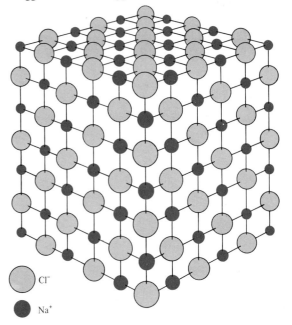

Cl^-

Na^+

Each sodium ion, Na$^+$, then attracts as many chlorine ions, Cl$^-$, as can fit around it, not just the one involved in the electron transfer. Similarly, each chlorine ion attracts and is surrounded by sodium ions.

The number of anions that can fit around a cation is called the **coordination number** and depends solely on the relative sizes of the ions involved (Figure 3.4). Each anion barely touches the surface of the cation, and their negative charges repel each other so that none of the anions come into contact with one another. In halite, six Cl$^-$ ions surround every Na$^+$ ion; they lie at the corners of an eight-sided solid called an octahedron, with the sodium at the center. This regular, geometrically perfect arrangement of ions is found in every sample of halite.

A different arrangement is found in quartz, SiO$_2$. The relative sizes of the Si^{4+} and O^{2-} ions are such that only four oxygens can fit around

Figure 3.4 The coordination principle. Every ion has as many neighbors of the opposite charge as can be packed around it. The number of neighbors (the coordination number) depends solely on the relative sizes of the cations and anions involved. *Coordination number 6:* sodium and six chlorines. *Coordination number 4:* silicon and four oxygens.

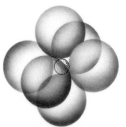

Coordination number 6
Example: 1 sodium and
6 chlorines

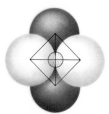

Coordination number 4
Example: 1 silicon and
4 oxygens

each silicon, forming a four-sided solid (a **tetrahedron**) with silicon at the center. This arrangement is found in all specimens of minerals containing silicon and oxygen.

Ionic charge

In halite, the six chlorine ions that surround sodium neutralize its attractive power; each expends one-sixth of its -1 charge to counter the $+1$ charge of the sodium. This leaves chlorine ions with a remnant negative charge ($-\frac{5}{6}$) that attracts more sodium ions. Every chlorine must be bonded to enough sodium ions (a total of six) to be electrically satisfied. Like any other compound, halite must be electrostatically neutral, and its formula, NaCl, shows that this is done by having equal numbers of $+1$ charged sodium and -1 charged chlorine ions.

Polymorphs: The Role of Temperature and Pressure

Diamond is the hardest mineral known. It is translucent and may either be colorless or appear in a variety of colors including yellow, blue, and green. Graphite, on the other hand, is one of the softest minerals. It is opaque and gray-black, and it has a greasy feel. Although their appearance and physical properties are drastically different, both minerals have exactly the same composition: native carbon. They are **polymorphs,** minerals that have identical compositions but different internal structures.

Diamond and graphite form under very different conditions, and their structures reflect this (Figure 3.5). The diamond structure is much more compact than that of graphite because diamond forms under extremely high pressures. Each carbon atom in diamond is bonded covalently to four others, forming a closely knit three-dimensional framework. In graphite, each carbon atom is bonded covalently to *three* others, forming a series of two-dimensional sheets. Weak van der Waals' forces hold the sheets together. As a result of the different structures, the density of diamond (the amount of mass per unit of volume) is greater than that of graphite [3.5 grams per cubic centimeter (g/cm^3) for diamond versus 2.3 g/cm^3 for graphite].

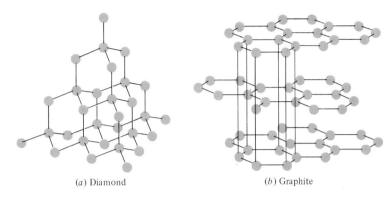

Figure 3.5 The internal structures of graphite and diamond: *(a)* In diamond, all carbon atoms are covalently linked to one another in a dense, tightly packed structure. *(b)* In graphite, sheets of covalently bonded carbon atoms are held together in a relatively open structure by van der Waals' forces.

(a) Diamond   *(b)* Graphite

Polymorphs such as diamond and graphite are useful for determining the physical conditions and locations in the earth where minerals form. The pressures needed to form diamond, for example, are greater than those found in the earth's crust so that rocks containing diamond must have originated in the mantle and somehow moved to the earth's surface.

VARIATIONS IN MINERAL COMPOSITION

Some minerals, such as graphite, C; quartz, SiO_2; and halite, NaCl, seem to have fixed chemical compositions, whereas other minerals, such as garnet, $(Ca,Fe,Mg,Mn)_3(Al,Fe,Cr)_2(SiO_4)_3$, have a range of possible compositions. Actually, nearly every mineral contains some impurities in its structure — ions trapped or included during growth. Quartz, for example, can be colorless, pink, purple, green, or yellow depending on the impurities present. Many minerals, such as the garnets, however, consistently show variations in composition that indicate a systematic substitution of ions rather than a random one. Several common ions can take the place of others in mineral structures (Table 3.1) in a process called **ionic substitution.**

Ionic Substitution

The olivine group of minerals, represented by the formula $(Fe,Mg)_2SiO_4$, provides an excellent example of ionic substitution. We saw earlier that the size and charge of ions determine mineral structures. Both magnesium, Mg^{2+}, and iron,

Fe^{2+}, ions are of the appropriate size to have a coordination number of 6 with oxygen ions, and both have the same charge. As a result, either can fit into the olivine mineral structure. Indeed, in many minerals, iron and magnesium are interchangeable and are said to *substitute* for one another. In general, if two ions have the same size and charge, they can substitute easily for one another.

The formula of the olivine group given above shows the possible substitution. The parentheses surround those ions which can substitute for one another, and indicate that the two ions can be present in any proportion. Pure Mg_2SiO_4

TABLE 3.1   Common ionic substitutions in minerals

Coordination number with oxygen ions	Approximate ionic radius, Å	Interchangeable cations
	Oxygen: 1.40	
4	Silicon: 0.34	Si^{4+}, Al^{3+}
6	Magnesium: 0.66	Al^{3+}, Mg^{2+}, Fe^{2+}, Mn^{2+}, Fe^{3+}, Cr^{3+}
8	Calcium: 0.99	Ca^{2+}, Na^+, Sr^{2+}
12	Potassium: 1.33	Na^+, K^+, Rb^+

(the mineral forsterite) or pure Fe_2SiO_4 (the mineral fayalite) can exist, and so can minerals with compositions intermediate between them. The olivine group is an example of a **solid-solution series,** a family of minerals that can have any composition between two extremes (called **end members**) and still maintain the same structure. Thus, forsterite and fayalite are end members of the olivine solid-solution series.

Coupled ionic substitution

Ions that do not have the same charge can substitute for one another if they are the same size and if electric neutrality can be maintained in some way. The plagioclase feldspar solid-solution series, one of the most abundant mineral groups in the crust, is an excellent example of this type of chemical variation. Its end members are albite, $NaAlSi_3O_8$, and anorthite, $CaAl_2Si_2O_8$. Ca^{2+} and Na^+ are so similar in size that they commonly substitute for each other in minerals, even though their charges are different. When a calcium ion substitutes for sodium in albite, there is an extra positive charge, but this problem is solved by simultaneously substituting an aluminum ion, Al^{3+}, for a silicon, Si^{4+}. This type of substitution is called **coupled ionic substitution**

because one change requires (or is coupled to) the other.

IDENTIFICATION OF MINERALS

Minerals must be identified easily if they are to be used to interpret earth processes and earth history. Fortunately, the unique combination of chemical composition and internal structure that defines a mineral results in a diagnostic set of physical properties by which the mineral can be identified. These properties make it possible for an unfamiliar mineral to be identified without the sophisticated and expensive equipment needed for chemical and structural analyses.

Physical Properties

Some of the properties of minerals, such as color and luster, require no testing and are simple to observe. Others, such as hardness or the manner in which breakage takes place, can be determined by very simple tests that require simple equipment or none at all. The physical properties discussed below are due either to the particular ions present in a mineral or to the nature of the bonding that holds them together.

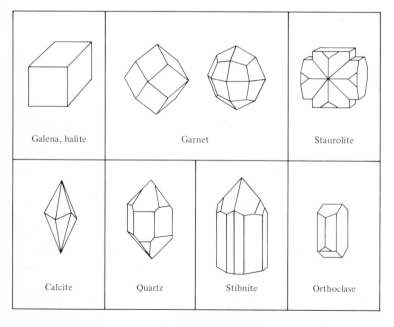

Figure 3.6 Crystal forms of some common minerals. Each of these minerals has a diagnostic crystal habit that can be used for purposes of identification.

Galena, halite   Garnet   Staurolite

Calcite   Quartz   Stibnite   Orthoclase

Crystal form (habit)

We have seen that a mineral's internal structure results in a regular crystal form if growth conditions permit. This crystal form is called a mineral's **habit** and is useful in identifying many minerals (Figure 3.6). Garnets, for example, form 12- or 24-sided equidimensional crystals but never the equidimensional cubes favored by halite or the mineral galena, PbS. Stibnite, quartz, and calcite commonly occur in elongate crystals, but the differences in the crystals allow us to distinguish the three easily. When unfavorable growth conditions prevent the formation of crystal faces, other properties must be used for identification.

Hardness

Hardness is the resistance of a mineral to being scratched and is an indication of the strength of the bonds between a mineral's constituent atoms or ions. Diamond is harder than quartz because the bonds holding diamond's carbon atoms together are stronger than the forces that bond silicon to oxygen in quartz. Geologists determine a mineral's hardness by comparing the mineral with a set of reference minerals that constitute a scale called **Mohs' hardness scale** (Table 3.2). A mineral will scratch any substance softer than itself but can be scratched by any harder substance. For example, minerals of the olivine group have hardnesses between 6.5 and 7. They can scratch the mineral orthoclase (hardness 6) but not quartz (hardness 7).

In certain mineral structures, bonds are stronger in some directions than in others, and so hardness must be tested carefully. A pocketknife (hardness 5.5) can scratch kyanite, Al_2SiO_5, if used parallel to the long dimension of the crystal but cannot scratch it if used parallel to the short dimension.

Breakage

When a mineral is struck with a hammer, it breaks along regions of relatively weak bonds. Some minerals break along smooth, flat planes, while others crumble into irregular fragments. The two different styles of breakage reflect different patterns of bonding within minerals.

TABLE 3.2   Mohs' scale for mineral hardness

Mohs' scale		Simple testing materials	
Hardness number	Mineral	Material	Hardness
10	Diamond		
9	Corundum		
8	Topaz, beryl		
7	Quartz	Streak plate	7
6	Orthoclase		
5	Apatite	Window glass	5.5+
		Pocketknife	5+
4	Fluorite		
3	Calcite	Penny	3
2	Gypsum	Fingernail	2.5
1	Talc		

Cleavage Graphite breaks along smooth, flat surfaces, and minerals of the mica family can be peeled into thin sheets (Figure 3.7). Feldspars characteristically break along smooth surfaces, but these are not all parallel to each other as in the micas. The feldspar breakage surfaces are at nearly right angles to one another. The tendency of these minerals to break along planar surfaces is called mineral **cleavage,** and the surfaces themselves are called **cleavage planes.** Cleavage occurs in minerals that have planar zones of weak bonding. In graphite, for exam-

Figure 3.7 Perfectly developed cleavage in one direction shown by a large mica crystal. *(Grant Heilman)*

ple, cleavage occurs because the weak van der Waals' bonds can be broken easily; as a result, the cleavage planes parallel the sheets of carbon atoms as shown in Figure 3.6.

In describing cleavage, it is necessary to note the number of different cleavage directions (two in feldspars, one in micas), the angles between cleavage directions, and the degree to which cleavage is developed (Figure 3.8). Two different minerals both might have cleavage in two directions, but the directions might be at right angles in one and at obtuse angles in the other. This is the case with the pyroxene and amphibole families of minerals, which have many similar properties but differ in their cleavage. Angles between cleavage planes in the pyroxenes are 87 and 93°, whereas the angles are 123 and 57° between amphibole cleavage planes. Some minerals cleave in three perpendicular directions (halite); others cleave in three nonperpendicular directions (calcite), four directions (diamond), or six directions (sphalerite).

Fracture Some minerals exhibit no cleavage because there are no planar zones of weak bonding in their structures, and such minerals break by what is called **fracture.** Some break along irregular, jagged, or splintery surfaces, but others favor smooth, curved surfaces in what is called **conchoidal fracture** (Figure 3.9). Quartz is an extremely common mineral that typically displays excellent conchoidal fracture.

Luster

The manner in which light is reflected from the surface of a mineral is called the mineral's **luster.** Luster depends on the composition of a mineral, the presence of impurities or defects in the structure, and pitting or chemical reactions on the outer surfaces. Pure, unflawed diamonds have a *brilliant* luster, but inclusions of tiny impurities or flaws in the internal structure can lead to a cloudy appearance. Most terms used to describe luster are self-descriptive. For ex-

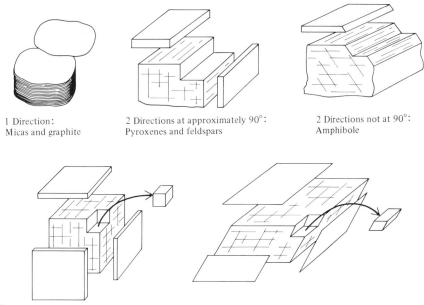

1 Direction:
Micas and graphite

2 Directions at approximately 90°:
Pyroxenes and feldspars

2 Directions not at 90°:
Amphibole

3 Directions at 90°:
Galena, halite

3 Directions not at 90°:
Calcite, dolomite

Figure 3.8 Different kinds of cleavage in common minerals.

ample, galena has a *metallic* luster; its surface looks like the metal of a new car. An *earthy* luster resembles dry, dull soil; a *glassy* luster is glasslike; a *pearly* luster is like the surface of a pearl, and so on.

Specific gravity

The **specific gravity** of a mineral is a comparison of its density with the density of water as shown in the relationship:

$$\text{Specific gravity} = \frac{\text{density of a mineral}}{\text{density of water}}$$

Density depends on the mass of the atoms in a mineral and the compactness with which they are bonded. Water at 25°C has a density of 1 g/cm^3, and graphite has a density of 2.4 g/cm^3. Graphite is 2.4 times as dense as water, and so its specific gravity is 2.4.

Most of the earth's crust is composed of minerals that contain silicon and oxygen (silicate minerals), and these have specific gravities that range between 2.40 and 4.50. However, many other types of minerals have much higher values. Gold (15.0 to 19.3) and platinum (14.0 to 19.0) have the highest specific gravities of known minerals. With practice, one can estimate specific gravity by holding a sample of an unknown mineral in one hand and comparing its weight with that of an equal-sized sample of a mineral with known specific gravity.

Color

It would seem that a mineral's color should be useful in its identification, and for many minerals it is. Galena, for instance, is always battleship gray; pyrite, brassy yellow; malachite,

Figure 3.9 Quartz crystal showing excellent conchoidal fracture. The dull faces at upper left are original crystal faces. (*Mira Schachne*)

green; azurite, blue; and cinnabar, red (Plate 1). Unfortunately, impurities and crystal defects cause many of the most common minerals to be highly variable in color. For example, quartz may be white, green, blue, gray, yellow, or colorless (Plate 2). Feldspars are colorless, white, gray, pink, or black. Sphalerite, ZnS, is white, yellow-brown, green, or black. Color *is* a diagnostic property for some minerals but not for others. It is the most easily determined physical property, but it must be used with extreme caution.

Streak

The color of a mineral's powder is called its **streak.** Streak is a more reliable property than surface color because surface effects are destroyed and impurities nullified during crushing. For many minerals, color and streak are the same, but often they are different, particularly for minerals that vary in color. Regardless of color, streak tends to remain constant. Thus, sphalerite yields a cream-yellow streak whether it is white, green, or black.

Pyrite, FeS_2, is often called fool's gold, but only a fool would not recognize it by its black streak (Plate 3). If a brassy-yellow mineral with a metallic luster gives a black streak, it can be thrown away because it is pyrite or one of its close relatives. If it streaks brassy yellow or gold, save the powder—it is probably gold.

Other properties

The preceding properties can be determined easily and are sufficient for identification of many common minerals, but there are several other properties that also prove helpful. Geologists often taste, smell, and rub their fingers across minerals because taste, odor, and feel are diagnostic properties for certain minerals. Halite obviously tastes salty, but sylvite, KCl, tastes bitter, and kaolinite sticks to the tongue. The streak of many sulfide minerals smells like rotten eggs, while that of arsenic minerals smells like garlic. Graphite and talc feel greasy. Geologists apply these tests with care. We tend to smell, in order to test for arsenic, before we taste!

Some minerals have one color in visible light but a very different appearance in ultraviolet light (Plate 4), a property called **fluorescence.** Other useful properties include **magnetism, malleability** (the ability of a mineral to be pounded into flat sheets), **ductility** (the ability to be pulled out into thin wires), and the flexibility or brittleness of a thin sheet of a sample. Indeed, any physical manifestation of composition and internal structure is useful.

One simple chemical test is often made because it requires no equipment and only a few drips of dilute (5 to 10 percent) hydrochloric acid. Minerals such as calcite that contain the carbonate anion complex $(CO_3)^{2-}$ effervesce ("fizz") in hydrochloric acid. A rapid chemical reaction takes place between mineral and acid, releasing carbon dioxide and water. The carbon dioxide bubbles out through the liquid to produce the effervescence. The reaction for calcite is:

$$CaCO_3 + HCl \rightarrow CaCl_2 + H_2O + CO_2$$

Calcite   Hydrochloric   Calcium   Water   Carbon
        acid   chloride   dioxide

Systematic Identification

Neither random guessing nor memorizing the appearance of individual mineral specimens is a successful method for mineral identification. The most effective method is to compile a complete profile of the physical properties of the unknown mineral and then compare this profile with those of minerals in determinative tables such as Appendix B.1. Very few minerals have exactly the same set of physical properties, and a look at some of the details in a table such as Appendix B.2 can help distinguish between them. The process is basically one of elimination—discarding those minerals whose properties do not match those of the unknown until only a single mineral remains.

Caution must be used in determining the properties. If a sample is not pure and contains grains of several minerals, the measured specific gravity will be an average value, not the true value for any of the minerals present. Minerals often occur in fine-grained aggregates that defy accurate testing for hardness because a scratch with a knife simply dislodges the grains; it does not test their hardness at all. A particular specimen might not show all its cleavage directions. Even though caution is needed, there is no need to panic because a few accurately determined physical properties are all you need to identify most common minerals successfully.

MINERALS AND ROCKS

How Do Minerals Form?

Minerals grow from small clusters of ions called **seed crystals,** or **crystal nuclei.** At first, a small group of ions forms a tiny **seed crystal** — a cluster of ions that possesses the appropriate proportions of elements and geometric configurations of the internal structure. More ions are attracted electrostatically to the seed crystal; as they are added in the correct structural positions, the seed grows into a visible grain. If seed crystals are far apart, each can grow without interference from the others, and well-formed crystals can be produced. If several seed crystals are close to one another, they interfere with each other as they grow, resulting in an interlocking aggregate of grains that displays no crystal faces, even though each grain has the appropriate crystalline structure.

Minerals grow in a wide range of physical conditions and in a variety of media. Some minerals grow from water. For example, ice forms by the solidification of water at 0°C and atmospheric pressure, and halite forms by precipitation as salt water evaporates at the same pressure but slightly higher temperatures. Other minerals (olivine and plagioclase) grow at much higher temperatures (550 to 1100°C) and pressures (10 kb) from a different kind of liquid — molten silicate material. Still others, such as garnet, form within solids at moderate temperatures (200 to 600°C) and pressures (2 to 10 kb) by a rearrangement of the ions already present.

Some minerals, such as quartz, can grow under a wide range of conditions, from those of the surface to those several kilometers down in the crust. Others, such as halite, form only within a restricted range of physical conditions. Minerals that form in a narrow range of conditions are used by geologists to interpret the conditions under which the rocks that contain them have formed.

Minerals and Rock-Forming Processes

Minerals do not form in isolation. Most are combined with other minerals or with many other grains of the same mineral in **rocks.** Rocks are divided into three types according to the nature of the processes that brought the minerals together; minerals can grow through each of the three types of processes.

Igneous rocks

Rocks that form by the crystallization of molten rock material called **magma** are known as **igneous rocks.** Seed crystals form as magma cools, and the igneous minerals grow in the hot fluid. Igneous minerals and rocks form wherever magma cools: at the surface during volcanic eruptions or deep within the crust and mantle. Olivine, plagioclase feldspars, and quartz can crystallize from magmas and thus may be igneous minerals.

Sedimentary rocks

The interaction of the atmosphere, hydrosphere, and solid earth causes the chemical and physical breakdown of rocks. Fragmented and dissolved rock material is transported at the surface by streams, glaciers, and wind; deposited or precipitated from solution; and finally recemented to produce a new rock. **Sedimentary rocks** are produced at the surface by these processes. They are composed of fragments of previously existing rocks so that igneous minerals may be reworked and become part of a sedimentary rock. **Sedimentary minerals** grow under surface conditions, usually by precipitation of dissolved ions from water. Halite, for example, is a common sedimentary mineral.

Metamorphic rocks

When igneous and sedimentary rocks are subjected to temperature and pressure conditions greatly different from those under which they first formed, they may respond by making mineralogic changes while still in the solid state.

How Can a Diamond Be Cut?

(a)

(b)

Figure 1 *(a)* **A diamond crystal (16.76 carats) showing its typical 8-sided habit.** *(b)* **Glass model of the Cullinan Diamond (3106 carats). The pencil is the same size in both photographs.** *(American Museum of Natural History)*

Diamonds are among the most beautiful and prized of all gems, and their multifaceted shapes reflect light brilliantly from rings, necklaces, and other jewelry. Diamond, however, is not found in nature in such magnificent shapes. Instead, its crystal habit is an octahedron (Figure 1*a*) and most of the largest specimens occur as irregularly shaped masses (Figure 1*b*). How can a diamond be "cut" into the intricate shapes we have come to enjoy? After all, diamond is the hardest natural substance. A brief look at the history of diamond cutting shows that it has changed over the past 500 years as we have learned more about the internal structure and physical properties of minerals. Ancient jewelers could not cut or modify the shape of diamond at all. Baffled by its hardness, they simply mounted the stone in a setting that enhanced its octahedral or irregular shape. Today, three different diamond-shaping methods are used, each taking advantage of the physical properties of cleavage and hardness.

Diamond Cleaving

Diamond cleaves in four directions, parallel to the faces of the octahedron, and jewelers can take advantage of this to alter the shape of a sample. A scratch is made on the surface of the diamond, using the only substance that can scratch it—another diamond. A sharp, flat blade is inserted in the scratch and oriented parallel to the cleavage plane. The blade is then tapped sharply with a mallet, and if all is oriented correctly, the mineral splits smoothly along its cleavage plane. If not, a precious gem shatters into tiny pieces. Irregularly shaped diamonds can be cleaved once the directions of the cleavage planes are located by modern x-ray methods.

Diamond Polishing, or Grinding

During the Middle Ages, jewelers found that certain faces and directions in diamond crystals are softer than others (kyanite has similar properties). Powder made from diamonds of nongem quality could be made into a paste by mixing it with olive oil; the paste was then used to grind

Figure 2 Diamond shapes produced by polishing.

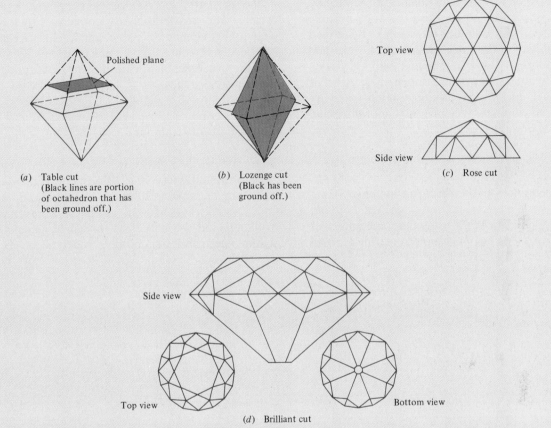

(*a*) Table cut
(Black lines are portion
of octahedron that has
been ground off.)

(*b*) Lozenge cut
(Black has been
ground off.)

(*c*) Rose cut

(*d*) Brilliant cut

away a gem-quality diamond along soft planes. The locations of these zones of softness became jealously guarded professional secrets. Two of the most commonly used planes resulted in the so-called table cut (Figure 2*a*) and lozenge cut (Figure 2*b*). In about 1520, more detailed knowledge of diamonds led to the more elaborate rose cut (Figure 2*c*). Finally, near the end of the seventeenth century, a new motif was created that modified the table cut by adding 57 more facets specifically designed to reflect light more brilliantly. This is now known as the brilliant cut (Figure 2*d*).

Diamond Cutting

None of the "cuts" described above was made originally by cutting because the brittleness of diamond makes it shatter very easily. Eventually, diamond cutting became a reality when wires were impregnated with diamond dust and then rubbed rapidly across a stone. Today, high-speed saw blades are impregnated with a diamond-dust paste, and they cut diamonds into desired shapes.

Rocks produced in this manner are called **metamorphic rocks,** and minerals that form during these changes are **metamorphic minerals.** Minerals of the garnet group are typical metamorphic minerals.

Some minerals, such as halite, are formed by only one of the three types of rock-forming processes; others, such as olivine, may be formed through two of the three processes (igneous and metamorphic). A few minerals with very wide stability ranges may form through all three. For example, quartz and plagioclase feldspars are produced through igneous, metamorphic, and sedimentary processes.

The Rock Cycle

It is apparent from the preceding statements that rocks are in a constant state of change. Nature is the original recycler, using the same ions over and over again to make new rocks in a broad scheme called the **rock cycle** (Figure 3.10). Igneous, metamorphic, and sedimentary rocks may weather at the surface to produce new sedimentary rocks and minerals. Sedimentary, metamorphic, and igneous rocks may melt deep in the crust to form new igneous minerals and rocks. The possibilities are complex. In the next four chapters we will examine closely the igneous, sedimentary, and metamorphic rock-forming processes and learn how to identify the different rock types.

ROCK-FORMING MINERALS

There are nearly 3000 known minerals, and new ones are found every year. Fortunately, only a small number of the 3000 are abundant, and a knowledge of these few is all that is necessary to understand the processes by which igneous, sedimentary, and metamorphic rocks form. In

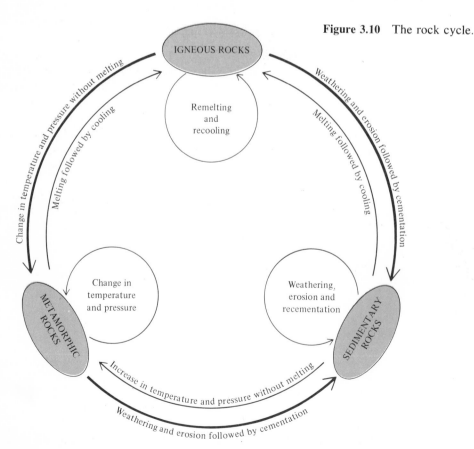

Figure 3.10   The rock cycle.

the pages that follow, we will examine some of the most important mineral groups.

Minerals are classified by the anion or anion complex they contain. For example, minerals that contain the silicon-oxygen tetrahedron are called silicates; those which contain the sulfur anion are called sulfides; and those which contain the carbonate group, $(CO_3)^{2-}$, are called carbonates. The following sections describe those groups which make up most of the earth's crust.

Silicate Minerals

We saw in Chapter 2 that the crust is composed predominantly of oxygen and silicon ions, and so it is not surprising that most of the crust is made of silicate minerals. The silicon-oxygen tetrahedron is the basic building block of all silicate minerals, but building blocks can be put together in many different ways. A brief look at the variety of silicate mineral structures gives further insight into the variety of mineral structures.

Types of silicate minerals

The silicon-oxygen tetrahedron, $(SiO_4)^{4-}$, forms, as we saw above, because the relative sizes of the Si^{4+} and O^{2-} ions permit only four oxygens to fit around a silicon. The tetrahedron has a net negative charge and acts as an anion complex, attracting cations to it. These cations satisfy the remaining negative charges of the oxygen ions and hold the tetrahedra together in a mineral structure. The specifics of this structure are governed by the precise size relationships of the ions involved and the temperature and pressure conditions of growth. This type of structure is called an **independent tetrahedron structure** and is found in olivines, $(Fe,Mg)_2$ (SiO_4); zircon, $(ZrSiO_4)$; and minerals of the garnet group.

Independent tetrahedron structures can form only when there is enough oxygen so that each silicon ion can fit into its own tetrahedron. This requires four oxygens for every silicon ion (note the SiO_4 in the mineral formulas above), but in many environments the oxygen/silicon ratio is less than 4:1. The coordination principle requires that each silicon be surrounded by four

oxygen ions; if there are not enough oxygens to go around, some must be shared between tetrahedra. The manner of sharing depends on the oxygen/silicon ratio, and the several different structural types are illustrated in Figures 3.11 through 3.16.

In some minerals, two tetrahedra share an oxygen ion and are joined together in a **twin-tetrahedral structure** (Figure 3.11). Notice that each Si^{4+} ion is surrounded by four oxygens, as required by the coordination principle, but one of the oxygens is part of both tetrahedra. The twin tetrahedron, with a composition of Si_2O_7 and a charge of -6, acts as an anion complex and attracts cations. The twin tetrahedron thus becomes the building block for all those minerals with an oxygen/silicon ratio of 7:2; and these units are held together by whatever cations are present.

Each silicon-oxygen tetrahedron in the twin-tetrahedra structure shares one of its oxygens with another tetrahedron. In other silicate minerals, each tetrahedron shares two of its four oxygens with adjacent tetrahedra. This can be accomplished in two ways. In the **ring silicates**, rings made of three, four, or six tetrahedra are formed, and *these* become the anion complexes that act as building blocks for the mineral structure (Figure 3.12). The six-tetrahedron ring shown in Figure 3.12 is the type of structure found in the gem minerals tourmaline and

Figure 3.11 Twin-tetrahedra structure. The $(Si_2O_7)^{6-}$ anion complex forms when two tetrahedra join by sharing an oxygen ion. The shared oxygen is dark gray; the unshared oxygens are light gray; the silicon is shown in color.

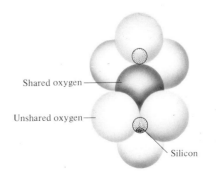

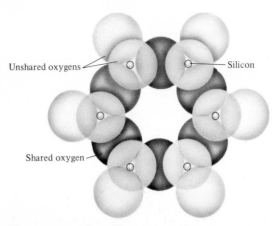

Figure 3.12   Ring silicate structures. Each tetrahedron shares two oxygens (dark gray) with neighboring tetrahedra.

beryl.   In other minerals, the tetrahedra line up to form long chains by sharing oxygens with their neighbors in what is called a **single-chain structure** (Figure 3.13).   The chains then act as anion complexes and are held together by cations.   In both the ring and single-chain silicates, the oxygen/silicon ratio is 3:1.

With progressively smaller amounts of oxygen, even more sharing takes place, resulting in yet more elaborate anion complexes. **Double chains** form when some tetrahedra share two oxygens and others share three, as shown in Figure 3.14. **Sheet silicate structures** form when every tetrahedron shares three of its four oxygens with neighbors (Figure 3.15).   This sheet structure is reflected clearly in the sheetlike cleavage found in minerals of the mica family.

When all four oxygens of every tetrahedron are shared with neighboring tetrahedra, a complex three-dimensional network called a **framework silicate structure** is formed (Figure 3.16). Quartz, plagioclase feldspar, and potassic feldspar—three of the most abundant minerals in the crust—have framework structures.

Rock-forming silicate minerals

Olivines   The **olivine** group consists of the solid-solution series forsterite-fayalite, $(Mg,Fe)_2SiO_4$, and has the independent tetrahedron structure.   Olivine minerals are characteristically green and have high specific gravities

for silicate minerals (3.27 to 4.37).   Occasionally olivine forms elongate crystals, but more often it occurs in the form of stubby, irregularly shaped grains.   Consequently, rocks composed almost entirely of olivine appear to be compact masses of grains and lack the elongate crystals common in rocks containing feldspars or amphiboles. Specimens with particularly brilliant luster are valued as the gemstone peridot.   Olivines are found throughout the crust in igneous rocks, but the magnesium-rich varieties are formed in some metamorphic rocks as well.   Olivines are very common in meteorites, and some geologists feel that they are important constituents of the earth's mantle.

Garnet   The **garnet** family consists of two solid-solution series, each of which has three end

Figure 3.13   Single-chain silicate structures.   Each tetrahedron shares two of its oxygens (dark gray) with adjacent tetrahedra, forming long chains.

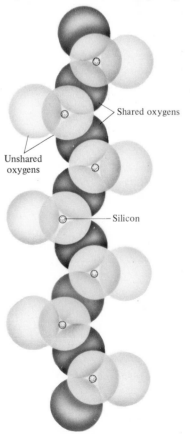

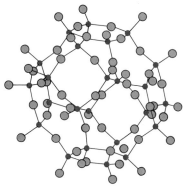

Figure 3.16 Framework silicate structures. All four oxygen ions (gray) of each tetrahedron are shared with neighboring tetrahedra in a complex three-dimensional framework.

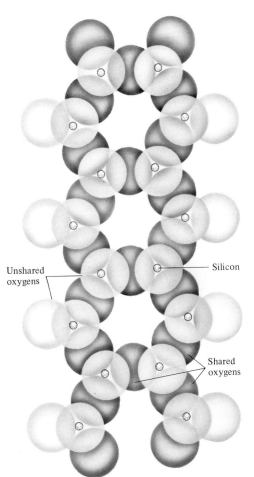

Unshared oxygens

Silicon

Shared oxygens

Figure 3.14 Double-chain silicate structures. Half the tetrahedra share two of their oxygens; the other half, three oxygens. Shared oxygens are shown in dark gray.

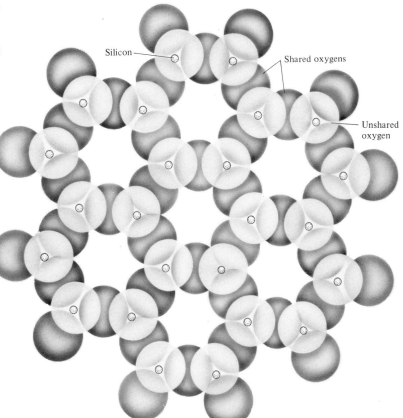

Silicon

Shared oxygens

Unshared oxygen

Figure 3.15 Sheet silicate structures. Each tetrahedron shares its three basal oxygens (dark gray) with neighbors in a silicon-oxygen sheet. The fourth, unshared oxygen (light gray) stands up above the plane of the shared oxygens.

members. The two series are $(Fe,Mg,Mn)_3Al_2$-$(SiO_4)_3$ and $Ca_3(Al,Fe,Cr)_2(SiO_4)_3$. With all the possible substitutions, it is not surprising that garnets appear in many colors. Red is the most common, but pink, brown, yellow, and bright green are also known. Garnets are independent tetrahedron silicates and are formed most commonly in metamorphic rocks, although a small fraction crystallize in igneous rocks as well.

Pyroxenes   The **pyroxene** family consists of several solid-solution series, all of which have single-chain structures. Some of these solid-solution series, such as $(Fe,Mg)SiO_3$ (the enstatite group), involve iron and magnesium substitution, whereas others also involve sodium or calcium. The augite group, $(Ca,Na)(Al,Mg)$-$(Si,Al_2)O_6$, and aegirine, $NaFe^{3+}Si_2O_6$, are examples of these. Some pyroxenes, such as augite, form under a wide range of conditions in igneous and metamorphic processes, but others are more restricted in their environment of formation. For instance, jadeite, $NaAlSi_2O_6$, forms in metamorphic rocks only at high temperatures and very high pressures. Most pyroxenes are medium to dark green, but some are brown, and one is beautiful enough to be semiprecious: jade. The weakest bonds in the pyroxene structure are always between the chains, and cleavage occurs in two directions at nearly right angles (87 and 93°). The cleavage planes parallel the chains.

Amphiboles   The **amphibole** family consists of several solid-solution series similar to those of the pyroxenes but with a double-chain structure in which the hydroxyl, OH^-, anion complex is present. Hornblende, the most common amphibole, is a major constituent of igneous and metamorphic rocks and constitutes a family of solid-solution series involving sodium, calcium, magnesium, and iron. Others, such as cummingtonite, $Mg_7Si_8O_{22}(OH)_2$; glaucophane, $Na_2Mg_3Al_2Si_8O_{22}(OH)_2$; and actinolite, Ca_2Fe_5-$Si_8O_{22}(OH)_2$, normally are produced only during metamorphism.

Amphiboles are difficult to distinguish from pyroxenes because both are chain silicates, contain the same elements, are the same color, and occur in the same types of rocks. In many in-

stances, only the oblique angle between the two cleavage directions in amphiboles (57 and 123°) permits distinctions to be made.

Micas   The **mica** family consists of several sheet silicate minerals in which large cations — K^+, Na^+, and Ca^{2+} — bond silicon-oxygen sheets together. The most abundant micas are the colorless mineral muscovite, $KAl_2(AlSi_3O_{10})$-$(OH)_2$, and the dark-brown or green mineral biotite, $K(Fe,Mg)_3(AlSi_3O_{10})(OH)_2$. Both are important constituents of metamorphic rocks and also form in many kinds of igneous rock. All micas have such excellent cleavage in one direction that they can be peeled into thin flexible sheets.

Plagioclase feldspars   The **plagioclase feldspars** are a solid-solution series of framework silicates between end members albite and anorthite, $NaAlSi_3O_8$ and $CaAl_2Si_2O_8$. Plagioclase minerals are the most abundant in the earth's crust and are found in nearly all igneous and metamorphic rocks. In addition, sodium-rich plagioclase (albite) can form in some sedimentary environments.

Plagioclases vary widely in color (white, gray, black, and colorless) but are recognized by their good two-directional cleavage at nearly right angles and their moderate hardness (6 on Mohs' scale). Some plagioclase grains exhibit very fine grooves called **striations** on one of the two cleavage directions, and these grooves are diagnostic for the entire family.

Potassic feldspars   The **potassic feldspars**, $KAlSi_3O_8$, are also framework silicates. There are three polymorphs of $KAlSi_3O_8$, each of which forms in different environments. **Sanidine** crystallizes in volcanic igneous rocks and in metamorphic rocks subjected to low pressures but very high temperatures. **Orthoclase** forms in igneous and metamorphic rocks, and to a much lesser extent in sedimentary rocks. **Microcline** has both igneous and metamorphic orgins.

Quartz   One of the most common minerals is **quartz**, SiO_2, which can form in igneous, metamorphic, and sedimentary rocks. It is a framework silicate, but unlike the feldspars there is no substitution for silicon. Quartz is identified readily in rocks by its vitreous (glassy) luster, high hardness (7), and conchoidal fracture. Quartz occurs in many colors. Its colorless variety is known as rock crystal, and each of its different-colored varieties has been given a different name: purple (amethyst), pink (rose quartz), gray or black (smoky quartz), yellow (citrine), and green (aventurine).

Clay Minerals

One important group of minerals — the **clay minerals** — does not fit into a simple chemical classification scheme because some clay minerals are silicates and others are nonsilicates. All form at the earth's surface by **weathering** — the interaction between previously existing minerals and the gases of the atmosphere. They commonly occur in very fine-grained mixtures that are plastic and can be molded when mixed with a little water. The most common clay minerals are the silicates kaolinite, $Al_2Si_2O_5(OH)_4$; montmorillonite, $(Al,Mg)_8Si_4O_{10}(OH)_{10} \cdot 12H_2O$; and illite (a mineral much like muscovite but

with less potassium and more silicon); and the nonsilicates gibbsite, $Al(OH)_3$, and boehmite, $AlO \cdot OH$.

Nonsilicate Minerals

There are more kinds of nonsilicate minerals than silicates, but the nonsilicates constitute only a very small part of the crust. Few of the nonsilicates are important rock-forming minerals. They occur most commonly as **accessory** (minor) minerals in rocks that are composed almost entirely of silicates. For example, the mineral apatite, $Ca_5(F,Cl,OH)(PO_4)_3$, is found in nearly every igneous rock, but it rarely amounts to more than 1 percent of the volume of any rock that it is in. However, nonsilicates are extremely important economically because they include most of the ore minerals from which we get metallic and nonmetallic mineral resources. A list of the most common nonsilicate minerals is given in Table 3.3, along with their modes of occurrence. We will focus here on the few non-

TABLE 3.3 Common nonsilicate minerals

Native elements	Oxides (O^{2-})	Sulfides (S)	Carbonates (CO_3)
Gold	Hematite: Fe_2O_3	Galena: PbS	Calcite: $CaCO_3$
Silver	Magnetite: $FeO \cdot Fe_2O_3$	Pyrite: FeS_2	Aragonite: $CaCO_3$
Platinum	Corundum: Al_2O_3	Marcasite: FeS_2	Dolomite: $CaMg(CO_3)_2$
Diamond (C)	Ilmenite: $FeTiO_5$	Pyrrhotite: $Fe_{1-x}S_x$	Siderite: $FeCO_3$
Graphite (C)	Chromite: $FeCr_2O_4$	Cinnabar: HgS	Malachite: $Cu_2CO_3(OH)_2$
Sulfur	Spinel: $MgAl_2O_4$	Sphalerite: ZnS	Azurite: $Cu_3(CO_3)_2(OH)_2$
		Orpiment: As_2S_3	
		Realgar: AsS	
		Stibnite: Sb_2S_3	

Halides (Cl^-, F^-)	Sulfates $(SO_4)^{2-}$	Hydroxides $(OH)^-$
Halite: NaCl	Gypsum: $CaSO_4 \cdot 2H_2O$	Goethite: $HFeO_2$
Sylvite: KCl	Anhydrite: $CaSO_4$	Gibbsite: $Al_2(OH)_6$*
Cryolite: Na_3AlF_6	Barite: $BaSO_4$	Boehmite: $AlO(OH)$*
Fluorite: CaF_2	Celestite: $SrSO_4$	

Phosphates $(PO_4)^{3-}$	Others
Apatite: $Ca_5(F, Cl, OH)(PO_4)_3$	Borax: $Na_2B_4O_7 \cdot 10H_2O$
Monazite: $(Ce, La, Y, Th)PO_4$	Wolframite: $(Fe, Mn)WO_4$
Turquoise: $CuAl_6(PO_4)_4(OH)_8$	Scheelite: $CaWO_4$
	Wolfenite: $PbMoO_4$
	Carnotite: $K_2(UO_2)_2(VO_4)_2 \cdot 3H_2O$

* In bauxite.

silicates that are important rock-forming minerals.

Carbonates

There are several **carbonate** minerals, each of which contains the carbonate $(CO_3)^{2-}$ anion complex. The two most abundant — **calcite, $CaCO_3$,** and **dolomite, $CaMg(CO_3)_2$** — are important constituents of sedimentary rocks and are also abundant in some metamorphic rocks. Both are relatively soft (around 3 on Mohs' scale) and display excellent cleavage in three nonperpendicular directions. Both react with HCl, although calcite effervesces vigorously and dolomite only weakly.

Sulfates

Gypsum, $CaSO_4 \cdot 2H_2O$, and anhydrite, $CaSO_4$, are the most common minerals containing the **sulfate** anion complex $(SO_4)^{2-}$. They are typically light gray to white in color and are among the softest of minerals (gypsum is 2 on Mohs' scale). They occur in massive aggregates or in well-developed crystals, some up to 1.5 m long. Gypsum and anhydrite are exclusively sedimentary in origin and form by the evaporation of seawater.

Halides

Elements in the periodic table next to the inert gases are called halogen elements, and their minerals are called **halides.** The only major halide rock-forming mineral is halite, NaCl, although fluorite, CaF_2, is an abundant accessory mineral. Halite occurs in clear cubic crystals and in granular masses, and it forms layers tens of meters thick in many areas. Halite, like gypsum and anhydrite, is a sedimentary mineral that forms by the evaporation of seawater.

MINERAL RESOURCES

Many minerals are useful, and we have gone to extraordinary lengths to discover and mine them. Today we dredge the deep ocean floors for manganese oxides, burrow thousands of meters into the earth for copper sulfides and diamonds, and pump steam underground to melt and recover native sulfur. Minerals are valuable resources only if they are concentrated in such large amounts that recovery is easy. In Chapters 4, 5, and 6 we will look at some of the igneous, metamorphic, and sedimentary processes that bring about these concentrations. Here we will look at those attributes of minerals which make them useful. A list of some of the important mineral resources is given in Table 3.4.

Some minerals are valuable because of the atoms they contain. Iron, aluminum, and copper — the metals we use the most — come from oxides (hematite and magnetite), sulfides (bornite and chalcocite), and hydroxides (bauxite). Uranium-bearing minerals such as uraninite, UO_2, and carnotite, $K_2(UO_2)_2(VO_4)_2$, are important sources of nuclear fuels.

Other minerals are useful because of their physical properties. The great hardness of diamond makes it an important industrial abrasive. The drills used to penetrate kilometers of rock in the search for oil are coated with diamonds that help grind through the softer rock-forming minerals. Emery cloth is coated with fine grains of corundum. Graphite, on the other hand, is so soft that it rubs off on paper. Remember, it is graphite that you write with when you use a pencil. In fact, pure graphite is too soft to be used in pencils because it smudges easily. It must be mixed with clay minerals to reach the desired hardness.

The greasy feel of talc and graphite is caused by very weak van der Waals' forces. Finger pressure alone breaks these bonds, causing sheets of silicon-oxygen tetrahedra (in talc) and carbon atoms (in graphite) to slide past one another in what we identify as greasiness. Their combined softness and greasiness make talc and graphite important lubricants. In contrast, very strong bonds give feldspars and the Al_2SiO_5 minerals andalusite, sillimanite, and kyanite very high melting points. These minerals are used in making porcelain and the refractory bricks that line the insides of steel-making blast furnaces.

Metallic bonding in gold, silver, platinum, and copper makes them superb electric conductors. Quartz responds differently to electricity,

TABLE 3.4 Minerals as natural resources (exclusive of gemstones)

AS ORES

Metal	Mineral ores
Iron	Magnetite, hematite, siderite, goethite, limonite
Copper	Cuprite, bornite, chalcopyrite, chalcocite
Zinc	Sphalerite
Lead	Galena
Aluminum	"Bauxite" (gibbsite, boehmite, diaspore)
Chromium	Chromite
Molybdenum	Molybdenite
Tin	Cassiterite
Titanium	Rutile, ilmenite

NONMETALLIC USES

Use	Minerals
Abrasives	Diamond, corundum, garnet, quartz
Building material	Gypsum, calcite, dolomite, asbestos (tremolite, serpentine)
Ceramics	Feldspars, clays, kyanite, wollastonite
Insulators	Muscovite
Optical lenses	Quartz, fluorite
Flux in metallurgy	Fluorite
Drilling muds	Barite
Medicinal	Sulfur (sulfa drugs), kaolinite, orpiment, realgar
To make acids	Fluorite (HF), pyrite, marcasite, sulfur (H_2SO_4)
Fertilizers	Apatite, potassic feldspars
Cleansers	Borax
Lubricants	Graphite, talc
Energy sources	Carnotite, uraninite, thorianite
Water purification	Zeolite family
Pigment	Limonite, orpiment, realgar

deforming slightly when subjected to a current. Thin plates of quartz vibrate in an alternating current and are used to tune the frequency of radio waves in transmitters and receivers.

The sheet structures of some clay minerals permit them to absorb large amounts of water and to swell to several times their normal volume. Such clays are used for a wide range of purposes: to thicken "thickshakes" in fast-food chains, to thicken fluids in humans (as a major ingredient in Kaopectate), and to prevent leaks in reservoirs or artificial lakes by swelling and plugging tiny cracks. The framework silicate structure of the zeolite minerals (Figure 3.17) permits small ions to pass through, but not large ones. Zeolites are used as the active ingredients in many water purifiers and softeners.

Figure 3.17 Zeolite mineral structures. Each corner represents the center of a silicon-oxygen tetrahedron. For simplicity, none of the oxygen ions has been shown.

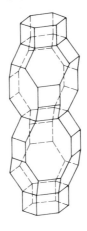

TABLE 3.5 Precious and semiprecious gemstones

Gemstone	Mineral name	Color	Hardness
Agate	Banded quartz	Varied	7
Alexandrite	Chrysoberyl	Emerald green	8.5
Amethyst	Quartz	Purple	7
Aquamarine	Beryl	Green-blue	8
Balas ruby	Spinel	Red	8
Carnelian	Quartz	Red	7
Citrine	Quartz	Yellow	7
Diamond	Diamond	Varied	10
Emerald	Beryl	Deep green	8
Garnet	Garnet family	Varied	6.5–7.5
Jade	Tremolite	Green	5–6
	Jadeite	Green	6.5–7
Lapis lazuli	Lazurite	Blue	5.5
Moonstone	Orthoclase	Translucent	6
	Albite	Translucent	
Opal	Amorphous silica	Varied	6.5
Peridot	Olivine	Green	6.5–7
Ruby	Corundum	Red	9
Sapphire	Corundum	Blue; varied	9
Topaz	Topaz	Varied	8
Tourmaline	Tourmaline	Varied	7
Turquoise	Turquoise	Blue-green	6
Zircon	Zircon	Varied	7.5

A combination of hardness, bright color, and brilliant luster is responsible for the use of minerals as gems (Table 3.5). Very few mineral specimens, even of diamond, have the right combination of properties to be gemstones; that is why gems such as those shown in Plate 5 are so expensive. Many gems are spectacular specimens of typical, everyday minerals. For example, ruby and sapphire are varieties of corundum, the same mineral we can buy cheaply at local hardware stores in emery cloth. Amethyst and agate are varieties of quartz.

There is literally no end to the uses to which minerals can be put, and new uses continue to be found. Rubies are now used to generate laser beams as well as to make jewelry, something unimaginable 25 years ago. The mixture of clay minerals we call bauxite was relatively unimportant until we learned to extract metallic aluminum from it. The more we learn about minerals, the more useful they become.

SUMMARY

Minerals are naturally occurring inorganic solids that make up most of the solid earth. Each possesses an ordered internal arrangement of atoms or ions and a specific chemical composition. A mineral's internal structure depends on the size and charge of its constituent ions and on the physical conditions under which it grows. Each ion in a mineral is bonded to as many ions of the opposite charge as can fit around it, and the total positive charge of the cations must equal the total negative charge of the anions. Polymorphs are minerals that have the same composition but different internal structures. Ions may substitute for one another in mineral structures if they are of similar size and charge.

Minerals grow by adding ions to seed crystals, and growth may take place as part of three different processes. Igneous minerals form from molten silicate materials at high temperatures. Sedimentary minerals form at the earth's surface by weathering or precipitation from water. Metamorphic minerals grow by ion migration in a solid medium at conditions intermediate between those of the surface and melting. Some minerals form through only one of the three processes, but others may form through all three.

A mineral's chemical composition, bonding, and internal structure result in a unique set of physical properties by which the mineral may be identified. Bond strength and distribution determine hardness, melting point, feel, and the manner in which the mineral breaks (cleavage or fracture). Chemical composition determines color, luster, streak, magnetism, reaction to acids, taste, and odor. Internal structure controls crystal form and, combined with composition, determines specific gravity.

Silicate minerals make up most of the crust, and all possess the silicon-oxygen tetrahedral anion complex. By sharing oxygen ions, tetrahedra may be joined in pairs, rings, chains, sheets, and elaborate three-dimensional frameworks. The silicon/oxygen ratio at the time of mineral formation determines which structures will form. A few silicate mineral families—olivine, plagioclase and potassic feldspar, pyroxene, amphibole, mica, and quartz—make up most of the rocks of the crust. There are many nonsilicate minerals, but most occur as accessory minerals in silicate rocks.

Economic uses of minerals depend on their chemical and physical properties. Many nonsilicates, particularly sulfides and oxides, are ores from which metals such as iron and copper are smelted. Minerals are used as abrasives, lubricants, electric conductors, tuning devices for radios, water softeners, jewels, and for many other purposes.

QUESTIONS FOR REVIEW AND FURTHER THOUGHT

1. What are the major differences between gaseous compounds and minerals?

2. What factors determine the internal structure of a mineral?

3. Lithium is in the same column of the periodic table of the elements as sodium and potassium and behaves as they do in chemical reactions. All three are cations with a $+1$ charge, but lithium does not substitute for sodium or potassium in minerals. Why not?

4. Andalusite, kyanite, and sillimanite are polymorphs of Al_2SiO_5 and have specific gravities of 3.16, 3.62, and 3.23, respectively. What can be inferred about the conditions of their formation?

5. Explain the reason for the existence of the following mineral properties: hardness, cleavage, specific gravity, and electric conductivity.

6. Why is color frequently less valuable than cleavage in mineral identification?

7. What controls the type of silicate structure that a mineral possesses?

8. All silicate minerals whose tetrahedra share three oxygens have sheet structures, but not all of those whose tetrahedra share two oxygens have the same kind of structure. Explain.

9. It is sometimes possible to determine a mineral's structure from its chemical composition. Compare the compositions of anthophyllite, $(Mg, Fe)_7Si_8O_{22}(OH)_2$, and pyrophyllite, $Al_2Si_4O_{10}(OH)_2$, with those of the other silicates and indicate the structural types present in each.

10. Why are minerals such as diamond more useful to geologists than minerals such as quartz in determining the conditions under which the rocks they are found in have formed?

ADDITIONAL READINGS

Beginning Level

Holden, A., and P. Singer, *Crystals and Crystal Growing,* Educational Services, Inc., Doubleday, New York, 1960, 320 pp.
(An inexpensive paperback written in highly interesting fashion for the layperson interested in mineral structures and crystals. Gives detailed instructions and "recipes" for growing crystals in the home)

Mott, N., "The Solid State," *Scientific American,* September 1967.
(An introduction to the ordered world of solids for the interested layperson)

Intermediate Level

Hurlbut, C., and C. Klein, *Dana's Manual of Mineralogy,* 19th ed., Wiley, New York, 1978.
(A sophomore- to junior-level textbook in mineralogy, and one of the best. It contains details of mineral structures, mineral crystal shapes, and the origin of the physical properties as well as descriptions of the most common rock-forming and ore minerals. It also contains superb determinative tables that permit identification of unknowns from their physical properties)

Vanders, I., and P. F. Kerr, *Mineral Recognition,* Wiley, New York, 1967.
(A book designed to help the amateur mineral collector identify minerals in the field)

CHAPTER 4

Igneous Processes and Rocks

Two hundred years ago, the scientific community was split into two groups over the origin of minerals and rocks. One group, called the Vulcanists, thought that all minerals and rocks are igneous, formed from cooling lava. Vulcanists pointed to similarities between ancient rocks and those found on the slopes of volcanoes such as Etna and Vesuvius to prove their hypothesis. The other group, the Neptunists, argued that all minerals and rocks are sedimentary, the result of precipitation from a worldwide ocean. Neptunists cited rocks containing fossil fish and clams as proof of their hypothesis, explaining that such creatures could not have lived in molten lava.

Both sides, of course, were partly right. In the mid-eighteenth century, however, no one realized that a third type of rock (metamorphic) exists. Today we recognize the three classes of rock because we understand the differences in the nature of their formation. Each of the three processes leaves a unique record in the resulting rocks that allows geologists to determine the manner in which the rocks were formed. In this chapter we will focus on igneous rocks, those formed by cooling of molten rock, or **magma.**

IGNEOUS ROCKS

Some igneous rocks form at the earth's surface during eruptions of volcanoes; they are called **volcanic** rocks. They are also called **extrusive** rocks because the magmas extrude (flow out) onto the surface from within the earth. When a magma reaches the surface of the earth, it is called **lava.** Extrusive rocks are forming today on the volcanoes of the Hawaiian and Aleutian Islands and make up the peaks of the Cascade Mountains of Washington, Oregon, and California. Rocks cooled from older lavas form the plateau through which the Columbia and Snake Rivers flow in Idaho, Washington, and Oregon, and still older cooled lavas underlie prominent ridges in Virginia, New Jersey, Connecticut, and Massachusetts. In Chapter 5 we will examine the causes of volcanic eruptions and the processes by which volcanic landforms are made.

Some magmas solidify within the earth, where we cannot observe their rock-forming processes directly. The resulting rocks are called **intrusive** because they commonly cut across or are injected into the surroundings, or **plutonic,** named after Pluto, god of the underworld. We see intrusive rocks at the surface today because uplift and erosion have exposed formerly deep-seated parts of the earth's crust. Intrusive rocks make up most of the Sierra Nevada Mountains of California and Mt. Katahdin and Stone Mountain in the eastern United States.

Igneous rocks can tell of volcanoes erupting explosively millions of years ago or of magma cooling quietly many kilometers beneath the surface—if we can learn to recognize the clues. The information is contained in the texture of every rock.

Texture

A rock's **texture** includes the grain sizes, the grain shapes, the degree of crystallinity, and the manner in which the individual grains are inter-

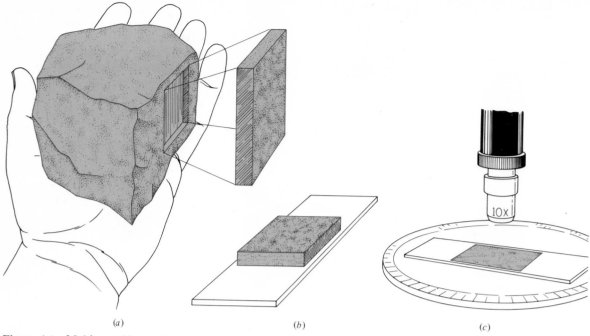

(a)   *(b)*   *(c)*

Figure 4.1 Making a thin section. *(a)* A hand sample has a representative chip cut from it. *(b)* The chip is mounted on a glass slide and is then ground thin enough (0.03 mm) so that light can pass through it. *(c)* The completed thin section is studied under a microscope, which reveals small-scale mineralogic and textural features.

grown. Certain textures immediately identify a rock as igneous, enable geologists to tell intrusive rocks from extrusive ones, and help us understand the cooling of magmas.

Many textural features are so small that they cannot be seen with the naked eye or even with a magnifying glass. To examine them, a small chip is cut off a rock with a diamond saw, and the chip then is cemented to a glass slide (Figure 4.1). The chip is ground with abrasive powders until it is so thin that light can pass through it. The slide is called a **thin section** and is studied with powerful microscopes that reveal even very small scale features. Plate 6 is a color photograph of a thin section of the fine-grained igneous rock called **basalt;** it shows how grains shorter than a millimeter can be studied with a microscope.

Interlocking grains

The mineral grains of plutonic rocks and many volcanic rocks interlock like pieces of a three-dimensional jigsaw puzzle because of the manner in which the minerals grow (Figure 4.2). The grains are intergrown intimately with their neighbors because each grain expands outward from its crystal seed until it encounters another grain, and both then grow to fill whatever space is available. Interlocking grains are not restricted to igneous rocks alone because numerous crystal seeds also form during sedimentary and metamorphic processes, but the *combination* of interlocking grains, mineral content, and the relationship of a rock to its surroundings (i.e., whether it is intrusive) is usually enough to identify a rock as igneous.

Grain size

There are igneous rocks in the Black Hills of South Dakota with crystals up to 10 m long and others in Hawaii with grains so small that they can be seen only in a thin section. Most igneous rocks have grains that are between these extremes. Grain size in an igneous rock de-

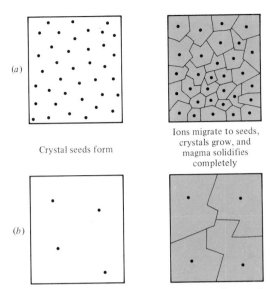

Crystal seeds form

Ions migrate to seeds, crystals grow, and magma solidifies completely

Figure 4.2 Interlocking texture, grain size, and number of crystal seeds. *(a)* In a magma with numerous seeds, crystals interfere with one another quickly as they grow, resulting in many fine-grained crystals. *(b)* Fewer crystals result from a magma in which there are a small number of seeds, but the crystals are coarser.

pends on two things: the number of crystal seeds in the magma and the amount of ionic migration to the seeds during crystallization.

If there are many closely spaced seeds, interference during growth leads to a large number of small crystals, as shown in Figure 4.2*a*. If there are only a few widely scattered seeds, a small number of large crystals will form if ions can migrate to them (Figure 4.2*b*).

In general, the more ionic migration there is, the larger each crystal can grow, and the fluidity and cooling rate of a magma determine how much migration there can be. Just as it would be more difficult for us to swim through molasses than through water, it is more difficult for ions to move through magmas with low fluidity than through highly fluid magma. If all other factors are the same, highly fluid magmas will yield the coarsest-grained igneous rocks. Even in a magma with very low fluidity, ionic migration can be extensive if it is allowed to continue for a long time. Conversely, the faster a magma cools, the less time there is for ions to move to

crystal seeds. Rapid cooling thus leads to fine grains, and slow cooling leads to coarse grains.

Volcanic rocks cool far more rapidly than plutonic rocks, just as pies cool more rapidly on a cool windowsill than in a warm oven. Consequently, volcanic rocks are fine-grained, even if the erupted lava had a high fluidity. Intrusive rocks cool more slowly and are coarser-grained. Those igneous rocks which are so fine-grained that their constituent minerals cannot be seen without a microscope are said to have an **aphanitic** texture, while those whose minerals can be seen with the naked eye are said to be **phaneritic.** Rocks with grains coarser than about 5 cm are called **pegmatitic** (Figure 4.3). Pegmatitic rocks are intrusive and probably formed from magmas that contained such large amounts of water that they were exceptionally fluid, thus permitting an unusually large amount of ionic migration.

A magma need not cool at a constant rate. Magma at depth will cool slowly and develop large crystals, but if it is driven rapidly upward before solidifying completely, the late-formed crystals will be finer-grained. This type of texture is called **porphyritic** (Figure 4.4) and indicates two cooling rates. The coarser, slowly cooled grains are called **phenocrysts;** and the finer, rapidly cooled grains are referred to as the rock's **groundmass.**

Figure 4.3 Pegmatitic texture. Granitic pegmatite composed of very large grains of muscovite, sodic plagioclase, and quartz. *(Mira Schachne)*

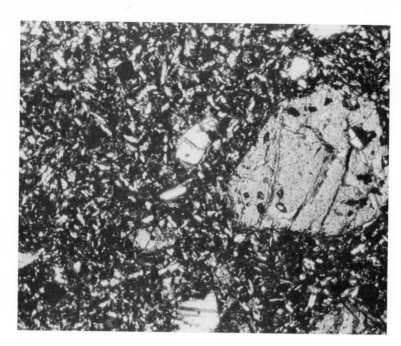

Figure 4.4 Porphyritic texture. Photomicrograph of a porphyritic basalt showing a fine grained groundmass of plagioclase feldspar and pyroxene crystals and a few large grains (phenocrysts) of plagioclase and pyroxene. *(Mira Schachne)*

Figure 4.5 A piece of obsidian, a volcanic glass. *(American Museum of Natural History)*

Some lavas cool so quickly that not even crystal seeds have a chance to form, and ions are frozen in place before they can become part of mineral structures. There are thus no minerals in the resulting rocks, and such rocks are said to be noncrystalline, or **amorphous.** Such rapidly cooled rocks are generally dark in color and glassy and are called **volcanic glasses.** Obsidian, the dark glassy rock used by the Aztecs and other early cultures for cutting implements and weapons, is a typical volcanic glass (Figure 4.5). Some magmas with two different cooling rates cooled so rapidly in their later stages that the groundmass became glassy. Such rocks have a **vitrophyric** texture. Vitrophyric rocks are partly crystalline and partly amorphous; the proportions of minerals and glass show how much of the magma cooled at each of the two different rates.

Textures produced by gas

Gases dissolved in magma play an important role in forming textures in volcanic rocks. Gases escape into the atmosphere during eruptions; the result is similar to what happens when a can of carbonated beverage is shaken vigorously and then opened. A froth composed of gas and magma forms and, if it cools quickly enough, is preserved as a **vesicular,** or **porous,** igneous rock (Figure 4.6).

The release of gases can be so explosive that the lava is broken into tiny spatters of liquid and

Figure 4.6 Porous texture in volcanic rock. *(J. C. Batte, U. S. Geological Survey)*

hurled into the air. The airborne lava solidifies to small glass fragments called **shards** that accumulate to form a rock when they fall back to the ground. When hot shards come into contact with one another on the ground, they may be welded together by their heat. Some rocks, called **tuffs**, are composed of millions of welded shards (Figure 4.7). Although individual shards are glassy, tuffs look very different from the homogeneous rock obsidian, and their texture is said to be **pyroclastic**, from the Greek roots *pyro* (meaning *fire*) and *klastos* (meaning *broken*).

Not all fragments that are shot into the air during volcanic eruptions are tiny shards. Geologists investigating the eruption of the volcano

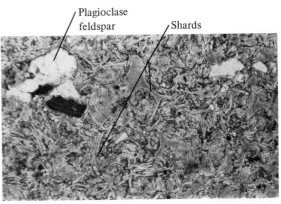

Figure 4.7 Welded Tuff. Photomicrograph of a rhyolitic tuff showing a few mineral grains embedded in a matrix of glass shards. Many of the shards show the characteristic narrow, curved shape. *(Mira Schachne)*

TABLE 4.1 Classification of tephra and pyroclastic rocks

Particle name	Size range	Name of consolidated rock
Dust	<0.062 mm	Tuff
Ash (shards)	0.062–4.0 mm	Tuff
Lapilli	4 mm–3.2 cm	Tuff
Bombs	3.2–25.6 cm	Agglomerate
Blocks	>25.6 cm	Agglomerate

Surtsey south of Iceland were pelted with blocks the size of automobiles. Material ejected from a volcano and thrown into the air is referred to as **tephra**. The names given to different size ranges of tephra are shown in Table 4.1.

It is apparent that the textures of igneous rocks supply information about the cooling histories of the magmas. These textural lines of evidence are summarized in Table 4.2.

TABLE 4.2   Summary of igneous rock textures and their interpretations

Texture	Description	Crystallization location	Interpretation
Pegmatitic	Very coarsely grained	Intrusive and/or extrusive	Crystallization from extremely fluid magma
Phaneritic	Grains visible to naked eye	Intrusive and/or extrusive	Relatively slow cooling; usually intrusive but may form in slowly cooled centers of thick lava flows
Aphanitic	Grains not visible to naked eye	Intrusive and/or extrusive	Relatively fast cooling; usually extrusive but may form in shallow intrusives or at margins of a pluton
Porphyritic	Some grains coarse; most fine	Intrusive and/or extrusive	Two cooling rates; phenocrysts cooled more slowly (deeper?) than finer-grained groundmass
Vitrophyric	Porphyritic with glassy groundmass	Intrusive and extrusive	Two cooling rates; slow at depth to produce phenocrysts, very rapid at surface to form glass
Glassy	No minerals formed	Extrusive	Extremely rapid cooling at surface
Vesicular	Porous; spongy	Extrusive	Rapid surface cooling with release of gases
Pyroclastic	Fragmental	Extrusive	Explosive eruption

Classification of the Igneous Rocks

There are many different igneous rocks and almost as many different ways of naming them. We will use a simplified classification scheme based on a rock's mineralogy and texture because both may be identified easily from hand samples, without the need for equipment other than a magnifying glass. In this scheme (Figure 4.8), texture is as important as mineral content in assigning a name to a rock. For example, a rock containing the appropriate proportions of potassic feldspar, plagioclase feldspar, quartz, and hornblende can be either a rhyolite (if fine-grained) or a granite (if coarse-grained). Similarly, a basalt can have the same mineral content as a gabbro. In each case, the finer-grained variety is volcanic and the coarser variety is plutonic.

Igneous rocks are classified in general by the proportions of ferromagnesian and feldspar minerals they contain. **Felsic** rocks (after *fel*dspar) are rocks with large proportions of potassic feldspar and sodium-rich plagioclase feldspar. They also often contain quartz. **Mafic** rocks (from *ma*gnesium and *F*e, the symbol for iron) contain calcium-rich plagioclase and large amounts of ferromagnesian minerals but little quartz or potassic feldspar. **Ultramafic** rocks are composed almost entirely of ferromagnesian minerals, with only minor amounts of feldspar. Because of these mineralogic differences, felsic rocks are usually lighter in color and have lower densities than mafic and ultramafic rocks.

The Most Abundant Igneous Rocks

Some igneous rocks are more abundant than others. The most commonly found igneous rocks are grouped into three categories: basaltic (including the intrusive equivalent gabbro), granitic (including the extrusive equivalent rhyolite), and andesitic (with the intrusive equivalent diorite). These three groups of rock are by far

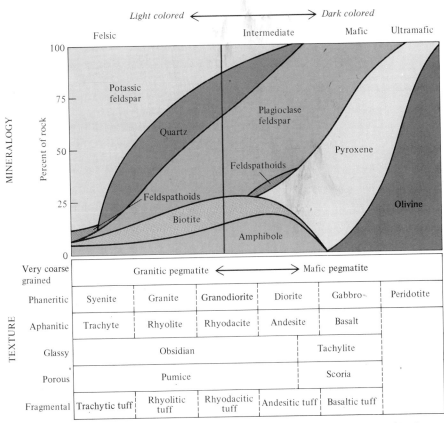

Figure 4.8 Classification of the igneous rocks based on mineral content and texture. Mineral content is indicated by drawing a vertical line in the upper part of the diagram. For example, the rock shown by the color line consists of 12.5% amphibole, 12.5% biotite, 40% plagioclase feldspar, 20% quartz, and 15% potassic feldspar. If phaneritic, it would be a granodiorite; if aphanitic, a rhyodacite.

the most abundant igneous rocks exposed at the earth's surface and play an important role in our attempt to understand how our planet works. We shall discuss them briefly here and will show how they can be used to interpret earth processes at the end of the chapter.

Basaltic rocks

Basalts and gabbros account for nearly 75 percent of the igneous rocks in the crust. They are mafic rocks composed mainly of plagioclase feldspar and one or two pyroxenes, with minor amounts of olivine or quartz. Basalts are generally dark gray to black and are fine-grained (Figure 4.9a), but thin sections clearly reveal elongate plagioclase feldspar crystals intergrown with irregularly shaped pyroxenes (see Plate 6 for color photographs of a hand sample and thin section of basalt). Porous basaltic rocks called **scoria** and basaltic tuffs are also common. Gabbros are coarser-grained (Figure 4.9b) and vary widely in color, depending on the amount and color of the plagioclase.

Granitic rocks

Granites and **rhyolites** are the next most abundant igneous rocks. They are felsic rocks composed of potassic feldspar, sodium-rich plagioclase feldspar, and quartz, with lesser amounts of a ferromagnesian mineral—usually biotite or hornblende. Some contain the colorless mica muscovite as well. Granites and rhyo-

Figure 4.9 Comparison of basalt and gabbro. *(a)* A medium grained basalt *(b)* coarser grained gabbro. The light minerals are plagioclase, the darker ones pyroxene. *(Mira Schachne)*

lites are generally light-colored rocks, occurring in shades of gray, pink, or white, but the color depends on the color of the potassic feldspar and some black varieties are known. The individual crystals of granites do not show the well-developed shape characteristic of many mafic rocks. This does not indicate different cooling processes but rather the tendency of some minerals to form good crystals more readily than other minerals. Plate 7 shows a pink granite in both a hand sample and a thin section.

Rocks with the chemical composition of granite display the widest range of textural variety among the igneous rocks. Obsidian is the

glassy variety, pumice the frothy, porous variety. Tuffs of rhyolitic composition are abundant, and the coarsest-grained of all the igneous rocks, the pegmatites, are generally of granitic composition.

Andesitic rocks

Andesites and **diorites** are intermediate in chemical composition, mineral content, and color between granitic and basaltic rocks. Most rocks of this composition occur as extrusive varieties, and intrusive masses of diorite are far less common than those composed of granite or gabbro. Intermediate plagioclase feldspar (30 to 50% anorthite), amphibole (usually hornblende), and pyroxene are the most important minerals, although quartz may be present in some varieties.

Shapes of Intrusive Rock Bodies

The general term **pluton** may be applied to any mass of intrusive rock regardless of its size or shape, but some plutons have been given special names based mainly on their shape (Figure 4.10). Flat, tabular (tabletop-shaped) plutons are called **dikes** if they cut across previously existing features such as layering in their host rocks, or **sills** if they are parallel to those features. Many dikes and sills are small, only a few centimeters across, but some, such as the Palisades Sill of New Jersey and New York, are immense. The Palisades Sill exhibits a feature common to many tabular plutons and lava flows—a systematic fracturing called **columnar jointing** (Figure 4.11). Columnar joints form during cooling, when the solidified rock contracts as heat leaves it. The columns generally form at right angles to the flat surfaces of the plutons or flows.

Many plutons are far less regular in shape than the dikes and sills. **Laccoliths** are flat-bottomed but have dome-shaped tops that arch the overlying rocks, and **lopoliths** are basically funnel-shaped. Irregularly shaped plutons are called **stocks, bosses,** or **plugs** if they are small, and **batholiths** if they underlie an area of over 75 km². For example, the Sierra Nevada batholith underlies nearly 40,000 km² of eastern California.

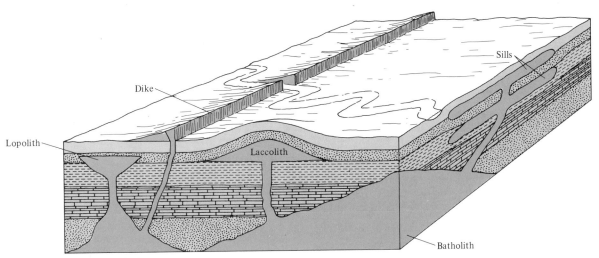

Figure 4.10   Shapes of plutons.

Figure 4.11   Columnar jointing in a basalt flow from the east wall of Grand Coulee, Washington. *(John S. Shelton)*

ORIGIN AND CRYSTALLIZATION OF MAGMA

The formation of an igneous rock involves two processes that are exact opposites of one another: melting and crystallization. In order to study these processes, geologists study igneous rocks and melt minerals and rocks in laboratories where temperatures, pressures, and chemical compositions are controlled carefully. The artificial melts are allowed to cool, and the results are compared with natural igneous rocks. The closer the agreement between artificial and natural materials, the more we can say about how igneous rocks form.

The Nature of Melting and Crystallization

The change from solid to liquid state caused by the addition of heat is called **melting.** We saw in Chapter 3 that ions are held rigidly in place in minerals. When a mineral is heated, its ions vibrate in their positions until enough heat is added to break the chemical bonds that hold them together. The ions are then free to move about in the liquid state. A magma thus consists of billions of ions, each moving freely at high temperature. Cations and anions frequently collide and form temporary bonds, but minerals cannot form until enough heat energy escapes from the magma into the surrounding rocks. When magma cools to the point at which there is no longer enough heat to counteract the attraction between ions, crystal seeds begin to form. Ions migrate to the seeds and are added in the geometric order required for mineral structures. In this way, rigidly ordered mineral structures grow from the chaotic disorder of a magma.

A mineral's **melting point** is the temperature at which it passes into the liquid state. Quartz, for example, melts at 1713°C, albite at 1118°C. A mineral's **crystallization** (or **freezing**) point, the temperature at which it becomes solid as magma cools, is the same as its melting point. Minerals with high melting points are the last to melt during heating but are the first to form when magma cools. Conversely, minerals with low melting points are the first to melt and the last to crystallize.

Melting and Crystallization of Minerals

Melting is such a common process that most people take it for granted. Geologists, however, cannot take melting for granted because experiments have shown that not all minerals melt in the same way that ice melts. Indeed, the minerals that make up igneous rocks can become part of magma in three ways: simple, continuous, or discontinuous melting. Since crystallization is simply the reverse of melting, minerals can form from magma by three different processes.

Simple melting and crystallization

Many common minerals, such as quartz, muscovite, albite, anorthite, and forsterite, melt in exactly the same way that an ice cube melts (Figure 4.12). When quartz, SiO_2, is heated, nothing happens until a temperature of 1713°C is reached. At that temperature, bonds between silicon and oxygen ions break, and the mineral melts to form a liquid with the composition SiO_2. This is called **simple,** or **congruent,** melting because it happens at a single temperature, and the solid and liquid have the same composition as soon as melting starts.

Crystallization is simply the reverse of this process. A magma with the composition SiO_2 remains molten until a temperature of 1713°C is reached during cooling. At that point, the ions are reunited to form quartz.

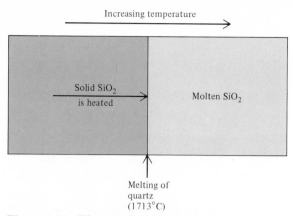

Increasing temperature →

Solid SiO_2 is heated

Molten SiO_2

Melting of quartz (1713°C)

Figure 4.12 The simple melting of quartz. Quartz melts at a single temperature and in a single-stage process. There is no reaction between quartz and magma.

Continuous melting and crystallization

Most minerals that are part of solid-solution series do not follow this simple melting behavior. For example, of the plagioclase feldspars, only the pure end members albite and anorthite melt simply, at 1118 and 1553°C, respectively. All other plagioclase feldspars are mixtures of these end members and melt in a more complex fashion.

As an example, consider the melting of a plagioclase feldspar composed of equal amounts of albite, $NaAlSi_3O_8$, and anorthite, $CaAl_2Si_2O_8$, as shown in Figure 4.13. Bonds between Na^+ and O^{2-} are weaker than those between Ca^{2+} and O^{2-} and thus require less heat to be broken. When heat is applied, all bonds do not break at the same temperature, as they would in simple melting. At the beginning of melting, more Na—O bonds break than Ca—O bonds. The first liquid formed thus has more of the sodium-rich component in it than calcium, even though the original mineral had equal amounts of the two. Because the sodium component is melted preferentially at low temperatures, the remaining unmelted mineral becomes richer in calcium. Only when more heat is added will the Ca—O bonds break. Plagioclase feldspars thus melt continuously over a range of temperatures, and melting ends when the last of the mineral adds its ions to the magma. Plagioclase feldspar and

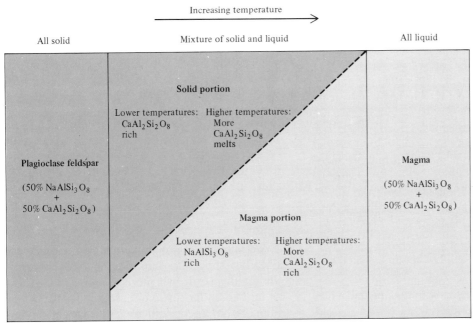

Increasing temperature →

All solid   Mixture of solid and liquid   All liquid

Solid portion

Lower temperatures: Higher temperatures:
$CaAl_2Si_2O_8$ More
rich $CaAl_2Si_2O_8$
                          melts

Plagioclase feldspar   **Magma**

($50\%\ NaAlSi_3O_8$ ($50\%\ NaAlSi_3O_8$
+   +
$50\%\ CaAl_2Si_2O_8$) $50\%\ CaAl_2Si_2O_8$)

Magma portion

Lower temperatures: Higher temperatures:
$NaAlSi_3O_8$ More
rich $CaAl_2Si_2O_8$
                          rich

Figure 4.13 Continuous melting of plagioclase feldspar. The first liquid that forms is richer in $NaAlSi_3O_8$ than the initial crystal. The remaining solid melts gradually over the melting interval, and the liquid continuously approaches the composition of the starting mineral. When melting is completed, the liquid achieves the same composition as the starting mineral.

other solid-solution minerals thus do not have a single melting point as quartz does, but rather a melting interval over which the solid is melting *continuously,* and both magma and mineral are changing compositions continuously.

Crystallization from a magma of this composition is also carried out over a range of temperatures. At high temperatures, calcium-rich plagioclase seed crystals form; but as temperature decreases, more and more of the sodium plagioclase component is added until all the magma is used up. The compositions of both mineral grains and residual magma change continuously over the crystallization interval, and the process is referred to as a **continuous reaction.**

Many plagioclase feldspar crystals exhibit a **compositional zoning** that reflects this change in composition (Figure 4.14). Such zoned crystals have calcium-rich cores and sodium-rich rims. However, many plagioclase crystals are not zoned. If plagioclase cools very slowly, ions from the crystal cores mix with those of the rims

Figure 4.14 Compositional zoning in plagioclase feldspar. The large gray crystals of plagioclase show a concentric structure indicative of compositional zoning. (*I. J. Witkind, U. S. Geological Survey*)

to produce a homogeneous crystal. Zoned plagioclase crystals thus indicate more rapid cooling than homogeneous plagioclase crystals.

Discontinuous melting and crystallization

Many of the common ferromagnesian silicates follow a third type of melting behavior called **discontinuous** melting. As an example, consider the melting of the calcium-poor pyroxene enstatite, $MgSiO_3$, as shown in Figure 4.15.

When enstatite begins to melt, there is a complete reshuffling of the ions involved. Some bonds break to free ions, but the liquid produced has the composition SiO_2. The ions in the remaining solid are rearranged to form the olivine mineral forsterite, Mg_2SiO_4. The process can be described by the equation:

$$\underset{2MgSiO_3}{\overset{\text{Enstatite}}{}} \xrightarrow{\text{Initial melting}} \underset{\substack{Mg_2SiO_4 \\ \text{Solid}}}{\overset{\text{Forsterite}}{}} + \underset{\substack{SiO_2 \\ \text{Melt}}}{\overset{\text{Melt}}{}}$$

Initial melting thus yields both magma and a new mineral, neither of which is at all similar in composition to the starting mineral. As more heat is added, forsterite gradually melts until there is no longer any solid and the magma has the same composition as the starting mineral. This type of melting is called discontinuous because the compositions of both solid and liquid change abruptly during the course of melting.

When magma with the composition $MgSiO_3$ crystallizes, the first crystals formed are of *forsterite*. Forsterite crystallizes until a reaction takes place with the remaining magma in which the forsterite is converted to enstatite. This type of reaction between mineral and magma is called a **discontinuous reaction** because the magma abruptly stops crystallizing one mineral and begins to form another.

In some rocks, the reaction between mineral magma does not go to completion. The result is a composite grain with forsterite in the core and enstatite in the rim (Figure 4.16). Rims like these are called **reaction rims.**

Melting of Rocks in Nature

Experiments tell how minerals melt and crystallize, but they also indicate that the processes within the earth are more complex than those described so far for two reasons: (1) Rocks that melt within the earth are made up of several minerals, not a single mineral, and (2) the presence of water in rocks has a significant effect on both melting and crystallization.

The effect of several minerals

A rock composed of several minerals or a magma made from such a rock behaves differently from the way single minerals behave be-

Figure 4.15 Discontinuous melting of enstatite. Enstatite melts over a range of temperatures in a two-stage process. When melting begins, enstatite is converted to molten SiO_2 and forsterite. Melting then continues as forsterite gradually becomes molten, until melting is completed and the magma has the composition $MgSiO_3$.

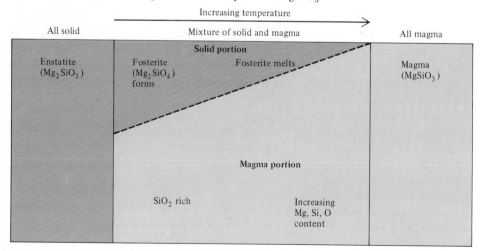

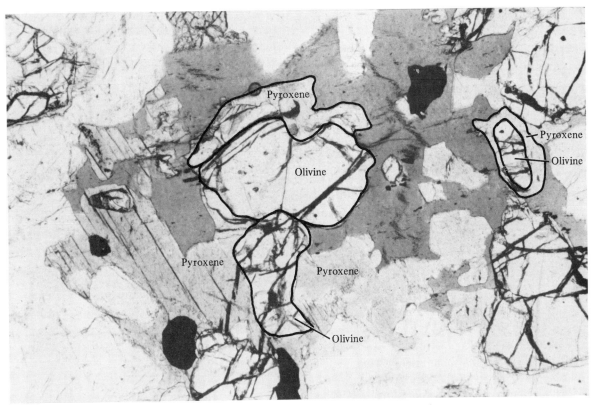

Figure 4.16 Photomicrograph showing reaction rims in a gabbro. Olivine crystals are surrounded and partially replaced by reaction rims composed of pyroxene. (*Mira Schachne*)

cause each mineral or ion has an effect on the behavior of others. For example, in a magma formed by melting a mixture of forsterite, Mg_2SiO_4, and anorthite, $CaAl_2Si_2O_8$, the calcium and aluminum ions get in the way of magnesium ions as they move toward potential olivine seed crystals. At the same time, the magnesium ions hinder the formation of anorthite seeds. The result is that the magma remains molten until lower temperatures are reached. The presence of several minerals in a rock tends to lower its melting point as well. Thus, a granite can melt at temperatures as low as 650°C, several hundred degrees lower than the melting points of its constituent minerals.

The role of water

In nature, heat is not the only agent that breaks bonds in minerals. Because of a peculiarity of its molecular structure, water exerts a force on ions in minerals that can help break them apart. Since water is present to some extent in nearly every rock, it can be a major factor.

Water is a **polar molecule** (Figure 4.17). This means that even though it is electrically neutral, it has positive and negative ends. The polarity is caused by the fact that both positively charged hydrogen ions are on one side of the molecule, leaving the other side with a very weak negative charge. Each water molecule acts as a magnet and exerts a weak electrostatic attraction on any cation or anion exposed at the surface of a mineral. This attraction partially counteracts the bonds that hold these ions in place, making it easier for them to be freed. Less heat energy is then needed to break the already weakened bonds, and so water lowers the melting points of most substances.

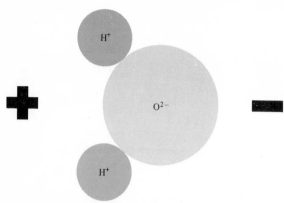

Figure 4.17 Water as a polar molecule. The position of both hydrogen cations on one side of the water molecule gives that side a weak positive charge. The other side has a weak negative charge. Both sides can attract ions in mineral structures and help dislodge them from the mineral.

Where does melting occur?

Melting takes place within the earth wherever there is sufficient heat energy from nuclear reactions. The most abundant radioactive elements—uranium, thorium, potassium, and rubidium—are concentrated in the crust and upper mantle, and most magmas probably form in these regions, close to the sources of heat. At one time, geologists thought that there was a continuous layer of magma in the mantle from which all igneous rocks were derived, but evidence from earthquake waves (see Chapter 19) shows that such a region does not exist. Melting takes place locally in small regions, called **magma chambers,** where heat is concentrated. The walls of the magma chamber are rocks that are hot but not hot enough to melt.

It is possible to estimate the depth at which melting can take place by comparing the melting temperatures of rocks with regional **geothermal gradients** (the rates at which temperatures increase with depth). Thus, experimental data show that at all pressures granitic rocks melt at lower temperatures than basaltic rocks, as shown by the black lines in Figure 4.18. The depth of initial melting can be estimated by plotting geothermal gradient curves (colored lines in Figure 4.18) on a diagram of the melting curve. Melting begins where the melting and gradient curves

intersect. Granitic magmas can form closer to the surface than basaltic magmas, but no closer than 20 km at an average geothermal gradient of 30°C/km. Basaltic magmas can form no closer than 25 km from the surface at the same gradient.

What Is Magma Like?

So far, we have treated magma as a substance that exists temporarily between the time a rock begins to melt and the time an igneous rock crystallizes from it. Certain properties of magma—notably temperature, composition, and viscosity—play important roles in the final nature of igneous rocks and must be considered here. Our knowledge of the properties of magma comes from those places where it reaches the surface and is then called lava and from deductions based on studies of igneous rock. Field geologists, some taking great risks (see Plate 8), have collected much of the data about magma from lava lakes and actively erupting volcanoes.

Temperature

Temperatures of basaltic lavas in Hawaii have been measured at 1000 to 1200°C, and melting experiments suggest a range of 650 to 1350°C for most silicate magmas. Typical magmas are

Figure 4.18 Estimated depths of melting. Melting begins where the geothermal gradients (colored lines) intersect the experimentally determined melting curves. At an "average" geothermal gradient of 30°C per kilometer, granite melts no closer than 20 km below the surface, and basalt no closer than 25 km.

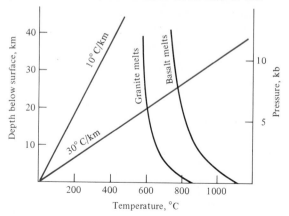

probably not heated much beyond their final melting points. Granitic magmas have the lowest temperatures, basaltic magmas the highest.

Composition

The composition of a magma may be estimated from the chemical composition of the rocks that form from it. Thus, all but a few magmas are composed mainly of ions that would give rise to silicate minerals. The chemical compositions of felsic, intermediate, and mafic rocks (granite, diorite, and basalt, respectively) are given in Table 4.3. The indicated range probably spans that of natural magmas because geologists believe that many ultramafic rocks form from mafic magmas by processes, called magmatic differentiation, which we shall discuss shortly. The differences in composition are shown most clearly by the values of SiO_2 and by corresponding values of iron and magnesium. Magmas high in SiO_2 crystallize quartz and feldspars. Magmas low in SiO_2 are generally high in FeO and MgO, and they crystallize ferromagnesian minerals such as olivine and pyroxenes.

Not all the constituents of magma end up in igneous rocks. All magmas contain dissolved gases, but most of these gases escape during cooling and are not incorporated in minerals. Water vapor and carbon dioxide are by far the most abundant gases in lavas, but significant amounts of sulfur dioxide, SO_2; sulfur, S; nitrogen; chlorine; hydrogen; methane, CH_4; and ammonia, NH_3, have been collected at volcanic vents. The earth's atmosphere probably evolved by the addition of volcanic gases from the interior.

Viscosity

The **viscosity** of a fluid is a measure of the sluggishness with which it flows; it is essentially the opposite of *fluidity*. For instance, water, molasses, and petroleum jelly are three fluids listed in order of increasing viscosity. Some lavas have very low viscosity and flow like molten rivers at several kilometers per hour, but others have high viscosity and move at only a few meters per hour.

Low viscosity is associated with high temperatures because each ion then has a lot of

TABLE 4.3 Compositions of representative igneous rocks (weight percent)

Oxide	Hornblende granite	Diorite	Olivine basalt
SiO_2	70.72	58.17	47.1
TiO_2	0.41	0.81	2.7
Al_2O_3	13.11	17.26	15.3
Fe_2O_3	1.85	3.07	4.3
FeO	1.97	4.18	8.3
MgO	0.50	3.24	7.0
MnO	—	—	0.17
CaO	1.36	6.93	9.0
Na_2O	3.35	3.21	3.4
K_2O	5.60	1.61	1.2
P_2O_5	—	0.21	0.41

MINERAL	COMPOSITIONS	
Hornblende granite	Diorite	Olivine basalt
Potassic feldspar	Ca-Na plagioclase	Ca plagioclase
Sodic plagioclase	Hornblende	Olivine
Quartz	Augite	Augite
Hornblende		
Biotite		

kinetic energy. This enables the ions to move rapidly and freely—the two requirements of a highly fluid substance. High water content also causes low viscosity because of the lubricating action of water and because the polar water molecule prevents ions from bonding to one another.

High SiO_2 contents, on the other hand, cause high viscosity and very sluggish behavior. This is because the attraction between Si^{4+} and O^{2-} ions is so strong that many temporary bonds form within the magma, restricting freedom of ionic movement. Basaltic magmas, with their low SiO_2 contents and high melting points, are typically the most fluid; granitic magmas are the most viscous.

Crystallization in Magma Chambers

The crystallization of millions of metric tons of magma in a magma chamber is far more complex than crystallization of a few grams of melt in a laboratory. However, features such as zoned plagioclase crystals and reaction rims found in igneous rocks indicate that our experiments are reasonable approximations of the processes involved in natural crystallization. The added complexities arise for two reasons: (1) Natural magmas are far more complex chemically than artificial melts, resulting in a greater number of minerals and more involved crystallization histories, and (2) experiments are controlled carefully so that there are no interruptions or changes in composition for the duration of the study, whereas natural crystallization may take tens of thousands of years, during which many changes in physical and chemical conditions are possible. Details of natural crystallization are preserved in rock textures so that we are able to follow the path of a natural magma through its evolution into the final igneous rock.

Sequence of crystallization: Bowen's Reaction Series

Granites contain several minerals—quartz, potassic feldspar, sodium-rich plagioclase feldspar, hornblende, and biotite—that display all three types of crystallization behavior. Gabbros also contain minerals that form through different processes. Some of these minerals have higher melting points than the others and crystallize early, but in most magmas several minerals crystallize simultaneously. It is conceivable that simple, continuous, and discontinuous crystallization can occur simultaneously in some magmas, whereas crystallization may be far simpler in others.

N. L. Bowen, one of the pioneers in the experimental study of melting and crystallization, reduced many of these complexities to a simple diagram that shows the sequence in which minerals generally form in magma (Figure 4.19). In his honor, the diagrammatic crystallization scheme is called **Bowen's Reaction Series.**

Figure 4.19 illustrates the sequence of the discontinuous reactions by which ferromagnesian minerals crystallize on the left side and the continuous reactions by which plagioclase feldspars crystallize on the right. Minerals that are on the same horizontal line crystallize simultaneously. Igneous rocks are composed of minerals that crystallize together or within a narrow range of temperature. Thus, minerals that are close to each other in Bowen's Reaction Series are commonly found in the same igneous rock, as shown by the shading. Minerals such as quartz and calcium-rich plagioclase that are far apart in the diagram are rarely found together in igneous rocks.

Bowen's Reaction Series does *not* indicate that olivine and calcic plagioclase are always the first minerals to form in magma. What crystallizes first depends on the magma's composition, temperature, and pressure. In some magmas, an amphibole such as hornblende is the first and only ferromagnesian mineral to form, and in others the first crystals may be of postassic feldspar. The diagram merely shows the sequence of mineral formation most commonly observed in nature and reproduced in experiments.

Magmatic differentiation: Changes in cooling history

A magma's crystallization is not always the reverse of its melting because there are several processes by which ions can be added or removed from a cooling magma. The more a

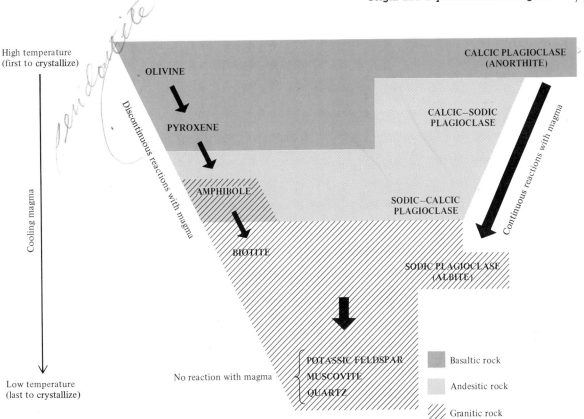

Figure 4.19 Bowen's Reaction Series. The order of crystallization of minerals from magma is shown by the arrows. Minerals that crystallize together are found together as shown by the three shaded areas that represent the three most common igneous rock types.

magma's composition changes, the more its cooling will deviate from that predicted from its melting behavior. A single magma may yield several different igneous rocks, depending on the extent to which these processes, called **magmatic differentiation,** have operated. Ions are added to magmas by the process of assimilation, while materials may be removed by gravity separation or by filter pressing.

Assimilation In melting experiments, the starting material is placed in a gold or platinum container so that the container will not react with the melt and yield erroneous results. However, in nature the walls of magma chambers are often melted locally by heat escaping from the magma, and the ions released from the wall rock are absorbed (**assimilated**) by the magma. Figure 4.20 shows blocks of wall rock incorporated in an

igneous rock. Such inclusions are called **xenoliths,** from the Greek roots *xenos* (meaning *stranger*) and *lithos* (meaning *rock*). Many xenoliths have rims from which ions have been removed, suggesting partial assimilation by the enclosing magma.

Once they are added to a magma, the new ions can change the magma's composition to the extent that minerals may crystallize that would not otherwise have formed, or the compositions of minerals such as pyroxene or plagioclase may be very different from those expected to form from the original magma. These mineralogic changes are usually most prominent near the contact between plutons and their wall rock.

Gravity separation Most crystals are denser than their parent magmas and therefore sink toward the floor of their magma chambers. This

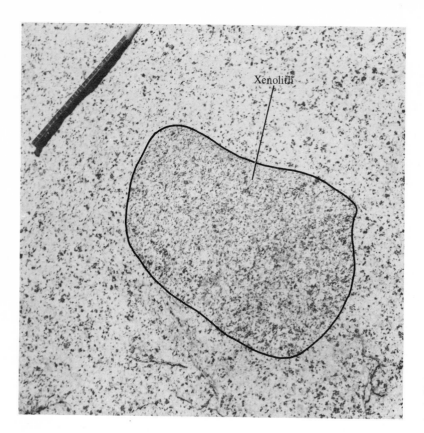

Xenolith

Figure 4.20 A xenolith in granite. The darker concentration of minerals in the center of the photograph is a partially assimilated xenolith incorporated within the granite. (*H. W. Smedes*)

process, called **crystal settling,** effectively removes early-formed crystals from reactions with residual magma (Figure 4.21). It is particularly effective in magmas with low viscosity because crystals can sink more easily. The magma behaves as if the crystals had never been present and starts crystallizing as if for the first time. The settled crystals form one rock, the residual magmas another.

One of the classic examples of crystal settling is in the Palisades Sill, a body of basaltic rock that forms cliffs nearly 300 m high along the west shore of the Hudson River in New Jersey and New York. Early-formed olivine crystals that normally would have entered into discontinuous reactions with the magma instead settled toward the bottom of the body and formed a rock composed largely of olivine. The remaining magma, depleted in $(Fe,Mg)_2SiO_4$, crystallized to produce rocks in which there is

very little olivine. The settling of mafic minerals such as olivine and pyroxene may thereby lead to the accumulation of an ultramafic rock from a magma of mafic composition.

Filter pressing The preceding processes assume that magma chambers remain undisturbed throughout cooling, but magmas commonly form in areas of great instability where rocks are squeezed, stretched, and broken. If a magma chamber is squeezed, the fluid in it is forced through small fractures into the surrounding rock, but the solid crystals are trapped against the small openings (Figure 4.22). This process is called **filter pressing** because the crystals are filtered out of the magma. The residual magma moves to a new location, where it begins crystallizing anew without the possibility of interacting continuously or discontinuously with the separated crystals.

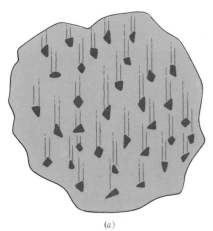

(a)   *(b)*

Figure 4.21 Differentiation by crystal settling. *(a)* Early-crystallized minerals with greater density than their parent magma sink toward the bottom of the magma chamber. *(b)* Crystals accumulate at the base of the magma chamber to form one type of rock; residual magma crystallizes above it to form another rock with a different composition.

Movement of magma

Some magmas may crystallize in the same magma chambers in which they formed, but most are forced from them and crystallize elsewhere. Some of the forces that cause the movement are external, such as those which squeeze a magma chamber and cause filter pressing. However, some of the forces are internal.

Magmas are not as dense as the rocks from which they have formed and as a result tend to rise upward in the crust until they come in contact with rocks of equal density. This upward movement is aided by gases dissolved in magma. Gases expand as magma rises into regions of lower pressure; as they expand, they carry the magma farther upward. In some instances the gas-powered rise may be both rapid and violent. Rapidly expanding gases are thought to bore upward rapidly through the crust, bringing with them magmas generated in the mantle. The gas-driven magmas incorporate blocks of the host rock as xenoliths as they rise, including some xenoliths from the mantle itself. We saw in Chapter 3 that diamond forms through temperatures and pressures of the mantle. Most dia-

Figure 4.22 Filter pressing. *(a)* A partially crystallized magma is disturbed as its chamber is squeezed, forcing the liquid out into cracks in the surrounding rocks. *(b)* The early-formed crystals are left in the old chamber, but the magma is separated from them and crystallizes in a new magma chamber.

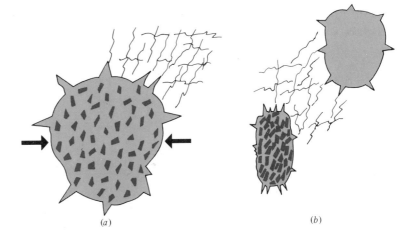

(a)   *(b)*

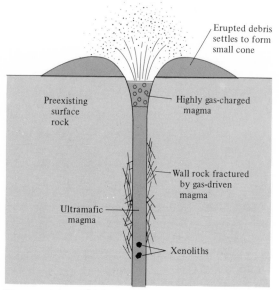

Figure 4.23 Origin of diatremes. Gas-powered magma drills through host rock and erupts at the surface.

IGNEOUS ROCKS AS A CLUE TO THE EARTH'S INTERIOR

Igneous rocks provide several important clues to the composition of the earth. For example, xenoliths found in diatremes are samples of the mantle. Most of the xenoliths are ultramafic rocks, indicating that at least the upper part of the mantle is of ultramafic composition.

The worldwide distribution of the three most abundant igneous rock types, shown in Figure 4.25, also proves valuable in determining the earth's composition. Basaltic rocks are found throughout the world. Basalts have erupted on every continent and form the floors of the world's ocean basins (see Chapter 16). However, granites and rhyolites are more restricted in their occurrence. Granitic plutons and rhyolitic volcanic rocks are confined almost entirely to the continents, and it has been shown that the rare rhyolites found in the oceans were differentiated from mafic magmas rather than having crystallized directly from a granitic magma. Thus, granitic magmas appear to be absent from the oceans. Andesitic rocks are even more restricted. Most are concentrated in islands situated near the margins of oceans, such as the Aleutian, Japanese, and Philippine Islands, or at the edges of continents. The west coasts of both North and South America, for example, contain many andesitic volcanoes in the Cascade Mountains and the Andes (from which andesite was named).

A simple model of the composition of the crust and upper mantle based on this distribu-

monds come from xenoliths in pipe-like plutons called **diatremes** (Figure 4.23), whose rock is thought to be emplaced by gas-driven uplift.

Few magmas rise so rapidly, and many simply move slowly upward through fractures in the host rock. A magma may make room for itself by a process called **stoping** (Figure 4.24). During stoping, projections of magma move along previously existing fractures until they engulf blocks of the host rock. The blocks are then detached from the surroundings and incorporated as xenoliths, while the magma fills the space from which they have been dislodged.

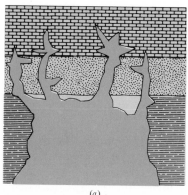

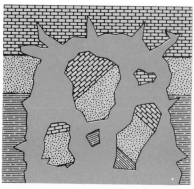

(a)   *(b)*

Figure 4.24 Upward movement of magma by stoping. *(a)* As magma rises, it extends projections into the host rock along fractures. The projections penetrate upward and grow, and begin to engulf blocks of the host rock. *(b)* Blocks of host rock are dislodged and incorporated as xenoliths in the magma, and the magma occupies the space from which they came.

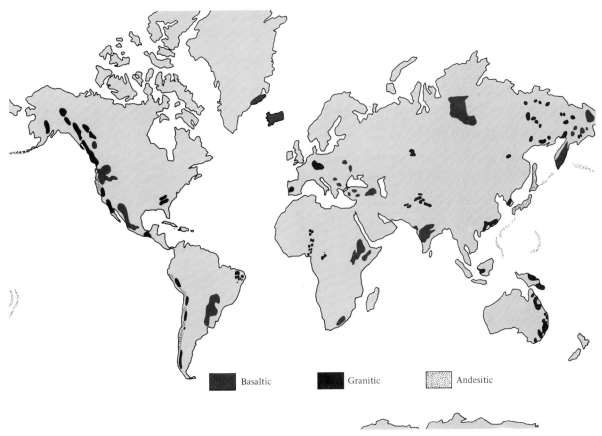

Figure 4.25   Distribution of the major igneous rock types on the continents.   The ocean basins are basaltic.   For simplicity, only those rocks younger than 225 million years are shown.   *(After Hyndman, 1972)*

tion pattern is shown in Figure 4.26; it indicates a major difference in composition beneath oceans and continents. The reasoning by which the model is constructed is as follows. We saw earlier in this chapter that granitic magma melts at lower temperatures than basaltic magma. If there is enough heat energy to form basaltic magma beneath the oceans, there must be enough to form granitic magma in the same place. The absence of granitic rocks from the oceans must be due to the absence of rocks that can melt to produce granitic magma. The abundance of granite and rhyolite on the continents indicates that rocks of appropriate composition are present there but not in the oceans. Basaltic rocks are found on the continents and in the oceans, indicating that rocks capable of melting to form basaltic magma underlie both regions.

Figure 4.26   Differences between oceans and continents based on distribution of igneous rock types. Rocks that melt to make granite are found only under the continents, but those which melt to make basalt are found under both continents and oceans.

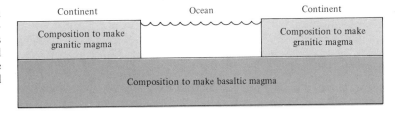

The andesites are something of a problem. In both composition and geographic location, they lie between the granites and basalts at the borders of oceans and continents. Some geologists have hypothesized that andesites form by the mixing of basaltic and granitic materials as basalt magma rises upward through a granitic layer of the crust. However, this does not explain why andesites are not abundant in continental interiors, where there are large basalt flows. The plate tectonic hypothesis states that andesites form in subduction zones in which two lithosphere plates collide and one is driven beneath the other, but none of the other four hypotheses deals specifically with the origin of the andesites. This problem will be reviewed in Chapter 21.

IGNEOUS ROCKS AS NATURAL RESOURCES

Many igneous rocks and the minerals of which they are made have become useful to our society, and in the hunt for energy resources we have tapped the heat given off by magmas cooling at relatively shallow depths. These energy resources are referred to as **geothermal** resources. A few of the igneous rock and mineral resources are described here as examples of the ways in which igneous processes and rocks are of economic importance.

Uses of Igneous Rock

Some igneous rocks are useful because of their beauty, strength, and durability. For example, granite has been used as a building stone and tombstone for thousands of years. Gabbro, particularly a variety with large, iridescent plagioclase crystals, makes a beautiful ornamental facing stone when polished. Some uses are less spectacular. Fine-grained basalts are crushed for use in road foundations. Pumice is used as an abrasive to smooth wood or remove calluses from hands and feet, and it is added to some soaps to abrade grease from the hands. In the past, finely pulverized pumice was used as a tooth powder because it abraded stains and dental plaque effectively from teeth. Unfortunately, the pulverized pumice abraded enamel

as well, and so other materials are now used instead.

Igneous Minerals as Resources

The processes that form igneous rocks produce several valuable mineral deposits. Gravity separation is particularly important because many ore minerals have extremely high specific gravities. Oxide and sulfide minerals such as hematite, Fe_2O_3; magnetite, Fe_3O_4; chromite, $FeCr_2O_4$; and bornite, Cu_5FeS_4, are important ores of iron, chromium, and copper and are present in many igneous rocks as accessory minerals. Gravity settling of early-formed ore minerals concentrates them effectively in abundant proportions and can make mining economically feasible. A single example illustrates the principle.

Chromite has a specific gravity of 4.62, far greater than that of the mafic magmas from which it crystallizes. Gravity settling of chromite duringing differentiation has produced layers of nearly pure chromite in many of the world's mafic plutons. The Bushveld igneous complex of South Africa contains several chromite layers, and one layer, the Merensky Reef, is a major source of platinum as well as chromite.

Hydrothermal Solutions

Water given off from magmas during the last stages of cooling often contains dissolved ions such as silicon, oxygen, sulfur, copper, lead, and zinc. These solutions easily penetrate fractures in host rock, and precipitate ore minerals in the fractures. Underground water not derived directly from magma may be heated when it passes through pores and fractures near a cooling magma. The heated water can then dissolve ions from the host rock, transport them to other locations, and precipitate them as ore minerals.

The heated water, whether from magma itself or from the local groundwater supply, is called a **hydrothermal solution,** from the Greek *hydro* (meaning *water*) and *therm* (the root for *heat*). Hydrothermal solutions were responsible for the copper deposits at Butte, Montana, and Chuquicamata, Chile, and for the extensive lead and zinc deposits of the midcontinental region centered on Joplin, Missouri.

SUMMARY

Igneous rocks are produced by the solidification of molten rock material called magma. Magma forms when heat energy from concentrations of radioactive elements in the crust and upper mantle overcomes chemical bonds in minerals and causes melting. Most magmas are of silicate composition, although the amount of silicon and oxygen relative to other elements varies considerably.

Crystallization of silicate minerals from magma is complex, reflecting three different ways in which minerals melt. Minerals such as quartz crystallize directly from magma at a single temperature—the melting point. However, plagioclase crystallizes over a range of temperatures and continuously changes from calcium-rich to sodium-rich composition as it forms. Ferromagnesian minerals such as enstatite undergo discontinuous reactions with magma as they crystallize. In these reactions, early-formed minerals react with residual magma to make new minerals.

The final products of crystallization cannot be predicted because of the possible assimilation of new material by melting the walls of the magma chamber and because of differentiation. Gravity settling of early-formed crystals and separation of crystals from magma by filter pressing may alter magma composition and hence final rock type(s) significantly.

Some magma moves to the surface, where it solidifies to form extrusive (volcanic) rocks. Some rises through the crust but solidifies beneath the surface to form intrusive (plutonic) rocks. The textures of igneous rocks reveal their cooling histories and enable us to determine whether they are intrusive or extrusive. Rapid cooling at the surface leads to fine-grained rocks, and slower cooling at depth leads to coarser-grained rocks. Extremely rapid cooling produces glassy-textured rocks devoid of minerals. Crystallization from extremely fluid magmas produces very coarse-grained rocks called pegmatites. The release of gases from extrusive rocks produces a highly porous texture.

A rock is identified as igneous by its texture, mineral content, and intrusive or extrusive relationships with surrounding rocks. It is classified and named by its texture and mineralogy. Basalts and gabbros are the most abundant igneous rocks and are found throughout the oceans and on the continents. Granites and rhyolites are restricted to the continents and continental margins, whereas andesite and diorite are abundant only at the boundaries between oceans and continents. This unequal distribution suggests that the material underlying the oceans lacks the correct composition to yield granitic magma but that such granite-forming material is abundant beneath the continents.

QUESTIONS FOR REVIEW AND FURTHER THOUGHT

1. What are the different types of melting exhibited by igneous minerals?

2. Contrast and compare the crystallization behaviors of enstatite, plagioclase feldspar, and quartz.

3. Discuss in as much detail as you can the melting and crystallization of a plagioclase feldspar composed of 30% anorthite and 70% albite and one composed of 90% anorthite and 10% albite. Refer to Figure 4.13 for details.

4. What factors determine the physical properties of a magma?

5. Explain how an igneous rock can be produced from a magma that does not have the same composition as the rock.

6. How can the textures of igneous rocks be used to distinguish between intrusive and extrusive rocks?

7. Why do some magmas reach the surface and become extrusive rocks while others crystallize deep within the earth?

8. What properties of accessory minerals make it possible for them to be concentrated during magmatic crystallization?

9. Forsterite, Mg_2SiO_4, and quartz, SiO_2, are almost never found together in igneous rocks. Refer to Figures 4.15 and 4.19 to explain why.

ADDITIONAL READINGS

Almost all readings about igneous rocks require some knowledge of mineralogy and the processes by which magmas cool. The following readings are all at a more advanced level than this text but are still readable and helpful in understanding igneous processes.

Intermediate Level

Bayly, Brian, B., *Introduction to Petrology,* Prentice-Hall, Englewood Cliffs, N.J., 1968.
(An introduction to all three rock types. It explains the details of the crystallization of minerals from magmas particularly well)

Bowen, N. L., *The Evolution of the Igneous Rocks,* Princeton University Press, Princeton, N.J., 1928. Reprinted by Dover Publications, New York.
(A classic discussion of melting and crystallization with detailed examples of magmatic differentiation. Written in an easy-to-read style that makes difficult concepts relatively simple to understand)

Williams, H., F. J. Turner, and C. M. Gilbert, *Petrography,* W. H. Freeman, San Francisco, 1954.
(An excellent introduction to all three rock types, with discussions and superb line drawings of typical igneous rock textures)

Advanced Level

Hyndman, D.W., *Petrology of Igneous and Metamorphic Rocks,* McGraw-Hill, New York, 1972.
(An excellent treatment of igneous rocks from a plate tectonic viewpoint. Each type of igneous rock is related to the processes involved in plate tectonics, and conflicting theories are evaluated. Some chemical and mineralogic background is necessary)

CHAPTER 5

Volcanoes

Few events are so terrifying as volcanic eruptions. The words alone conjure up an image of thunderous explosions, poisonous fumes, choking clouds of hot, glowing ashes, and swift-flowing rivers of molten lava. Few geologic processes are so violent or so awesome in terms of the amount of energy expended in a short period of time. But not all volcanic activity is violent. Not all volcanoes are the same. Some extrusive igneous activity does not even produce a volcanic cone.

Why not? What are the different types of volcanoes? What are they made of? Why are there volcanoes in Oregon, Washington, Hawaii, and Alaska but not in Kansas, Florida, or Maine? Are volcanoes dangerous? Can they be beneficial? In this chapter we will try to answer these questions.

WHAT ARE VOLCANOES MADE OF?

In Chapter 4 we saw that when magma reaches the surface, it may remain fluid and flow along the ground as lava or may become solid, break into small fragments, and erupt violently into the air as pyroclastic material, or tephra. Both types of material—lava and tephra—are used to build volcanoes and other surface features.

Lava

Magma that has reached the surface of the earth is called **lava.** It contains the same ions, has the same properties, and is at the same range of high temperatures as intrusive magmas. There are as many different compositions of lava as there are of intrusive igneous rocks. Different compositions and different temperatures result in lavas with a range of physical properties, most notably viscosity.

Pahoehoe lava

Highly fluid lava behaves much like water in a stream, except for the fact that it is much hotter—around 1000°C. It flows rapidly down steep slopes and more slowly along gentle surfaces; sometimes it falls much as waterfalls do (Figure 5.1). Much lava in the Hawaiian Islands is of this type, and the Hawaiian word **pahoehoe** (pronounced pa-hoy'-hoy) is used to describe highly fluid lava. As pahoehoe cools, it forms a crust where it meets the air and is chilled. This crust commonly retains a "ropy" appearance that indicates the high fluidity of the lava (Figure 5.2). The escape of gases from nearly solidified pahoehoe lava results in a highly vesicular texture in the resulting rock.

Aa lava

Highly viscous lava is known by the Hawaiian word **aa** (pronounced ah-ah). Its advance across the countryside is more like that of a glacier than a stream in that it is nearly solidified and moves very slowly (Figure 5.3). The movement of aa is not affected so much by ground slope as is the movement of pahoehoe, and its advance is somewhat like that of a bulldozer

Figure 5.1 River of molten lava flowing over partially solidified pahoehoe. *(D. B. Jackson, U. S. Geological Survey)*

tread (Figure 5.4). The outer surface of an aa flow is a rubbly crust composed of solid sharp, angular chunks of hot volcanic rock. Nearly solidified, highly viscous lava within the flow pushes slowly but doggedly onward, dragging the brittle crust with it and plowing chunks under as it moves.

Features associated with lava flows

Lava tubes and tunnels Many pahoehoe flows contain long tubelike cavities called **lava tubes** or **lava tunnels.** These form when the margins of a flow solidify, leaving a central passageway through which lava continues to flow (Figure 5.5). During the waning stages of an eruption, the amount of lava coming from the source area may decrease to the point where this passageway is no longer completely filled with lava; when final cooling occurs, a tunnel is left within the flow. Lava tunnels on earth may be hundreds of meters long and 10 to 20 m high (Figure 5.6). Sinuous valleys on the moon called **rilles** have long puzzled astronomers,

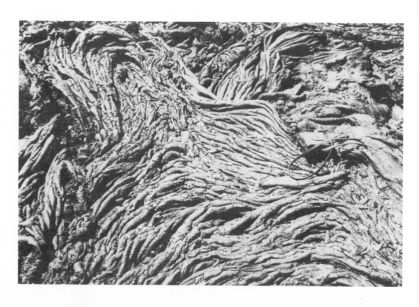

Figure 5.2 Ropy surface of a solidified pahoehoe lava flow. *(G. A. MacDonald, U. S. Geological Survey)*

Figure 5.3 Blocky surface of an aa lava flow. Compare with pahoehoe in Figure 5.2. *(G. A. MacDonald, U. S. Geological Survey)*

and one was visited early in the Apollo program. Rilles resemble lava tunnels on earth whose roofs have collapsed, and they may have been produced by the same mechanism.

Pillows   When lava flows into a river, lake, or ocean, there is often an explosive meeting of magma and water as the heat from the lava converts water rapidly to billowing clouds of steam.

In some instances, particularly at great depths where the pressure of the overlying water prevents the rapid escape of steam, the confrontation between lava and water is far less violent. As lava erupts on the sea floor, small tongues of pahoehoe break through the already solidified lava front; upon contact with the seawater, the pahoehoe forms globular masses called **pillows** (Figure 5.7).

Figure 5.4 Movement of an aa flow.

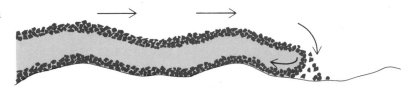

Figure 5.5 Origin of lava tunnels.

1. Lava flow cools on outer surfaces but flow continues in molten material in central passageway.

2. During waning stages of eruption, supply of molten material does not completely fill passageway and, when lava solidifies, a tunnel is formed.

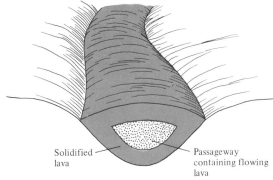

Solidified lava

Passageway containing flowing lava

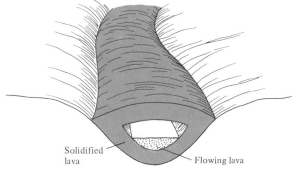

Solidified lava

Flowing lava

Figure 5.6 Interior of a lava tunnel in Lava Beds National Monument, California. The floor of the tunnel shows the texture of the flowing lava. *(J. Boucher, National Parks Service)*

Figure 5.7 Pillows in a basalt flow. Several pillows broken by weathering show typical radial structures formed during cooling. Some broken and unbroken pillows are circled. *(P. D. Snavely, U. S. Geological Survey)*

Tephra

Many volcanic eruptions are accompanied by the explosive ejection of particles of lava into the air. These particles chill in the air and fall as solidified or partly solidified **tephra** particles. Tephra ranges in size from microscopic volcanic dust particles to "bombs" the size of trucks (see Table 4.1). There is perhaps no better way to visualize the enormous amounts of energy involved in volcanic eruptions than to try to imagine how much energy is needed to propel hundreds of truck-sized blocks of lava hundreds of meters into the air. When the particles fall back to earth, they are incorporated in volcanoes or may form tephra blankets whose extent is controlled by the wind.

When tephra falls through the air, it is affected by the same currents as nonvolcanic, wind-transported particles and eventually is deposited on the ground or in a body of water. The particles are of igneous origin, but the process that brings them together is sedimentary. Is the resulting rock igneous or sedimentary? Perhaps a little of each? These processes cross the arbitrary boundaries between rock types that we set up in Chapter 3, reminding us that natural processes do not necessarily obey the rules established by scientists.

What is tephra made of?

Tephra may be composed of several kinds of particles. Some tephra is made of volcanic glass–lava spatters that chill almost instantly in the air so that minerals do not have a chance to form. However, crystals or parts of crystals are also found in many tephra deposits. These crystals form within the earth as the magma rises upward and are carried into the air along with the unsolidified magma. Fragments of igneous and other rock types may also be ejected from a volcano. When magma rises toward the surface, it may dislodge fragments of the rock through which it is passing, and these fragments, like early-formed crystals, can be carried with the magma during the eruption.

Shapes of tephra particles

The shapes of individual tephra fragments tell of their history. The largest blocks tend to be irregular, but many blobs of lava hurled into the air cool during flight and show a streamlining that indicates aerodynamic control of shape (Figure 5.8). Many of the smallest fragments, called **shards,** have curved surfaces and sharp, angular corners. Shards form from nearly solidified frothy lava, a mixture of gas and lava similar to the foam on a glass of beer (Figure 5.9), When the bubbles burst at the earth's surface, the fragments of their walls become the shards. The curved surfaces represent the bubble surfaces, and the sharp angular corners represent the places where the walls broke.

Figure 5.8 Volcanic bomb showing aerodynamically streamlined shape. *(Mira Schachne)*

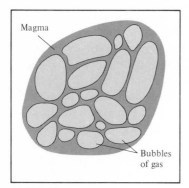

(a) Magnified view of frothy magma. Bubbles are formed in nearly solidified magma.

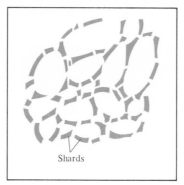

(b) Bubbles burst, gases escape, and walls of bubbles are shattered. These pieces fall back to earth to become the shards of tuffaceous rocks.

Figure 5.9 The origin of volcanic glass shards.

Lava or Tephra?

What determines whether magmas will extrude as lava or as tephra? Composition plays an important role. Basaltic magmas extrude as pahoehoe, aa, and tephra; andesitic magmas extrude as aa or tephra; and rhyolitic magmas extrude as either tephra or an exceptionally viscous lava. The silicon and oxygen content of the magma determines to a great extent the type of material extruded. Andesite and rhyolite have much more silicon and oxygen than basalt. Thus, in the magmatic state there are many more temporary bonds formed between these ions than there are in basaltic magma. This cuts down on ionic mobility and makes andesitic and rhyolitic magmas much more viscous than basalt. As a result, only basalt forms pahoehoe.

Temperature is also an important factor. The hotter a magma is when it reaches the surface, the freer the ions will be to move and thus take part in pahoehoe flows. Basaltic magmas have higher temperatures than andesitic or rhyolitic magmas, again explaining their occurrence as pahoehoe. Rhyolitic magmas are higher in silicon and oxygen (hence more viscous) and lower in temperature than basalts. In general, they also contain more water that can escape explosively, and as a result they erupt most commonly as tephra.

VOLCANOES AND OTHER VOLCANIC LANDFORMS

The accumulation of lava and tephra on the surface of the earth produces topographic features. The most familiar of these are the cone-shaped mountains called **volcanoes.** Other landforms composed of extrusive igneous rock include **lava domes,** extensive **lava plateaus** (broad uplands underlain by thick sequences of lava flows), and sheets of volcanic ash.

Volcanoes

Geologists recognize three types of volcanoes. Each kind is characterized by its shape and by the materials from which it is constructed. In the initial stages of the growth of a volcano, lava or tephra is brought to the surface through a central **feeder conduit,** a roughly cylindrical, pipelike passageway. The end of the conduit at which lava and tephra are erupted is called the **volcanic vent.** Eruptions at later stages may be through conduits that branch off from the original one or through elongate fractures called **fissures.**

Cinder cones

A **cinder cone** is a volcano made of tephra. It is shaped like an inverted cone with moderately steep slopes, and it generally has a bowl-shaped depression called a volcanic **crater** at the apex. The shape is controlled basically by gravity. After particles are ejected from the volcanic vent, they fall back to the ground, with most landing near the vent itself (Figure 5.10). This produces a mound of debris that grows higher and higher as more tephra is erupted. The slope of the volcano's flanks is determined by how steep a pile of particles can be made without being flattened by the effects of gravity. The effect is similar to that in the lower half of an hourglass; however, the ability of tephra particles to be welded to each other enables cinder-cone slopes to be somewhat steeper than would be possible for grains of sand. The crater is kept free of tephra by the explosive activity that accompanies eruptions.

Cinder cones vary widely in size but in general constitute the smallest kind of volcano. Some are small hills only a few tens of meters high, but the largest, the Mexican volcano Parícutin, measures approximately 400 m from base to top (Figure 5.11).

Cinder cones can be built very quickly, and geologists were fortunate to be able to watch Parícutin through almost its entire growth cycle. In February 1943, a farmer plowing a cornfield noticed hot fumes rising from a small crack in the earth. The fumes were sulfurous and were joined by cinders, steam, and small soil fragments hurled upward from the crack. The farmer was soon chased from his field by an assault from two directions. Tephra and nauseating gases cascaded down onto him while the earth beneath his feet began to rumble and shake. When he returned to the cornfield the next day, he found a small cinder cone nearly

Figure 5.10 Formation of a cinder cone. (a) Origin of the cone. (b) A cross section through a cinder cone showing layers of tephra.

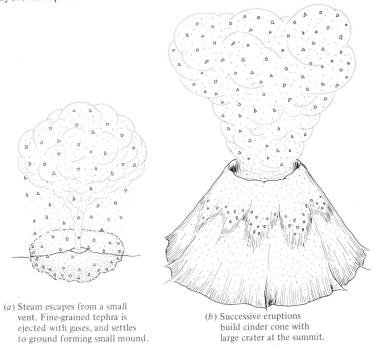

(a) Steam escapes from a small vent. Fine-grained tephra is ejected with gases, and settles to ground forming small mound.

(b) Successive eruptions build cinder cone with large crater at the summit.

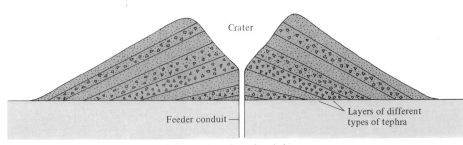

(c) Cross section through a cinder cone.

Figure 5.11 Mt. Paricutin erupting. The trees in the foreground were killed by a blanket of ash from earlier eruptions. *(W. F. Foshag, U. S. Geological Survey)*

10 m high built over the crack. It was to grow to nearly 130 m in the first week of its existence. Within a month the formerly flat upland fields of the area were the home of a 300-m-high, extremely active cinder cone. Eventually Parícutin also erupted lava, not from its crater but from vents along its flanks.

Shield volcanoes

Some volcanoes are extremely broad, with very gently sloping flanks made up of lava. They are called **shield volcanoes** because their shape resembles a shield lying flat on the ground. Lava extruded from a central vent flows outward in all directions over the surrounding land surface or sea floor (Figure 5.12). Because it is a fluid, the lava cannot be piled up into a steep-walled structure like that of a cinder cone. To demonstrate this, try pouring maple syrup onto a flat surface and see how steep a cone you can produce. Successive flows increase the size of the volcano and accumulate to form a gently sloping structure that is quite unlike that of a cinder cone and also much larger.

The Hawaiian Islands are a chain of shield volcanoes of truly enormous size. The island of Hawaii contains three individual cones: Mauna Loa, Mauna Kea, and Kilauea. The summit of Mauna Loa is nearly 4000 m above sea level, and it constitutes less than half the volcano because the base sits on the floor of the Pacific Ocean, nearly 6000 m below sea level. The total height of the lava pile, nearly 10,000 m, makes Mauna Loa far larger than Mt. Everest in height and total volume of material. It is the largest mountain on earth.

What are eruptions of shield volcanoes like? In the early stages of growth, most outpourings of lava come from the central vent. A crater commonly forms by subsidence after lava has been extruded from the region beneath the volcano. The ground subsides because the lava has been removed from a subterranean chamber, leaving it empty and incapable of supporting the huge mass of the volcano above it. During later central eruptions, lava first fills this crater with a lake of molten, glowing material, and beautiful lava fountains several tens of meters high are created as eruptions continue through the lava lake (see Plate 9). The lake fills until it overflows or breaks through its walls, and rivers of lava flow down the flanks of the volcano, adding to its height and extent.

Not all eruptions are from the summit. **Flank eruptions** occur when lava creates a new conduit for itself and reaches the surface at a new vent along the flanks of the volcano (see Figure 5.12). Late-stage eruptions may be from fissures rather than vents, resulting in sheets of lava pouring out on the volcano flank. Pahoehoe, aa, and even small amounts of tephra contribute to the immense volume of shield volcanoes. Most of the great shield volcanoes, by far the largest volcanoes on our planet, are made up almost entirely of basaltic lava.

Composite cones (stratovolcanoes)

Most of the best-known volcanoes of history and mythology are neither cinder cones nor shield volcanoes. They are made up of alternating layers of lava and tephra and are thus a composite of the two types discussed already. Not surprisingly, they are called **composite cones.** The mountains Etna, Vesuvius, and Vulcano in Italy; Pelée in Martinique; Katmai in Alaska; Fujiyama in Japan; Lassen, St. Helens, Shasta, Rainier, and Baker of the Cascade Range; and the cones of Krakatoa Island (between Java and Sumatra) are all composite cones. Mt. Mayon in the Philippines and Mt. Shishaldin in the Aleutian Islands (Figure 5.13) are reputedly the most perfectly symmetrical cones of this type. Composite cones are intermediate in size between shield volcanoes and cinder cones.

Figure 5.12 Formation of a shield volcano.

(a) Lava extrudes from vent, spreads out, and solidifies.

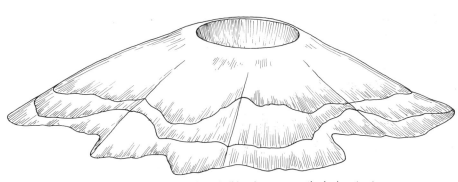

(b) Successively overlapping flows build an immense, gently sloping structure.

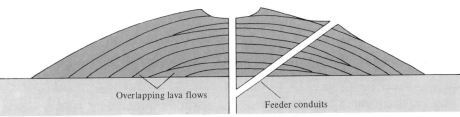

Overlapping lava flows   Feeder conduits

(c) Cross section through a shield volcano.

Figure 5.13 Mt. Shishaldin in the Aleutian Islands shows the classic profile of a composite cone. *(National Archives)*

Many cinder cones, such as Parícutin, contain small amounts of lava, and many shield volcanoes, such as Mauna Loa, erupt small amounts of tephra, but composite cones are made of large amounts of both. Periodic alternation of lava with tephra produces the classic profile of the composite cone (Figure 5.14). Because the volcanoes are composed of **strata** (layers) of different types of extrusive rock, they are sometimes referred to as **stratovolcanoes.**

What are eruptions of stratovolcanoes like? Eruptions involving lava are not terribly unlike shield volcano eruptions, except that andesitic and rhyolitic materials are often ejected as highly viscous flows. Eruptions involving tephra are among the most violent ever observed, and in many cases the eyewitness reports are from the few survivors of towns destroyed by the eruptions. These reports, coupled with reconstruc-

tions based on geologic field evidence, reveal the existence of a terrifying process unique to this type of volcano.

The 1902 eruption of Mt. Pelée on the island of Martinique in the Caribbean Sea gives a frighteningly clear picture of such an eruption. In late March 1902, a group of hikers noticed sulfurous fumes in a small crater near the summit of Pelée. For about a month, activity increased gradually, with minor explosions of tephra and the formation of a small cone within the crater. Residents of the surrounding countryside flocked into St. Pierre, the largest town on the island, located nearly 6 km from the crater. The population of St. Pierre swelled from some 20,000 to nearly 30,000. Just before 8 A.M. on May 8, 1902, nearly all these people were killed as St. Pierre was destroyed almost instantaneously by the eruption of Pelée. Fewer than 10 survived, and

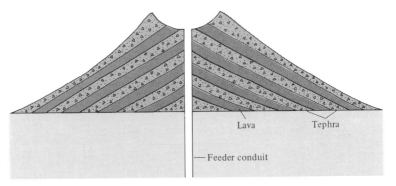

Figure 5.14 Structure of composite cones.

from them the following sequence of events has been determined.

After a series of loud explosions, a large black cloud rose upward from the crater and then sank toward the ground and rolled downslope toward and through St. Pierre at very high velocity. The cloud was a hot (several hundred degrees Celsius) mass of tephra and poisonous gases. Thousands died by asphyxiation or burning as they inhaled the searing gases. One survivor was a prisoner in the local dungeon. He was spared because his cell was partly below ground level, out of the direct path of the glowing cloud, but even he was affected. Gases and ash penetrated through ventilation ducts to his cell and burned him severely in a matter of seconds.

The sudden and violent onslaught of the hot, glowing cloud of gas and tephra is nature's version of the surge of material outward from the base of a nuclear mushroom cloud. The fiery cloud is called *nuée ardente.* It sinks toward the ground because it is denser than the surrounding air and rolls downslope under the influence of gravity. On Martinique, *nuée ardente* from Mt. Pelée knocked down stone walls, set St. Pierre on fire, twisted steel beams into knots, and took nearly 30,000 human lives in a matter of seconds. Figure 5.15 shows a *nuée ardente* erupted from Mt. St. Helens in its catastrophic 1980 eruption.

Calderas

Many volcanic peaks are characterized by huge, irregularly shaped craters called **calderas** at their summits. Some calderas form when a volcano literally blows its top during extremely violent eruptions. Other calderas form by inward collapse as the ground adjusts to the removal of large volumes of lava or tephra from within the volcanic structure. Some calderas form by a combination of these processes.

Crater Lake, Oregon Crater Lake is an example of a caldera formed by a combination of explosion and collapse (Figure 5.16). Originally the site of a large stratovolcano (named Mt. Mazama by geologists), this volcanic center exploded, showering the surrounding countryside with ash and debris for several hundred kilometers. So much material was ejected that large-scale subsidence within the crater followed, producing the final shape of the caldera. Later, less violent eruptions formed a small cone within

Figure 5.15 Nuée ardente from Mt. St. Helens. *(J. Rosenbaum, U. S. Geological Survey)*

Figure 5.16 Crater Lake, Oregon fills the caldera of an extinct volcano called Mt. Mazama. Wizard Island, in the center of the photograph, is a volcanic cone built after the caldera formed. *(H. R. Cornwall, U. S. Geological Survey)*

the caldera, and this cone stands up above the present level of the lake as Wizard Island.

Krakatoa Krakatoa Island lies between Java and Sumatra in the Pacific Ocean, and in 1883 it was composed of a series of volcanic centers. After nearly 200 years of inactivity, Krakatoa went through a series of explosive eruptions that culminated in several titanic blasts heard clearly many hundreds of kilometers away. When the smoke and ash had cleared enough for the first courageous surveyors to approach the island, they found some surprises (Figure 5.17). What had been the largest of a small group of islands was considerably shrunken. A huge crater over 300 m deep and mostly below sea

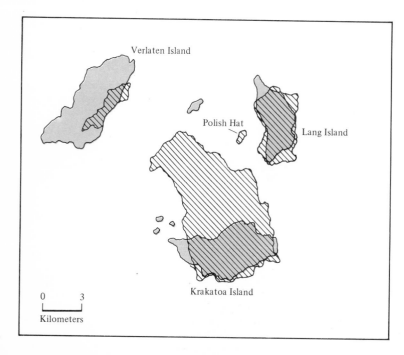

Figure 5.17 Krakatoa Island and vicinity before (lined area) and after (shaded area) the 1883 eruption.

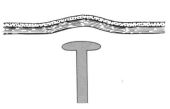

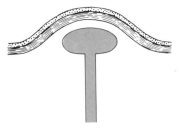

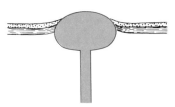

(a) Conduit brings magma toward surface and highly viscous magma bulges upward from conduit.

(b) Magma arches overlying strata as it moves upward.

(c) Magma may break through strata to surface.

Figure 5.18   The origin of lava domes.

level replaced the prominent volcanoes on Krakatoa Island, where formerly peaks had reached hundreds of meters above sea level. In this case the caldera formed explosively beneath the water.

Many geologists believe that the submarine caldera formed by the explosion of the Mediterranean island of Santorin in about 1400 B.C. is the basis for the legend of Atlantis, the sudden disappearance from the surface of the earth of a heavily populated island empire. The combination of earthquake and ashes that accompanied the eruption destroyed the flourishing Minoan civilization on Crete.

Lava Domes

Exceptionally viscous lavas, generally of a rhyolitic composition, are rare but produce rather oddly shaped **lava domes** when they are erupted. These lavas are so viscous that they do not flow far from their vents, and so they produce only small hills. The eruption is very much the opposite of a *nuée ardente* type and is much like a blob of material (such as toothpaste) being squeezed upward from a tube (Figure 5.18). Many lava domes form late in the eruptive history of a volcano and fill craters produced during earlier eruptions. Mono Craters of eastern California, for example, are lava domes composed almost entirely of obsidian (Figure 5.19).

In some instances, the upward-moving magma is so viscous that instead of bursting through the surficial materials, it arches them upward. In direct contrast to violent eruptions such as those of Pelée and Krakatoa, the growth of lava domes is very slow and usually quite safe

Figure 5.19   A lava dome (in the center of the photograph, Mono Craters, California. *(F. E. Matthes, U. S. Geological Survey)*

to watch. The Showa-Shinzan dome in Japan bulged upward for nearly a year before lava finally reached the surface. The Santiaguito dome in Guatemala has been growing quietly for decades, but so slowly that geologists actually can climb over the surface and measure changes as the dome arches over the rising magma.

Lava Plateaus

Large upland regions underlain almost entirely by lava are called **lava plateaus.** They are found throughout the world, and in all instances the lava is of basaltic composition. Some of the better-known examples include the Karroo basalts of South Africa, the Deccan basalts of India, the Giant's Causeway of Ireland, and the Paraña basalts of South America. North America also has its share of plateau basalts. The Columbia River Plateau of Oregon, Washington, and Idaho consists of basalts that originally covered some 300,000 km^2 (Figure 5.20).

Lava plateaus form by the extrusion of highly fluid lava from large fissures. The lava spreads quietly outward from the fissures in fissure flows that may cover thousands of square kilometers. Successive flows build up immensely thick masses of lava. The Columbia River Plateau, for example, contains hundreds of flows that today amount to over 600 m in thickness (Figure 5.21). However, individual

Figure 5.21 A sequence of basaltic lava flows in the Columbia Plateau, exposed in the wall of the Grand Coulee, Washington. Note columnar jointing in several layers. *(John S. Shelton)*

flows may range in thickness from a few centimeters to tens of meters. Lava tubes and tunnels occur in the thicker flows and are particularly well developed in the Lava Beds National Monument in northern California, where tunnels up to 20 m in diameter form an intricate network. Columnar jointing produced by contraction during cooling is developed spectacularly in some flows (see Figure 4.11).

Ash Falls and Ash Flows

Extensive deposits of fine-grained tephra from explosive eruptions such as those of Vesuvius and Mt. St. Helens form blankets that cover the countryside surrounding the vent. Coarse particles settle closest to the vent, but the finest ones may be carried great distances. Those from the Krakatoa eruption circled the world in the upper atmosphere for over a year before falling back to earth; during that year, sunsets were spectacularly beautiful because of diffraction of light through the ash. The thickness of the ash blanket depends on the amount of tephra erupted and on the area over which it is dispersed. Some ash sheets are only a few centi-

Figure 5.20 Geographic extent of the Columbia River Plateau basalts.

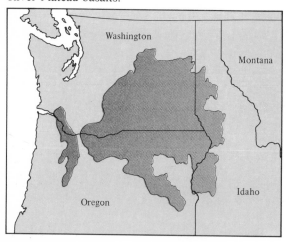

meters thick but cover thousands of square kilometers. Others, such as that produced by the Katmai eruption of 1922, are over 1000 m thick. Ash-flow deposits from *nuée ardente* eruptions surround many major volcanoes such as Pelée and Etna. Plateaus underlain by alternating layers of lava flows and ash flows may cover thousands of square kilometers, as in the San Juan Mountains of Colorado.

Erosional Remnants of Volcanic Landforms

Many lava flows are more resistant to erosion than the sedimentary rocks with which they are interlayered. As a result, the flows act as a protective cap rock and form topographic highs overlooking areas underlain by the less-resistant rock types. For example, basalts of the Holyoke Range of central Massachusetts stand high above the red sedimentary rocks of the Connecticut River Valley (Figure 5.22).

However, aa lava and tephra may be eroded more easily than the associated rocks if they are poorly consolidated or highly fragmental. The cylindrical feeder conduits that brought the magma to the surface commonly are preserved as dense, crystallized igneous rocks that stand up as isolated remnants when the rest of the volcano has been eroded away (Figure 5.23). Some of these remnants, called volcanic **necks,** are famous features of the world's landscape: Arthur's Seat overlooking Edinburgh in Scotland, Ship Rock in New Mexico, and steep-sided pinnacles in the Auvergne, France.

VOLCANOES AND HUMAN BEINGS

Volcanic eruptions often interfere with human activities, as the 1980 eruption of Mt. St. Helens vividly reminded us (see the perspective on Mt. St. Helens). However, not all effects of volcanism need be hazardous. Indeed, some volcanic products are economically valuable.

Hazards

Most of the destruction caused by volcanic activity is centered on the flanks of the erupting volcano, but activity associated with volcanism can also affect areas hundreds of kilometers away. The damage in outlying areas is often more severe because the inhabitants have no warning that anything abnormal has happened.

Figure 5.22 The Holyoke Range, Massachusetts. The ridge is held up by a thick basalt flow that is more resistant to erosion than the surrounding sedimentary rocks. The contact between the basalt and the underlying sandstones is close to the abrupt slope break at the right side of the photograph. *(Marshall Schalk)*

Mt. St. Helens, Washington, 1980
It Can Happen Here

Until 1980, most North Americans associated volcanism with such exotically named far-off places as Fujiyama, Krakatoa, Pelée, and Surtsey. The eruptions of Mt. St. Helens, one of the composite cones of the Cascade Range of the northwestern United States, reminded us that our continent too has its share of volcanic activity. The eruptions also gave geologists an unparalleled view of the workings of a stratovolcano, providing a natural laboratory much as Parícutin and Mauna Loa had done years before for cinder cones and shield volcanoes, respectively.

After nearly 100 years of dormancy, a week-long series of low- to moderate-intensity earthquakes began on March 20, 1980. On March 27, the volcano erupted a large volume of steam and ash, and observers reported a new crater and two large cracks forming near the summit. A group of explosive eruptions occurred the next day. Ash and steam rose to more than 1.5 km above the volcano, and several avalanches composed of ash-laden snow cascaded down the east flank. By March 30, a huge tephra cloud had deposited materials as far as 250 km away.

On April 1 and 2, indications began of yet more violent activity. Continuous harmonic tremors suggestive of the upward movement of lava were recorded by seismologists for the first time; they were to continue sporadically until May 8, ten days before the major eruption. The shape of the volcano was monitored carefully by geologists of the U.S. Geological Survey. By late April, part of the upper north face of the volcano had bulged outward by an estimated 100 m, suggesting imminent eruptions and prompting hazard warnings by the local and state governments.

On May 18, Mt. St. Helens blew its top in an explosive eruption heard as much as 300 km away. The upper 400 m of the cone was removed completely (Figure 1), and there was nearly complete failure of the north face in the vicinity of the bulge. Approximately 1 km³ of material was erupted in this one blast, and a tephra cloud billowed upward to over 20,000 m shortly afterward (Figure 2). This cloud was followed by satellite cameras for the next few days as it spread out over the American Midwest (Figure 3). The blast blew down trees as far as 25 km from

Figure 1   Mt. St. Helens on October 22, 1980. The original symmetrical cone was destroyed by the catastrophic eruptions, leaving this large, asymmetric crater at the top of the mountain.   *(Max Gutierrez, UPI)*

Figure 2   Aerial view of Mt. St. Helens erupting at 11:30 A.M. on May 18, 1980. A large ash-laden plume rises directly upwards from the summit crater, while steam and ash clouds in the background are being blown to the east.   *(U. S. Geological Survey)*

Figure 3 Satellite photograph of western United States and Canada taken on May 18, showing spread of ash cloud from Mt. St. Helens across adjacent states. (NOAA)

the volcano. A *nuée ardente* flow and landslide moved down the breached north face and along the nearby Toutle River for up to 360 km, filling the valley with as much as 60 m of debris. A tephra flow followed from the breach, building a dam that impounded 30 to 50 m of water and created an instant flood hazard. Fully grown trees were scattered like children's pickup sticks in the debris slides (Figure 4).

After its initial explosive activity, the volcano became less violent. Tephra continued to be erupted, but a dome, or plug, of highly viscous lava formed a cap over the vent, reducing visible activity greatly. Once gas pressures had built up beneath the plug, it was blasted off in a small-scale version of the eruption of May 18. Several cycles of quiescence and activity followed, and at this writing (mid-September 1980), the volcano is once again relatively peaceful.

We have learned many things from the eruptions of Mt. St. Helens and will spend years analyzing the data collected from our instruments. We have also been reminded of what a volcanic eruption can do to human lives. Even though the eruption was twice as powerful as

Figure 4 The explosive May 18, 1980 eruption of Mt. St. Helens leveled 185 square miles of prime timber, scattering full-grown trees like matchsticks. (U. S. Geological Survey)

Figure 5 Ash clouds from the Mt. St. Helens eruption passing over Ephrata Airport, Washington, on May 19. Such clouds disrupted air and ground traffic in the northwest for days. (*Associated Press*)

that of Mt. Pelée, fewer than 100 lives were lost, principally because the volcano is in a sparsely populated area. Warnings by geologists helped keep the death toll low, but it requires little imagination to picture what would happen to Seattle if Mt. Rainier were to erupt in a similar way.

The effects of the eruption were felt hundreds of kilometers away. In some areas the light dusting of fine tephra was simply a nuisance, but in other areas it was far more dangerous (Figure 5). Tephra clogged carburetors and stalled vehicles on roads throughout the affected region, and residents of Washington resorted to gas masks or surgeon's masks to walk down the street. Speed limits were lowered so that cars would not stir up tephra that had already been deposited. Respiratory problems were caused by the fine dust. Airline traffic in the region had to be rerouted. Millions of dollars worth of prime timber was destroyed, and the fragile ecology of the region was disrupted severely. The loss of animal life was far greater than the loss of humans. Longer-term effects, such as the change in stream and groundwater composition, the change in the soil, and so forth, remain to be seen. A natural laboratory for volcanologists turned part of Washington into a disaster area.

Figure 5.23 Devils Tower, Wyoming, is a volcanic neck. The remainder of the volcano has been eroded. (*L. C. Huff, U. S. Geological Survey*)

Damage due directly to volcanic eruption

We have already seen how explosive tephra-producing eruptions can destroy a modern city (Pelée) or wipe much of a large island off the face of the earth (Krakatoa). Heavy ash falls can also be deadly. For example, in A.D. 79, Mt. Vesuvius buried the nearby town of Pompeii beneath a thick pile of ash, killing hundreds but also preserving a Roman city for future archaeologists (Figure 5.24).

Although most of the catastrophic eruptions described so far have been of tephra, lava flows can also prove disastrous. Pahoehoe flows from flank eruptions of Kilauea have advanced several times on the Hawaiian city of Hilo (see Plate 10). Slow but inexorably moving aa flows from the 1973 eruption of Helgafell in Iceland entered the important fishing port of Heimaey but stopped before engulfing the town.

Sometimes the damage does not occur instantly but follows the actual eruption. Herculaneum was a Roman town situated on the flanks of Mt. Vesuvius, about 15 km northwest of Pompeii. It was affected only slightly by ash falling from the eruption that engulfed Pompeii, but it did not escape destruction for long. Loose cinders piled up on the slopes of Vesuvius after the major eruption. Heavy rains saturated the slopes with water, soaking the unconsolidated tephra. This decreased cohesiveness between particles to the point where a massive volcanic mudslide was generated. A sea of water-saturated mud swept down the side of Vesuvius and demolished Herculaneum. Heavy rains often accompany volcanic eruptions, caused in part by the addition of large amounts of heat to the atmosphere and in part by the tephra itself. Small volcanic dust particles seed clouds much

Figure 5.24 Excavations in the southwest part of Pompeii, a Roman city buried by ash from Mt. Vesuvius. *(Wide World)*

as meteorologists do, and bring about the rainfall. The year A.D. 79 sounds like ancient history but, as illustrated by Mt. St. Helens in 1980, volcanic mudslides are not.

Indirect damage

The side effects of volcanic eruptions extend the range of destruction by hundreds of kilometers. Submarine and coastal eruptions may trigger **tsunamis**—large-scale waves that travel at hundreds of kilometers per hour across the ocean. When these waves (often mistakenly called tidal waves) strike a shoreline, they can do tremendous damage. Most of the loss of life associated with the eruption of Krakatoa was caused by tsunamis crashing onshore elsewhere in the Indonesian islands, where fishing villages had no advance warning.

Have we any defense?

Most deaths associated with volcanic eruptions occur because people are caught by surprise. The worst effects could thus be alleviated if we were able to predict volcanic eruptions so that people could be evacuated from the danger areas. But how predictable are eruptions? The Romans did not predict the eruption of Vesuvius, and St. Pierre was taken completely by surprise. Are we any better now at prediction? To a great extent, yes, but we still failed to predict the eruption of Mt. St. Helens. Despite the complexities of the problem, we have made significant steps toward being able to predict when certain volcanoes will erupt.

Volcanoes have long been classified as active, dormant, or extinct. **Active** volcanoes are those which are erupting now or have erupted in the very recent past. **Dormant** volcanoes are those which have not erupted within the past few hundred years but have a history of eruption within human record. **Extinct** volcanoes are those which have not erupted in historic times. The most thoroughly studied active volcanoes are probably those of Japan, Iceland, and Hawaii —areas where large population centers are situated on or close to the volcanoes. Volcanologic institutes in these three regions have made careful measurements of active volcanoes to try to determine whether there are any clues to eruptions, any physical indications that an eruption is about to occur.

The shift of magma toward the surface just before an eruption can be detected in several ways. Ground temperatures rise, ground slopes are tilted, the land surface rises slightly, and the frequency of small earthquakes increases. There is increased activity from gas vents surrounding the major volcanic center. If all these parameters can be monitored carefully, major eruptions can be predicted. But heat-flow sensors, seismic networks for earthquake detection, and the extremely sensitive tiltmeters and surveying instruments needed for the job are very expensive, and some of the methods are still experimental. The expense, however, is minimal compared with the savings that would result if we could forecast eruptions accurately. Thus, in the future major population centers situated near eruptive centers may be ringed by the needed apparatus.

Is there any defense once an eruption has begun? Can lava flows be deflected away from cities or tephra falls and *nuées ardentes* kept out of populated areas? At this time, such methods are more on the order of science fiction than fact, but some experiments have been made. After inhabitants of Heimaey had been evacuated safely from the advance of aa from Helgafell, observers watched the slow but relentless advance of lava toward the city. Possessed by a fierce determination not to stand by and watch their homes be destroyed without a fight, several Icelanders trained high-pressure hoses on the front of the advancing lava in the hope that the cold North Atlantic water would chill the lava enough to make it solidify completely and thus lose its mobility. The flow front did indeed slow and eventually stop, but whether this was due to the efforts of humans or the waning of a natural event remains unknown.

Benefits

Lava and tephra brought to the surface of the earth help create new land in the oceans. After all, the Hawaiian Islands would not exist if not for volcanoes. The same lava that nearly des-

troyed Heimaey flowed into the North Atlantic and created a natural breakwater that improved Heimaey harbor greatly. Sometimes the growth process is rapid and geologists are fortunate enough to have a front-row seat for the growth of a new volcanic island. Such an event took place off the south coast of Iceland in 1963. After first building a submarine cone, tephra broached sea level and began to accumulate on what became a small island called Surtsey. North Atlantic waves immediately began breaking against the poorly consolidated tephra and eroded it quickly. Eventually Surtsey erupted lava that was much more resistant to the waves than the tephra, and this protected the rest of the island.

New land often means fertile land because many of the ions that make up volcanic rock are important nutrients for plants. Volcanic glass and porous tephra weather rapidly and yield soils that can support abundant plant life. The luxuriant plant growth on Hawaii reflects chemical weathering that produces fertile soil in an ideal tropical environment; but even in the inhospitable North Atlantic, plants rooted within a year of the emergence of Surtsey. Coffee in Central America, rice and vegetables in Indonesia and the Philippines, and heather in Scotland all grow on volcanically derived soil.

Mineral and energy resources are also associated with volcanic activity. Sulfurous gases such as SO_2, H_2S, and S are exuded by many volcanoes, and pure sulfur coats the cracks and fissures of volcanoes, particularly when they cease erupting lava or tephra and give off steam and gases instead. In such a state they are called **fumaroles.** Nearly perfect sulfur crystals have been collected from Italian fumaroles for thousands of years. Volcanic sulfur accounts for only a small percentage of the world's sulfur supply, but it is an important resource in Chile and Bolivia.

Copper in the Chilean Andes and the Keeweenaw Peninsula of Michigan is mined from volcanic rock, as is iron in Argentina and gold in Colorado. Tephra itself can be useful. Pumice is used as an important lightweight aggregate material in plaster, stucco, and cement and is found in most cinder blocks. Volcanic ash sometimes alters to **bentonite,** a group of clay minerals that swell enormously by absorbing water into their structures. Bentonite is used as a sealer for artificial lakes, reservoirs, and other human-built bodies of water. Bentonite is also an important constituent of drilling muds — the mixture of minerals and water pumped into oil wells during drilling operations. These muds are prepared scientifically to have a precise combination of viscosity, density, and lubricating ability; and bentonite acts as a thickening agent.

High heat flow associated with active and dormant volcanic centers helps produce hot spring and geyser activity and is potentially a source of energy that we can tap. It is a form of geothermal energy and is one of the most recently developed alternatives to fossil fuels such as coal and oil. Most of the heating in Reykjavík, the capital city of Iceland, is already carried out by tapping geothermal energy of this type.

VOLCANOES AS CLUES TO INTERNAL PROCESSES

Volcanoes and lava plateaus not only are features of the earth's surface, they also are important clues to the nature of its interior. The extrusive igneous rocks supply two kinds of data about the planet. The rocks themselves are samples that tell us about the composition of parts of our planet that we have never seen, while the geographic distribution of volcanoes tells us about deep earth processes active today. Both types of data help evaluate the five hypotheses outlined in Chapter 1.

Composition of the Mantle

We learned in Chapter 4 that basaltic magma forms deeper in the earth than granitic magma. Xenoliths in some basalts enable geologists to estimate the depths at which the magma formed. These inclusions were picked up by the rising magma and carried to the surface during eruptions. Some xenoliths are peridotite, an ultramafic rock composed of olivine and pyroxene; others are eclogite, a garnet-pyroxene rock; and a few contain diamond. Experiments show that

rocks such as these probably form at extremely high pressures, under conditions unlike those which exist in the crust. The xenoliths are thought to be pieces of the upper mantle, suggesting that the mantle is composed essentially of ultramafic rocks made up of ferromagnesian minerals formed under high pressures.

Origin of Basalt

If the mantle is made of ultramafic rocks, where does the basalt come from? Once again we turn to experimental evidence. Melting experiments have shown that basalt is probably a product of the *partial melting* of an ultramafic rock rather than the complete melting of something with the composition of basalt. As heat is applied, those minerals in the ultramafic source rock which have the lowest melting points melt. The resulting magma rises through the crust and crystallizes as basalt, while the unmelted part of the rock has the composition of peridotite or eclogite. Bear in mind the fact that nobody has ever seen the postulated source rock. This is merely a hypothesis based on the composition of basalt and the xenoliths and on laboratory experiments.

Geographic Distribution of Volcanoes

Active volcanoes are not distributed evenly on the earth. Instead, they occur in elongate groups in some areas and are completely absent from others. For example, the west coasts of North and South America contain several active volcanoes, whereas there are none on the east coast of either continent. The Pacific Ocean Basin is surrounded by active volcanoes, called the "Ring of Fire" (Figure 5.25), whereas the edges of continents bordering the Atlantic Ocean are nonvolcanic. A long chain of volcanoes can be found near the centers of most oceans, producing what are called the midocean ridges.

What can be determined from this distribution? There are two conditions that must be met for a volcano to form. First, there must be a source of heat in the mantle beneath the volcanic center. Second, there must be an avenue of access for the magma to follow toward the surface. Long chains of volcanoes imply elongate heat sources and access routes. This in turn indicates

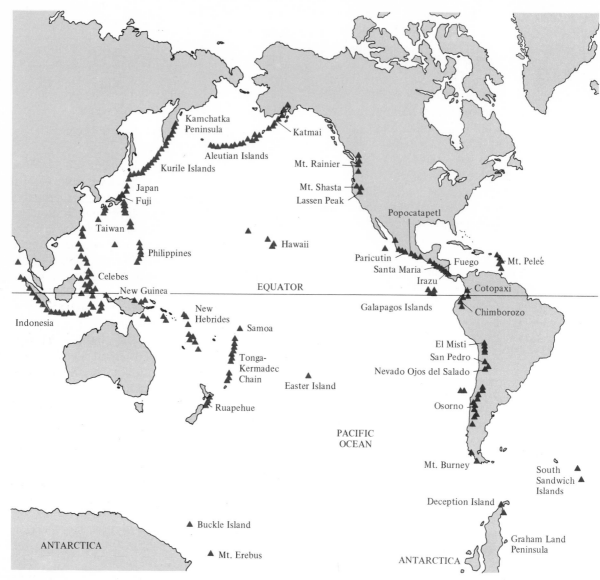

Figure 5.25 Location of active volcanoes (▲) in the Pacific Ring of Fire.

that there are elongate zones of instability and intense activity within the earth as well as zones of relative quiet.

Volcanoes and the five hypotheses

Four of the five hypotheses outlined in Chapter 1 attempt to deal with the location of volcanoes. Once geologists began to study the earth in detail and learned how to determine the age of rocks (see Chapter 8), they found that the earth is anything but static. New volcanoes, such as Parícutin and Surtsey, develop almost overnight; and areas that had been sites of active volcanism in the past are very quiet today. For example, had we been able to sail along what is now the coast of Maine some 400 million years ago, we would have had to dodge tephra from an extensive chain of volcanoes. Today there are no active volcanoes within thousands of kilometers of that coast.

Compare Figures 1.8 and 5.25. According to the plate tectonic hypothesis, the boundaries between lithosphere plates should be the most active regions on earth, whereas the centers of plates should be relatively quiet. The chains of active volcanoes in Figure 5.25 correspond closely with the plate boundaries sketched in Figure 1.7, and areas devoid of volcanoes seem to coincide with plate centers. Does this prove that plate tectonics is the only viable hypothesis?

Not necessarily. Elongate chains of volcanoes could represent equally well cracks in the crust of an expanding earth or wrinkles in a shrinking planet. However, only the plate tectonic model tries to explain the concentration of andesitic volcanic rock in the Ring of Fire and similar environments in Europe. Plate tectonicists argue that andesites can form only in subduction zones, by the melting of water-rich subducted igneous and sedimentary rocks.

SUMMARY

Volcanic landforms are made of fluid lava, particles of tephra, or a combination of the two. Different types of lava result from differences in temperature, composition, and content of gases in magma. Solidified lava flows retain features that indicate whether they were highly fluid or highly viscous during eruption. Tephra forms by the explosive ejection of solidified or nearly solidified magma into the air, where final solidification takes place. Solid but still hot tephra particles may become welded to one another when they are in contact on the surface. Silica and water content determine whether a magma will extrude as fluid lava (pahoehoe), blocky viscous lava (aa), or tephra.

Volcanic landforms are constructional features. Volcanoes are cone-shaped mountains that may be of three types. Cinder cones are relatively steep-sided volcanoes composed almost entirely of tephra. Shield volcanoes are made of lava; because much lava is quite fluid, shield volcanoes have broad, gentle slopes. Composite cones (stratovolcanoes) are made of both lava and tephra, often in alternating bands.

Volcanic eruptions vary considerably in degree of violence. Eruptions of shield volcanoes are the "quietest," with little explosive activity. Stratovolcano eruptions, on the other hand, are often extremely violent. They produce a cloud of hot tephra particles mixed with gases (nuée ardente) that flows rapidly down the flanks of the volcano. Eruptions of extreme violence, sometimes accompanied by subsurface subsidence, result in huge irregular craters called calderas.

Lava plateaus and ash-fall plateaus are built up from material extruded from linear fractures called fissures rather than from individual vents, as is the case with volcanoes. Although individual flows or ash falls may be thin, hundreds of meters of material may be accumulated in sequences of hundreds of separate eruptive events.

Volcanism may be both destructive and beneficial. Lava flows, ash falls, and nuées ardentes have in the past destroyed population centers. Rain-generated volcanic mudslides, famine caused by ash falling in agricultural and grazing areas, and tsunamis caused by submarine volcanism have also resulted in extensive loss of life and property. Recent advances in seismology and surveying suggest that prediction of eruptions may be possible in the near future.

Benefits from volcanism include creation of new land in oceans, supply of new material to be converted by weathering to fertile soil, and a variety of mineral deposits. Geothermal energy from areas of recent volcanism supplies a local alternative to fossil fuels for heating and generation of electricity.

Volcanic activity provides clues to the earth's internal processes and supplies data that act as checks of global-scale models of how the earth works.

QUESTIONS FOR REVIEW AND FURTHER THOUGHT

1. What are the different types of material produced in volcanic eruptions?

2. Why are there different types of lava?

3. Why is it more likely that steam tubes and pillows will be associated with pahoehoe than with aa flows?

4. Explain why there are few rhyolitic flows but many rhyolitic tuffs.

5. Explain how gravity ultimately controls the shape of volcanoes.

6. What are the roles of water in volcanic eruptions?

7. Volcanoes rarely erupt the same type of material throughout their active stages. Suggest an explanation for this phenomenon.

8. Why are some eruptions more violent than others?

9. What are the dangers inherent in living near a volcano on a continent? What other problems are associated with living near island volcanoes?

10. Several global models for earth processes are compatible with the observed distribution of volcanoes, but some of these models are in conflict with one another. How is this possible? Suggest detailed methods for resolving the conflicts.

ADDITIONAL READINGS

Beginning Level

Francis, Peter, *Volcanoes,* Penguin Books, Baltimore, 1976, 368 pp.
   (A very well written explanation of volcanic activity, with case histories of some classic eruptions and their effects on people)

Green, J., and N. M. Short, *Volcanic Landforms and Surface Features: A Photographic Atlas and Glossary,* Springer-Verlag, New York, 1971.
   (A collection of photographs showing features of volcanoes and volcanic features from around the world. Some of the discussion is written for those already conversant with the nomenclature discussed in this chapter)

Oakshott, G. B., *Volcanoes and Earthquakes: Geologic Violence,* McGraw-Hill, New York, 1976, 143 pp.
   (A short, well-written book that deals with types of eruptions and causes of volcanic activity in a plate tectonic setting)

Thorarinsson, S., *Surtsey: The New Island in the North Atlantic,* Viking Press, New York, 1967.
   (An interesting, well-written view of the evolution of the volcano Surtsey by a geologist who watched it grow and was one of the first to set foot on it)

U.S. Geological Survey, *Atlas of Volcanic Phenomena,* 1972.
   (A set of 20 posters showing the origin, composition, location, and tectonic significance of volcanic features)

CHAPTER 6

Sedimentation and Sedimentary Rocks

Many of the great mountains of the earth's past are long gone, but evidence of their existence is preserved in the sedimentary rocks derived from their breakdown. Exposed to the atmosphere, hydrosphere, and biosphere for millions of years, the rocks in those ancient mountains were weathered and broken down slowly to form unconsolidated debris called **sediment** along with dissolved ions. This material was transported into lower areas by gravity and ancient streams, glaciers, and winds. The sediments, along with the remains of any plants and animals living in the areas, accumulated slowly, layer by layer. Subsequent physical and chemical changes transformed the sediments into sedimentary rocks.

Sedimentary rocks are unique in a number of ways. Unlike metamorphic rocks and many igneous rocks, they formed originally near the earth's surface. The successive layers of sedimentary rocks thus provide a detailed picture of surface conditions in the past. The conditions for life are not generally compatible with the conditions present during the formation of metamorphic rocks and most kinds of igneous rocks. However, plants and animals live in many of the same areas in which sediments accumulate, and so their remains are incorporated commonly into sedimentary rocks. The plant and animal remains in successive layers of sedimentary rocks provide a partial record of the progressive changes in life at the earth's surface.

Sedimentary rock layers are much like the pages in a book — they tell a fascinating story to those who can read their clues. Features within sedimentary rocks give clues about the climate in the source area, the medium (air, water, or ice) that transported the original sediments, how fast the medium was moving, and even in which direction and for how long it traveled. Changes in sedimentary rocks from layer to layer record changes in surface conditions such as the invasion of the land by the sea, the onset of cold climates during which glaciers were more extensive than today, and the elevation of great mountain chains.

SEDIMENT

Sediments are composed of three major components produced by physical, biologic, and chemical processes: rock and mineral particles derived from the breakdown of older materials, particles produced through the life activities of plants and animals, and crystals which have precipitated out of concentrated solutions at or near the earth's surface. The sediment at any one place may be composed of only one component or may be a mixture of two or more of them (see Figure 6.1).

Breakdown of Older Sediments and Rocks

Sediments derived from the physical, biologic, or chemical breakdown of other materials (rocks and sediments) are called **clastic** sediments (Figure 6.1). Many clastic particles come from the breakdown of rocks by a process called **weathering.** Weathering refers to the changes that occur in materials as they are exposed to the atmosphere, hydrosphere, and the plants and animals that make up the biosphere. Weathering processes and products are described in detail in Chapter 9. The following discussion summarizes the role of weathering in producing clastic sedimentary particles.

In many areas, physical processes *disintegrate* the rocks into fragments. For example, water freezing or tree roots growing into the crevices in a granite exposure can break the outer surface into fragments. At the same time, chemical processes may be *decomposing* the rock through reaction of the component minerals with fluids and gases. In many cases, the decomposition forms new minerals which are more stable at the earth's surface. For example, the feldspars in granite may react with water and carbon dioxide to form clay minerals. The iron-bearing minerals, such as biotite, may react with water and oxygen to form limonite and hematite.

Weathering reduces massive rock exposures to clastic particles ranging in size from boulders to clay. These particles can then be removed through the process of **erosion** by gravity, water, ice, or wind and deposited elsewhere as sediment accumulations.

Biologic Activity

Sediments that were, at some point, actually manufactured by plants and animals are called **biogenic** (Figure 6.1). The most abundant biogenic sediment component is the shells of aquatic animals. Some shells can be found along the edges of freshwater lakes. However, shells form a much higher percentage of the particles making up the sediment along oceanic shorelines. Shells are also important components in the sediments on the deep ocean floors. Large areas of the ocean bottoms are covered by the microscopic calcium carbonate or silica shells of amoebas

Figure 6.1   Sediments are composed of various mixtures of clastic and biogenic particles and crystals.

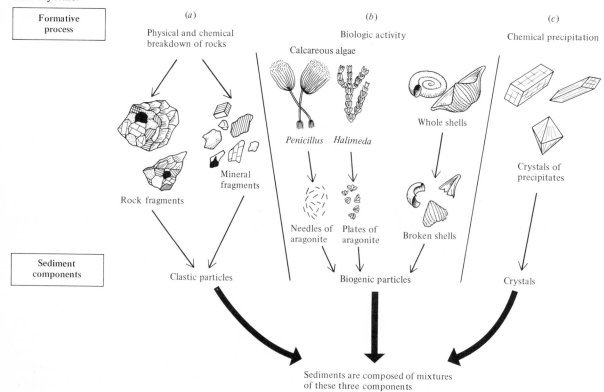

Sediments are composed of mixtures of these three components

Figure 6.2 Scanning electron micrograph of aragonite needles secreted by calcareous algae. The crystals are about 10 μm long. (*Courtesy of D. S. Marszalek*)

and algae that formerly lived in the surface waters and later settled to the ocean floors after death.

In southern Florida and in many parts of the Bahamas, most of the submarine sediment is of biogenic origin. The shallow, warm, and clear waters in these areas provide optimum conditions for the growth of many types of calcium carbonate–secreting organisms. The calcium carbonate may be in the form of calcite or in the form of its polymorph aragonite. Muds composed of needle-shaped crystals of aragonite accumulate in the quieter parts of the area (Figure 6.2). These fine aragonite crystals were originally thought to have been precipitated directly from seawater, but recent research has shown that certain calcareous algae remove the calcium carbonate from seawater and incorporate it into their skeletons. When the alga dies, the aragonite crystals are deposited on the ocean floors, forming a silt or clay-sized sediment. Other types of calcareous algae produce plates of aragonite which form coarser biogenic sediments. In warm seas, coral reefs are another common source of biogenic sediments. The reefs are composed of calcium carbonate that is secreted by the little animals called corals. Waves smashing against the reefs erode pieces of the coral

structure and break up the shells of other animals that live within the reef to produce a coarse sand or gravel-sized sediment.

An economically important biologic activity that occurs on land is the formation of peat. **Peat** is a biogenic sediment formed from plant material which accumulated in ancient swamps. Ideal conditions for peat formation are abundant plant debris, absence of clastic sediments that would dilute the organic debris, and stagnant water conditions which prevent the oxidation of the plant material. After the peat is buried, it begins a slow transformation into coal—a process which we will return to later in this chapter.

Chemical Precipitation

Chemical sediments are composed of crystals precipitated out of concentrated solutions. A common example of chemical sediment is halite, NaCl, crystals, which form on the edges of bodies of water undergoing extensive evaporation. At White Sands National Monument, New Mexico, evaporating waters precipitate crystals of gypsum, $CaSO_4 \cdot 2H_2O$. (The dot in the formula means that water is incorporated into the calcium sulfate crystal.) These gypsum crystals are picked up subsequently by desert winds and accumulate in the white sand dunes for which the area is famous.

A major chemical component in some calcium carbonate sediments is concentrically banded spherical particles of aragonite called **ooliths.** Ooliths form in shallow, wave-agitated marine waters where calcium carbonate is deposited around a nucleus of a shell fragment, quartz grain, or any other "seed." The concentric layering in ooliths is the result of equal deposition of calcium carbonate on all sides of the oolith as it is tossed around in the agitated water.

CHARACTERISTICS OF SEDIMENT

Now that we have considered the different ways in which sediments form, we can look at their distinguishing characteristics. These characteristics provide the clues by which geologists can determine the environmental conditions in which the sediments accumulated.

Composition

A number of factors govern the mineralogic composition of a sediment. The primary control on clastic sediment composition is the mineralogy of the source rocks. For example, a stream draining an area in which only basaltic rocks are exposed (see Chapter 4) would not have any quartz in its sediments because quartz is absent in the source rocks. The duration and intensity of weathering in the source area also affect the mineralogy. Minerals formed at conditions closest to those at the earth's surface have structures which enable them to survive weathering best. Some original rock minerals, such as quartz, are stable under most weathering conditions and increase in relative proportions in the weathered debris. Because of its resistance to weathering, quartz is one of the most abundant minerals in clastic sediments. Sedimentary minerals undergo continuous changes until they are buried by younger sediments and become protected from further weathering. Rate of burial also affects mineralogy. If the rate is fast, it is possible to preserve more of the minerals which weather easily. The

mineralogic composition of a sediment does not always remain fixed after burial. Later in this chapter we will discuss the further mineralogic changes that can occur as the buried sediment is transformed into sedimentary rock.

While there are many minerals in igneous and metamorphic rocks, the great majority of clastic particles in sedimentary rocks are composed of quartz, feldspar, calcite, iron oxides, clay minerals, and rock fragments. Thus, just as fractional crystallization can produce many different types of igneous rocks from one magma (see Chapter 4), differences in weathering intensity and duration, changes during transport, and differing rates of burial can produce several different types of sediments from a single source rock.

Size and Sorting

Walk along a beach in northern New England and feel the coarse-grained beach sand beneath your feet. Walk through the dunes at Cape Hatteras, North Carolina, or along the Oregon coast and feel the finer-grained wind-transported sediment. Compare these with the large boulders at the bases of cliffs in the Rocky Mountains or with a handful of mud scooped up from the bottom of a lake. Obviously, the size of sedimentary particles is highly variable.

The subdivisions of the **Wentworth scale** (Table 6.1) are used to describe the size of sedimentary particles. Each of these common terms —**boulder, cobble, pebble, granule, sand, silt,** and **clay**—describes sedimentary particles whose diameters lie between the appropriate limits.

TABLE 6.1 The Wentworth scale

Particle size range, mm	Name of particle	Name for sediment composed of that particle size
> 256	Boulder	Boulder gravel
256–64	Cobble	Cobble gravel
64–4	Pebble	Pebble gravel
4–2	Granule	Granule gravel
$2-\frac{1}{16}$	Sand	Sand
$\frac{1}{16}-\frac{1}{256}$	Silt	Silt
$< \frac{1}{256}$	Clay	Clay

Terms such as fine, medium, and coarse are used to subdivide each of the size classes. These names are related *only* to particle size; they tell nothing of the composition of the material. For example, we can describe the composition of sand-sized particles on beaches in different areas by using terms such as quartz sands (New England), calcium carbonate sands (southern Florida), and olivine sands (Hawaii). In a similar fashion, the term clay refers only to particles smaller than $\frac{1}{256}$ mm. To avoid possible confusion, sediments composed of the layered silicate *clay minerals* described in Chapter 3 will be referred to as clay-rich sediments in subsequent chapters.

What accounts for the diversity in sizes of sediment particles? The nature of the source rock exerts some influence on grain size. Weathering of a coarse-grained granite will result in larger quartz particles than weathering of a fine-grained granite would. The kinetic energy of the transporting agent determines what size sediments can be carried. For example, slowly moving (low-energy) streams transport fine particles, whereas fast-moving (high-energy) streams transport much larger particles. Collisions between particles of sediment as they are being transported result in abrasion of the particle edges, reducing particle size further.

The range of particle sizes in a sediment is called the **sorting.** *Well-sorted* sediments have particles of similar sizes, while *poorly sorted* sediments have a wide range of particle sizes. Why are some sediments better sorted than others? One of the main reasons is that transporting agents have different densities and thus are able to transport only certain sizes. Air is much less dense than water and is able to transport only sand-sized and finer particles. Winds blowing across a mixture of different sizes of sedimentary particles can move only the finest sand particles, depositing them later as a well-sorted sand. In contrast, glacial ice is much more dense and viscous than air, enabling the ice to transport extremely large particles along with extremely small ones. Consequently, sediments carried and deposited by glaciers have very poor sorting (Figure 6.3). Stream sediments, by comparison, are better sorted than

Figure 6.3 Glacial till exposure in the Adirondack Mountains, New York. *(Courtesy of Walter S. Newman)*

those of glacial ice but not as well sorted as wind-blown sediments.

Roundness

One major factor distinguishing the grains in most sedimentary rocks from those in igneous and metamorphic rocks is that the originally sharp crystal boundaries of the sedimentary particles have been rounded by abrasion during transport. The degree of development of rounded edges on sedimentary particles is called **roundness.** Roundness increases with the number of collisons the sedimentary particles have as they are being transported. Softer minerals, such as the gypsum in the dunes at White Sands, New Mexico, become rounded at a faster rate than harder quartz particles do.

Some general correlations can be made between degree of particle roundness and agent of transport. Particles carried by wind have the greatest opportunity for collision with each other and with the ground. These collisions smooth off rough edges, giving wind-blown sand particles a high degree of roundness. The presence of highly rounded particles in a sediment implies that, in at least part of the past history of the sediment, it was transported under conditions in which sedimentary particles were in continual collision with each other. Examples of such types of transport occur in reworking of beach sands by waves and in wind transport of dune sands. Particles carried within glacial ice, on the other hand, rarely collide with each other, and they may retain their initial angularity even though they are transported great distances. Sedimentary particles carried by streams have a lower roundness than air-transported particles but a greater roundness than glacially transported ones.

Color

An impressive display of colored sedimentary rocks can be seen from the edge of the Grand Canyon or in Bryce and Zion national parks. Some rocks are dull gray or black, while others are brilliant white, tan, orange, or red.

Why is there such a diversity of color in sediments and sedimentary rocks? The colors are due to three major factors: the color of the original mineral particles, the color of weathering products of the original minerals, and the percentage of organic material in the sediment or rock. In some rocks, the color reflects the color of the original minerals. For example, a feldspar-rich rock derived from a granite may have a pink color which reflects the salmon-pink color of orthoclase feldspar in the rock. Similarly, rocks composed entirely of quartz may appear white.

In many rocks the color is not a reflection of the original minerals in the rock but of the weathering products of those minerals. For example, weathering of iron-bearing minerals in the rock forms the brightly colored iron oxides hematite (red and reddish brown) and limonite (yellow and orange). Most sedimentary rocks contain less than 5% iron; however, weathering of only a portion of this iron is sufficient to impart a brilliant color to the rock. In general, finer-grained rocks have more intense colors than coarser-grained rocks of the same composition. Sedimentary rocks with high percentages of organic matter tend to be gray, bluish gray, or black.

Changes in sedimentary mineralogy, weathering, and organic content with time result in the deposition of a succession of multicolored rock layers, such as the spectacular sequences exposed in numerous national parks in the western United States (Plate 11).

TRANSPORT AND DEPOSITION OF SEDIMENT

A desert windstorm, a mountain stream, and a slowly moving glacier can each move large quantities of sediment. However, each of these transporting mediums — air, water, and ice — has a different viscosity and density so that each carries sedimentary materials in a different way. The sedimentary material carried by each is called the **load**. In water transport, some of the load is carried physically as particles (**physical load**) and some is transported chemically as ions in solution (**chemical load**). In air transport, all the material is carried physically as particles. However, in a stream, sedimentary particles may be bouncing along the bottom at the same time that dissolved ions are moving in the waters above. In this chapter we will discuss only those aspects of sediment transport and deposition which are necessary to an understanding of the formation of sedimentary rocks. Specific aspects of sedimentation in different environments will be discussed in detail in Chapters 9 through 16.

Physical Transport and Deposition

Physical transport is controlled by the amount of kinetic energy and the viscosity of the transporting agent. Sedimentary particles begin to move when the force of the airflow or water flow overcomes the gravitational and cohesive forces holding the sedimentary particles together.

Plate 1   Minerals come in nearly every color imaginable. All rows are labeled from left to right. Top row: azurite and malachite (Bisbee, Arizona); crocoite (Dundas, Tasmania); dioptase in calcite (Guchab, Namibia). Second row: fluorite (Cumberland, England); rhodochrosite (Alma, Colorado); quartz crystals (Dauphiny, France). Third row: wulfenite (New Mexico); mangantantalite (Minas Geraes, Brazil); calcite (Joplin, Missouri). Bottom row: uraninite and curite (Katanga, Zaire); hematite (Cumberland, England); carnotite (no source given). *(Identification by Dr. Fred Pough, American Museum of Natural History)*

Plate 2   Some minerals can occur in a variety of colors, as shown by these samples of quartz. *(Courtesy of Mira Schachne)*

Plate 3 The streak of some minerals is the same color as that of the mineral (as shown for cinnabar, right), but for others (like pyrite, left) color and streak are very different. *(Courtesy of Mira Schachne)*

A

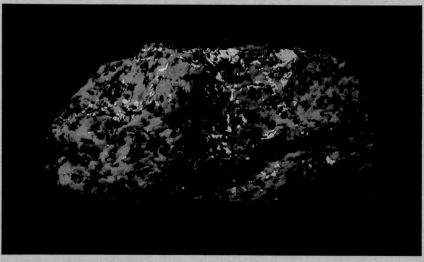

B

Plate 4 Some minerals fluoresce; that is, they have different colors when viewed under normal and ultraviolet light. Calcite, willemite, and franklinite in this specimen from Franklin, New Jersey are pale in normal light *(a)* and fluoresce brightly under ultraviolet light *(b)*. *(Courtesy of Mira Schachne)*

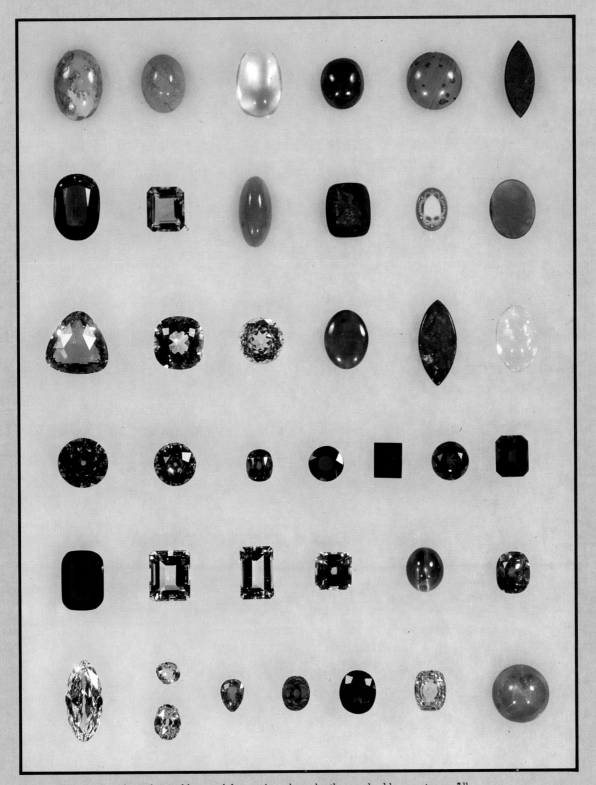

Plate 5 The spectacular color and luster of these minerals make them valuable gemstones. All rows are labeled from left to right. Top row: turquoise matrix, turquoise, moonstone, azurite and malachite, malachite, lapis lazuli. Second row: amethyst, citrine topaz, chrysoprose, sandonyx, carnelian cameo, carnelian. Third row: precious topaz, blue topaz, purple spinel, jade, blue precious opal, Mexican precious opal. Fourth row: blue zircon, gold zircon, green zircon, garnet, green tourmaline, blue tourmaline, red tourmaline. Fifth row: imitation emerald, aquamarine, morganite, chrysoberyl, cat's eye, alexandrite. Bottom row: diamond, pink diamond and green diamond, ruby, sapphire, green sapphire, yellow sapphire, star sapphire. *(Identification by Dr. Fred Pough, American Museum of Natural History)*

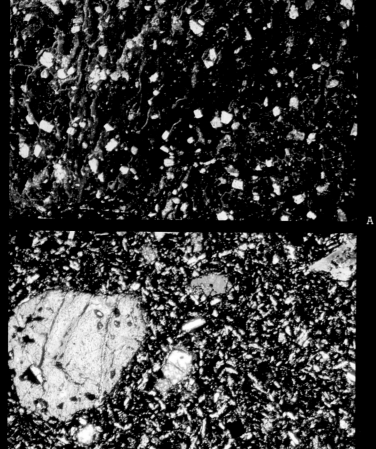

A

Plate 6   Basalt. *(a)* Hand sample of basaltic porphyry with coarse feldspar and olivine grains in a finer-grained feldspar-rich matrix. *(American Museum of Natural History)* *(b)* Thin section showing coarse olivine (bright colored) and finer plagioclase (gray) crystals. *(Courtesy of Allan Ludman)*

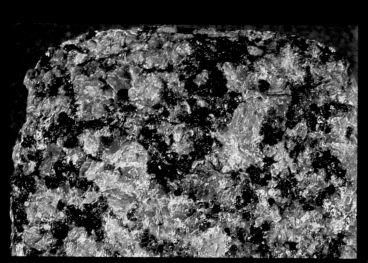

B

Plate 7   Granite. *(a)* Hand sample of granite composed of pink potassic feldspar, white plagioclase, and translucent quartz. *(American Museum of Natural History)* *(b)* Thin section of granite. The plaid mineral is potassic feldspar, the gray plagioclase, the white quartz, and the brown biotite. *(Courtesy of Allan Ludman)*

A

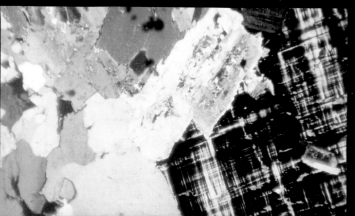

B

Plate 8   Geologist taking lava sample from lava lake in crater of Mauna Ulu, Hawaii, January 1974. (J. C. Ratté, U. S. Geological Survey)

Plate 9   Lava fountaining at night in the Hawaiian Islands. Red-hot molten lava contrasts sharply with darker solidified lava. (Cecil W. Stoughton, National Park Service)

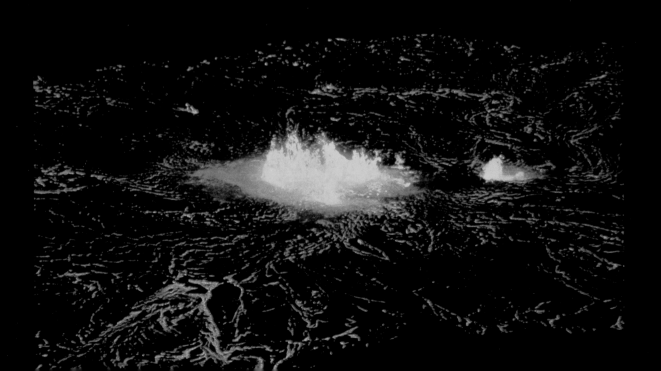

Plate 10　Destruction of forest caused by Hiiaka pahoehoe lava flow, Hawaii. *(Hawaii Volcanos National Park)*

Plate 11　Variations in color in the sedimentary rocks in Zion National Park, Utah. Differential weathering and erosion of the underlying red shales isolate the masses of the overlying red and white sandstone as the great monoliths for which the park is famous. *(Aerial photograph by Nicholas K. Coch)*

A

Plate 12   Sedimentary environments. *(a)* Kaskawuish Glacier, Yukon Territory,
Canada. View looking up-valley showing merging of several valley glaciers to
form a trunk glacier. *(b)* Coastal barrier beach and tidal wetlands, Nauset
Harbor, Cape Cod, Massachusetts. *(c)* Carbonate coast, southwest of North
Eleuthera Island, Bahamas. The white areas are sheets of carbonate sands.
(Courtesy of John S. Shelton)

B

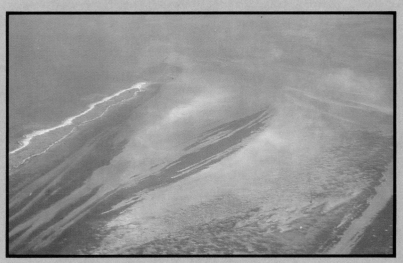

C

Plate 14 Saprolized hornblende gneiss in Jackson County, Georgia. Note the soil on top and the
ock underneath. (Courtesy of Vernon J. Hurst)

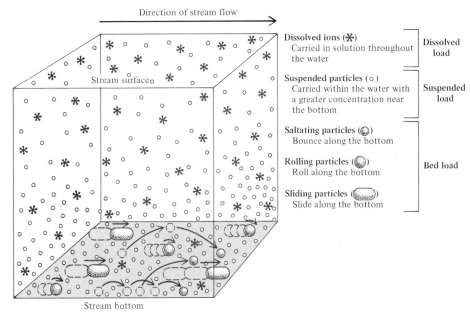

Direction of stream flow

Stream surface

Dissolved ions (✳)
Carried in solution throughout
the water

Dissolved
load

Suspended particles (o)
Carried within the water with
a greater concentration near
the bottom

Suspended
load

Saltating particles (◐)
Bounce along the bottom

Rolling particles (◑)
Roll along the bottom

Sliding particles (◢)
Slide along the bottom

Bed load

Stream bottom

Figure 6.4 Phases of sediment transport in fluids such as streams and wind.

The ability of a transporting agent to carry material, as measured by the amount carried at a given point per unit of time, is called the **capacity**. The **competence** of a transporting agent is a measure of the largest particles that can be carried. An increase in the kinetic energy can increase both the capacity and competence of a transporting agent. For example, a slow-moving stream may be transporting only clay, silt, and fine sand particles. The same stream in flood has a greater kinetic energy and can carry a greater load (increased capacity) and also particles coarser than fine sand (increased competence).

Particles can be physically transported in a number of different ways (Figure 6.4). The larger grains roll (if round) or slide (if flat) along the ground surface or streambed. The movement of particles in constant contact with the bottom is called **traction**. The smaller particles may bounce along the bottom. Transport of particles by bouncing along the ground or streambed is called **saltation** (from the Latin *saltere* meaning *to jump* or *leap*). All of the particles moving by traction and saltation make up the **bed load**. The finer particles may be transported continuously within the fluid in **suspension**. Suspended particles give a cloudy appearance to many streams and account for the poor visibility in windstorms. Glacial ice can transport particles within and below the ice, but a glacier also has a density, viscosity, and strength that enable it to transport material on its upper surface.

Flowing air and water are rarely tranquil but are characterized by numerous "whirlpool-like" internal movements called eddies in which the local flow may be downwards, upwards, sidewards or even upcurrent. This type of churning fluid flow is called **turbulent flow** and is the type of flow that occurs in most streams and winds. As the velocity of the flow increases so does the degree of turbulence. Turbulence is important in the transport of sedimentary particles because the upward movement in turbulent eddies acts against the gravitational force which tends to make the particles settle out of suspension.

Deposition of bed load and suspended load occurs when the transporting medium loses kinetic energy, resulting in a decrease in velocity and turbulence. When a stream flows into a quiet lake, its velocity drops abruptly and the sediment it is carrying is deposited in the lake. A windstorm moves fine sand until an obstacle, such as a house or a rock ledge, blocks the airflow and the sand piles up against it. Organic structures also aid sediment deposition. Aquatic grasses slow down water flow resulting in deposition of sediment around their roots.

Whether a suspended particle will continue to be transported or will be deposited depends on its settling velocity and the turbulence of the flow. The **settling velocity** is the speed at which a particle of a given size, density, and shape settles through still water or air. Settling velocity increases with increasing particle size and density, and varies according to particle shape (spherical particles tend to settle fastest). As the density of the transport medium increases, the settling velocity of a given particle decreases. For example, spherical particles of a given size and density have a greater settling velocity in air than in water.

As long as the upward velocities of the turbulent eddies exceed the settling velocity of a particular particle that particle remains in sus-

pension. When the flow velocity decreases, the turbulence drops and those particles with the highest settling velocities are deposited first.

The process of suspension settling can be illustrated by what happens to the suspended load as the velocity and turbulence of a stream drop, in a series of steps, over a period of time (Figure 6.5). At first the stream energy is sufficient to transport *all* of the particles (Figure 6.5a) in suspension. When the stream energy decreases, the particles with the greatest settling velocity (coarse sand in this case) fall out of suspension and are deposited as a layer on the bottom of the stream (Figure 6.5b). Later, the stream's energy drops even more, and medium sand-sized particles are deposited in a second bed on the stream's bottom (Figure 6.5c). The accumulation of sedimentary particles in distinct layers is called **stratification**. Stratification is one of the diagnostic characteristics of sedimentary rocks and will be discussed in more detail later on in this chapter.

Figure 6.5   Formation of a series of beds along the bottom of a stream. The kinetic energy of the stream is decreasing in a series of steps. Each time the kinetic energy of the stream decreases, the suspended particles with the highest settling velocity are deposited first.

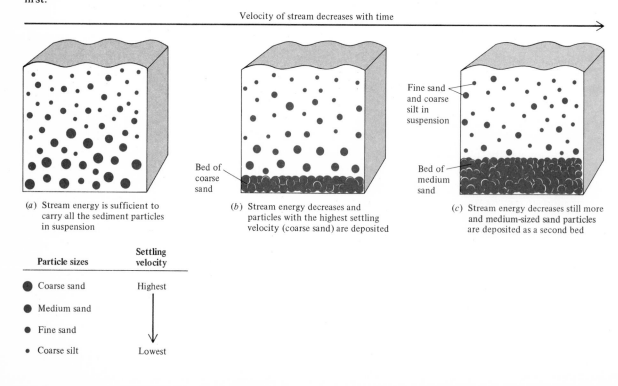

Velocity of stream decreases with time

(a) Stream energy is sufficient to carry all the sediment particles in suspension

(b) Stream energy decreases and particles with the highest settling velocity (coarse sand) are deposited

Bed of coarse sand

(c) Stream energy decreases still more and medium-sized sand particles are deposited as a second bed

Fine sand and coarse silt in suspension

Bed of medium sand

Particle sizes	Settling velocity
● Coarse sand	Highest
● Medium sand	
● Fine sand	
● Coarse silt	Lowest

Chemical Transport and Deposition

The dissolved ions present in all bodies of water are just as important in the formation of certain types of sedimentary rocks. Water bodies differ in both the total amount of dissolved ions that they carry and in the concentration, or the amount of dissolved ions per unit volume. The kind and amount of ions in a body of water is determined by the solubility of the minerals with which the water has been in contact.

The dissolved load may be deposited when chemical changes make certain ions less soluble or when a physical change, such as evaporation, increases the concentration of certain dissolved ions beyond the concentration permitted by the temperature and solubility of those ions in water. When ions are present in low concentrations, water molecules separate them and prevent them from coming into contact with each other. If evaporation of the water increases the concentration of ions, the chance of ions meeting one another increases. As the ions collide in these concentrated solutions, they may be bonded to one another and precipitate out as crystals which accumulate as a layer of chemical sediment. Precipitation will continue until the concentration of the remaining ions reaches a low enough value for them to remain in solution.

Chemical sediments deposited from highly concentrated solutions are called **evaporites.** The high ionic concentrations needed to precipitate evaporites occur in hot and arid climates in a variety of geographic areas. Major areas where evaporites are accumulating today include isolated desert basins, coastal areas, and (sporadically) the slopes and floors of isolated marine basins such as the Dead Sea, the Red Sea, and the Mediterranean. Streams carry dissolved ions down into hot, arid basins, such as Death Valley, where the water evaporates and the dissolved loads are deposited as evaporites. Evaporites are also forming along the arid coasts of the Persian Gulf and along the Trucial Coast in the Middle East. Seawater trapped within the pores of the sediments undergoes extensive evaporation, resulting in deposition of crystals of gypsum, $CaSO_4 \cdot 2H_2O$, and anhydrite, $CaSO_4$, within the sediments. The characteristics of this recently deposited gypsum and anhydrite resem-

ble those in ancient evaporite rocks, suggesting that at least a portion of the older evaporites may have had a similar origin.

While most present-day evaporites are accumulating on the continent or along arid coasts, the thick deposits of ancient evaporite rocks probably had a different origin. Such evaporites are believed to have accumulated in deep basins which were separated from the ocean by a submerged barrier. This barrier, a coral reef in many cases, allowed only a minimal exchange of waters with the ocean (Figure 6.6). Ocean water entered through and over the submerged barrier. This water became enriched in dissolved salts by evaporation, thus increasing its density.

When a basin reaches saturation of dissolved ions, the dense saline water may settle to the bottom of the basin, depositing evaporite crystals on the basin floor. In other instances, evaporite crystals form at the air-water contact points in saturated surface waters. These crystals then settle down and accumulate on the basin floor. In yet other instances, deep basins may actually evaporate completely, and shallow-water or coastal-type evaporite deposits may accumulate on formerly deep basin floors.

Laboratory studies show that as evaporation proceeds, different salts are precipitated at different concentrations. Theoretically, $CaCO_3$ precipitates first, followed by $CaSO_4$, then NaCl, and finally salts of magnesium and potassium. The high concentration needed to deposit magnesium and potassium salts accounts for the relative rarity of these types of evaporites.

Special conditions can give rise to exceptional thicknesses of evaporites. Deep drilling beneath the floor of the Mediterranean disclosed that great thicknesses (an average of 2000 m) of evaporites underlie the basin. From this and other evidence, it was deduced that a barrier developed between the Atlantic and the Mediterranean, restricting circulation between the two bodies of water. Nearly complete evaporation of the Mediterranean waters began about 7.2 million years ago and resulted in the deposition of great thicknesses of evaporites. Evaporite deposition ceased about 5 million years ago with the subsidence of the barrier separating the Atlantic and the Mediterranean and the forma-

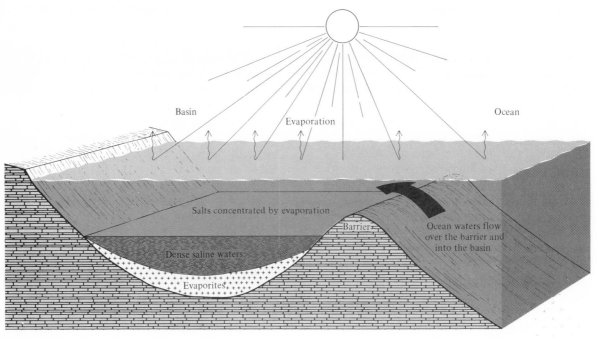

Figure 6.6 Deposition of evaporites in a basin partially separated from the ocean by a barrier.

tion of the Straits of Gibraltar. This was followed by a rapid refilling of the Mediterranean basin by Atlantic waters.

SEDIMENTARY ENVIRONMENTS AND FACIES

The sediments on the earth's surface and on the ocean floors are deposited under a wide variety of conditions. The term **sedimentary environment** is used to describe an area with distinctive physical, chemical, and biologic conditions. A number of common sedimentary environments are shown in Figure 6.7. Each of these environments contains a distinctive type of sediment that reflects conditions in that environment. The fast-flowing water of the river carries sands and gravels along its bed, while the water above may be cloudy with suspended sediment. When the stream reaches the quiet waters of the lake, velocity decreases and the stream deposits sand and coarser sediment in a delta. Finer silts, clays, and organic particles are carried in suspension into the lake. These settle slowly through the lake waters to form the dark muds on the lake bottom. Occasional river floods may transport fine sand particles in suspension out into the lake, leaving thin sand layers between dark mud layers normally deposited in the lake (Figure 6.7). Some of the coarse material supplied by the river is moved along the lakeshore to form sandy beaches. Winds blowing across the beach remove the finer sand, building it into dunes. On the opposite shore, dense growths of vegetation trap fine-grained sediments in marshes.

Each of these sedimentary environments (stream, delta, beach, dune, lake, and marsh) is underlain by a distinctive type of sediment that reflects the physical, chemical, and biologic conditions prevailing in that environment. In this example we have considered briefly only a few of the great number of sedimentary environments, some of which are shown in Plate 12. In Chapters 9 through 16 we will describe these environments and others in greater detail. Geologists use the term **facies** to describe sediment (or rock) with distinctive characteristics. The area illustrated in Figure 6.7 contains a number of different sedimentary facies. Each of the different facies is being deposited at the same time and grades laterally into other facies. In a similar fashion, sedimentary *rock* facies that grade laterally into one another must have been deposited at the same time. This principle is an extremely important tool for establishing the geologic history of an area, and we shall return to it in Chapter 8. The number of different sedimentary environments is another factor accounting for the wide variety of types of sediment found on the earth's surface and on the ocean bottom.

Sediments deposited in one environment may be eroded and reworked again in a different environment. For example, ancient stream deposits in cliffs may be eroded by coastal waves, and streams may erode banks composed of older glacial deposits. Thus, grains of sediment may bear the "imprint" of many previous sedimentary environments. These cycles of erosion, transportation, and deposition may be repeated many times over periods of millions of years, with each reworking changing the sediment more and more from its initial character.

Figure 6.7 (*a*) Sedimentary environments, and (*b*) sedimentary facies. Each sedimentary environment is underlain by a distinct sedimentary facies.

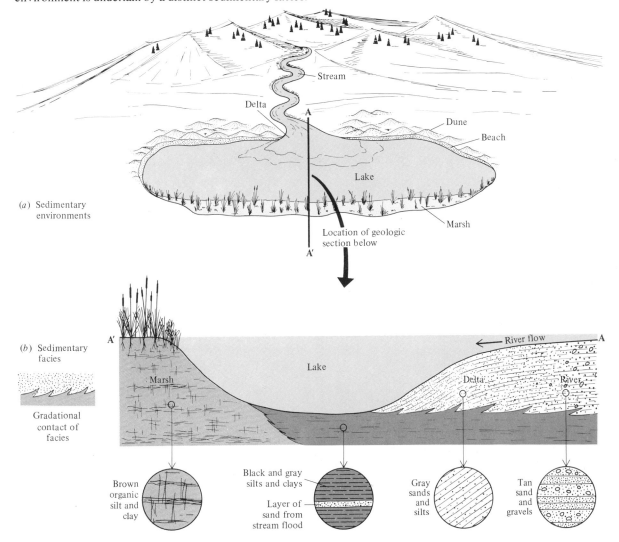

LITHIFICATION

Eventually, the sediment reaches an area where it is protected from subsequent erosion and accumulates layer by layer. Some sediments accumulate on the continents in river valleys, lakes, and desert basins. However, for many sediments the ultimate depositional area is the ocean basins. As the sediment becomes more deeply buried, it is subjected to higher temperatures and pressures. The conditions then are present for the transformation of the sediment into sedimentary rock. This transformation is known as **lithification.** The physical, chemical, and biologic changes that occur after deposition and during lithification are referred to as **diagnenesis.** There are a wide variety of diagenetic changes (Figure 6.8), some of which lithify the sediment while others change the mineralogy and characteristics of its component grains.

Compaction

The spaces between sedimentary particles are called **pores** (Figure 6.8a). The volume of pore space compared with the total volume of the sediment is called the **porosity.** Initially, the sediment may have a high porosity, and its pore spaces may be filled by air or by liquids. As the sediment is buried progressively by younger sediments, the pressure on its component grains

Figure 6.8 Diagenetic changes that occur in the lithification of sedimentary rocks: (*a*) compaction; (*b*) particle intergrowth; (*c*) cementation; (*d*) overgrowths on particles; (*e*) matrix formation; (*f*) solution.

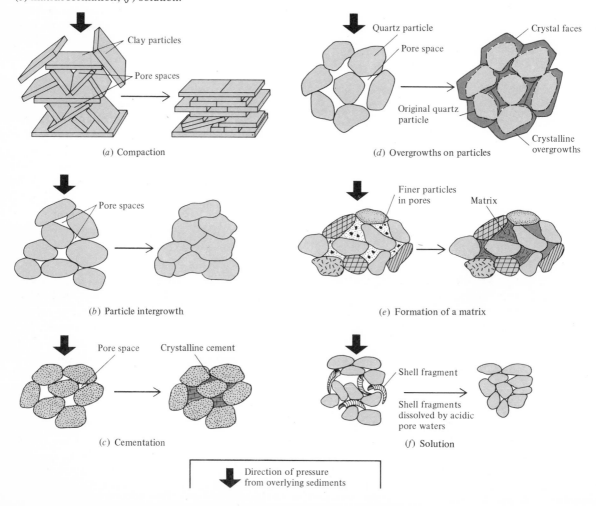

(*a*) Compaction

(*b*) Particle intergrowth

(*c*) Cementation

(*d*) Overgrowths on particles

(*e*) Formation of a matrix

(*f*) Solution

Direction of pressure from overlying sediments

increases. As this pressure increases, the particles become rearranged, resulting in a decrease of porosity, and the expulsion of part of the liquids that originally filled the pore spaces. The process of rearrangement of particles owing to pressure from above is called **compaction.** Compaction results in a decrease in porosity and in bed thickness (Figure 6.8a). The decrease is greatest in fine-grained sediments such as silts and clays because they initially have a greater porosity than sands and coarser sediments. Extreme compaction may push the grains together to such a degree that the original pore spaces are destroyed. The edges of the particles may even penetrate one another, giving the rock a "sutured" look (Figure 6.8b).

Cementation

Sediments are also lithified by the precipitation of crystalline materials between the particles in a process called **cementation** (Figure 6.8c). Common cementing agents include silica, SiO_2; calcite, $CaCO_3$; and iron oxide, Fe_2O_3. The cementing materials in sedimentary rocks have a number of different origins. They can be derived from the liquids trapped within the pores at burial, from fluids circulating through the sedimentary pores after deposition, or from the solution of some of the sedimentary particles. Sometimes the dissolved material is similar in chemical composition to the original particles, and it is deposited around the rims of the particles as **overgrowths** (Figure 6.8d).

Mineralogic Changes

Diagenetic changes also can alter the mineralogy of the sediment during lithification. These mineralogic changes occur by the complete breakdown of some minerals, the partial replacement of the ions in other minerals, the change from one mineral structure to a more stable structure without a change in composition, or even the formation of a completely new mineral.

The particles in some sedimentary rocks are not lithified by visible cementation but by solidification of a fine-grained interparticular material called the **matrix.** In some sedimentary rocks the matrix represents fine-grained sedimentary particles incorporated within the sediment during its burial. At least part of the matrix in other rocks is believed to form *after* burial by the breakdown of some of the smaller particles to form the fine-grained matrix. New minerals that formed at the same time, or subsequent to the formation of a sedimentary rock, are referred to as **authigenic minerals** (Figure 6.8e). Some of these authigenic minerals form a matrix while others form nodules, concretions, or even isolated crystals within the sedimentary rock.

Some of the original sedimentary particles may be removed by solution during lithification. For example, acidic waters circulating through sedimentary pores may dissolve shell material composed of calcium carbonate (Figure 6.8f). Many present-day aquatic organisms secrete shells made of aragonite. The shells found in sedimentary rocks are not composed of aragonite but rather of calcite, because aragonite commonly inverts to its more stable polymorph, calcite, during diagenesis.

Changes in Organic Materials

The changes that occur in organic materials are well illustrated by the diagenetic transformation of highly organic freshwater swamp sediments into coal. As this organic sediment is buried under the stagnant swamp waters, increased pressure and temperature begin to break down the vegetal debris. The organic material breaks down slowly, releasing gases in the process. These released gases include water, carbon dioxide, and methane.

The transformation of organic material into coal involves a series of intermediate steps with the progressive release of more and more gaseous components, enriching the residue in carbon. The first stage of this transformation is peat, a soft, spongy, wet fibrous mass with a low carbon content and a low heating value. In many areas of the British Isles today, peat is cut out of bogs, stacked to dry, and used as a low-grade fuel. As temperature and pressure increase, peat is transformed successively into **lignite; bituminous,** or soft, coal; and finally **anthracite,** or hard coal. Anthracite is high in carbon and low in volatiles—this gives it the greatest heat value.

SEDIMENTARY ROCK TYPES

The wide variety of sedimentary rocks reflects differences in sources along with differences in mode of transport and in depositional conditions. Sedimentary rocks are classified into three major groups based on the origin of their major component: clastic, biogenic, and chemical. Texture (the size and arrangement of the grains) and mineralogic composition are used to further subdivide the three major rock groups (Figure 6.9). **Clastic rocks** are composed of particles from preexisting rocks. These rocks may have a crystalline cement (Figure 6.9a) or a fine-grained matrix between the particles (Figure 6.9b). **Biogenic rocks** are composed of materials derived from the activities of plants and animals. The biogenic particles are surrounded by either crystalline material or fine-grained matrix (Figure 6.9c). **Chemical rocks** form from the precipitation of crystals from concentrated solutions and consist of tightly interlocking crystals (Figure 6.9d).

Most sedimentary rocks consist of more than one of these three components but are named for the dominant component. For example, a sandstone composed largely of quartz particles, containing some shell fragments, and cemented by crystalline calcite would be classified as a clastic rock because it is composed largely of clastic particles. There are a number of sedimentary rock classifications, each of which emphasizes different criteria. The simplified classification used here emphasizes those points which are readily visible in sedimentary rock samples.

Figure 6.9   Textures of major types of sedimentary rocks.

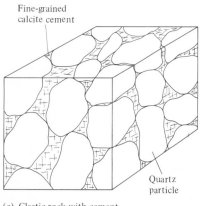

(a) Clastic rock with cement

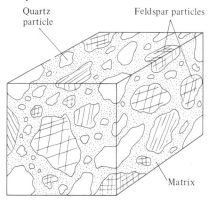

(b) Clastic rock with matrix

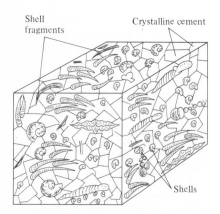

(c) Biogenic rock with cement

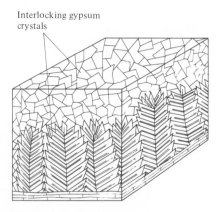

(d) Chemical rock

Clastic Rocks

Clastic rocks composed of gravel-sized particles are called **conglomerates** if the particles are rounded (Figure 6.10*a*), and **breccias** if the particles are angular (Figure 6.10*b*). Rocks composed of sand-sized particles are called **sandstones.** If the sandstone contains less than 15% matrix, it is called an **arenite.** If it has more than 15% matrix, it is called a **wacke** (pronounced wack–ie) (Figure 6.10*c*). Rocks composed of silt-sized particles are called **siltstones** (Figure 6.10*d*), and those composed of clay-sized particles are called **shales** (Figure 6.10*e*). If a rock is composed of two or more different-sized particles, the one occurring in the larger percentage is given last in naming the rock. For example, a rock in which rounded gravel predominates over sand would be called a sandy conglomerate.

The composition of the particles is given by a term preceding the textural term (Table 6.2). For example, a clastic rock composed of quartz grains cemented together is called a **quartz arenite,** whereas another rock composed of quartz grains and more than 15% matrix is called a **quartz wacke.** If the sandstone is composed of rock fragments and matrix, it is referred to as a **lithic wacke.** Sandstones composed of 10–15% feldspar are called **feldspathic sandstones;** those with higher percentages of feldspar are referred to as **arkoses.** All the clastic rocks may be classified by using combinations of terms denoting the texture and composition of the component particles.

Biogenic Rocks

Biogenic rocks are classified according to their composition, texture, and in the case of crystalline limestones, crystal size. The two major biogenic rocks are **limestone,** which is composed of calcite, $CaCO_3$, and **dolomite,** which is composed of calcium-magnesium carbonate, $CaMg(CO_3)_2$. Dolomite is actually a diagenetically altered crystalline or biogenic limestone, and as such it is more of a chemical than a biogenic rock. Field studies in warm and shallow intertidal environments have determined how the change occurs. As calcium carbonate is deposited, calcium is removed from the seawater, and the seawater becomes relatively enriched in magnesium. Evaporation increases the magnesium concentration further. When the magnesium-enriched seawater circulates through the pores of the underlying limestone, the magnesium replaces some of the calcium in the rock, converting it to dolomite.

Biogenic rocks such as limestone may have many different textural components, such as ooliths, shells, and shell fragments along with sand, silt, and clay-sized particles derived from the breakdown of calcareous algae (Figure 6.2). Some limestones are composed largely of shells and shell fragments with a crystalline calcite cement. **Chalk** is a soft, white rock composed mostly of the skeletons of microscopic marine plants and animals. **Coquina** is a limestone composed entirely of the shells and shell fragments of marine animals (Figure 6.11). Some fine-grained limestones are formed from the lithification of lime muds derived from the breakdown of calcareous algae (Figure 6.2). However, most limestones are composed of a mixture of textural elements and are classified

TABLE 6.2 Terms used to classify clastic sedimentary rocks

TEXTURE

Particle size	Rock name
Gravel	
Rounded particles	Conglomerate
Angular particles	Breccia
Sand	Sandstone
Less than 15% matrix	Arenite
More than 15% matrix	Wacke
Silt	Siltstone
Clay	Shale

COMPOSITION

Component	Rock name
Quartz	Quartzose
Feldspar	
10–25% feldspar	Feldspathic
More than 25% feldspar	Arkosic
Iron oxide	Ferruginous
Calcite	Calcareous
Rock fragments	Lithic

Figure 6.10 Types of clastic rocks. *(Nicholas K. Coch)*

(a) Conglomerate.

(d) Siltstone laminae (dark) in very fine sandstone.

(b) Breccia.

(e) Shale with fossil plant impressions.

(c) Sandstone.

by using terms that describe their major component. For example, a rock containing ooliths of calcium carbonate cemented by calcite is referred to as an **oolitic limestone.**

Lignite, bituminous, and anthracite coal are biogenic rocks formed from organic materials by the processes described earlier in this chapter.

Chemical Rocks

Chemical rocks are classified according to the composition of the crystals that compose the rock. The most common chemical rocks are **chert,** SiO_2; **gypsum,** $CaSO_4 \cdot 2H_2O$; **anhydrite,** $CaSO_4$; **halite,** NaCl; and **phosphorite** (com-

Figure 6.11 Coquina. *(Nicholas K. Coch)*

plex phosphates of calcium) (Figure 6.12). Crystal size in chemical rocks ranges from crystals too small to be seen without magnification (**cryptocrystalline**) to coarse-grained crystalline rocks.

SEDIMENTARY STRUCTURES

We mentioned before that sedimentary rocks provide a record of surface conditions in the past. Many of the clues to these former environments are derived from features within the rock called **sedimentary structures.** These are structures which form within the sediment during its accumulation and before it is lithified. They can be formed by any combination of the physical, chemical, and biologic processes acting in the environment in which the sediment accumulated. **Primary** sedimentary structures form as the sediment is being deposited. Examples of common primary sedimentary structures that you might see include the layering of different types of sand in a beach or river bar, the ripple marks on the surface of sand along a shoreline or river, and the tracks that organisms make as they move along the sedimentary surface. **Secondary** sedimentary structures form after sediment deposition. Examples of such structures include nodules, concretions,

and the spherical, mineral-filled structures called **geodes** which are so prized by mineral collectors.

There are a great number of primary sedimentary structures. In this chapter we can consider only a few of them, specifically those which tell us something about conditions at the time of sediment deposition. When properly deciphered, these primary sediment structures can be used to answer questions as diverse as: Which way was the current flowing? What was the climate like? Are these rock layers in the same order that they were deposited in, or have they been overturned since deposition?

Stratification

Nearly all clastic sedimentary rocks are produced by sedimentary particles settling out of fluids such as water and air and accumulating for the most part in originally horizontal layers. These layers are called **beds** if they are thicker than 1 cm and **laminae** if they are thinner than 1 cm. Only some of the sedimentary layers deposited may be preserved because some beds may have been eroded away partially or completely by later currents. Sometimes pieces of the older bed may be incorporated as fragments in the younger one.

(a) Gypsum crystals.

(c) Halite — large crystalline mass.

Figure 6.12   Chemical sedimentary rocks. *(Mira Schachne)*

(b) Chert (dark layer) interbedded with two limestone beds.

Each bed is deposited under a given set of environmental conditions. Adjacent beds are distinguished because they are separated by a discrete physical break in the rock called a **bedding plane.** When environmental conditions change or when deposition begins again after ceasing for a while, another bed begins to form. The various beds can be distinguished from each other based on differences in particle size, composition, sorting, color, and shell content. Since each bed represents a particular set of environmental conditions, study of a vertical sequence of beds allows a geologist to determine the successive changes in environmental conditions in the area.

Thicker beds generally indicate a longer time of sediment accumulation. However, some thick beds can be deposited quickly, while some thin ones accumulate over a long period of time. For example, a storm sweeping across a beach may deposit a $\frac{1}{2}$-m bed of gravel and broken shell in a few hours, whereas sedimentation of a 1-cm-thick bed of oceanic clay may take thousands of years.

There are several major types of stratification, each of which forms under a different set of environmental conditions. Thick and uniform beds are called **massive** and imply that conditions were constant during the deposition of the bed. Beds in which the grain size decreases from the base to the top are called **graded** (Figure 6.13*a*). Graded beds are produced where sediment-laden fluids undergo a fairly rapid decrease in kinetic energy. As the energy decreases, finer and finer particles are deposited on the bottom. For example, as a flooding stream spills out of its channel, there is an abrupt decrease in the stream's energy, and a graded bed may be deposited over the adjacent area. Stratification consisting of an alternation of

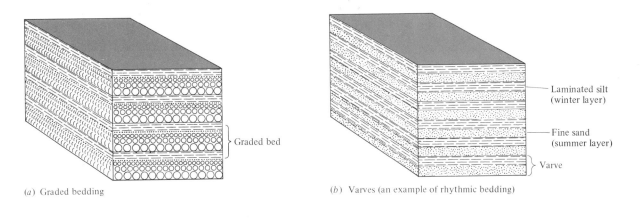

(a) Graded bedding

(b) Varves (an example of rhythmic bedding)

Graded bed

Laminated silt (winter layer)

Fine sand (summer layer)

Varve

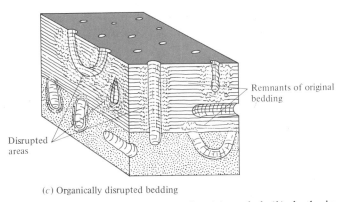

Disrupted areas

Remnants of original bedding

(c) Organically disrupted bedding

Figure 6.13 Major types of stratification in sedimentary rocks: (a) graded; (b) rhythmic; (c) organically disrupted bedding.

two different types is called **rhythmic** bedding (Figure 6.13b). Rhythmic bedding implies a sequential alternation of two depositional conditions. One environment in which rhythmic bedding occurs is the lakes that develop at the borders of melting glaciers.

In the summer, streams from the melting glaciers deposit their bed load and coarser suspended load into the lake as graded beds. In the winter, the stream and lake freeze over and bed load is excluded. At this time, the suspended organic matter and silt- and clay-sized particles settle out of suspension and are deposited as a dark, fine-grained layer across the lake bottom. As the ice melts in the spring, bed load again reaches the lake and another cycle of deposition begins. Such a yearly couplet of a coarsely graded bed (summer) and a fine bed (winter) constitutes a special type of rhythmic bedding called a **varve.**

Some beds that look massive at first glance show remnants of an original stratification upon closer examination. Usually the bedding has been modified by organisms that live on and within the accumulating sediments, burrowing through, disrupting, and homogenizing them (Figure 6.13c).

Cross-bedding is a type of stratification in which the layers *within* the bed are not horizontal but are inclined (Figure 6.14). Cross-bedding forms by sand and coarser sediment being moved by air or water along a surface in stream-lined mounds such as ripples or dunes. A section cut through a ripple or dune shows that they are composed of sand layers which are usually inclined in the down-current direction (Figure 6.14a). As the ripples or dunes move across the surface, they leave behind beds of cross-bedded sediments. The cross-bedding deposited by migrating high mounds of sand such

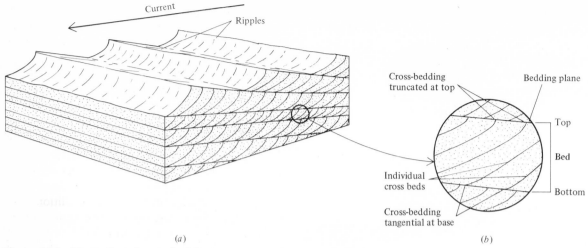

Figure 6.14   Formation of cross-bedding by the migration of sandy ripples.

as dunes can reach great thicknesses (Figure 6.15). Since the cross-bedded layers are usually inclined down-current, cross-stratification can be used to determine the direction of the ancient current, or **paleocurrent,** which formed the deposit. The cross-stratification within any one

Figure 6.15   Large scale dune cross-bedding in the Navaho sandstone, Zion National Park, Utah. *(D. Carroll, U. S. Geological Survey)*

bed either makes a small angle or is tangential to the bottom of the bed (Figure 6.14*b*). The individual cross-strata may be concave upward and cut off or truncated at the top of the bed. A geologist can therefore determine whether a bed is in the normal position or has been overturned by examining the orientation of the cross-bedding. Sedimentary structures that can be used to determine the tops and bottoms of beds are called **top and bottom structures.**

Ripple Marks

You may have noticed regular undulations in the sandy material near the shore of a lake, river, or ocean, or on the sides of a sand dune. These regular undulations are called **ripple marks.** They form as currents move over a sandy surface. There are two major kinds of ripple marks, each of which forms under different conditions. **Oscillation ripples** are symmetrical and form by wave action (Figure 6.16*a*). The pointed crests of the ripple marks always point upward, and this makes oscillation ripple marks a useful top and bottom structure. **Current ripples** are asymmetrical ripples formed when the current moves along the bottom (Figure 6.16*b*). The steeper side of the current ripple is always inclined down-current, making it useful for determining paleocurrent directions.

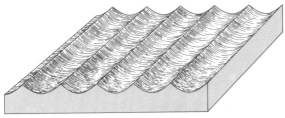

(a) Oscillation ripple marks (symmetrical)

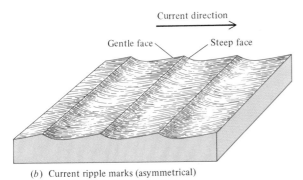

Current direction

Gentle face   Steep face

(b) Current ripple marks (asymmetrical)

Figure 6.16 (a) Oscillation and (b) current ripple marks.

Mud Cracks

When a wet, clayey sediment dries, it contracts, forming a series of polygonal **mud cracks** on the surface (look back at Figure 1.4). Sediment deposited later fills in the cracks and preserves them. Mud cracks are a useful indicator of exposure to air drying and give some clues to ancient climatic conditions, or **paleoclimates.** They record a wet time during which the clayey sediment accumulated, followed by a dry time during which the mud cracks formed.

Sedimentary Structures and Geologic History

The geologic history of an area is determined by studying the sequence of exposed sedimentary rocks. This task can be quite complicated, especially if postdepositional deformation has tilted, and in some cases overturned, the rock layers. How do we know whether this has occurred? If the sedimentary rocks in the exposure contain any top and bottom structures, such as mud cracks, oscillation ripple marks, or

cross-bedding, it is possible to determine the original tops of the beds. Geologists examining the rock exposure shown in Figure 6.17 might think at first that they were dealing with 10 different layers. After a more careful examination of the top and bottom structures in these beds, it would be apparent that the upper part of the exposure is an overturned version of the lower part. For example, the cross-bedding in layer C_2 is tangential to what appears to be the top of the bed and truncated at what appears to be the base of the bed. (Refer back to Figure 6.14 to see the normal orientation.) In addition, unit C_2 contains rock fragments which appear to be from unit B_2. The rock fragments would not have been available if B_2 had not formed before C_2 was laid down. These are not the normal relationships to be expected and they indicate that layer C_2 has been overturned. Once it became apparent that a portion of the rock sequence had been overturned, the reconstruction of the area's geologic history could be revised accordingly.

FOSSILS

Fossils are any traces or imprints of plant or animal life that have been preserved by natural processes. Fossils are generally found only in sedimentary rocks, although in rare cases they have been preserved in some types of extrusive igneous and metamorphic rocks. What parts of these ancient life forms become "fossilized"? In general, only the hard parts such as shells and bones are preserved because the soft flesh decomposes quickly. There are a few examples of exceptional fossilization in which both the hard and soft parts have been preserved. One of these exceptional preservations is the Ice Age mammoths found in frozen soil in Alaska and Siberia. These mammoths apparently fell into cracks in the ice, and their carcasses froze before their flesh could decompose (Figure 6.18). Remains of Arctic vegetation were even found in their stomachs! Many complete insect fossils are preserved in the hardened tree sap called **amber.** The insects originally were caught by the sticky sap oozing from trees and were incorporated into drops of sap which

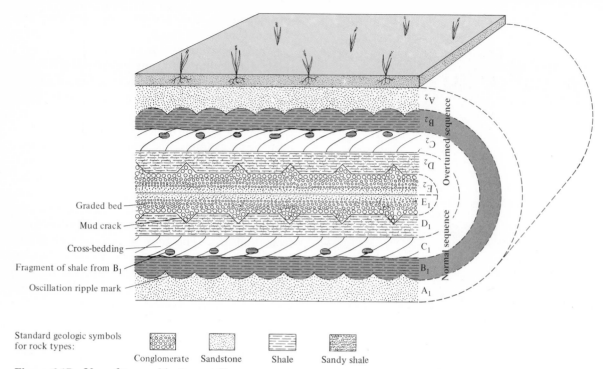

Graded bed
Mud crack
Cross-bedding
Fragment of shale from B_1
Oscillation ripple mark

Standard geologic symbols for rock types:

Conglomerate   Sandstone   Shale   Sandy shale

Figure 6.17 Use of top and bottom sedimentary structures to determine the depositional order in a sequence of overturned rocks. Sequence is composed of the following beds: (A) sandstone with oscillation ripple marks; (B) shale; (C) sandstone with cross-bedding and pieces of underlying shale; (D) shale with mud cracks filled in with material from layer E; (E) sandstone with graded bedding and chips of layer D at base. Dashed lines show inferred reconstruction of bed sequence.

solidified into amber. The amber protected the insects from decomposition and prevented them from being crushed during compaction of sediment (Figure 6.19).

The fossils found in sedimentary rocks form in different ways. Organic structures such as tree trunks incorporated in sediments may have their soft parts replaced by minerals precipitated from solutions circulating through the enclosing sediment. Such a mineral transformation is called **petrifaction.** The fossil wood at the Petrified Forest National Monument was formed when silica replaced the original woody tissues of the tree trunks, forming a hard, resistant fossil (Figure 6.20). A similar type of fossilization is **replacement,** in which the *hard* parts of former animals are replaced by different minerals. For example, in some rocks the original calcium carbonate in fossil marine shells has been replaced by silica, forming **silicified** fossils.

Leaves and other plant material falling into the muds of stagnant lakes and swamps are preserved from oxidation. In the transformation of the muds to shale, the plant material is transformed to carbon imprints, preserving the form of the original material. The process of forming such carbon imprints is called **carbonization** and is part of the process of coal formation. Thus, carbonized plant remains are especially common in coal beds and the shales associated with them. Such imprints can be seen on the shale shown in Figure 6.10*e.*

A buried organism or plant may be dissolved during diagenesis. This leaves a cavity with the impression of the organism on the sides (Figure 6.21). Such impressions are called **molds.** This cavity may be filled subsequently with sediment which forms a three-dimensional **cast** of the original organism. Thus, the shape of the original shell is preserved.

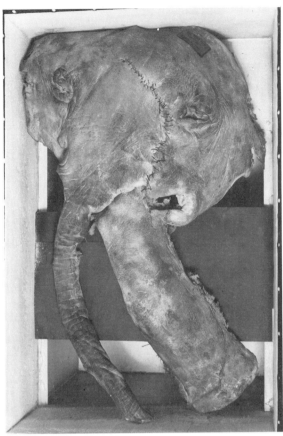

Figure 6.18 Fossilized baby mammoth dug out of frozen ground in Alaska. *(American Museum of Natural History)*

Figure 6.19 Ant preserved in amber. *(American Museum of Natural History)*

Figure 6.20 Petrified wood in the Petrified Forest National Monument. *(N. H. Darton, U. S. Geological Survey)*

If the material is fine-grained and deposited slowly, remarkably detailed impressions may be formed. The most famous impression is that of the primitive bird *Archaeopteryx* preserved in very fine grained limestone in Bavaria. The impression is so good (Figure 6.22) that the teeth and feathers are clearly visible.

The arrangement and preservation of fossils tells us something about the environment in which the original organisms lived, died, and were buried. The complete preservation of a delicate-shelled fossil indicates that there was little reworking of the deposit by currents. On the other hand, a coquina, composed entirely

Figure 6.21 Formation of molds and casts.

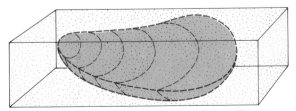

After a dead organism is buried by accumulating sediments, it is dissolved by waters circulating through sand. An open cavity forms with the impression of the shell on sides.

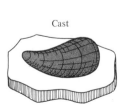

Cast

Cavity filled by crystals precipitating from circulating waters, forms a cast with shape similar to original shell.

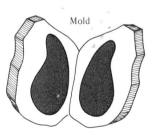

Mold

Original fossil dissolved giving void with impression of fossil organism.

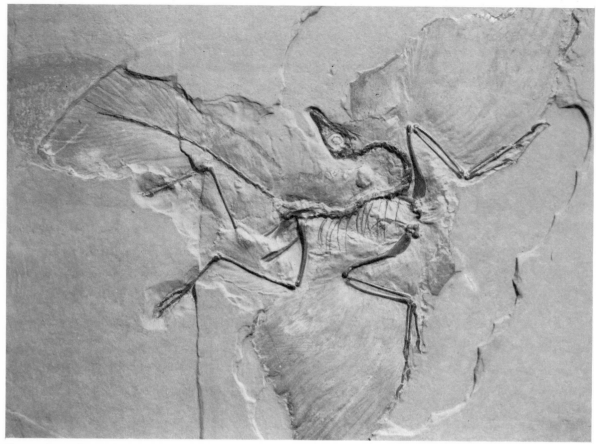

Figure 6.22 Cast of *Archaeopteryx* in fine-grained limestone. *(American Museum of Natural History)*

of broken shell fragments (see Figure 6.11), indicates intensive reworking of the deposit before burial. In some deposits, elongate fossils are aligned much as a weathervane aligns with the wind. These aligned fossils can be used to determine the paleocurrent direction at the time of sediment deposition.

Some types of fossils do not represent the remains of the organisms themselves but rather their *effects* on the sediment during deposition. As sediments accumulate, organisms live on and within the sediment, leaving a record in tracks and trails as they move across the surface and in burrows as they bore through the bedding. Sedimentary structures resulting from the life activities of animals are called **trace fossils** and can be seen in the organically disrupted bedding in Figure 6.13*c*. Trace fossils are especially important in determining the origin of sedimentary rocks that do not contain any fossilized remnants of original life. The presence of trace fossils in such rocks shows that animals did live in that area as the sediments accumulated.

SUMMARY

Sediments are produced from the breakdown of older rocks, from organic activity, or by the precipitation of crystals from solutions.

Sedimentary materials are carried physically as particles in bed load and suspended load, and chemically as ions in solution. Bed load and suspended load are deposited when the kinetic energy of the fluid decreases. Dissolved load is deposited when chemical conditions change or when evaporation concentrates the dissolved ions that are crystallized as evaporite minerals.

Mineralogic composition of sediments is a function of mineralogy of the source rock, weathering, changes during transport, and diagenetic changes after burial. Particle size in clastic sediments increases with increases in the kinetic energy, density, and viscosity of the transporting medium. Particle sorting is best in windblown sands because air has a low density and can carry only a narrow range of particle sizes. The greater density and viscosity of glacial ice enable it to carry a wide range of particle sizes. Consequently, glacial deposits have the poorest sorting. The rounding of particle edges increases with both the energy and the distance of transport.

Once sediment is deposited, it commonly is eroded and transported once again, passing through a number of different sedimentary environments. This constant recycling of the sediment, along with its dilution with other sediments, can change it greatly from its original composition, size, roundness, and sorting.

Eventually, the sediment reaches an area on the continent or in the ocean where it is protected from further erosion and accumulates layer by layer. The sediment then undergoes a series of diagenetic changes which result in its lithification. Major changes include solution, replacement, compaction, cementation, formation of authigenic minerals, and decomposition of organic tissues and plants.

Sedimentary rocks are classified into three types based on the dominant origin of their component particles or crystals. They are subdivided further on the basis of particle composition and texture. The three major types of sedimentary rocks are clastic, biogenic, and chemical. However, many sedimentary rocks are composed of a mixture of these components. Stratification and sedimentary structures provide information about the environmental conditions that existed during accumulation of sediment. Some of these sedimentary structures enable geologists to determine the tops and bottoms of beds. This is important in determining whether the rock layers have been overturned since their deposition. The fossils in sedimentary rocks enable geologists to determine the successive changes in life forms with time.

QUESTIONS FOR REVIEW AND FURTHER THOUGHT

1. What factors determine the mineralogy of a sediment?

2. Why do sediments deposited by wind, streams, and glaciers have such different characteristics?

3. *a.* What factors determine grain size in clastic rocks?
   b. What factors determine size in chemical rocks?

4. What are the diagenetic changes that result in lithification? How do they modify the original characteristics of the sediment?

5. Under what conditions would it be possible to have the following?
   a. A very young lithified sediment
   b. An older unlithified sediment

6. What information do the following sedimentary structures provide about the environmental conditions under which the sediments accumulated?
   a. Graded bedding
   b. Rhythmic bedding
   c. Mud cracks
   d. Asymmetrical ripple marks
   e. Cross-stratification

ADDITIONAL READINGS

Beginning Level

Hsu, K. J., "When the Mediterranean Dried Up," *Scientific American*, vol. 227, no. 6, December 1972, pp. 26–36.

(Describes how the great thicknesses of evaporites were deposited in the Mediterranean Basin when it dried up between 7 and 5.5 million years ago)

Kuenen, P. H., "Sand," *Scientific American,* vol. 202, no. 4, April 1960, pp. 94–110.
(Reviews the origin, transport, and deposition of clastic particles)

Seilacher, A., "Fossil Behavior," *Scientific American,* vol. 217, no. 2, August 1967, pp. 72–80.
(Describes different kinds of trace fossils and what geologists can deduce about the organisms that made them)

Intermediate Level

Laporte, L. F., *Ancient Environments,* 2d ed., Prentice-Hall, Englewood Cliffs, N.J., 1979, 163 pp.
(Detailed discussion of the interrelationships between organisms and sediments. Excellent examples of how geologists determine ancient environments by studying sediments and fossils. A good reference for an overall study of sedimentary geology)

Pettijohn, F. J., *Sedimentary Rocks,* 3d ed., Harper & Row, New York, 1975, 628 pp.
(Deals extensively with properties of sediment and sedimentary rocks but has relatively little on sedimentary processes and environments. A good beginning reference for the study of sedimentary rocks)

Advanced Level

Friedman, G., and J. E. Sanders, *Principles of Sedimentology,* Wiley, New York, 1978, 792 pp.
(An up-to-date discussion of sedimentary particles and processes of sedimentation, along with an overview of sedimentary environments. Has a detailed glossary and bibliography)

Reading, H. G. (ed), *Sedimentary Environments and Facies,* Elsevier, New York, 1978, 557 pp.
(Written by a number of experts on different sedimentary environments. Contains a great number of diagrams and geologic sections showing sedimentary facies in different environments)

Reineck, H. E., and I. B. Singh, *Depositional Sedimentary Environments,* Springer-Verlag, New York, 1973, 439 pp.
(An overview of sedimentary environments with an emphasis on their sedimentary structures. Numerous excellent photographs supplement the text)

CHAPTER 7

Metamorphism and Metamorphic Rocks

An iceberg floating southward from the Arctic Ocean melts gradually as it comes into contact with warm air and water. A balloon rising into the upper atmosphere expands and may burst as it adjusts to the gradually decreasing pressure around it. Rocks also change when forces within the earth subject them to temperatures and pressures greatly different from those under which they first formed. The changes are called **metamorphism,** from the Greek roots *meta* (meaning *change*) and *morph* (meaning *form*), and the rocks they produce are metamorphic rocks.

The realm of metamorphism begins below the near-surface environment where sedimentary processes take place and ends wherever melting begins in the lower crust and mantle. Metamorphic rocks thus contain a record of the conditions, processes, and compositions of the crust and mantle that is not preserved in igneous or sedimentary rocks. In this chapter we will examine the causes and kinds of metamorphic changes, the metamorphic rocks themselves, and the information they provide about the earth.

METAMORPHISM

What Is Metamorphism?

Not all changes that take place in a rock after it forms are considered metamorphism. Metamorphism must occur *within* the earth so that weathering at the surface is not a metamorphic process, even though the change from rock to sediment or soil is a major one. Furthermore, metamorphism occurs only in the solid state so that the onset of melting marks the end of metamorphism and the beginning of igneous activity.

Distinctions between metamorphism and the processes of melting and weathering are clear, but the boundary between metamorphism and the sedimentary process of diagenesis is not nearly so precise. In Chapter 6 we saw that diagenesis during burial brings about changes in both the mineralogy and texture of sedimentary rocks, much as metamorphism does at greater depths. Some geologists select specific temperatures or pressures to define the boundary between metamorphism and diagenesis, while others use the formation of particular minerals to mark the transition. We shall arbitrarily choose 250°C and 2 kb to mark the beginning of metamorphic conditions.

Agents of Metamorphism

Metamorphism occurs when the environment surrounding a rock becomes significantly different from the environment in which the rock first formed. In general, three factors in the environment bring about metamorphic changes: heat, pressure, and solutions of ions circulating through the rock. These are referred to as the **agents of metamorphism.**

Heat

Most minerals expand when they are heated so that ions are separated from one another, stretching and weakening the bonds that hold

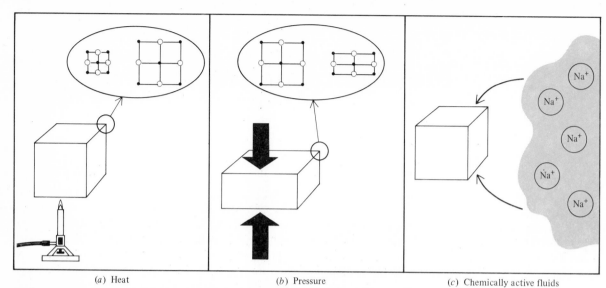

(a) Heat   (b) Pressure   (c) Chemically active fluids

Figure 7.1 The agents of metamorphism. *(a)* Heat causes most substances to expand, increasing interionic distances. This weakens bonds and makes some minerals unstable. *(b)* Pressure: Compression forces ions closer to one another, breaking some bonds and producing new mineral structures. *(c)* Chemically active fluids: New ions are transported to a rock, changing its composition and creating new minerals.

the minerals together (Figure 7.1a). Heat breaks some of the bonds (but not enough to cause melting) and enables the freed ions to migrate through the rock to form the seeds of metamorphic minerals. Most ions probably move along boundaries between grains, aided by whatever fluid is present, but some pass through minerals, either through defects in crystalline structures or through relatively open structures, as shown in Figure 7.2. This movement is slower than that of ions in magma or seawater, but it is just as effective in creating minerals.

Pressure

The effect of pressure is opposite to that of heat because pressure compresses rocks and minerals, forcing their ions closer together (Figure 7.1b). This puts such a strain on some bonds that they break, and the ions are rearranged into a more compact mineral structure. This is basically how scientists convert graphite into diamond in laboratories. Two different types of pressure are involved in metamorphism, and each type causes a different kind of change in rocks (Figure 7.3).

Lithostatic pressure (from *lithos* meaning *rock,* and *status* meaning *position*) is the pressure exerted on a rock as it is buried deeper and deeper in the earth (Figure 7.3a). The greater the thickness of overlying rocks, the greater the lithostatic pressure. Lithostatic pressure is of equal intensity from all directions, and as a result it compresses a rock into a small volume but cannot flatten it. The ultimate cause of lithostatic pressure is the earth's gravitational pull, and the geobaric gradient of 3.3 km/kb given in Chapter 2 is the gradient of lithostatic pressure.

Directed pressures differ from lithostatic pressures in that they are greater in some directions than in others (Figure 7.3b) and can therefore deform and flatten rocks. Directed pressures are not the result of gravity alone but result from the same internal forces that are responsible for large-scale folding and fracturing of rocks and the formation of mountains like the Alps and Himalayas.

Chemically active fluids

Changes in the composition of a rock and the formation of minerals during metamorphism

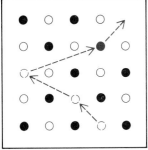 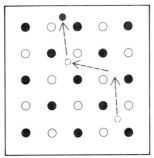 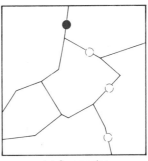

(a) Ions migrating from one vacant site to another

(b) Small ions moving between ions of a mineral structure

(c) Ions moving along grain surfaces

Figure 7.2 Migration of ions through solid rock. *(a)* Defects in crystal structures often result in some vacant structural positions. Ions may migrate from one vacant site to another. *(b)* Small ions may move between the ions in a mineral structure if the mineral has a relatively open framework structure. *(c)* Ions may move along the surfaces of the grain, aided by whatever fluid is present.

can be brought about by **chemically active fluids** (Figure 7.1*c*), solutions containing large quantities of dissolved ions. Some are hydrothermal solutions that emanate from cooling magmas, and others result from metamorphism occurring at great depths in the crust. The solutions can produce major changes in rock composition because the ions they carry interact with the minerals already present to form new minerals. Sometimes the water itself combines with minerals. For example, olivine can resist extensive changes in heat and pressure but reacts rapidly with water to form talc or serpentine.

Types of Metamorphism

Heat, pressure, and chemically active fluids are involved in all metamorphism to some degree. However, in some areas, lithostatic pressure may be the most important agent; in others, the agent may be heat, directed pressures, or chemically active fluids. Geologists can reconstruct the roles played by the agents of metamorphism because the manner in which a rock changes depends on which agents are dominant.

Several different types of metamorphism are recognized, based on the manner in which the metamorphic agents were applied. They are summarized in Table 7.1 and discussed in the following sections.

Figure 7.3 Lithostatic and directed pressure. *(a)* Lithostatic pressures are equal in all directions. They compress but do not greatly deform original grains. *(b)* Directed pressures are greater in some directions than others. They compress *and* change the shapes of original grains.

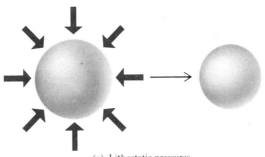

(a) Lithostatic pressures

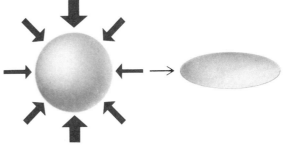

(b) Directed pressures

TABLE 7.1 Types of metamorphism

Agent	Type of metamorphism	Description
Heat	Thermal (contact) metamorphism	Areas around intrusive igneous rocks
Pressure		
Lithostatic	Burial metamorphism	Base of a thick pile of accumulating sedimentary rocks
Directed	Dynamic (cataclastic) metamorphism	Zones of faulting, intense fracturing
Chemically active fluids	Metasomatism	Transport of ions to solid rock by fluids from cooling magma
Combinations		
Heat, directed and lithostatic pressures, and chemically active fluids	Regional metamorphism	Large areas subjected to intense compression at depth; mountain ranges such as the Alps
Directed pressure and heat	Impact (shock) metamorphism	Meteorite craters

Contact metamorphism

The intrusion of magma into host rocks provides a source of heat for metamorphism without altering pressure significantly. Heat given off by a cooling magma is conducted to the host rock, where it causes changes in texture and mineralogy. This type of metamorphism is called **thermal metamorphism** because heat is the dominant agent, or **contact metamorphism** because its effects are most intense at the contact between pluton and host rock. Contact-metamorphic intensity decreases with distance from the pluton until distances are reached at which the effects of heat from the pluton are not felt at all. The region within which contact metamorphism occurs is known as the **contact aureole** of the pluton. Contact aureoles around small dikes may be only a few centimeters across, while those surrounding plutons like the Sierra Nevada batholith may be a few kilometers wide.

Burial metamorphism

The application of lithostatic pressure alone takes place during burial of sedimentary or vol-canic rocks beneath later deposits and flows. A little heat is involved also because the overlying rocks act as an insulating blanket and prevent heat from radioactive decay from escaping, but it is the pressure that dominates in this process, called **burial metamorphism.** The intensity of burial metamorphism increases with depth and with the density of the overlying rocks.

Dynamic metamorphism

Directed pressures are applied to rocks in major fracture zones in the earth called **faults,** where large blocks of rock grind past one another (Figure 7.4). Some frictional heat is added also, but it is the directed pressure that controls the changes in the affected rock. This type of metamorphism is called **dynamic metamorphism** because it is caused by movement, or **cataclastic** (from the Greek words meaning *broken at depth*) **metamorphism,** in reference to the fragmentation that accompanies the grinding. Dynamic-metamorphic intensity decreases rapidly away from the actual zone of grinding and crushing, and the width of the affected zone depends on the

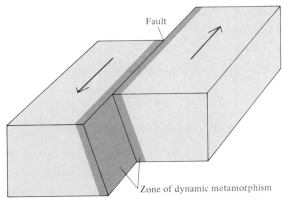

Figure 7.4 Dynamic metamorphism accompanies faulting in the earth. The arrows indicate directions of movement on opposite sides of the fault. Crushing and grinding of dynamic metamorphism is concentrated in the region of faulting.

duration and intensity of faulting. Along such major faults as the San Andreas and Garlock Faults of California, the zone of dynamic metamorphism may be hundreds of meters or even kilometers wide. In smaller faults, it may be only a few centimeters or meters across.

Metasomatism

If a rock's chemical composition changes markedly during metamorphism, the rock is said to have undergone **metasomatic metamorphism,** or **metasomatism.** Many metamorphic processes drive water or carbon dioxide from rocks, which suggests that metasomatism is a common event. However, many geologists restrict the use of the term to processes in which large amounts of cations such as K^+, Na^+, and Ca^{2+} are added by chemically active fluids, and we will follow this usage here.

Regional metamorphism

Heat, lithostatic and directed pressures, and chemically active fluids combine to produce **regional metamorphism.** During regional metamorphism, deeply buried sedimentary and volcanic rocks (lithostatic pressure) are squeezed (directed pressure), intruded by igneous rocks (heat), and permeated by chemically active fluids. The intensity of regional metamorphism is greatest in the deepest and most highly deformed parts of the affected area.

This type of metamorphism is called regional metamorphism to indicate the broad scale on which it operates. As an example, a single episode of regional metamorphism about 400 million years ago affected all of New England and Maritime Canada from New York City to Newfoundland. Forces that can cause such a widespread event are on the same order of magnitude as those involved in mountain building. Regional metamorphism provides important clues to the origin of mountains and to the acceptability of the five hypotheses outlined in Chapter 1. We will return to these clues after discussing the nature of metamorphic changes and rocks.

Impact metamorphism

The rarest form of metamorphism on earth occurs when a meteorite collides with the surface of the planet. During such a collision, the enormous kinetic energy of the meteorite is converted to heat, and tremendous pressures are built up at the moment of impact. This type of metamorphism is called **impact,** or **shock, metamorphism** and is rare on earth because most meteors burn up in our protective atmosphere. However, it is the most common type of metamorphism on the moon, Mars, and Mercury and will be discussed in greater detail in Chapter 23.

Metamorphic Changes

Nearly every aspect of a rock can change during metamorphism, including texture, mineralogy, and chemical composition. In many instances the changes are so great that the metamorphic rock bears little resemblance to its parent. In order to determine the history of a metamorphic rock, we must know what kinds of changes can take place.

Textural changes

All textural features of a rock—grain size, grain shape, and relationships between adjacent grains—can be altered during metamorphism. During burial and regional metamorphism, pores in sedimentary rock and vesicles in volcanic rock are closed by pressure. Other textural changes are brought about by the processes of recrystallization and reorientation.

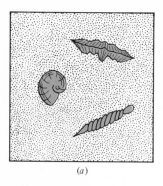
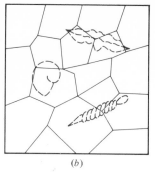

(a)   (b)

Figure 7.5   Recrystallization. Recrystallization may destroy original rock textures. *(a)* This fine-grained limestone contains a few large fossils. *(b)* The small calcite grains recrystallize until nearly all traces of the original grain size and organic content have been destroyed.

During **recrystallization**, original grains grow by merging with one another or are broken into smaller grains, thus obscuring or obliterating the original texture. For example, recrystallization of clasts, matrix grains, and fossils in sedimentary rocks yields interlocking grains of nearly uniform size (Figure 7.5). Grain size generally increases during regional, contact, and burial metamorphism. For example, when fine-grained limestones recrystallize, they commonly are transformed into coarse-grained aggregates of interlocking calcite grains—the metamorphic rock called **marble** (Figure 7.6). However, grain size decreases during the grinding and crushing that accompany dynamic metamorphism.

Grain shapes also may change. This is most pronounced when strong directed pressures are involved because such pressures can stretch or flatten grains. Ooliths or rounded clasts in sedimentary rocks can undergo extreme changes in shape as shown in Figure 7.7, and crystals in igneous rocks can be affected in a similar way.

Reorientation of original grains also occurs because of directed pressures, and it may take one of two forms (foliation or lineation) depending on the shapes of the grains (Figure 7.8). **Foliation** is the name given to a parallel alignment of platy minerals such as micas or chlorite or of flattened grains. **Lineation** is the parallel alignment of rodlike grains such as amphibole crystals. Reorientation takes place by the rotation of grains in response to directed pressure. Platy minerals and flat grains change position so that their flat surfaces are at right angles to the great-

Figure 7.6   Limestone and marble. The fine-grained limestone on the left can be metamorphosed to produce the coarser grained marble on the right. *(Mira Schachne)*

Figure 7.7 Deformed clasts in a metamorphosed conglomerate. The elliptical light and dark particles were originally spherical, and were flattened during cataclastic metamorphism. *(Mira Schachne)*

est directed pressures. This enables geologists to determine the direction of the greatest pressures active during metamorphism simply by measuring the orientation of the foliation. Foliation and lineation are not found in all metamorphic rocks because directed pressures are not important in all kinds of metamorphism. For example, contact-metamorphosed rock generally displays no foliation at all.

Mineralogic changes

The textural changes described above can occur without changes in the mineralogy of a rock, but metamorphism commonly involves the disappearance of original minerals and the formation of new ones. This happens because whether minerals form in igneous, sedimentary, or metamorphic processes, they represent the most stable grouping of ions for a particular

Figure 7.8 Preferred reorientation of grains. Randomly oriented grains may rotate because of directed pressure to produce *(a)* foliation, the parallel alignment of platy minerals or flattened grains, or *(b)* lineation, the parallel alignment of rodlike minerals.

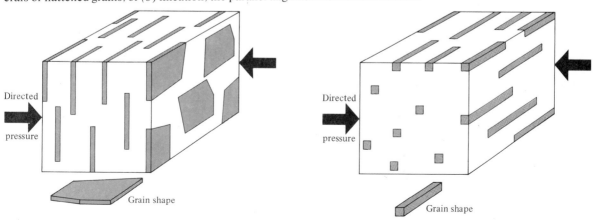

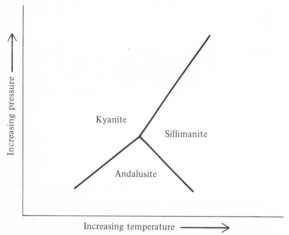

Figure 7.9 Stability diagram for the polymorphs of Al_2SiO_5. Each of the minerals shown is stable within its particular region of the pressure-temperature diagram. With changes in condition, one mineral may be converted to another.

environment. For example, clay minerals such as kaolinite have open, sheetlike structures that are stable at the earth's surface, where they form during weathering. They cannot survive either lithostatic pressures or the temperatures found at great depth, and they are converted to more compact micas during the early stages of metamorphism.

Transformations from one polymorph to another represent the simplest mineral changes since no ions are added or removed but rather are rearranged to fit the new structures. For example, calcite, one of the most common sedimentary minerals, is transformed to its denser polymorph aragonite when limestones are metamorphosed at very high pressure. The compound Al_2SiO_5 exists as one of three polymorphs in metamorphosed shales (andalusite, kyanite, and sillimanite), depending on the conditions of temperature and pressure (Figure 7.9).

Other mineralogic changes are more complex because the new minerals form by a recombination of ions from several original minerals. Examples of this type of reaction are given in Table 7.2. Because of the importance of pressure as a metamorphic agent, the metamorphic minerals tend to be denser, with their ions packed more tightly than in the original minerals. The mineral changes that occur in the metamorphism of shale and limestone are mainly **dehydration** (water is given off) or **decarbonation** (carbon dioxide is given off) reactions. The loss of these compounds decreases the volume of the remaining rock, thus increasing its density. If directed pressures are active during mineral reactions, the new metamorphic minerals generally become foliated or lineated.

TABLE 7.2 Types of metamorphic reactions

DEHYDRATION REACTIONS (WATER IS GIVEN OFF)

Muscovite + quartz $\longrightarrow$ sillimanite + potassic feldspar + water

$KAl_2(AlSi_3O_{10})(OH)_2 + SiO_2 \longrightarrow Al_2SiO_5 + KAlSi_3O_8 + H_2O$

DECARBONATION REACTIONS (CARBON DIOXIDE IS GIVEN OFF)

Dolomite + quartz $\longrightarrow$ diopside + carbon dioxide

$CaMg(CO_3)_2 + SiO_2 \longrightarrow CaMg(Si_2O_6) + CO_2$

REACTIONS INVOLVING BOTH WATER AND CARBON DIOXIDE

Dolomite + quartz + water $\longrightarrow$ tremolite + calcite + carbon dioxide

$5CaMg(CO_3)_2 + 8SiO_2 + H_2O \longrightarrow Ca_2Mg_5Si_8O_{22}(OH)_2 + 3CaCO_3 + 7CO_2$

REACTIONS INVOLVING NEITHER WATER NOR CARBON DIOXIDE

Jadeite + quartz $\longrightarrow$ albite

$NaAlSi_2O_6 + SiO_2 \longrightarrow NaAlSi_3O_8$

How Much Does Rock Change during Metamorphism?

The amount of change that a rock undergoes during metamorphism depends on the (1) intensity of the metamorphic agents, (2) duration of metamorphism, and (3) type of rock involved. The importance of the first two factors is understandable because the greatest changes logically should accompany the greatest heat, pressure, and chemical alterations, and the longer the agents operate, the more sweeping the changes should be. However, the type of rock is equally important. Some rocks, such as quartzose sandstones and granites, are made of minerals that are stable over nearly the entire range of metamorphic conditions. These rocks change very little, while rocks such as shale or lime-stone undergo textural and mineralogic changes readily.

Rocks subjected to low-intensity metamorphism generally retain many of their original characteristics, such as bedding, texture, and fossils, and they are called **low-grade** metamorphic rocks. In contrast, **high-grade** rocks generally have lost most of their original features because of the intensity of the metamorphic agents. Figure 7.6 shows the change from unmetamorphosed limestone to a high-grade marble, but some changes are even more drastic, as shown in Figure 7.10. Unmetamorphosed shale becomes slate (Figure 7.10*a*) but can be converted to schist (Figure 7.10*b*) or gneiss (Figure 7.10*c*) at higher grades. Extreme metamorphism culminates with partial melting of the rock, resulting in an intimate mixture of igneous

Figure 7.10 Degrees of metamorphic change. Slate *(a)* represents a very low grade change of an original shale. With progressively higher grade changes, it can become schist *(b)*, gneiss *(c)*, or migmatite *(d)*. (a, *Mira Schachne;* b, c, *American Museum of Natural History;* d, *B. H. Bryant, U.S. Geological Survey*)

a

c

b

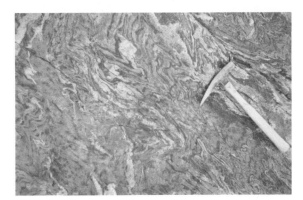

d

rock and high-temperature minerals in the un-melted rock (Figure 7.10*d*).

The Progressive Nature of Metamorphism

With the exception of impact metamorphism, metamorphic intensity increases gradually over hundreds or millions of years. Rocks respond by changing continuously to meet the altered conditions and do not pass from an unmeta-morphosed state to a high-grade condition over-night. Original minerals are converted to low-grade metamorphic minerals in the early stages of metamorphism, and the low-grade minerals themselves are converted to higher-grade min-erals as metamorphic intensity increases. This step-by-step adjustment of rocks is referred to as **progressive metamorphism,** and the changes that occur as the grade (intensity) of metamor-phism increases are called **prograde** meta-morphic changes.

Retrograde Metamorphism

Metamorphic intensity eventually peaks and then gradually diminishes. At first thought, we might expect a metamorphosed rock to readjust to the conditions of lower temperature and pres-sure and perhaps even to return to its original unmetamorphosed state. However, the existence of garnets and diamonds in jewelry and the use of slate for roofs on houses indicates that this does not happen. There are two basic reasons why a metamorphic rock does not return to its original state as metamorphic intensity wanes: absence of a driving force and lack of appropriate ions.

Heat is the driving force behind ionic migra-tion. Thus, the more a rock cools, the more its ions tend to remain in place rather than recom-bining to form new minerals. Similarly, when the grinding of dynamic metamorphism stops, crushed grains do not revert to their original sizes and shapes because no force exists to make them do so. Second, some metamorphic changes that might be reversed are not because some of the necessary ingredients are no longer available. Dehydration and decarbonation re-actions such as those listed in Table 7.2 can be reversed only if water and carbon dioxide are still present in the rock. However, both are extremely mobile and tend to escape through fractures. Once gone, they cannot interact with the metamorphic minerals to recreate a rock's original mineralogy.

Some mineralogic adjustments do occur during the waning stages of metamorphism, and these are called **retrograde** (backward) changes. Most retrograde changes involve the small amounts of water or carbon dioxide that have not escaped from the rocks and as a result re-place the prograde minerals only *partially* as shown in Figure 7.11.

Figure 7.11 Retrograde metamor-phism. The large andalusite crys-tal in the center of the photomicro-graph is rimmed by muscovite flakes produced during retrograde metamorphism. *(Mira Schachne)*

METAMORPHIC ROCKS

Each metamorphic rock is a record of all the processes that have affected it and all the changes it has undergone. It is also a geologic thermometer and barometer—a measure of the temperature and pressure at which metamorphism took place. The record of these conditions, as in igneous and sedimentary rocks, lies in a rock's texture and mineralogy.

Metamorphic Textures

The texture of a metamorphic rock is potentially more difficult to interpret than that of an igneous or sedimentary rock because it can contain features produced both during metamorphism and during the process by which the unmetamorphosed rock first formed. The textural features inherited from the original rock are called **relict features.** They are often modified during metamorphism, but even such delicate features as vesicles and fossils may be recognizable in metamorphic rocks. However, in high-grade metamorphism, most original textural features are destroyed and are replaced by metamorphic textures.

Regardless of the grade of metamorphism, metamorphic textures can be used to determine what type of metamorphism has taken place and can give an indication of the grade of metamorphism. Grain size and relationships between grains are the major clues used to interpret the history of a metamorphic rock.

Grain size

As in igneous rocks, the amount of ionic migration determines the size of mineral grains in a metamorphic rock. Because heat enables ions to move, we would expect the coarsest-grained metamorphic rocks to be those which remained hottest for the longest time. In general this is correct; high-grade regionally metamorphosed rocks are coarser than low-grade rocks. Compare, for example, the grain size of the low-grade slate and high-grade schist in Figure 7.10. Where heat can escape rapidly or metamorphism lasts only a short time, however, grains tend to be small regardless of the intensity of the metamorphic agents. Thus, even high-grade impact-metamorphic rocks are fine-grained, as are many contact-metamorphosed rocks.

Remember also that dynamic metamorphism generally leads to *decreased* grain size. As a rock is subjected to grinding in a fault zone, its grains are crushed, stretched, and flattened, resulting in what is called a **mylonitic texture** (Figure 7.12). In extreme cases, grains are so pulverized that the resulting rock looks almost like a volcanic glass. The texture of such a rock is said to be **ultramylonitic,** and the rock can be distinguished from a true glass only with the aid of a high-powered microscope.

Some minerals in a metamorphic rock may grow much larger than the others either because the ions did not have far to travel or because the ions involved migrate more easily than others. The result, shown in Figure 7.13, is called a **porphyroblastic** texture, one in which a few large grains (called **porphyroblasts**) are embedded in a finer-grained matrix. Garnet, staurolite, andalusite, kyanite, and sillimanite are common porphyroblasts in schists, whereas actinolite, hornblende, and garnet commonly form porphyroblasts in metamorphosed basalts.

Figure 7.12 Mylonitic texture. This photomicrograph shows a smeared out, strongly foliated appearance and the suturing of individual quartz grains. (*W. A. Braddock, U. S. Geological Survey*)

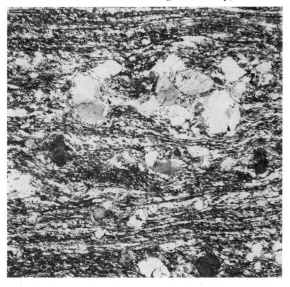

Figure 7.13 Porphyroblasts. The large outlined crystals are garnet porphyroblasts. *(Mira Schachne)*

Relationships between grains

Interlocking grains are characteristic of metamorphic rocks, and the presence of specific metamorphic minerals can be used to distinguish metamorphic rocks from igneous or sedimentary rocks that also have interlocking textures. For details of the metamorphic processes, it is important to note whether minerals in a metamorphic rock have a preferred orientation or a random orientation with respect to one another and whether a compositional layering has been developed.

Preferred orientiation We have seen that directed pressures produce foliation and lineation in metamorphic rocks. When foliation is defined by the parallel alignment of platy minerals such as muscovite, biotite, or chlorite, it is given a special name — **schistosity.** Foliation may also be defined by alignment of flattened grains resulting from dynamic metamorphism, and as a result many mylonitic rocks are strongly foliated. The presence of well-developed lineation or foliation indicates the operation of strong directed pressures, suggesting regional or dynamic metamorphism. Weakly developed foliations are characteristic of burial metamorphism.

Random orientation Metamorphic rocks formed without directed pressures, particularly during contact metamorphism, exhibit neither foliation nor lineation. Their random, unaligned mineral orientation is called a **granoblastic** texture, and a rock exhibiting this texture is called a **granofels** (Figure 7.14).

If platy and elongate minerals are part of the granoblastic texture, directed pressures were certainly absent during metamorphism because such minerals align themselves readily. However, it is important to realize that a granoblastic texture need not form only during contact metamorphism. This is proved clearly by outcrops made of alternating layers of granofels and strongly foliated schist. Directed pressures must have been involved in the metamorphism of such rocks in order to produce the schistosity, and it is not logical to assume that these pressures affected only some of the layers. The granofels is unfoliated simply because it contains no platy minerals. Granofels in such exposures typically consists of quartz and feldspars, minerals that tend to be equidimensional in metamorphic rocks.

Figure 7.14 Granofels. Sketch of a thin section of marble showing roughly equidimensional grains that exhibit no preferred orientation.

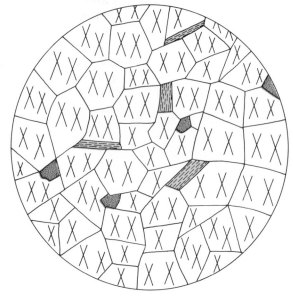

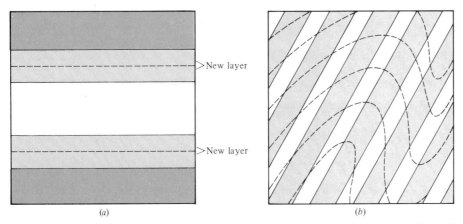

Figure 7.15 Layering in metamorphic rocks. *(a)* Layering may form by migration of ions between original layers. Original layer boundaries (shown as dashed lines) are destroyed and replaced by new layering (solid lines). In these cases, new layers (light color) parallel the old ones. The remains of the old layers are shown as dark color and as white. *(b)* Layering may form by movement of material along fractures associated with deformed rock. The dashed lines indicate the original layering and show how it has been folded, while the new layers (solid lines) cut across and almost obliterate all traces of the original boundaries.

Layering Compositional **layering** is common in metamorphic rocks. In some instances it may be a relict feature—preserved traces of original bedding or igneous layering. However, metamorphic processes also can produce layering by the segregation of ions during migration. In some rocks, layering forms by mixing of ions from original layers (or beds), and the metamorphic layering parallels the original layering (Figure 7.15a). When directed pressures are involved, a layering often develops that is parallel to foliation and may cut across or completely obliterate the original layers (Figure 7.15b).

Gneissosity Some metamorphic rocks consist of alternating layers of foliated and granoblastic-textured materials (look back at Figure 7.10c). The layering commonly is indicated by alternation of light and dark bands, with foliation restricted to one of the two, usually the dark. This combination of layering and foliation/nonfoliation is called **gneissosity** (pronounced nice-ossity), and rocks displaying it are called **gneisses** (pronounced nices). Most gneisses are high-grade rocks, and their layering represents a metamorphic feature rather than relict bedding.

Mineralogy

Unless there has been significant addition or removal of materials during metamorphism, the ions present in the original rock determine the minerals that can be found in the metamorphosed equivalent. Metamorphic rocks can be subdivided into four broad compositional groups—pelitic, calcareous, mafic, and quartzofeldspathic—and each has its particular group of minerals that can form during metamorphism.

Pelitic rocks are those which originally contained a large proportion of clay minerals and were mostly shales or sandstones and siltstones with clay-rich matrixes. Clay minerals are rich in aluminum so that metamorphosed pelitic rocks contain highly aluminous minerals such as muscovite, biotite, and chlorite. **Calcareous rocks** were originally limestones and dolomites, and their metamorphosed equivalents contain minerals with high calcium and magnesium contents such as diopside, $CaMgSi_2O_6$, or wollastonite, $CaSiO_3$. Mafic rocks were originally basalts and gabbros rich in ferromagnesian minerals and plagioclase feldspars. The quartzofeldspathic rocks were either felsic igneous rocks, such as granite or rhyolite, or sedimentary rocks, such as arkose.

TABLE 7.3 Metamorphic minerals and original rock composition

Rock type	Original rocks	Metamorphic minerals*
Pelitic	Rocks rich in clay minerals: Mostly shales and some silt-stones and sandstones with clay-rich matrixes	Aluminum-rich minerals: Andalusite Cordierite Kyanite Muscovite Sillimanite Staurolite Biotite† Chlorite† Garnet (almandine)†
Calcareous	Limestones and dolomites	Calcium and calcium-magnesium silicates: Garnet (grossularite) Idocrase Wollastonite Actinolite† Diopside† Epidote† Talc†
Mafic	Mafic igneous rocks: basalts and gabbros	Garnet (almandine) Hornblende Pyroxene (calcium-poor) Actinolite† Chlorite† Diopside† Epidote†
Quartzo-feldspathic	Felsic igneous rocks (granite, rhyolite) and sedimentary rocks (arkose, sandstone)	Plagioclase feldspar Potassic feldspar Quartz

* In addition to the minerals listed, quartz, plagioclase, and potassic feldspar may be found.

† These minerals can be formed in rocks of different original compositions. In such cases, the entire group of minerals found in the rock must be used to identify the initial rock composition.

Table 7.3 lists some of the common metamorphic minerals found in the four rock types. Note that some minerals, such as sillimanite and wollastonite, may occur only in one type of rock, whereas others, such as almandine garnet, may occur in more than one. This happens because the ions needed to make a mineral may be found in more than one type of initial rock.

Classification of Metamorphic Rocks

The classification and naming of a metamorphic rock are based on its two most readily identifi-able properties: mineralogy and texture. A simple classification scheme based on these criteria is outlined in Table 7.4. Most of the names in Table 7.4 are descriptive and say nothing about either the intensity or the type of metamorphism, but there are three exceptions. Extremely high-grade rocks composed of an intimate mixture of igneous material formed by partial melting and residual metamorphic rocks that did not melt are called **migmatites.** This is the only name in the table that indicates a particular grade of metamorphism.

TABLE 7.4 Classification of the metamorphic rocks

Texture	Composition			
	Pelitic	Calcareous	Quartzofeldspathic	Mafic
Foliated	Slate Phyllite Schist	Calcareous schist		Greenschist Amphibole schist
Foliated and Layered	Gneiss Migmatite	Calc-silicate gneiss	Quartzofeldspathic gneiss Migmatite	Mafic gneiss
Cataclastic	Mylonite and ultramylonite			
	Hornfels (if produced by contact metamorphism)			
Nonfoliated	Granofels	Calc-silicate granofels	Quartzite Quartzofeldspathic granofels	Greenstone Amphibolite Mafic granofels

The other exceptions involve the type of metamorphism. Cataclastic textures formed during dynamic metamorphism are similar regardless of the mineralogy of the affected rocks. Rocks with these textures are called mylonites or ultramylonites. Rocks subjected to contact metamorphism commonly become dense, very fine grained granoblastic rocks regardless of their composition, and they are called **hornfels.** The term hornfels is used only when it is absolutely certain that contact metamorphism is involved. When there is doubt, the descriptive term granofels is preferred because it does not imply any particular type of metamorphism.

We will now take a brief look at the progressive metamorphism of the four major rock types by following the changes that occur in representative sedimentary and igneous rocks. This will show how the different rock types listed in Table 7.4 are related to each other and clarify the changes that occur during metamorphism.

Metamorphism of pelitic rocks

Shales are fine-grained rocks composed mostly of clay minerals and varying amounts of quartz and feldspars. The changes that shales undergo during metamorphism are summarized in Figure 7.16. During early stages of regional and burial metamorphism, pressure completes the diagenetic process of compaction.

Platy clay minerals are realigned parallel to one another and pores are closed, resulting in the compact metamorphic rock called **slate.** Mineral changes take place with increased intensity, and the clays break down in dehydration reactions to produce very fine flakes of muscovite and chlorite. Foliation of the muscovite and chlorite produces a shimmery surface in which individual grains are too small to be seen with the naked eye. Rocks with this mineralogy and texture are called **phyllites.** In still higher grades, grain size increases by recrystallization and biotite replaces chlorite by mineralogic reactions. These strongly foliated, coarse-grained rocks composed mostly of mica are called **schists.** With further metamorphism, aluminous minerals form as porphyroblasts in the schist, and these minerals should be mentioned to complete the rock's name. For example, a coarse-grained, well-foliated rock composed of muscovite and biotite with a few garnet porphyroblasts should be called a garnetiferous muscovite-biotite schist.

If layering develops, a high-grade pelitic rock can become a **gneiss** by segregation of quartzose and feldspathic minerals in light-colored layers and micas in the darker layers. In the absence of layering, the final product can be a coarse granoblastic rock composed of quartz, potassic feldspar, plagioclase feldspar, and aluminous minerals such as sillimanite and

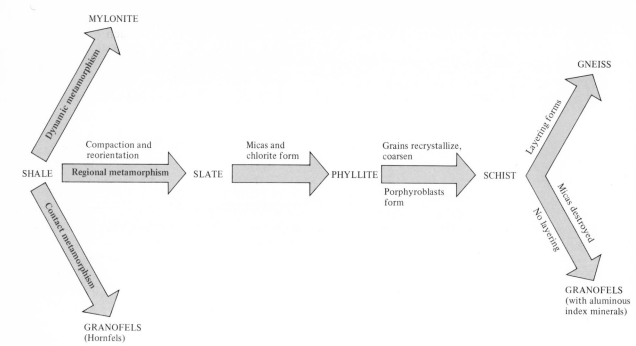

Figure 7.16 Flow diagram for metamorphism of pelitic rocks.

garnet. If partial melting occurs, quartzofelds-pathic magma forms and cools within a residual micaceous rock to produce migmatite.

Metamorphism of quartzose and quartzofeldspathic rocks

A quartz sandstone with a silica cement undergoes little change during regional meta-morphism. Matrix silica and clasts recrystallize until there is no distinction between them, and a tough, dense granofels called **quartzite** is formed. Quartzofeldspathic rocks such as arkose and arkosic wacke undergo similar tex-tural changes, and a homogeneous grain size is the result. There are no mineralogic changes because quartz and the feldspars are stable throughout metamorphism and do not react with one another. However, if there has been a clay matrix, as there is in some wackes, alum-inous metamorphic minerals can form as they do in pelitic rocks.

Metamorphism of calcareous rocks

If a pure calcite limestone or pure dolomite is metamorphosed, it generally recrystallizes to the granofels called marble. If the original lime-stone or dolomite contained quartz and clay mineral impurities, reactions take place between these minerals and the carbonates, as illustrated in Table 7.2. Minerals made of calcium, silica, and various amounts of magnesium result, and they are called calc-silicate minerals. If layering develops, the result is a calc-silicate gneiss.

Metamorphism of mafic rocks

Basalts commonly are interlayered with sedimentary rocks near volcanic islands such as Japan and the Aleutians, and can be metamor-phosed along with the sedimentary rocks. Stages in the metamorphism of basalt are shown in Figure 7.17.

In the early stages, water driven out of the sedimentary rocks by dehydration reactions combines with the anhydrous minerals of basalt (pyroxene, olivine, and plagioclase) to produce chlorite, epidote, and actinolite along with sodium-rich plagioclase. Because all these minerals except the plagioclase are green, low-grade metamorphosed basalts are called **green-stones** if granoblastic and **greenschists** if foliated.

At higher grades, the greenstones undergo dehydration reactions like the surrounding pelitic rocks and progressively lose the water that had been added during the low-grade reactions. Hornblende forms from epidote, chlorite, and actinolite, and rocks composed mostly of hornblende and plagioclase result. They are called **amphibolites** if granoblastic and **hornblende schists** if foliated. Increased metamorphism completes dehydration by converting hornblende to pyroxene, and a granoblastic pyroxene-plagioclase rock is produced that has almost the same mineralogy as the original basalt, although with very different texture. If layering develops by the segregation of feldspar from pyroxene, a mafic gneiss can be produced.

INTERPRETATION OF METAMORPHIC ROCK HISTORY

A metamorphic rock preserves a record of complex geologic history that includes both metamorphic and premetamorphic events. Our goal is to read this record in as much detail as possible and ideally to learn the following things about every metamorphic rock: (1) the type of rock from which it was produced, (2) the type of metamorphism involved, and (3) the intensity of the metamorphic agents.

We have already seen several types of evidence that can be used to accomplish the first two objectives. The composition of the original rock (called the **protolith**) can be determined from its metamorphic minerals, as shown in Table 7.3. Even more helpful in this regard are **mineral assemblages**—the complete combination of minerals present in a rock. For example, biotite may form during metamorphism of shales, basalts, or clay-rich dolomites, but the assemblage biotite-muscovite-chlorite-quartz-plagioclase in a rock is so highly aluminous that it could have come only from a pelitic protolith. Similarly, almandine garnet may form in pelitic

Figure 7.17 Flow diagram for metamorphism of mafic rocks.

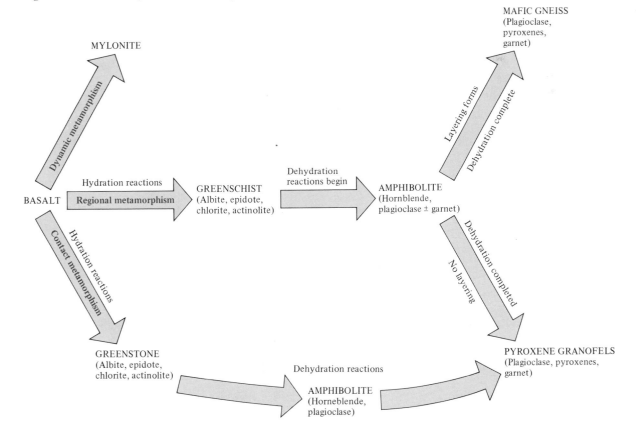

or mafic rocks, but the assemblage almandine-hornblende-diopside-plagioclase is so rich in iron and magnesium that it indicates a mafic protolith.

Type of metamorphism is indicated by metamorphic texture, mineralogy, and field relationships. Well-developed schistosity implies regional metamorphism, cataclastic textures indicate dynamic metamorphism, and the association of a meteorite crater with extremely high-pressure minerals clearly indicates impact metamorphism. The rapid decrease of metamorphic intensity with increased distance from a pluton would, of course, indicate contact metamorphism.

Determining the Intensity of Metamorphism

Ideally, we would like to know the precise temperature and pressure at which every rock was metamorphosed. Unfortunately, the determination of absolute metamorphic intensity (the exact temperature and pressure) is not always possible. Indeed, it has been accomplished satisfactorily only in recent years, and even then only in a small number of cases.

Even if the absolute conditions cannot be learned, it is not difficult to establish a scale of *relative intensity*—a scheme in which one rock can be described as being more or less intensely metamorphosed than another. We have already used some terms of relative intensity (low-grade and high-grade) because they are almost intuitive. We will now look at a more precise system for describing relative metamorphic intensity.

Index minerals and metamorphic zones

The contact aureole surrounding a pluton is an ideal place to study relative metamorphic intensity because metamorphic grade decreases away from the pluton in all directions. We will focus on the aureole surrounding a small granitic pluton to show how individual metamorphic minerals can be used to describe relative intensity.

The Hartland pluton is a 365-million-year-old granite that intrudes tightly folded sedimentary rocks just northeast of the town of Skowhegan in central Maine. The pluton underlies an area of about 175 km^2 and has produced a contact aureole in surrounding pelitic and calcareous rocks that ranges from 4 to 10 km in width (Figure 7.18).

We will concentrate on the pelitic rocks, which are shown in color in Figure 7.18. As one follows any single band of pelitic rock toward the granite, several mineralogic changes are noted. Green, chlorite-rich sandstones and shales far from the pluton are replaced by purplish-gray, biotite-rich slates and granofels closer to the contact, and these rocks are followed in turn by others containing porphyroblasts of andalusite. Within a few hundred meters of the pluton, small sillimanite crystals are present in the pelitic rock. Enough relict textures are preserved to show that the rock within a single color band is all of the same type so that the appearance of the different minerals must reflect different contact-metamorphic intensities. Sillimanite represents the highest grade of metamorphism in the aureole since it is produced only near the pluton, while chlorite represents the lowest grade. Sillimanite cannot be found far from the granite because the amount of heat needed to produce it was not supplied to rocks far from the heat source. It could form only close to the pluton, where the magma's heat was concentrated. Similar reasoning explains the geographic distribution of andalusite and biotite.

The contact aureole can be divided into **metamorphic zones,** each characterized by the presence of a mineral called a metamorphic **index mineral** that indicates a particular grade of metamorphism. Thus, the chlorite zone around the Hartland pluton encompasses the area in pelitic rocks where metamorphism was intense enough to produce chlorite but not intense enough to make biotite. Each index mineral represents a metamorphic intensity greater than that of the next mineral farther from the granite but less than that indicated by the next mineral closer to the pluton.

Metamorphic zones are also found in the calcareous rocks but are identified by different minerals because the calcareous rocks do not have the appropriate compositions to form chlorite, biotite, andalusite, or sillimanite. Instead, the sequence of increasing metamorphic intensity is given by the index minerals actino-

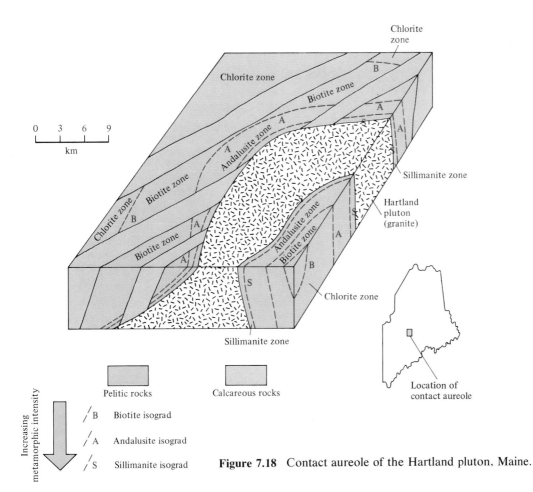

Figure 7.18   Contact aureole of the Hartland pluton, Maine.

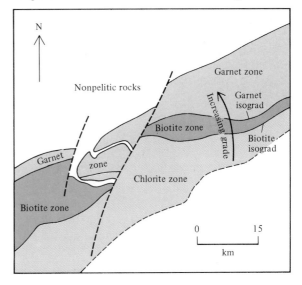

Figure 7.19   Metamorphic mineral zones and isograds for pelitic rocks in the southern Scottish Highlands.

lite, diopside, grossularite garnet, and wollastonite. Had basalts been intruded by the pluton, they would also have shown a zonation, but one defined by *mafic* index minerals.

Metamorphic zones are not restricted to contact metamorphism. At the beginning of the twentieth century, George Barrow described mineral zones in regionally metamorphosed pelitic rocks of the Scottish Highlands (Figure 7.19). Although there was not an obvious heat source as there is in a contact aureole, Barrow was able to deduce the direction of increasing intensity by studying the degree to which primary sedimentary features were preserved. He eventually recognized the zones of chlorite, biotite, almandine garnet, staurolite, kyanite, and sillimanite in order of increasing intensity. This sequence of index minerals is different from that found around the Hartland pluton because the metamorphic conditions, particularly the pressure, were different. However, the principle of metamorphic zones is identical.

Isograds   The boundaries between mineral zones have special significance. For example, on the low-grade side of the boundary between the chlorite and biotite zones, no biotite formed because there was not enough heat. There obviously was enough heat on the high-grade side. Everywhere *along* the boundary, the minimum conditions necessary for the formation of biotite must have been attained. Thus, metamorphism must have been of equal intensity, or equal grade, everywhere along the boundary. For this reason, the boundaries are called **isograds,** from *iso* (meaning *equal*) and *grad* (meaning metamorphic *grade*). An isograd is named after the index mineral that appears on its high-grade side. Thus, the biotite isograd separates the chlorite and biotite zones.

Metamorphic facies

The metamorphic zone system of describing metamorphic intensity has the major drawback that it is restricted to a single type of rock at a time. For example, "staurolite zone metamorphism" has no meaning to a geologist studying calcareous rocks because limestones have neither the aluminum nor the iron needed to make staurolite, even when the temperature and pressure are correct. Conversely, a geologist studying pelitic rocks has little use for diopside or wollastonite zones. What is needed is a system that can be used for all rocks regardless of their composition.

The concept of metamorphic facies was proposed by P. Eskola to fill this need. Just as a sedimentary facies contains all rocks deposited in a single environment, a **metamorphic facies** includes all rocks metamorphosed at the same general temperature and pressure conditions. Furthermore, rocks of one metamorphic facies pass into those of another at the point where metamorphic intensity changes, just as one sedimentary lithofacies passes into another at the point where depositional environments change. Figure 7.20 illustrates the metamorphic facies that are recognized currently and plots them in their appropriate relative positions in a pressure-temperature diagram. The lines that separate the facies represent mineral reactions. For example, the boundary between the amphibolite and granulite facies coincides roughly with reactions marking the final dehydration of mafic rocks by conversion of hornblende to pyroxene.

The facies to which a rock belongs is determined by its mineral assemblage. At a particular set of temperature and pressure conditions, calcareous rocks would have one set of assemblages, pelitic rocks another, and mafic rocks a

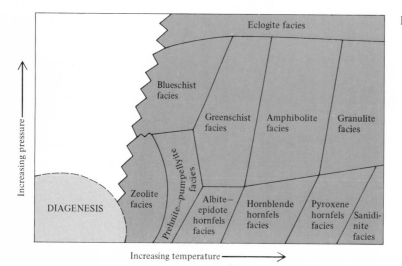

Figure 7.20   Metamorphic facies.

third. All would belong to the same facies. The most intensely metamorphosed rocks in the contact aureole of the Hartland pluton belong to the hornblende hornfels facies because the pelitic assemblage sillimanite-muscovite-biotite-quartz-plagioclase feldspar and the calcareous assemblage wollastonite-garnet-calcite found at the contact are typical of that facies throughout the world. If there had been mafic rocks in the contact aureole of the Hartland pluton, the assemblage probably would be something like hornblende-plagioclase-diopside because this is a typical assemblage found in hornblende hornfels facies rocks of mafic composition elsewhere.

Absolute Metamorphic Intensity

In order to estimate actual values for the temperature and pressure conditions of metamorphism, geologists combine field studies such as those of the Hartland pluton with laboratory work. Powdered minerals are placed in an inert container and then heated and squeezed to the desired temperature and pressure conditions. The contents of the container are then examined to see whether a reaction has taken place. By trial-and-error experiments such as this, the minimum conditions needed for particular reactions to occur can be determined. It is also from experiments such as this that the stability diagram of the polymorphs of Al_2SiO_5 shown in Figure 7.9 was constructed. The assumption is then made that the conditions for a reaction in nature are approximately the same as the conditions for a laboratory reaction.

This assumption is not necessarily correct, but it is essential if we are to arrive at any conclusions about absolute metamorphic conditions. Natural metamorphism may be somewhat different from laboratory-produced metamorphism in that it can operate over long periods of time whereas we cannot wait a few centuries in the laboratory for a reaction to run. Results of the experimental work are generally consistent with qualitative data from the field so that to a first approximation, the laboratory data can be applied successfully to natural assemblages.

REGIONAL METAMORPHISM AS A KEY TO EARTH HISTORY AND PROCESSES

Contact metamorphism can happen anywhere in the world at any time. Intrusive magma is all that is needed, and we saw in Chapter 4 that magmas are found throughout the world. Regional metamorphism is an entirely different matter. The tremendous areal extent of regional metamorphism requires that large-scale forces be involved—forces of the magnitude required by the different models for the earth proposed in Chapter 1. While the textures and mineralogy of metamorphic rocks record the conditions and types of metamorphism, the distribution of regionally metamorphosed rocks in space and time provides important clues to the causes of metamorphism.

Geographic Distribution of Regionally Metamorphosed Rocks

Regionally metamorphosed rocks are found on every continent. Some lie in elongate belts in the centers of mountain ranges such as the Himalayas, Alps, and Appalachians, and were formed during the mountain building itself. Others occupy broad, flat terrains covering hundreds of thousands of square kilometers at the centers of the continents. Separating the elongate belts from the broad metamorphic terrains are extensive regions of unmetamorphosed sedimentary and volcanic rocks. For example, in North America, the central metamorphic terrain is called the Canadian Shield (Figure 7.21), and it is separated from the metamorphic rocks of the Rocky, Appalachian, and Ouachita Mountains by the sedimentary rocks of the Great Plains, the midcontinent platform, and the Appalachian Plateau.

The ocean basins, in striking contrast, contain almost no regionally metamorphosed rocks but are underlain instead by sedimentary and volcanic rocks (see Chapter 16). The regionally metamorphosed rocks of the oceans are restricted to a few island chains, such as the Japanese Islands, that lie near the continents. Mid-ocean islands such as Hawaii, Iceland, and the

Figure 7.21 Distribution of regionally metamorphosed rocks exposed at the surface in North America.

Azores contain no regionally metamorphosed rocks at all.

In Chapter 4 we saw that igneous activity in the oceans differs from that on the continents, and now we find that metamorphism is also drastically different. Compositional differences alone can account for the distribution of igneous rocks described in Chapter 4, but metamorphism affects all rocks regardless of composition. The scarcity of regionally metamorphosed rocks in the oceans indicates significant differences in the *processes* affecting continents and oceans as well as compositional differences.

The difference seems to be an absence (from the oceans) of strong directed pressures and intense squeezing, forces that are required for regional metamorphism. This lack can be explained with varying degrees of success by each of the hypotheses described in Chapter 1. Either a shrinking earth or the shrinking phase of a pulsating earth can produce strong directed pressures. But if this is the case, why is there so

little regional metamorphism in the oceans that cover nearly 75 percent of the planet's surface? Intense squeezing is not at all compatible with an expanding earth, and this would fit the near absence of regional metamorphism in the oceans; but the continents have certainly undergone squeezing that could not have happened as the planet expanded. Neither the expanding nor the contracting earth model fits the observed metamorphic histories of the oceans and continents.

The plate tectonic model, on the other hand, does fit the metamorphic data very well. In this model, ocean floors form by eruption of basalt at central ridges and grow as the basalt is moved away from the ridges as part of large lithosphere plates. The basalts are transported passively, with none of the intense squeezing that accompanies regional metamorphism; this explains the absence of regional metamorphism in most of the oceans. Where plates collide in subduction zones, however, enormous directed pressures are produced. Islands such as Japan would have experienced their regional metamorphism as a result of plate collision.

Metamorphism and Time

The distribution of regional metamorphic events in geologic time also gives insight into the way in which the earth operates. Regional metamorphism does not take place simultaneously everywhere throughout the world; while some areas are being subjected to the intense heat and pressures of regional metamorphism, others are completely unheated and unsqueezed. This is difficult to reconcile with a shrinking earth, in which simultaneous worldwide compression might be expected. However, it is perfectly compatible with the plate tectonic model, in which plate collisions can result in strong directed pressures at subduction zones while midplate regions can be undeformed and unheated.

Regional Metamorphism and Ancient Geothermal Gradients

Consider what geothermal gradients would be in a subduction zone. Close to the subduction zone itself, directed pressures are most intense but there is little added heat, resulting in a relatively low geothermal gradient. Farther from the subduction zone, the effects of directed pres-

sures decrease and heat rises from the subducted plate, causing a relatively high geothermal gradient. Metamorphism in the two different areas should reflect the different geothermal gradients. Close to the subduction zone, directed pressures dominate and we would expect a type of metamorphism called **high-pressure–low-temperature metamorphism;** farther from the subduction zone, heat is dominant and **high-temperature–low-pressure metamorphism** should occur (Figure 7.22). If we could determine the ancient geothermal gradients in metamorphic terrains, we conceivably could locate ancient subduction zones.

This is done by studying the sequence with which rocks of one metamorphic facies pass into those of another. For example, in New Zealand and California, rocks of the zeolite facies are followed with increasing grade by those of the prehnite-pumpellyite facies and blueschist facies (Figure 7.23). However, in parts of Japan, increasing metamorphism is indicated by the sequence of albite-epidote hornfels through sanidinite facies; and in Barrow's classic area of Scotland, the sequence is greenschist, amphibolite, granulite. The colored lines on Figure 7.23 show the different paths of temperature and pressure change during metamorphism of the different regions. In effect, each line represents the

Figure 7.22 Plate tectonics and metamorphism. Regions of no regional metamorphism and of different types of regional metamorphism are predicted by the plate tectonic model for subduction zones and spreading centers.

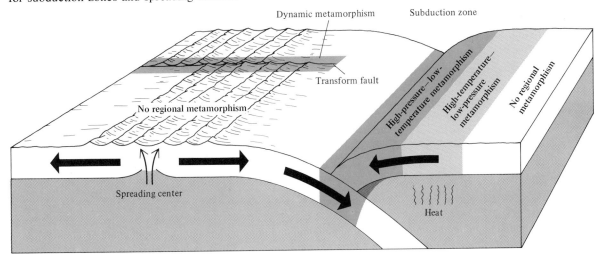

geothermal gradient active during metamorphism.

The examples of California and New Zealand shown in Figure 7.23 represent high-pressure–low-temperature metamorphism, and the example of Japan represents high temperature–low-pressure metamorphism. The Scottish Highlands is intermediate in gradient between these two extremes and is an example of moderate-temperature–moderate-pressure metamorphism. In most areas where the metamorphic record is well preserved, regional metamorphism seems to follow a consistent pattern. As we shall see in Chapter 21, some geologists use this pattern to locate ancient subduction zones.

METAMORPHISM AND NATURAL RESOURCES

Metamorphism is important to industry as well as to the geologist because it has created several economically valuable natural resources. As with igneous and sedimentary resources, both specific metamorphic minerals and entire metamorphic rocks prove useful.

Metamorphic Rock Resources

The two most widely used metamorphic rocks are probably marble and slate. Marble has been used as a decorative facing stone for thousands of years. Because of its almost endless variety of color (pink, tan, white, green, black, and so forth), texture, and layering, people from ancient Babylonia to modern times have not tired of seeing it where they work and live. Marble is also a favorite medium of sculptors because its relatively low hardness (calcite is 3 on Mohs' scale) allows it to be carved and chipped easily.

Slate also comes in a range of colors but has very different uses. Its dense, impermeable nature and the ease with which it can be split into flat slabs make it an excellent roofing shingle. It also forms the bases of many of our best fish tanks and billiard tables. We walk on it (as flagstones) and set our hot cups and pots on it (as trivets).

Metamorphic Mineral Resources

Metamorphic mineral resources come from rocks of nearly every composition and metamorphic grade. Graphite forms by metamorphism of carbonaceous shales, coal, or limestones, while corundum is formed by metamorphism of highly aluminous rocks. Low-grade metamorphism of either ultramafic rocks or impure dolomites yields talc for our talcum powder. Low-grade ultramafic rocks also yield serpentine, a mineral whose fibrous variety is known as asbestos.

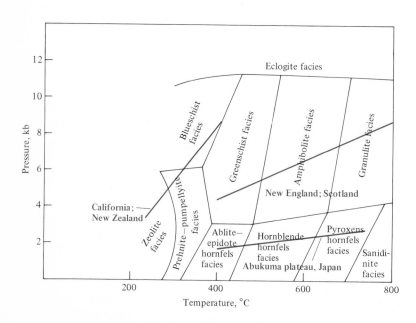

Figure 7.23 Determining ancient geothermal gradients. The sequence of regional metamorphic facies in an area indicates the geothermal gradient active during metamorphism.

Asbestos is used as a major component in automobile brake linings and to make heat-resistant cloth. High-grade minerals such as andalusite, kyanite, and sillimanite are used to make high-temperature porcelains found in spark plugs, and garnet, an abundant index mineral in pelitic rocks, is used both as an abrasive and as a gemstone.

Metasomatism is important in the formation of industrial minerals because it often concentrates the ions of ore minerals at the contacts between igneous and sedimentary rocks. These deposits are called **contact metasomatic ores** and include tungsten minerals in Nevada and California, iron minerals in Utah and Pennsylvania, lead in New Mexico, and gold in Montana.

SUMMARY

Metamorphism consists of the solid-state changes that take place when a rock is subjected to environmental conditions different from those under which it first formed. Changes in rock texture, mineralogy, and chemical composition take place in response to heat, lithostatic pressure, directed pressure, and changes in the chemical composition of the environment.

Different types of metamorphism are caused by different combinations of these agents. Contact metamorphism occurs in response to heat energy from intrusive igneous rocks. Burial metamorphism is caused by lithostatic pressure in a pile of accumulating sedimentary and volcanic rocks. Dynamic metamorphism accompanies movement along faults in the earth and is caused by directed pressure. Large-scale application of heat and pressure during broad regional deformation causes regional metamorphism. Shock (impact) metamorphism results from collisions of meteorites with the earth.

Textural changes in rocks include reorientation of grains in response to directed pressure; flattening, stretching, and crushing of grains; and recrystallization to increase or decrease grain size. Mineralogic changes produce new minerals from those which were originally present. Changes from one polymorph to another involve no interaction between different minerals, but other types of mineralogic change involve interaction between different minerals by recombination of their ions. Many of these changes result in the removal of water or carbon dioxide from the parent rock.

Some rocks, called low-grade metamorphic rocks, are changed only slightly during metamorphism because of the low intensity of the metamorphic agents, but others (high-grade rocks) may undergo drastic changes. Metamorphism is a series of progressive changes that occur as rocks readjust constantly to increasing metamorphic intensity in a series of steps.

The final mineralogy and texture of a rock depend on the original mineralogy; the intensity, duration, and type of metamorphism; and the addition or removal of ions. Rock textures give clues to the type of metamorphism involved. Foliation and lineation are produced only by directed pressure, and hence during regional or dynamic metamorphism. Granoblastic textures result from metamorphism in the absence of directed pressure (contact metamorphism) or the absence of grains with shapes that are appropriate to exhibit foliation or lineation. The final mineral assemblage indicates the intensity of metamorphism, and either index minerals (mineral zones) or complete assemblages (metamorphic facies) can be used to determine relative metamorphic intensity. The absolute intensity of metamorphism is estimated by comparing laboratory-produced mineral assemblages with natural ones, assuming that the laboratory reactions occur at approximately the same conditions as those in nature.

Metamorphic rocks underlie large areas of continental interiors and also occur in linear belts in major mountain ranges. However, they are rare in the ocean basins, being restricted to island chains near ocean margins. Metamorphism is not restricted in time, having occurred throughout geologic history. The distribution of metamorphism and metamorphic rocks both geographically and in time is most compatible with the plate tectonic model of the earth.

QUESTIONS FOR REVIEW AND FURTHER THOUGHT

1. Compare the roles of heat and pressure in causing metamorphic changes in a rock.

2. What are the different types of metamorphic changes?

3. Explain why the different kinds of metamorphism can result in unique textures, unique mineral assemblages, or a unique combination of mineralogy and texture that can be used to determine which kind of metamorphism has occurred.

4. Why are some minerals useful as metamorphic index minerals while others are not?

5. Why are metamorphic mineral assemblages more valuable than index minerals for assigning relative metamorphic intensities to rocks?

6. Why are metamorphic facies more useful than metamorphic zones as indicators of relative metamorphic intensity?

7. Discuss the factors that determine the mineral content of a metamorphic rock.

8. A sheet of basalt sandwiched between layers of shale may be either a sill or a lava flow. Suggest a method for distinguishing between these possibilities by using the principles of metamorphism.

9. A sequence of interbedded sandstones and shales is subjected to intense regional metamorphism. What differences would you expect in the responses of the two rock types to the same metamorphic conditions?

ADDITIONAL READINGS

There are very few readings available on an introductory level that deal in detail with metamorphism or metamorphic rocks. The readings suggested below are divided into two categories: an intermediate level in which most of the material is understandable and useful to the introductory student, and an advanced level recommended for students with strong chemistry and physics backgrounds and a knowledge of minerals.

Intermediate Level

Bayly, Brian, 1968, *Introduction to Petrology,* Prentice-Hall, Englewood Cliffs, N.J., 371 pp.
  (A very readable treatment of the types of changes that occur in metamorphic rocks)

Williams, H., F. Turner, and C. Gilbert, 1954, *Petrography,* W. H. Freeman, San Francisco, 406 pp.
  (A comprehensive review of all rock types, with drawings of typical metamorphic rock textures and a brief review of metamorphism)

Advanced Level

Hyndman, Donald, *Petrology of Igneous and Metamorphic Rocks,* McGraw-Hill, New York, 1972, 533 pp.
  (An excellent in-depth view of metamorphic rocks with specific application to interpretation of the plate tectonic model)

Spry, Alan, *Metamorphic Textures,* Pergamon Press, New York, 1969, 350 pp.
  (An excellent treatment of the origin and appearance of metamorphic textures, with superb photomicrographs of actual rocks. Some of the material is very complex, but the photographs and descriptions are useful)

CHAPTER 8

Geologic Time

Most people think of time in relation to human life spans and consider the passing of several generations as a long time. However, geology deals with much greater intervals of time. Geologic time is measured in relation to the origin of the earth—an event which is believed to have occurred around 4.5 *billion* years ago. The human perspective of time makes it very difficult to grasp the vastness of geologic time. To understand and appreciate the material in subsequent chapters of this book, you should think from now on in terms of geologic rather than historic time. This is especially important because advanced humanlike creatures began to appear only in the last 0.1 percent of geologic time.

Geologists express the geologic age of earth materials (rocks, fossils, and sediments) or events in two different ways. One way is to determine whether a particular earth material or event is younger or older than another, although we may not know the *actual* age of either. For example, a sedimentary rock layer containing dinosaur bones is relatively older than volcanic sediments containing human bones. The **relative age** of earth materials or events is their age in *comparison with the ages of others.*

Another way to express the age of earth materials or events is by determining *their actual age in years* using the radioactive and other dating methods discussed later in this chapter. Ages obtained by radioactive dating are called **radiometric dates.** If we apply one of the radioactive dating methods to the volcanic rock containing human bones that was just described, we may obtain a radiometric date of 1.2 million years. Under the best of conditions, this radiometric date would be very close to the actual age of the layer dated.

We will start our discussion of geologic time by considering the sequence of discoveries that allowed early geologists to determine the relative ages of different types of rock layers. Then we will be able to discuss how actual ages are obtained for these layers, thus working out the geologic history of the earth.

RELATIVE TIME

By the end of the nineteenth century, geologists were able to establish the relative ages of many rocks by utilizing a combination of physical and biologic concepts. The rocks in individual areas were then assigned a chronologic order to make up a local section. A **local section** is a series of rocks, from oldest at the bottom to youngest at the top, that records the history of deposition in an area. As geologic studies increased, the local sections in different areas were compared with each other to establish **regional** sections recording the history of rock deposition (and erosion) over wider geographic areas. The concepts used to establish the relative ages of the rocks in local and regional sections seem obvious to us in hindsight, but each was revolutionary when it first was proposed.

Physical Methods for Determining Relative Ages

Several of the basic principles geologists use to determine the relative ages of rocks come from studies made by Nicolaus Steno in the latter half of the seventeenth century. Steno observed that sediments accumulate by the deposition of individual particles from a medium such as air or water. This deposition builds up a series of horizontal layers, with the oldest at the bottom and the youngest at the top. Steno's observations led him to propose three basic principles in 1669. Perhaps the most important is the principle of **superposition,** which states that in any sequence of rocks that has not been overturned subsequently, the oldest rock is at the bottom and the youngest at the top. His principle of **original horizontality** states that sedimentary rocks normally are deposited in horizontal layers; if rock layers are inclined, we can attribute this to some postdepositional force which tilted or folded the originally horizontal layers. The last concept is the principle of **lateral continuity,** which states that as a sedimentary rock layer is deposited, it extends outward horizontally until it thins out and disappears or until it terminates against the boundaries of the basin in which it accumulates. This last principle suggests that if a rock layer appears in one local section, the same layer, or its facies equivalent (see Chapter 6), probably would appear in adjacent sections as well. The principle is illustrated in the geologic section shown in Figure 8.1. Rock layer A underlies the

whole area, but it undergoes a facies change from sandstone in the west to shale in the east. Rock layer B appears in the western part of the area. but it thins eastward and is not present in the eastern part of the area. We will discuss the other aspects of this diagram later in the chapter.

Two additional physical methods for determining relative age were proposed by James Hutton in 1788. These principles are illustrated in Figure 8.2 The principle of **crosscutting relationships** states that an igneous rock that cuts across a rock layer must be younger than that rock layer. For example, the igneous intrusion (layer E) shown in Figure 8.2 cuts across layers A, B, C, and D, and so it must be younger than they are. The principle of **inclusions** states that a rock containing fragments of another rock must be younger than that rock. For example, layer F in Figure 8.2 contains fragments of rocks eroded from layers D and E, and so layer F must be younger than layers D and E.

How does a geologist make use of these physical criteria to determine the relative age of a rock layer in a local section? The techniques used are described in the analysis of the igneous and sedimentary rock layers that is shown in Figure 8.2 First we must use top and bottom sedimentary structures (Chapter 6) to determine

Figure 8.1 Rock layers can be traced laterally by their distinct physical characteristics and fossils, even though the individual rock exposures are separated widely. Rock layer A undergoes a facies change from sandstone in the west to shale in the east. Rock layer B thins eastward and does not appear in the eastern part of the area.

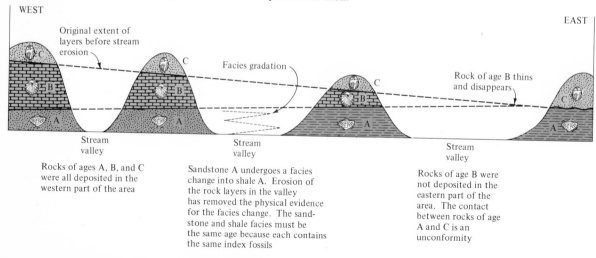

WEST

Original extent of layers before stream erosion

Facies gradation

Rock of age B thins and disappears

EAST

Stream valley

Stream valley

Stream valley

Rocks of ages A, B, and C were all deposited in the western part of the area

Sandstone A undergoes a facies change into shale A. Erosion of the rock layers in the valley has removed the physical evidence for the facies change. The sandstone and shale facies must be the same age because each contains the same index fossils

Rocks of age B were not deposited in the eastern part of the area. The contact between rocks of age A and C is an unconformity

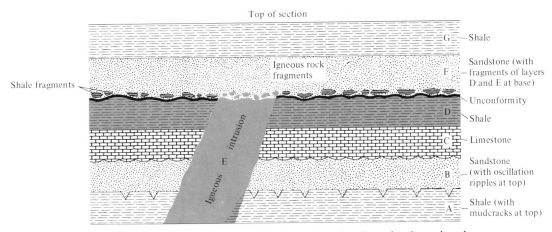

Figure 8.2 Determining the relative ages of rock layers in a local section by using the principles of Steno and Hutton.

whether the rock layers are in the same order in which they were deposited originally or whether they have been overturned since deposition. The mud cracks at the top of the shale (layer A) open upward, and the oscillation ripple marks at the top of the sandstone (layer B) point upward; both factors indicate that the rock layers have the same orientation in which they were deposited (see Chapter 6). The principle of superposition indicates that the oldest layer (A) is at the bottom of the section and that layers B, C, and D are successively younger. The igneous intrusion (layer E) cuts across layers A, B, C, and D, and so it must be younger than all of them. The sandstone (layer F) contains rock fragments (inclusions) of *both* layer D (below) and the igneous intrusion (layer E), and so it must be younger than both. Superposition indicates that the uppermost layer (G) is younger than layer F. Thus, the physical principles of Steno and Hutton have been used to establish the relative age of the rocks in this section and show that layer A is the oldest and layer G the youngest.

Biological Methods for Determining Relative Ages

Fossils—the preserved remains, traces, or imprints of plants and animals described in Chapter 6—are extremely important in establishing the relative age of sedimentary rocks. As soon as the chronologic succession of rock layers in local

sections had been determined using the physical criteria described earlier in this chapter, it became apparent that sedimentary rocks of the same relative age had the same kinds of fossils. Once that association was made, it became possible to determine the relative age of a sedimentary rock in another area by its fossils *alone*.

William Smith, a British engineer working in the late 1700s and early 1800s, made the first detailed studies showing that certain rock layers had the same fossils. After examining many geographically isolated sections of sedimentary rocks exposed in mines and canal excavations, he was able to show the following:

1. Rock layers with distinct physical characteristics and the same overlying and underlying rock layers occurred over a wide area.

2. Each of these sedimentary rock layers contained a type of fossil which was not found in the other layers.

3. The rock layers in geographically isolated exposures were identifiable on the basis of their fossil contents alone.

Smith's observations were very significant because they provided the means by which rock layers in widely separated areas could be correlated with each other. This principle is illustrated by Figure 8.1 Layer A is a sandstone in the western part of the area that undergoes a facies change into a shale in the eastern part. However, the part of the layer showing the gradational facies change has been eroded away, and a geologist studying the area would see two different types of rock at either end of the area. However, examination of the sandstone and shale would show that they have the same fossils and thus the same age, even though they have different physical characteristics.

At about the same time Smith was working in England, important discoveries were being made in the Paris Basin of France by Georges Cuvier. In 1812, he showed not only that each of the rock layers in the vicinity of Paris had a distinct group of fossils but also that the most primitive fossils were found in the lower layers and that fossils closer to present-day life forms occurred in the upper layers. Such an upward sequential change in fossil types is called **faunal succession.** Cuvier explained the differences between fossils in the adjacent rock layers as resulting from the extinction of one fossil organism in a catastrophic event and the appearance of a somewhat different form by the time the next rock layer was deposited.

Charles Darwin provided the explanation for faunal succession in his classic work, *Origin of Species,* in 1859. He showed that fossil forms evolved over long periods of time, resulting in progressive changes in appearance. Thus the different appearances of fossils in adjacent rock layers need not have been brought about by a catastrophic event as Cuvier had suggested.

As more and more field studies were carried out, it became apparent that the rock record was incomplete *at any one locality.* A substantial break in the rock record where there has been a break in the continuity of deposition of a sequence of sedimentary rocks, or a break between eroded igneous or metamorphic rocks and younger sedimentary rocks, is called an **unconformity.** A dramatic unconformity in the inner gorge of the Grand Canyon can be seen in Plate 13. Figure 8.2 shows how an unconformity develops and the relationships between the beds above and below it. After the deposition of layer D and the subsequent intrusion by layer E, there was a break in sedimentary deposition, and portions of layers D and E were eroded. When

sedimentary deposition resumed again with layer F, pieces of layers D and E were incorporated in the base of layer F. The actual surface of the unconformity is shown by the thick line cutting across the tops of layers D and E.

The Relative Geologic Time Scale

Even though the fossil record at any one place is incomplete, geologists were able to piece together the total fossil record by examining rocks in different areas. By the middle of the nineteenth century geologic time had been subdivided into **eras** based on major differences in fossils in the sedimentary rock record (Table 8.1).

The oldest era, the Precambrian, extends from the earth's beginning to the first appearance of primitive life (bacteria), followed by simple marine plants such as algae and then by soft animals such as worms and jellyfish. The next era, the Paleozoic (meaning *ancient life*), is dominated by invertebrates (animals without backbones) such as clams and by the first appearance of fish, land plants, and amphibians. The next era, the Mesozoic (meaning *middle life*), is dominated by reptiles such as the dinosaurs. The most recent era, the Cenozoic (meaning *recent life*), represents the time when mammals and flowering plants became dominant and humans first appeared.

Eras are subdivided into **periods** and the periods into **epochs** based on evolutionary changes that are less dramatic than those used to subdivide eras. Not all fossils are useful in determining geologic time subdivisions because many of these plants and animals lived for great lengths of time and may be found in rocks deposited during different epochs or even periods. Only certain fossils, called index fossils, are useful for identifying rocks of a specific age. An **index fossil** is one with unique features that lived over a wide geographic area for only a short period of time and that has been preserved abundantly in the rock record. Several index fossils are shown in Figure 8.1. The index fossil in layer A enables a geologist to determine that the sandstone in the west is the same age as the shale in the east. The establishment of a time equivalence between two rock units, such as the

TABLE 8.1 Relative and radiometric geologic time scales

Era	Period	Epoch	Radiometric time, millions of years before present	Major life forms in each era
Cenozoic	Quaternary	Holocene	— 0.01 (10,000 yr.) —	Mammals, including humans, in the latest part. Flowering plants
		Pleistocene		
			1.5	
	Tertiary	Pliocene	— 12 —	
		Miocene	20	
		Oligocene	35	
		Eocene	55	
		Paleocene		
			65	
Mesozoic	Cretaceous	Numerous epochs recognized	130	Reptiles. Conifers and cycad plants. Advanced marine invertebrates
	Jurassic		185	
	Triassic			
			230	
Paleozoic	Permian	Numerous epochs recognized	265	Marine invertebrates. Fish (amphibians in later part). Marine plants (land plants in later part)
	Pennsylvanian*		310	
	Mississippian*		355	
	Devonian		413	
	Silurian		425	
	Ordovician		475	
	Cambrian			
			570	
Precambrian			600	Simple marine animals without mineralized skeletons, such as worms and jellyfish
			3500	Oldest fossils (bacteria)
			3800	Oldest known rocks
			4500	Formation of the solid earth
			5000	Formation of the protoearth

* These two periods are known collectively in Europe as the Carboniferous Period.

sandstone and shale in this example, is called **correlation.** The absence of index fossil B in the eastern part of the area suggests that rocks of age B were not deposited there; if they were, they must have been eroded away before deposition of layer C. The absence of rocks or fossils of age B in the eastern part of the area indicates an unconformity between layers A and C. This is indicated by a wavy line in Figure 8.1.

ACTUAL TIME

The biologic and physical methods used in the eighteenth and nineteenth centuries to determine the relative age of rocks could not provide answers to fundamental questions such as: How old is the earth? When did life begin? At what time did humans evolve? Answers to these questions had to wait until accurate actual measurements of geologic time became possible in the early part of the twentieth century. Before considering these methods, we should first ask, how *do* we measure time?

We can use a number of different devices to measure the passage of time, such as a sand-filled hourglass or a clock with a swinging pendulum. What do all such time-keeping devices have in common? They *operate at a fixed rate.* Therefore, the time elapsed can be measured by the amount of sand falling through the hourglass or the number of swings of the pendulum. In order to keep accurate time, however, our time-keeping device must not be interrupted and restarted.   For example, in order for an electric clock to keep accurate time, it must not be affected by a power failure.   These same conditions must be satisfied by any of the "geologic clocks" used by geologists to determine the ages of rocks.

Early Attempts

Geologists of the nineteenth century first used earth processes as clocks to obtain absolute dates of past geologic events.

In 1897, Lord Kelvin calculated that the earth took 20 to 40 million years to cool from an assumed initial molten state.   Kelvin's calculations now are known to be erroneous because he assumed that all the earth's heat energy dates from its formation and that the rate of heat loss has been constant since that time.   We now know that the breakdown of radioactive elements within the earth's crust produces heat and that this slows the earth's rate of cooling (see Chapter 2).   Therefore, the earth's age must be much greater than the age proposed by Kelvin.

In 1899, John Joly estimated the age of the earth from the salt content of its oceans.   He assumed an initial freshwater ocean on a primitive earth into which streams were pouring salts at a fixed rate.   After measuring the average salt content of present rivers and comparing it with the salt content of the oceans, he arrived at a figure of 80 to 90 million years for the age of the earth.   One of the problems with Joly's estimate was that he assumed a fixed rate of supply of salts to the ocean—a highly improbable event. In addition, fossil animals and other evidence suggest that the oceans reached a maximum content of salts, similar to today's levels, in the Precambrian Era.

Each of these early dating attempts was doomed to failure because of the assumption that these geologic processes acted at fixed rates.   It was only after the discovery of radioactivity by A. H. Becquerel in 1896 that geologists had a fixed-rate process that could date past geologic events accurately.

Radioactivity as a Dating Tool

Nuclear reactions proceed at a constant rate that is not affected by changes in temperature, pres-

sure, and chemical reactions.   In 1905, Ernest Rutherford made the first clear suggestion that radioactivity could be used to estimate the age of the earth.   Shortly thereafter, B. B. Boltwood utilized the radioactive breakdown of uranium and obtained radiometric dates for some minerals that ranged up to about 2 billion years.   Thus, **geochronology**, the science of obtaining radiometric dates for geologic materials, was born.

Types of Nuclear Reactions Used in Geochronology

Each of the radioactive dating methods utilizes one of three different types of nuclear reactions. A product of a nuclear reaction is called **radiogenic**.   In each reaction, the starting element is referred to as the **parent** and the element produced by decay of the parent is referred to as the **daughter**.   The three types of nuclear reactions are alpha decay, beta decay, and electron capture.

Alpha decay

Alpha decay occurs in several isotopes of uranium.   Recall from Chapter 2 that all isotopes of an element have the same atomic number—which means that they have the same number of protons.   What differs from isotope to isotope is the number of neutrons.   Uranium has atomic number 92 (it has 92 protons).   One of its more common isotopes is $^{238}_{92}U$, which contains 146 neutrons and thus has an atomic mass of $92 + 146 = 238$.   Another isotope is $^{235}_{92}U$, which has 143 neutrons and an atomic mass of 235.   Both uranium 238 and uranium 235 are radioactive.   Both break down into an isotope of thorium, Th, as follows:

$$^{238}_{92}U \rightarrow {}^{234}_{90}Th + {}^{4}_{2}\alpha + energy$$
$$^{235}_{92}U \rightarrow {}^{231}_{90}Th + {}^{4}_{2}\alpha + energy$$

Notice that the uranium has lost two neutrons and two protons in the form of an **alpha particle,** α.   A nuclear reaction that releases an alpha particle is called **alpha decay.**

In some cases the daughters, such as $^{234}_{90}Th$ and $^{231}_{90}Th$, are *also* radioactive.   The radioactive daughter then continues to break down into a succession of radioactive isotopes until a nonradioactive daughter finally is produced.   In the case of $^{234}_{90}$Th, the stable daughter is an isotope of lead, $^{206}_{82}$Pb. This type of decay, in which there is a series of steps between the parent and a stable daughter, is called **series decay.**

Beta decay

The type of nuclear decay in which the parent-to-daughter transition is accompanied by release of a small, negatively charged particle (an electron) is called **beta decay.** The particle is called a **beta particle,** β^-.

An example of beta decay is the transition of the radioactive isotope of rubidium, $^{87}_{37}$Rb, to strontium, $^{87}_{38}$Sr:

$$^{87}_{37}\text{Rb} \rightarrow \, ^{87}_{38}\text{Sr} + \underset{\substack{\text{Beta} \\ \text{particle}}}{\beta^-} + \text{energy}$$

Notice that the daughter has gained a proton, but there is essentially no change in atomic mass. Where does the additional proton come from? How can it be added without changing the mass of the daughter from that of the parent? Furthermore, the beta particle has the physical properties of an electron, but how can an electron be ejected from a nucleus which is composed of only protons and neutrons? These questions can be answered if we think of a neutron as consisting of a proton joined to an electron. A review of Chapter 2 shows that the mass of the neutron is nearly exactly the mass of a proton and an electron, and the neutral charge also fits this model. If one neutron in the $^{87}_{37}$Rb nucleus is converted to a proton and an electron, the atomic mass remains unchanged but the atomic number is increased by 1. The electron is then driven off as a beta particle.

Electron capture

Electron capture is involved in the radioactive decay of potassium 40, $^{40}_{19}$K, to the gas argon 40, $^{40}_{18}$Ar:

$$^{40}_{19}\text{K} \rightarrow \, ^{40}_{18}\text{Ar} + \text{energy}$$

In this case, the atomic number decreases by 1 and the atomic mass remains the same. This decay scheme differs from the other types of radioactivity in that it involves an electron from the electron shells as well as a nuclear particle. One electron, usually but not always from the innermost shell, combines with one of the pro-

tons in the nucleus to form a neutron. Hence the name **electron capture.** No particle is ejected in this process, but energy is emitted.

Obtaining Radiometric Dates

The basis for radiometric dating can be described as follows. As minerals crystallize, they may incorporate atoms of some radioactive isotopes into their structure. These atoms then continue to decay at a constant rate called the **decay rate.** The decay rates for all the radioactive isotopes used in geochronology have been determined from laboratory experiments. The **half-life** of a radioactive isotope is the time it takes for one-half the total number of parent atoms to break down. This process is shown in Figure 8.3. When the mineral diagrammed in Figure 8.3 crystallized, it incorporated some radioactive isotopes with a half-life of 1000 years. One thousand years later (one half-life) 50 percent of the original parent atoms have been converted into daughter atoms. After another 1000 years have passed, only 25 percent of the parent atoms are left, and so on. Note that as the percentage of parent decreases, the percentage of daughter increases (Figure 8.3). However, the sum of the parent and daughter atoms remains the same at any given time.

Radiometric ages for minerals are obtained by crushing the rock and separating the appropriate minerals. These minerals are then analyzed to determine the proportion of parent and daughter atoms. Since the half-life is known, the age of the mineral can then be obtained. For example, if the ratio of parent atoms to daughter atoms is 1:1, the time elapsed since the formation of the mineral is one half-life.

Not all kinds of rocks are useful for radiometric dating. Igneous and metamorphic rocks are best for determining dates because the minerals in them crystallized at the time of rock formation. In contrast, most of the minerals in sedimentary rocks are derived from the breakdown of a number of older rocks (see Chapter 6) and thus would not indicate an accurate date for the time of formation of the sedimentary rock. Some success has been obtained in dating the authigenic potassium-rich mineral glauconite in sedimentary rocks because it forms when the sediments are accumulating.

Accuracy of Radiometric Dates: Resetting the Geologic Clocks

In any of the radiometric dating methods involving a parent-to-daughter breakdown, three criteria must be satisfied in order to obtain an accurate radiometric date for the crystallization of the mineral. The mineral (or rock) must have measurable parent and daughter contents. The system must have been a closed one in which there was no parent or daughter gain or loss except for the parent-to-daughter decay. If there was any daughter product trapped in the mineral at the time of formation, it must be possible to determine the amount and adjust the radiometric age accordingly. Otherwise, the calculated radiometric age would be older than the actual age. Additional errors may result from sample and analytic procedures. *Every* radiometric date involves some uncertainty, but under ideal conditions the error in the radiometric date can be as small as ±2 percent.

If a rock is metamorphosed, the products may be driven out of the crystals. If these crystals are then dated, the lower daughter content will give an apparent date *younger* than the true date. When the daughter product is a gas, such as argon in the potassium-argon dating method, the date obtained is not that of the original crystallization of the mineral but that of the later metamorphic change. Where the daughter product is not a gas, metamorphic changes may drive it out of the crystals in which it formed, but it still may be retained within the *rock*. If none of the premetamorphic daughter products have been lost from the rock as a whole during metamorphism, the ratio of parent to daughter in the *whole rock* will give a radiometric date for the original formation of the rock. A radiometric date obtained in this manner is called a **whole-rock date.**

Radiometric Dating Methods

The major radioactive isotopes used to date earth materials (sediments, organic debris, and rocks) are indicated in Table 8.2. Most of them are used to date minerals and rocks. One method

Figure 8.3 Change in the proportion of parent and daughter atoms in a crystal with time. Half-life of the parent is 1000 years. As time increases, the percentage of parent decreases as more and more atoms decay to form the daughter. The number of atoms shown here is for illustrative purposes only; in reality great numbers of atoms are involved.

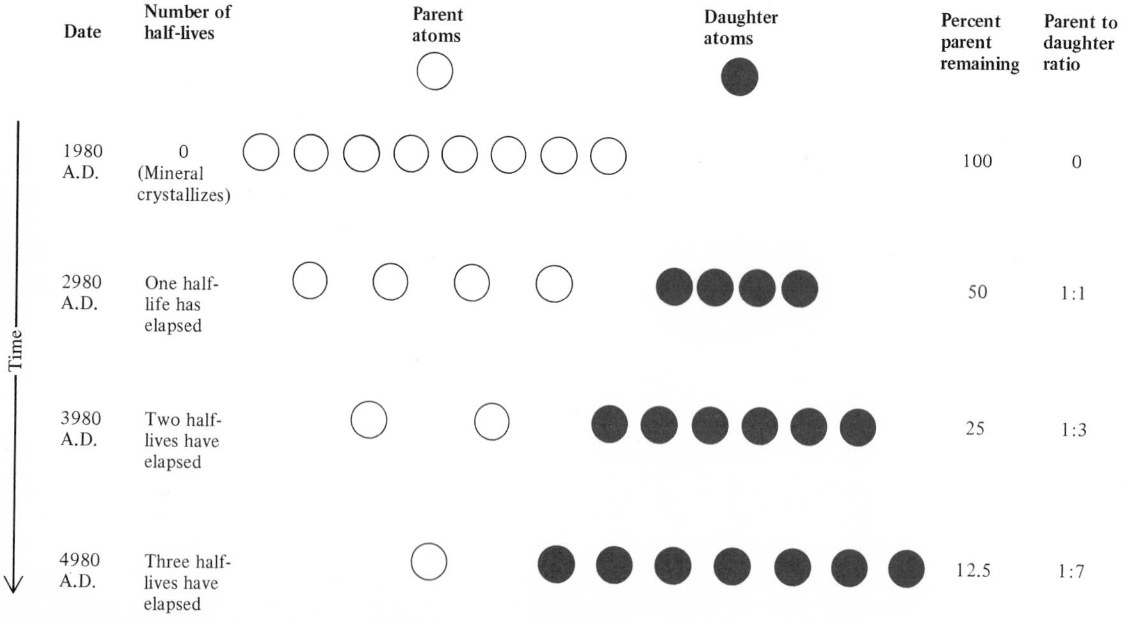

TABLE 8.2 Major radioactive isotopes used to obtain absolute dates

Method	Parent	Daughter	Decay scheme	Half-life of parent, yr	Effective age range, yr	Materials commonly dated	Comments
Uranium-lead	$^{238}_{92}U$	$^{206}_{82}Pb$	A chain of alpha and beta decays	4.5 billion	100 million to 4.5 billion	Zircon Uraninite Whole rocks	Uranium 238 and uranium 235 are found together in all uranium-bearing deposits. A radiometric date obtained from one isotope can be cross-checked against one obtained from the other.
Uranium-lead	$^{235}_{92}U$	$^{207}_{82}Pb$	A chain of alpha and beta decays	0.71 billion	100 million to 4.5 billion	Zircon Uraninite Whole rocks	
Rubidium-strontium	$^{87}_{37}Rb$	$^{87}_{38}Sr$	Beta decay	47 billion	100 million to 4.5 billion	Muscovite Biotite Potassic feldspar Whole igneous rock Metamorphic rock	Very useful in studies of metamorphic rocks. The oldest earth rock dates were obtained by this method.
Potassium-argon	$^{40}_{19}K$	$^{40}_{18}Ar$	Electron capture	1.3 billion	100,000 to 4.5 billion	Muscovite Biotite Hornblende Glauconite Sanidine Whole rock	Potassium-argon dating of volcanic deposits interlayered with sediments containing fossil human bones has provided dates for successive stages of human evolution. If the rock subsequently is heated, radiogenic argon may be driven off, giving an age younger than the true one.
Carbon 14	$^{14}_{6}C$	$^{14}_{7}N$	Beta decay	5730	100–50,000; up to 70,000 with special techniques	A wide variety of carbon-bearing materials including wood, charcoal, bone, cloth, paper, cave deposits, underground water, and oceanic water	Extremely versatile over its age range. Major advances in archaeology, anthropology, and glacial geology have been made by using carbon-14 dating. Particle accelerators will greatly expand the uses for carbon-14 dating by extending the age range covered and by allowing much smaller samples to be analyzed.

(the carbon-14 method) can be used to date carbon-containing materials less than 70,000 years old. Each of these parent-daughter pairs has a different half-life and thus a different effective age range. Most of the radiometric dates obtained for rocks younger than those from the Precambrian Era have been obtained by the potassium-argon and rubidium-strontium decay schemes. Most Precambrian dates were obtained by the rubidium-strontium and lead-uranium decay schemes, while some were obtained by the potassium-argon method. Only one of the major dating methods works on materials less than 100,000 years old; if the material contains carbon and is less than 70,000 years old, it can be dated by the carbon-14 method.

Uranium-lead dating

This was the first method used to date minerals. It is based on the radioactive decay of two uranium isotopes—uranium 235, $^{235}_{92}U$, and uranium 238, $^{238}_{92}U$. The $^{235}_{92}U$ ultimately breaks down to form lead 207, $^{207}_{82}Pb$, while the $^{238}_{92}U$ breaks down to form lead 206, $^{206}_{82}Pb$. All naturally occurring uranium deposits contain both uranium-235 *and* uranium-238 isotopes. Therefore, a rock can be dated by both isotopes. The date determined from the second isotope serves as a check on the date obtained from the first. At first, only uranium minerals could be dated by this method. However, uranium minerals are not found commonly in rocks, and this constituted an early limitation on this method. The increasing sophistication of modern laboratory techniques has made it possible to analyze small amounts of uranium and lead in the mineral zircon, $ZrSiO_4$, and in other minerals that are more common in many igneous rocks. Recent developments in uranium-lead dating enable geochronologists to determine (1) whether lead or uranium has been lost or added since the rock was formed, (2) the time when this change occurred, and (3) the correct age of the rock in spite of the alteration.

Rubidium-strontium dating

Rubidium 87 undergoes beta decay, forming strontium 87. A potential problem in using this method is that some *nonradiogenic* strontium 87 may be present in the minerals being tested. However, the amount of nonradiogenic strontium 87 can be determined because it occurs in a constant ratio with strontium 86, which is *completely* nonradiogenic. Therefore, the amount of nonradiogenic strontium 87 can be determined and subtracted from the total amount of strontium 87 to determine the amount of radiogenic strontium 87 (daughter). Metamorphic rocks commonly are dated by the rubidium-strontium whole-rock method.

Potassium-argon dating

The radioactive isotope potassium 40 decays to form two radiogenic products as indicated below:

Most (89 percent) of the potassium 40 undergoes beta decay to form radiogenic calcium 40. This decay branch is not useful for radiometric dating because the radiogenic calcium 40 cannot be distinguished from any nonradiogenic calcium 40 originally present in the crystal. About 11 percent of the potassium 40 undergoes decay by electron capture, forming argon 40, a chemically unreactive gas which accumulates within the crystal. This decay branch is useful in radiometric dating because no nonradiogenic argon 40 is incorporated in minerals at their formation. Because argon 40 is a gas, subsequent heating of the rocks during metamorphism may drive off part or all of the argon 40. Consequently, the radiometric age of the rock after metamorphism would be *younger* than its true age and more an indicator of the time of metamorphism.

The "shorter" half-life of potassium 40 relative to the other major isotopes used in dating permits it to be used to date materials formed between 100,000 and 10 million years ago (and beyond). This range includes the time when major human evolutionary changes occurred. Potassium-argon dating of volcanic sediments interlayered with sedimentary beds containing human fossils has allowed anthropologists to date successive stages of human evolution.

Carbon-14 dating:
Window on the earth's recent past

The great majority of radiometric dates used in geology are obtained from radioactive isotopes other than carbon 14. However, the carbon-14 method is probably the best-known isotopic dating method because of its extensive application in archaeology, anthropology, and geology in the dating of glacial and other events in the last 70,000 years. This dating technique is described here in more detail than the other methods because we will be referring to it in Chapter 13. Unlike the other radiometric dating schemes, the carbon-14 method rarely is used to date rocks. The short half-life of carbon 14 (5730 years) makes it possible to date only those materials less than 70,000 years old, and the great majority of sedimentary rocks are older than that. W. F. Libby, who developed the carbon-14 dating technique, received the 1960 Nobel Prize in chemistry.

Carbon has six protons (atomic number 6). Two of its isotopes are carbon 12, $_6^{12}C$, which has six neutrons, and carbon 14, $_6^{14}C$, which has eight neutrons. Carbon 12 is stable, and carbon 14 is radioactive. Carbon 14 forms continuously in the upper atmosphere when cosmic-ray-produced neutrons bombard stable nitrogen atoms, $_7^{14}N$, releasing a proton and forming the radioactive isotope $_6^{14}C$ (Figure 8.4). The atoms of carbon 14 along with atoms of stable carbon 12 combine with oxygen to form carbon dioxide, CO_2. This carbon dioxide mixes rapidly through the atmosphere and through the oceans, lakes, underground water, and glaciers of the hydrosphere. Plants manufacture sugars and starches out of this carbon dioxide, and when animals eat the plants, the carbon 14 becomes incorporated into their tissues as well.

Carbon 14 reverts very quickly back to nitrogen through beta decay. However, *as long as the organism lives*, new carbon 14 enters the organism's tissues by exchange with the atmosphere, hydrosphere, or both. The carbon-14 level in the organism thus reaches a concentration equal to the concentration of carbon 14 in the atmosphere and hydrosphere. But as soon as the organism dies, decayed carbon 14 can no longer be replaced, and the amount of radioactive carbon in the organism decreases steadily. A radiometric age for the material is obtained by measuring the number of beta particles emitted by a given sample as it decays. For example, after about 5730 years has elapsed (one half-life), the concentration of carbon 14 is only one-half the concentration in the atmosphere.

The carbon-14 dating method, while restricted in age range, is highly versatile. It has been used to date wood, charcoal, peat, shells, charred bones, paper, cloth, pollen, leaves, and other carbon-bearing objects in archaeological and anthropological studies. The age of the famous Dead Sea Scrolls was obtained by dating the linen in which the scrolls had been wrapped. The concentration of carbon 14 in campfire charcoal and other objects has been used to date the cultures of ancient peoples. Even though carbon 14 rarely is used to determine the age of rocks, it is very useful in dating certain relatively recent geologic events. For example, the carbon-14 dates of ancient forests in Wisconsin that were overridden by glacial ice provide a date for the glacial advance.

Carbon-14 dating currently is being revolutionized by the use of high-energy particle accelerators to determine the number of carbon-14 atoms in a sample *directly*. Refinements of the method may soon push the range of carbon-14 dating to 100,000 years before the present. This would close the time gap between the potassium-argon and carbon-14 methods. In addition, particle-accelerator dating speeds up the analytical process greatly (hours rather than days for old samples) and utilizes samples one one-thousandth of the size needed in present analyses.

Carbon-14 dates are subject to error. If the material being dated has been contaminated with "recent" carbon, perhaps carried by underground waters, the age will be in error (too young). Another problem is that comparison of ages determined by counting annual growth rings of trees with carbon-14 dating of the same rings indicates that some carbon-14 dates are

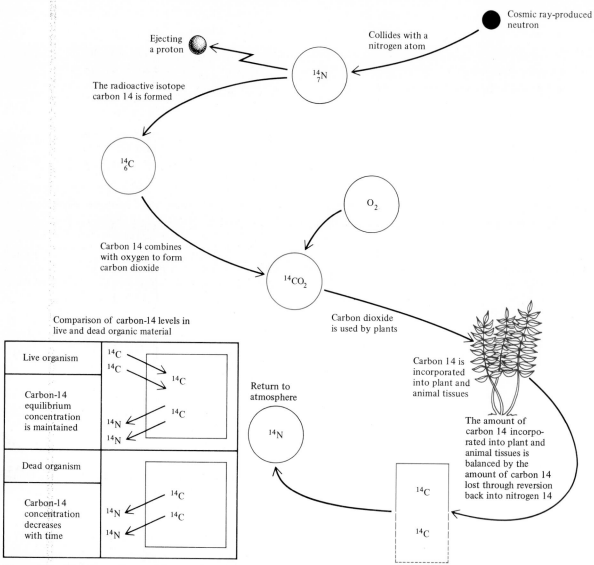

Figure 8.4 The formation and decay of carbon 14.

less and some greater than the "true" ages obtained by counting the annual rings. Since the decay rate of $^{14}_{6}C$ is constant, it is believed that this age discrepancy is caused by fluctuations in carbon-14 *productivity*. These fluctuations could have resulted from a number of causes, including changes in the earth's magnetic field, sunspot cycles, burning of fossil fuels, and even nuclear bomb tests. Carbon-14 dates back as far as 6273 B.C. can now be corrected through tree-ring data. What of the older carbon-14

dates? The possible uncertainty in these older dates has led some geologists to use the term *radiocarbon years* in giving carbon-14 dates. This is done to avoid the assumption that the years we have measured are necessarily the same in length as calendar years.

Other Dating Methods

While most geologic dates are obtained by the radiometric methods we have described, other methods are used in more specialized situations.

Varves

In Chapter 6 we described a type of annual rhythmic stratification called a varve. Counting the number of varves in an exposure can give an idea of the length of time it took for them to accumulate. This assumes that only one varve formed per year and that all the varves originally formed are preserved. For example, varve counting provides information on the length of time that lakes that were fed from glacial melting existed in a given area.

Tree-ring dating

In temperate climates, the trunk and branches of a tree usually increase in thickness by one layer during each growing season. If a tree is cut, these layers can be seen as concentric rings (Figure 8.5). Since each layer usually represents a year's growth, it is called an **annual ring.** The pattern of rings provides information about the area's past climate. For example, the rings laid down in wet years are thicker than those laid down in dry years. Additionally, by

Figure 8.5 Cross section of a tree trunk showing the annual rings used to determine the age of the tree. *(Laboratory of Tree Ring Research, University of Arizona)*

counting the rings, one may be able to estimate how long the tree has lived.

After a number of years, the trees in any one region have a characteristic sequence of annual rings of different widths. If we examine trees of overlapping ages, a continuous record of tree-ring variation can be built up. This record is extended back in time by study of long-lived species such as the bristlecone pine. The age of a wood sample containing a number of annual rings can be dated by comparing the ring patterns with a "standard" for the area. Tree-ring dating in the United States currently can be extended back in time to about 8254 years before the present.

Amino acid dating

As fossil bones "age," there is a known rate of change in the ratio of two different forms of amino acids (the D and L isomers). After a long period of time the two amino acid types equalize in abundance. An approximate age may be determined by calculating the D/L amino acid ratios. This dating scheme works for materials older than about 200 years and less than 1,000,000 years old.

Measurement of obsidian hydration layer

Fresh obsidian exposed to the atmosphere reacts with the humid air, becoming hydrated. When the moisture-absorption speed of a particular kind of obsidian is known, the depth of the hydrated glass layer gives an approximate date for the glass sample. This date records when the glass first was exposed to the air, perhaps when primitive humans first broke the obsidian to fashion a spear point or other tool. This dating method works for material older than about 500 years to younger than 1,000,000 years old.

Fission-track dating

As the nucleus of a radioactive isotope such as uranium 238 decays, high-energy particles are given off. These particles damage the surrounding areas, forming narrow and continuous **tracks** through the crystal structure. When the crystal surfaces are polished and etched with a strong solvent, the tracks become more visible and can be counted. The first tracks start forming shortly after the mineral crystallizes, and the number of tracks per unit area, called the **track density,** increases with time. The track density also increases with uranium content, and so it is necessary to measure uranium content before a fission-track age can be calculated. A wide variety of crystals and glasses have been dated, and this method can be used for materials from less than 100 years old to materials as old as the oldest rocks.

Magnetic dating

One of the newest methods of dating rocks and sediments is based on changes over time in the intensity of the earth's magnetic field. We know now that at different times in the past, the earth's magnetic field reversed (North Pole became South Pole and vice versa). These changes occurred over unequal time periods. Every time the magnetic field reversed, this was recorded in the deposition of iron-rich sedimentary particles and in the cooling of crystals from magmas. By dating the rock layers with radioactive isotopes, a magnetic dating scale was constructed. This magnetic dating scale can be used to obtain the ages of sediments or rocks with given magnetic properties. We will return to the topic of magnetic dating in more detail in Chapter 20.

The Radiometric Time Scale

When geologists want to determine the ages of the rock layers in an area, they generally employ a variety of methods. The igneous and metamorphic rocks can be dated directly by a number of radioactive isotopes (Table 8.2). The sedimentary rocks, the most abundant on the earth's surface and generally the only ones that contain fossils, can be dated directly only in the rare cases in which they contain authigenic minerals such as glauconite. However, ages for sedimentary rocks can be obtained if the rocks are either cut across or interlayered with igneous rock. The technique can be illustrated by analyzing the rock layers shown in Figure 8.6. The rock sequence is composed of a number of fossiliferous and unfossiliferous sedimentary rocks interlayered with volcanic ash and lava flows and cut across by a granitic dike. The relative age of the various layers is established by the principles of superposition, crosscutting relationships, inclusions, and fossils (see the relative age column in Figure 8.6). Radiometric dating of the volcanic ash (bed F) is 300 million years, and for the lava flow (bed H) it is 230 million years. By means of relative dating, we have already found out that the fossiliferous sandstone (bed G) is younger than F and older than H, and so we can assume that the sandstone is between 230 and 300 million years old.

Layers A through E can be analyzed in the same manner.

Once geologists have dated the rock layers in several areas, they often can use the information to date still other areas. For example, many sedimentary beds are not associated with datable igneous or metamorphic rocks. However, if we are trying to date a sedimentary bed that has the same index fossils as those found in layer G of Figure 8.6, we know that the bed correlates with layer G.

Figure 8.6 Determination of the ages of a series of sedimentary rocks interbedded with and cut across by igneous rocks, which can be dated radiometrically.

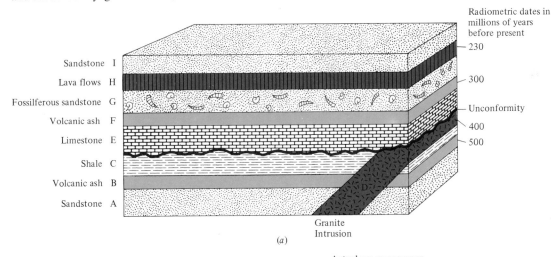

(a)

	Layer	Relative age	Actual age or age range (millions of years before present)
Youngest	I	Younger than H	Younger than 230
	H	Younger than G Older than I	230
	G	Younger than F Older than H	230–300
	F	Younger than E Older than G	300
	E	Younger than D Older than F	300–400
	D	Younger than A, B, and C Older than E	400
	C	Younger than B Older than D	400–500
	B	Younger than A Older than D and C	500
Oldest	A	Older than D Older than B	Older than 500

(b)

Answers to Some Basic Questions

As more and more field studies and radiometric dating were carried out, the absolute ages of relative time spans became subdivided more finely (see Table 8.1). Radiometric dating finally has enabled geologists to answer many important questions about the time of origin of the earth and of life and to date the sequence of life changes on the earth.

How old is the earth?

Geologists believe that the meteorites hitting the earth today formed at or near the time of the formation of the earth. Radiometric dating of meteorites gives ages around 4.5 billion years. Radiometric ages for the lunar rocks collected on the Apollo missions range from 3.8 to 4.2 billion years. Other lines of evidence suggest that the earth is slightly older than 4.5 billion years. However, the age of the *solid* earth is presently believed to be no older than 5 billion years. The oldest rocks found so far on earth are from southeast Greenland. Rubidium-strontium and uranium-lead dating of these rocks established their ages at about 3.8 billion years. If the earth is indeed 4.6 to 5.0 billion years old and the oldest earth rock found so far is about 3.8 billion years old, a good part of the earth's early rock record remains unaccounted for. Will we ever find the original "genesis" rock dating back to the earth's beginning? Probably not. The earth recycles its rock material constantly (see Chapter 3), and the elements in that genesis rock very probably have been recycled completely into other rocks.

When did life evolve on the earth?

In 1980, geologists reported primitive bacterialike cells from 3.5-billion-year-old chert deposits in northwestern Australia. The cells contained no nuclei and were linked together like pearls in a necklace. Some scientists believe that possible organic remains may occur in rocks as old as 3.8 billion years.

The subsequent evolution of life forms has been dated accurately by radiometric methods. The first recognizable primitive soft-shelled animal fossils occur in the late Precambrian Era, about 1 billion years ago. Abundant animal fossils with mineralized skeletons appear in the Cambrian Period, about 600 million years ago, and modern humans appeared between 1.5 and 2.0 million years ago.

Dating human evolution

The dating of human evolution is one of the most fascinating applications of geochronology. Geologists, working hand in hand with anthropologists, are providing radiometric dates for successive stages of human evolution. These researchers have been fortunate because many

Figure 8.7 Fossil footprints in volcanic ash. (*John Reader, copyright National Geographic Society*)

layers of sediment containing human and humanlike fossils also occur interbedded with layers of volcanic ash. Radiometric ages can be obtained for the ash layers, thus bracketing the age of the intervening fossil-bearing layers. We will close our discussion of geologic time by describing some recent research which has great importance in the study of human evolution.

In 1979, anthropologist Mary D. Leakey startled the scientific community with her discovery in Tanzania, east Africa, of sets of fossil foot prints made by early humanlike forms (hominids). This find indicated that hominids, with feet similar in form to ours, were walking erect half a million years earlier than indicated by previous fossil evidence. These primitive humans had walked across freshly deposited volcanic ash, leaving sets of footprints. The footprints (Figure 8.7) were preserved by burial under sediments and later volcanic ash deposits.

Potassium-argon dating of biotite crystals in overlying and underlying volcanic ash deposits established the age of the footprints as being between 3.6 and 3.8 million years! Leakey believes that when these hominids walked fully upright, they were free to use their forelimbs for other purposes. This presented new opportunities, which the brain subsequently evolved to take advantage of. As Leakey says:

> . . . those footprints out of the deep past, left by the oldest known hominids, haunt the imagination. Across the gulf of time I can only wish them well on that prehistoric trek. It was, I believe, part of a greater and more perilous journey, one that — through millions of years of evolutionary trial and error, fortune and misfortune — culminated in the emergence of modern man.*

* M. D. Leakey, "Footprints in the Ashes of Time," *National Geographic,* vol. 155, no. 4, April 1979, p. 457.

SUMMARY

Geologic time differs from historic time because it is reckoned in millions and billions of years back to the origin of the earth some 4.5 billion years ago. Geologic events or materials (sediments, rocks) may be dated relatively by their relation to other events or materials and their fossil contents, or radiometrically by using radioactive isotope methods. Relative dating is accomplished through superposition, cross-cutting relationships, and other physical criteria. Index fossils are used in relative dating to identify rocks of a certain age. The relative geologic time scale is based on differences in evolution of fossil animals. The major subdivisions, called eras, are defined by major evolutionary changes, while less dramatic evolutionary changes are used to subdivide eras into periods and periods into epochs. Detailed studies have shown that the rock record at any one place is incomplete, but a fairly complete record of earth history can be obtained by piecing together the rock record from different areas. Gaps in the rock record, called unconformities, result from erosion of older rocks prior to deposition of the overlying rocks.

Radiometric dating is based on the known and constant decay rates of certain radioactive isotopes (parents), which break down to form radiogenic products (daughters). Accurate radiometric dates require the following: The mineral (or rock) must have measurable parent and daughter; there must be no parent or daughter loss or gain except for the parent-to-daughter decay; and there must be no daughter present initially, or there must be some way of correcting for it. When an igneous or sedimentary rock is metamorphosed, there may be a loss of daughter from the minerals, especially if the radiogenic product is a gas such as argon 40. If the daughter has been driven out of the minerals but is still preserved within the rock, the age of the initial mineral formation may be obtained by the whole-rock method.

The radiometric time scale provides dates in years for the subdivisions of relative geologic time. Igneous and metamorphic rocks are dated directly by one or another of the radiometric methods. Sedimentary rocks can be dated if they are cut across or are interlayered with igneous rocks or if they grade laterally into other sedimentary rocks containing index fossils. Radiometric dating has provided information about the age of the earth (4.5 billion years), the oldest rocks (3.8 billion years), the oldest fossils (3.1 to 3.5 billion years), and the stages in hominid evolution which culminated in modern humans.

QUESTIONS FOR REVIEW AND FURTHER THOUGHT

1. How can geologists use physical methods to determine the chronologic sequence of rock layers in an area?

2. How can fossils be used to indicate the relative ages of sedimentary rocks?

3. If the rock record at any one place is incomplete, how is the complete record of relative geologic time assembled?

4. How can geologists correlate sedimentary rocks in widely separated areas?

5. What are the necessary conditions and assumptions associated with radiometric dating?

6. *a.* Describe the method used to obtain the radiometric age of a rock by the potassium-argon method.

b. What assumption is made when presenting the date?

c. Can you think of a few possible ways to check the accuracy of the radiometric date independently?

7. Under what conditions can fossils be used to approximate closely the real age of sedimentary rocks?

8. How were actual ages finally determined for the different kinds of rocks that make up the regional geologic sections?

ADDITIONAL READINGS

Beginning Level

Dott, R. H., Jr., and R. L. Batten, *Evolution of the Earth*, 3d ed., McGraw-Hill, New York, 1981.
(An excellent, well-illustrated, and detailed account of earth history. The methods used for relative and radiometric analyses are described, along with data on sedimentary geology which supplement the data in Chapter 6 of this book. A good place to look for the geologic history of a particular area of the United States)

Gannon, R., "How Old Is It? — The Elegant Science of Dating Ancient Objects," *Popular Science*, November 1979, pp. 76–81.
(A well-written and readily understandable treatment of a wide variety of dating methods. Includes some fascinating stories of applications of these dating methods in a variety of fields)

Hume, J. D., "An Understanding of Geologic Time," *Journal of Geological Education*, vol. 26, 1978, pp. 141–143.
(The vastness of geologic time is made clear through several easy-to-understand discussions)

Leakey, M. D., "Footprints in the Ashes of Time," *National Geographic*, vol. 155, no. 4, April 1979, pp. 446–457.
(Fascinating account of how fossil footprints made by hominids were radiometrically dated and the significance of the find in terms of human evolution)

Newman, W. L., *Geologic Time*, U.S. Geological Survey, No. 0-261-226 (9), 1978, 20 pp.
(A readable and well-illustrated discussion of how geologists determine the relative and radiometric ages of earth materials)

U.S. Geological Survey, *Tree Rings — Timekeepers of the Past*, Information Booklet 73-15, No. 1976-211-345/82, 1976, 16 pp.
(An extremely well-illustrated discussion of how tree-ring dating works and how tree-ring studies provide valuable data on past climates)

Intermediate Level

Barghoorn, E. S., "The Oldest Fossils," *Scientific American,* vol. 224, no. 5, May 1971, pp. 30–42.
(A well-illustrated discussion of the occurrence and description of the earliest known fossils in Precambrian rocks)

Bennett, C. L., "Radiocarbon Dating with Accelerators," *American Scientist,* vol. 67, July-August 1979, pp. 450–457.
(Describes the method by which carbon-14 dating is carried out by counting the carbon-14 atoms directly. Describes how accelerator dating can be used to date materials, such as cave paintings and individual tree rings, which cannot be dated at the present time)

Eicher, D. L., *Geologic Time,* 2d ed., Prentice-Hall, Englewood Cliffs, N.J., 1976, 150 pp.
(Describes the development of relative and absolute dating and the methods by which rock layers in different areas are correlated with fossils. Includes a good review of sedimentary geology which extends our discussion in Chapter 6 of this book)

Faul, H., "A History of Geologic Time," *American Scientist,* March-April 1978, pp. 159–165.
(A description of the developments that led to relative and radiometric dating of earth materials)

Advanced Level

Faure, G., *Principles of Isotope Geology,* Wiley, New York, 1977, 464 pp.
(A detailed treatment of all the major radiometric dating methods)

CHAPTER 9

Weathering and Soils

The blurring of the inscription on a marble tombstone, the crumbling of the sandstone facing on a building, and the cascading of rock debris onto a highway all show that rocks exposed at the earth's surface are subject to continual breakdown. The changes that occur in rocks and sediments as they come in contact with the atmosphere, the hydrosphere, and the biosphere are called **weathering.** Most of the minerals in rocks and sediments were formed at higher temperatures and pressures than those at the earth's surface. Weathering processes transform them into new minerals which are more stable under surface conditions. Rocks at the earth's surface are usually covered partially or completely by a layer of loose rock and sediment called **regolith.** Regolith that is derived in place from the breakdown of underlying materials is called **residual regolith** (see Plate 14). In some cases, the regolith covering rock exposures is not derived from the rocks below but is deposited on top of the rocks by streams, wind, glaciers, or mass movements of materials down slopes. This type of regolith is called **transported regolith** and will be dealt with in more detail in Chapters 10, 11, and 13 to 15.

Residual regolith may be derived from the breakdown of rocks or sediments. The general term **parent material** is used to describe the material being weathered. Parent material is broken down gradually by a number of different weathering processes. These weathering processes can be grouped into two major categories: physical and chemical. The *disintegration* of the parent material into smaller and smaller pieces with no change in mineral structure or composition is called **physical weathering.** The *decomposition* of the original parent material, forming new minerals and soluble ions, is called **chemical weathering.** Chemical and physical weathering processes work together to form residual regolith almost everywhere on the surface of the earth. However, at any one place the relative importance of chemical or physical weathering varies with factors such as temperature, vegetation, and the abundance of water.

Weathering is a vital part of the recycling of rock materials in the geologic cycle (Chapter 6). This recycling produces new sediments, agricultural soils, and important deposits of iron and aluminum. Adjacent rocks may weather at very different rates. This is called **differential weathering** and is an important factor in the development of landscapes on the earth's surface.

PHYSICAL WEATHERING

Physical weathering breaks down rocks into smaller pieces that retain most of the characteristics of the parent material. Grain size becomes smaller, but compositions are not changed and no new materials are formed.

Processes of Physical Weathering

A number of physical and biologic processes are active in physical weathering. Some processes, such as the freezing of water within rock fractures, are entirely physical. Others, such as the wedging apart of rocks by the growth of tree and plant roots along rock fractures, are biologic.

Frost wedging

Failure to drain water pipes in an unheated building prior to the winter season causes extensive damage when water freezes within the pipes. The trapped water expands about 9 percent with freezing. This can create enough pressure on the pipe to rupture it. A similar process, called **frost wedging**, occurs when water fills the crevices and pores of a rock and then is subjected to freezing temperatures. The water near the rock surface, exposed to the cold air, freezes first, exerting pressure on the water remaining in the rock crevices and pores. This produces the mechanical effect of a wedge on the rocks. The pressure resulting from the growth of ice crystals within rocks is sufficiently high to crack the rocks apart (Figure 9.1). Frost wedging is most extensive in areas where temperatures fluctuate periodically above and below 0°C, forming meltwater which freezes as the temperature drops. Frost wedging is less extensive in very cold regions where freezing temperatures prevail year-round.

Animals and plants

Sidewalks are buckled and underground sewage pipes broken by the expansive forces caused by growing tree roots. In similar fashion, the growth of plant and tree roots into rock crevices pries apart the rock in a process called **root wedging** (Figure 9.2). Burrowing animals and extensions of plant roots help in the formation of regolith by physically displacing rock particles and increasing the volume of openings in the regolith. These new openings allow more oxygen and water to penetrate the regolith, increasing the effects of chemical weathering.

Release of confining pressure

Many of the rocks at the earth's surface were formed at considerable depths and later were uncovered by erosion of the overlying rock material. Deeply buried rocks are subjected to high, and equal, pressures on all sides. The

Figure 9.1 Frost-wedged rocks, Uinta Mountains, Utah. *(W. R. Hansen, U. S. Geological Survey)*

Figure 9.2 Rock wedged apart by the growth of tree roots, Sierra Nevada Mountains, California. *(G. K. Gilbert, U. S. Geological Survey)*

vertical pressure lessens progressively as more and more of the overlying rocks are removed by erosion. This release of pressure allows the compressed rock to expand in the vertical direction, forming curved sheets parallel to the earth's surface. The formation of curved sheets of rock by release of pressure is called **exfoliation.** Exfoliation results in the formation of rounded features called **exfoliation domes.** Excellent examples of such features can be seen in Yosemite National Park in California (Figure 9.3).

Salt crystal growth

We noted earlier in this chapter that ice crystals forming within openings in rocks exert sufficient pressure to break off pieces of the rock. In a similar fashion, salts crystallizing in the pores and crevices of surface rock exert a force that is sufficient to break off small chips of the rock, exposing new areas to weathering. This process occurs in dry areas where evaporation of surface water results in salt crystallization in surface openings (Figure 9.4), in coastal areas where rocks are exposed constantly to salt spray, and along some highways where salt used

Figure 9.3 Exfoliation dome in Yosemite National Park. *(F. C. Calkins, U. S. Geological Survey)*

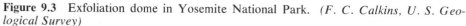

Figure 9.4 Wood post shattered by the growth of salt crystals, Inyo County, California. The original ground level was at the lower edge of the bulge. Water rising in the wood evaporated at ground level and the post was shattered by the salt crystallization. *(C. B. Hunt, U. S. Geological Survey)*

for deicing splashes onto adjacent rocks. Salt crystal growth commonly results in a cavernous and honeycombed rock surface.

Thermal expansion and contraction

Rocks are poor conductors of heat. When the outside surface of a rock is heated by the sun, the ions in the minerals start to vibrate around their structural positions, and the outer few centimeters of the rock expands. The inner part of the rock does not receive much heat, and therefore stresses are built up between the daily expanding and contracting exterior and the unchanging interior. These differences in stress may result in the fracturing and "peeling off" of the outer layers if the outer surface of the rock is heated to very high temperatures, such as in a forest fire.

In addition, each mineral in the rock expands to a different degree upon heating. For example, quartz expands three times as much as feldspar. The different expansive stresses set up between individual minerals may result in the crystals breaking off at the rock's surface.

These thermal effects would be most marked in areas of considerable daily changes in temperature, such as deserts. Shattered angular rock fragments do litter many desert areas, and in places these angular pieces can be reassembled into larger cobbles and boulders. However, in laboratory studies, experimental heating and cooling of rocks in thousands of cycles failed to shatter them, and so currently it is not clear how the process works in nature.

Characteristics of Physically Weathered Materials

Physical weathering processes result in regolith that has coarse grains with angular edges and a composition similar to that of the unweathered rock. These processes also produce a very large increase in the surface area of the weathered material. For example, a rock cube 2 m on each edge has a total surface area of 24 m^2 (Figure 9.5). If physical weathering breaks the rock along three mutually perpendicular fractures, eight cubes, each with 1-m edges, are formed. The total surface area of the rock cubes is 48 m^2, twice the surface area of the original rock cube. In each successive halving of a cube edge, the surface area doubles while the volume remains the same. The smaller rock fragments weather even faster since there is more surface area exposed to both physical and chemical weathering.

CHEMICAL WEATHERING

Chemical weathering consists of a number of reactions in which the ions that make up minerals combine with the ions of the atmosphere and hydrosphere. Some of the ions recombine to form new minerals that replace those originally present; other ions are dissolved and removed in solution from the parent rock or sediment. The factors governing the extent and rate of chemical weathering are the same as those which play a role in increasing the rates of chemical reactions in the laboratory—namely, the presence of water

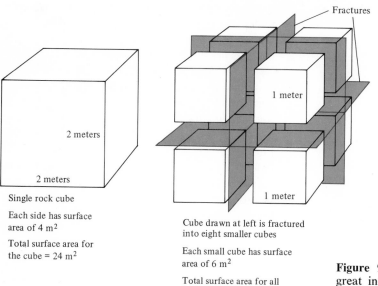

Single rock cube

Each side has surface area of 4 m²

Total surface area for the cube = 24 m²

Cube drawn at left is fractured into eight smaller cubes

Each small cube has surface area of 6 m²

Total surface area for all eight cubes = 48 m²

Figure 9.5 Physical weathering results in a great increase in the surface area of the rock particles, shown here as cubes.

or other solutions, mixing of the reactants, increase in temperature, and increase in the surface area of the reactants.

Water is the most important factor in chemical weathering. In fact, every reaction in chemical weathering involves water either as a reactant or as a carrier of reaction products such as soluble ions. Water is an especially effective weathering agent because it can exist as a fluid, gas, or solid over the temperature ranges at the earth's surface. In humid climates, water may partially or completely fill the fractures and pores in a rock, whereas in drier climates, it may exist only as a thin film coating individual grains. Subsurface water serves both to carry the gases and fluids that attack the rock chemically, and to remove soluble weathering products, enabling chemical reactions to continue toward completion.

Physical weathering increases the effectiveness of chemical decomposition by increasing the surface area exposed to chemical reactions (Figure 9.5). Chemical weathering is more effective in warmer climates because heat increases the rate of most chemical reactions. All other factors being equal, reaction rates at normal temperatures approximately double for each 10°C rise in temperature. Animals and

plant and tree roots make openings through the regolith which allow agents of weathering such as solutions, carbon dioxide, and oxygen to penetrate below the surface. Burrowing animals also move particles of rock and mineral around, exposing new surfaces to chemical reactions.

Chemical Weathering Processes

The most common reactions in chemical weathering involve the addition of water, carbon dioxide, and oxygen to rocks and sediments at the earth's surface, although sulfurous fumes produced by industry recently have been recognized as important factors in chemical weathering near urban centers. The reactions involving water are hydrolysis and hydration, while those involving carbon dioxide and oxygen are carbonation and oxidation, respectively.

Hydrolysis

In Chapter 3 we saw that water is a polar molecule and therefore facilitates melting of silicate minerals to form magma. The polar nature of water also makes it an excellent solvent during chemical weathering, as shown in Figure 9.6. The charged ends of the molecule help break the bonds between sodium and chlorine ions in the halite structure and enable these ions to be removed in solution.

In a given volume of water, some of the molecules can be thought of as having dissociated into their constituent ions, H^+ and OH^-, by the following reaction:

$$H_2O \rightarrow H^+ + OH^-$$

Both these ions exert an electrostatic attraction on ions in mineral structures that is far greater than the attraction exerted by the neutral water molecule; as a result, relatively insoluble minerals can be decomposed. The reaction of the hydrogen, H^+, and hydroxyl, OH^-, ions with minerals is called **hydrolysis**; it is the most important weathering reaction in the decomposition of silicate minerals. When hydrolysis is combined with carbonation, the decomposition of the mineral is accelerated. The weathering of potassium feldspar (orthoclase) is an example of such a reaction:

$$2KAlSi_3O_8 + 2H_2CO_3 + 9H_2O \rightarrow Al_2Si_2O_5(OH)_4 +$$

Orthoclase   Carbonic   Water   Kaolinite
acid

$$4H_4SiO_4 \; + \; 2K^+ \; + \; 2HCO_3^-$$

Silicic   Potassium   Bicarbonate
acid   ion   ion

└────── IN SOLUTION ──────┘

The end products of this reaction are soluble potassium and bicarbonate, HCO_3^-, ions; dissolved silica in the form of silicic acid; and the residual mineral kaolinite. Similar hydrolysis of plagioclase feldspars yields kaolinite and dissolved Na^+ and Ca^{2+} ions. Kaolinite is one of a group of fine-grained sheet silicate clay minerals produced by weathering. Clay minerals are important constituents of the regolith, and some are important in industry. For example, kaolinite is used as a filler and a coating agent in the manufacture of paper, as a thickener for foods, and as a raw material for ceramics.

Hydration

Chemical weathering of some minerals takes place by the addition of water molecules to the mineral structure, a process called **hydration**. For example, the evaporite mineral

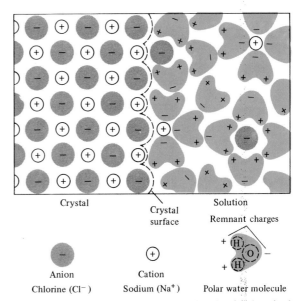

Crystal   Solution
          Crystal
          surface   Remnant charges

Anion   Cation
Chlorine (Cl^-)   Sodium (Na^+)   Polar water molecule

Figure 9.6 Role of polar water molecules in chemical weathering. Because of the remnant charges on the water molecule, water is able to neutralize charges at the mineral surface and take ions into solution.

anhydrite, $CaSO_4$, is converted by hydration to gypsum, $CaSO_4 \cdot 2H_2O$, by the following reaction when water is available:

$$CaSO_4 + 2H_2O \rightarrow CaSO_4 \cdot 2H_2O$$

Anhydrite   Gypsum

This transformation involves an increase in volume to accommodate the water in the mineral structure; this change commonly causes distortion and deformation in the resulting gypsum crystals. The opposite, but related, process of **dehydration** also occurs in nature. Dehydration involves the removal of water from the mineral structure and sometimes leads to the formation of anhydrite from gypsum.

Carbonation

Carbon dioxide in the atmosphere combines with water to form the weak acid H_2CO_3 (carbonic acid) by the reaction

$$H_2O + CO_2 \rightarrow H_2CO_3$$

The regolith around plant roots is enriched in carbon dioxide by plant respiration to levels 10 to 1000 times the normal atmospheric concentration. As rain soaks through this regolith, large quantities of carbonic acid are formed.

The decomposition of minerals by reaction with carbonic acid is called **carbonation.** It is particularly effective in the solution of limestones because calcite, the major mineral in limestone, is dissolved by carbonic acid by the reaction

$$CaCO_3 \;+\; H_2CO_3 \rightarrow Ca^{2+} \;+\; 2HCO_3^-$$

Calcite   Carbonic   Calcium   Bicarbonate
(calcium carbonate)   acid   ion   ion

Whenever limestone occurs at the surface in areas of high rainfall and organic activity, it dissolves rapidly.

Dissociation of carbonic acid to form the hydrogen and bicarbonate ions also frees some H^+ ions for hydrolysis reactions as described earlier:

$$H_2CO_3 \rightarrow H^+ \;+\; HCO_3^-$$

Carbonic   Hydrogen   Bicarbonate
acid   ion   ion

Oxidation

Anyone who leaves tools outdoors in moist air will soon notice that they rust. The originally dark steel becomes covered by a yellowish brown or reddish substance that we call rust, which actually is a hydrated iron oxide. The process of rusting is basically a combination of oxidation and hydration, and it is a common weathering reaction.

Oxidation is the combination of oxygen ions with cations such as iron, magnesium, calcium, sodium, and potassium. Among the most common oxidation reactions are those involving the ferromagnesian minerals in which a reddish or yellow-green coating replaces the original mineral partially or completely. The oxidation of brass-yellow pyrite $-FeS_2$, iron sulfide $-$ in the presence of water is an example of this process. Pyrite first combines with oxygen and water to form ferrous iron, Fe^{2+}, ions:

$$2FeS_2 + 7O_2 + 2H_2O \rightarrow 2Fe^{2+} + 4SO_4^{2-} + 4H^+$$

Pyrite   Ferrous
(iron sulfide)   ion

If more oxygen is available, the ferrous iron ions are oxidized further to form ferric iron, Fe^{3+}, ions:

$$4Fe^{2+} + O_2 + 4H^+ \rightarrow 4Fe^{3+} + 2H_2O$$

Ferrous   Ferric
ion   ion

Figure 9.7 Spheroidal weathering in basalt, Puerto Rico. *(C. A. Kaye, U. S. Geological Survey)*

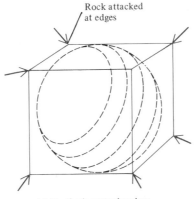
Rock attacked at edges

(a) Weathering attacks edges of cube rounding them quickly to form a sphere

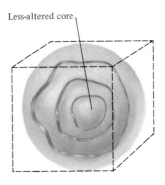
Less-altered core

(b) Successive layers of the sphere slake off as chemical and physical weathering continue

Figure 9.8 Formation of spheroidal weathering. *(a)* Chemical weathering starts on the rock edges first because they have three surfaces to attack. *(b)* Progressive weathering reduces the rock cube to a sphere, the geometric form with the lowest surface area for a given volume.

The ferric ion then combines with oxygen and water to form the mineral limonite, $HFeO_2$, the substance we normally call rust.

Spheroidal Weathering

Massive and uniform rocks such as granite and basalt commonly weather into curving "peels" of crumbly altered minerals. This is sometimes referred to as "onion-skin" weathering because of its appearance (Figure 9.7), but it is more properly referred to as **spheroidal weathering** because the weathering processes tend to form spheroidal boulders of relatively unweathered material. Spheroidal weathering is similar in appearance to the exfoliation that we described previously but differs from it both in scale and in origin. Exfoliation is a physical process. The exfoliation sheets are relatively unweathered and quite large, sometimes covering whole mountains (see Figure 9.3). In contrast, spheroidal weathering is produced by chemical decomposition of the rock; it creates highly weathered crumbly outer sheets, and it operates on a much smaller scale (less than 10 m). Chemical weathering of a block of rock is most effective at the corners and edges. At these locations, weathering proceeds from two or more directions, and the edges and corners become rounded quickly. This is analogous to the

melting of an ice cube, in which the corners become rounded early in the melting process (Figure 9.8). As spheroidal weathering proceeds from the outside inward, new minerals form and the resulting curved weathering "rind" breaks away, exposing new rock to weathering.

Mineral Stability in Chemical Weathering

Minerals have differing susceptibilities to weathering under equivalent conditions. The relative weatherability of common rock-forming minerals has been determined from numerous field and laboratory studies and is known as the **Goldich Stability Series** (Figure 9.9). A comparison of the Goldich Stability Series with Bowen's Reaction Series (Figure 4.19) shows the relationship between conditions of formation of minerals and their weathering susceptibilities. Minerals formed at higher temperatures and pressures tend to be less stable in a weathering environment than those formed at lower ones. Thus, if an igneous rock includes both olivine and pyroxene, the olivine crystals would be more susceptible to decomposition than the pyroxene crystals under the same weathering conditions. Because of its high susceptibility to weathering, olivine very rarely occurs in regolith. The higher chemical stability of quartz, on the other hand, along with its abundance, makes it the most common mineral in regolith derived from quartz-

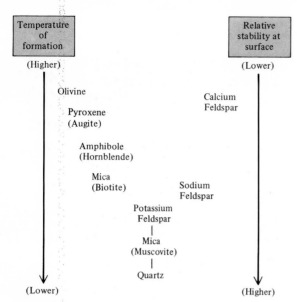

Figure 9.9   Relative stability of minerals in chemical weathering. Those minerals which form at conditions most different from those at the earth's surface are the most easily weathered.

bearing rocks and one of the major minerals in sedimentary rocks (see Chapter 6).

Products of Chemical Weathering

Rocks subjected to chemical weathering yield a number of different products. Some minerals, such as quartz, are quite stable and in temperate areas survive weathering virtually unaltered. Unaltered minerals present in the original rock are referred to as **primary minerals.** Most other minerals are broken down to form new minerals. For example, feldspars break down to form clay minerals, while biotite mica breaks down to form iron oxides, such as hematite and limonite, and clay minerals. New minerals formed by weathering processes are called **secondary minerals.** A considerable portion of the original rock is broken down into dissolved ions.

What happens to these different weathering products? Some of the primary and secondary minerals may remain in the weathering area and accumulate as residual regolith. The rest may be transported out of the area by wind, water, glacial ice, or gravity. These transported sediments eventually accumulate in a basin of

deposition and ultimately are transformed into sedimentary rocks (Chapter 6). Primary and secondary minerals such as quartz, feldspar, clay minerals, and iron oxides make up most of the minerals in clastic sediments and sedimentary rocks. Dissolved ions are not so visibly obvious a product of weathering as primary and secondary minerals, but they are extremely important in both geologic and biologic processes. Dissolved ions may be utilized within the source area or transported out of it by surface and subsurface waters. One of the major effects of chemical weathering is that it frees elements from rocks so that they can be utilized by plants as essential mineral nutrients. Plants remove a portion of the dissolved ions through their root systems. In dry areas, dissolved ions may be deposited as a result of the evaporation of surface and subsurface water (Chapter 6). The remaining dissolved ions are carried out of the area and accumulate elsewhere, commonly in the ocean basins. These dissolved ions are utilized in the formation of the shells of marine organisms, chemical sedimentary rocks, and the cements holding many clastic rocks together (Chapter 6).

Weathering reduces the large number of minerals in igneous, metamorphic, and sedimentary rocks to a relatively small number of minerals that are stable under conditions at the earth's surface (Figure 9.10). Residual regolith is made up largely of organic material, quartz, clay minerals, and iron oxides along with partially weathered rock. The changes in atomic structure that accompany the formation of secondary minerals result in changes in physical properties. The secondary minerals are generally less dense and have a different luster, and many are softer than their unweathered predecessors.

Sometimes "unstable" high-temperature minerals are found in significant amounts in sediments or sedimentary rocks. These rare occurrences are the result of any or all of the following conditions:

1. The mineral was very abundant in the source rocks so that more mineral grains managed to survive chemical and physical attack during weathering, erosion, and transport.

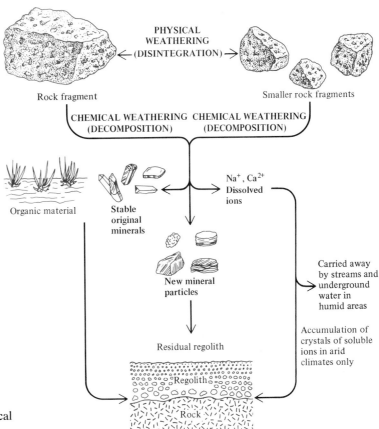

PHYSICAL
WEATHERING
←(DISINTEGRATION)→

Rock fragment   Smaller rock fragments

CHEMICAL WEATHERING CHEMICAL WEATHERING
(DECOMPOSITION)   (DECOMPOSITION)

Organic material

Stable
original
minerals

Na^+, Ca^{2+}
Dissolved
ions

New mineral
particles

Carried away
by streams and
underground
water in
humid areas

Accumulation of
crystals of soluble
ions in arid
climates only

Residual regolith

Regolith

Rock

Figure 9.10 Products of chemical
and physical weathering.

2. The source rocks were nearby so that there
was not enough time to alter their minerals
during erosion and transportation.

3. The source rocks were not nearby, but the
weathered material was so rapidly eroded,
transported, deposited, and buried that
complete weathering could not take place.

4. The rock or sedimentary layers were exposed for an insufficient time for appreciable
weathering to occur.

5. The climate was dry or cold so that chemical
weathering was not effective.

Weathering of Granite and Basalt

As an example of differences in weathering
products we can consider what happens to two
common rock types, granite and basalt, when

they weather in a warm humid climate (Figure
9.11). Both granite and basalt contain feldspars
and dark ferromagnesian (iron- and magnesium-
bearing) minerals, whereas only granite contains
significant quantities of quartz. The residual
regolith on the granite will contain abundant
quartz, clay minerals from the decomposition
of sodium and potassium feldspars, and iron
oxides derived from the weathering of ferromagnesian minerals such as amphibole and
biotite. The residual regolith on the basalt will
also contain clay minerals derived from the calcium feldspars, and abundant iron oxides derived from the decomposition of pyroxene or
amphibole, but will not contain significant
quantities of quartz. After a long period of
weathering, the residual regolith developed on
the two rocks will appear to be a similar reddish
clayey material but will differ in that the granitic

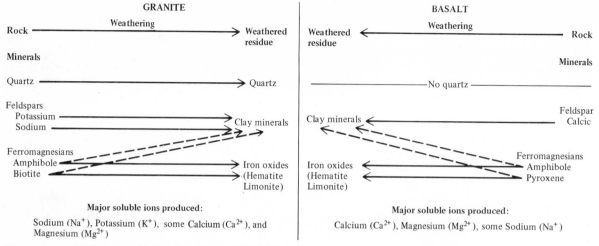

Figure 9.11   Differences in the weathering of granite and basalt in a warm humid climate. Although quite different in original mineral composition, both rocks weather to form the same secondary minerals. The major difference between the two residual regoliths is that quartz occurs only in the granite regolith. This is because quartz is present only in the granite and is also resistant to weathering.

regolith will contain quartz, whereas the basaltic regolith will not since quartz is usually absent in the original basalt.

Differential Weathering and the Molding of Landscapes

Weathering, along with winds, rivers, ocean currents, and glaciers, plays a major role in carving the landscape. Differential weathering of two adjacent rocks leads to significant differences of elevation. In a temperate humid climate such as that of Pennsylvania, limestone and sandstone areas once had similar summit elevations. But, with time, the limestone areas became progressively lower as the calcite in the limestone dissolved in the humid climate. Therefore, relatively resistant quartzose sandstones now are viewed as ridges above the less-resistant limestone lowlands (Figure 9.12).

In a dry climate, limestone deposits may underlie topographically high areas because the scarcity of water greatly inhibits the solution of the limestone. For example, in the southwestern United States limestone stands up in steep cliffs because chemical weathering is minimal in this dry climate (Figure 9.13).

SOIL

The term **soil** is used to describe loose material at the earth's surface that has been weathered sufficiently so that it can support the growth of rooted plants. Soils are derived from the physical and chemical breakdown of a variety of parent materials such as rocks, transported regolith, or even an older soil.

Soil Formation and Zonation

How does a soil begin to form on bare rock? To illustrate the process, we will describe the formation of a soil on a granite exposed in a humid temperate climate such as that of southern New England. Only the most primitive plants, such as lichens, can grow on a bare rock surface because the lichens have limited needs for mineral nutrients and can obtain them *directly* from the rock. Secretions released by the lichens etch the rock surface, loosening mineral particles and producing small amounts of mineral nutrients. These mineral particles, along with atmospheric dust and organic debris from dead lichens, accumulate in rock crevices as small pockets of soil. Spores and seeds of higher plants then may gain a foothold in these thin soil pockets be-

Figure 9.12 Differential weathering in Pennsylvania. The ridge in the distance is underlain by sandstone, whereas the low areas are underlain by limestone and shale. The limestone is especially susceptible to solution in such a humid and temperate climate. *(G. W. Stose, U. S. Geological Survey)*

Figure 9.13 Limestone underlying a ridge, Nye County, Nevada. In a dry climate limestone is less susceptible to solution. (Compare with Figure 9.12.) *(E. H. McKee, U. S. Geological Survey)*

cause weathering has released the mineral nutrients necessary for their growth. The growth of the plants and trees accentuates the physical and chemical weathering of the rock, and the soil thickness grows steadily with time so long as none of the soil is removed by surface erosion.

With the passing of time, weathering products accumulate in some parts of the regolith and are depleted in others. This leads to the development of layers called **soil horizons,** which can be distinguished on the basis of organic content, color, mineralogy, grain size, and percentage of unweathered parent material. Soils which have been forming for a sufficiently long time to develop distinct horizons are called **mature soils.**

Five major soil horizons are recognized. Starting with the uppermost layer, these are referred to as the O, A, B, C, and R horizons (Figure 9.14).

The uppermost layer is the **O horizon,** characterized by the accumulation of organic material and the absence of mineral matter. The upper portion of this organic horizon is made up of fresh plant matter, whereas the lower part contains plant matter that is more highly decomposed. The organic accumulations in the O horizon give it a dark brown color. The next layer is the **A horizon,** composed largely of mineral particles and organic material. In humid climates, percolating waters charged with carbon dioxide dissolve any calcium carbonate in the A horizon. The removal of soluble materials such as calcium carbonate is called **leaching,** and for this reason the A horizon is sometimes referred to as the **leached horizon.** The soluble carbonates, along with clay minerals and iron oxides, are transported by water moving down through the soil and accumulate in the underlying **B horizon,** sometimes referred to as the **accumulation horizon.** The concentrations of iron oxide in the B horizon commonly give it a yellowish to red color. The **C horizon** is composed of partially weathered parent material. At the top of the C horizon is highly weathered rock or sediment which may still preserve some of the characteristics of the parent material. The lower portion of the C horizon contains distinct particles of the parent material. The **R horizon,** the unweathered parent material, forms the base of the soil profile.

Factors in Soil Formation

Soils in adjacent areas may have very different characteristics because of the numerous factors that determine what type of soil may develop over a given area. In general, soils develop much more quickly on unconsolidated sediments than on bedrock. The major factors that determine soil characteristics are parent material, climate, biologic activity, topography, and elapsed time. All these factors operate simultaneously, and one can offset the effects of the others in some cases. For purposes of discussion we will consider each as if it acted alone.

Parent material

The parent material controls soil mineralogy to the extent that all the soil minerals either must come directly from the parent material (primary minerals such as quartz) or must be derived from those original minerals by weathering (secondary minerals such as iron oxides and clay minerals).

The relative percentages of minerals in a soil may be very different from those in the original rock. The weathering of an impure limestone in a humid climate provides an example of this. Many limestones are not pure calcium carbonate but contain small percentages of quartz sand and clay minerals. Limestones exposed in humid areas are dissolved readily by water containing carbon dioxide (carbonic acid), yielding soluble ions that are removed by surface and subsurface waters. Consequently, if the limestone is impure, the residual regolith is enriched greatly in percentages of minerals resistant to weathering, such as quartz and clays, which make up only a small percentage of the original rock.

Climate

Climate ranks with parent material as one of the most important factors in soil development. Climate determines the availability of the water which is essential to chemical weath-

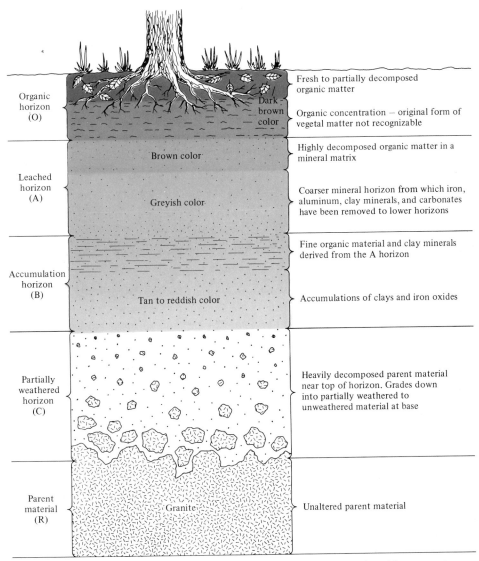

Figure 9.14 Generalized soil profile developed on a granite in a humid temperate area such as the middle Atlantic states or southern New England.

ering and also the temperature at which chemical reactions occur. In areas where rainfall is greater than 50 to 60 cm per year, calcium carbonate and other soluble salts are dissolved by carbonic acid and removed from the soil, while clays and iron oxide minerals accumulate in the B horizon. A soil that does not contain calcium carbonate and that has accumulations of iron oxide and clay minerals in its B zone is called a **pedalfer.** A typical pedalfer is shown in Figure 9.14. Pedalfers are common in the eastern United States (Figure 9.15). When rainfall is less than 50 to 60 cm per year, there is insufficient water to form the carbonic acid necessary to dissolve calcium carbonate and remove it from the soil. A soil in which there are accumulations of calcium carbonates is called a **pedocal.** An example of a pedocal is desert soil, which is enriched in a number of soluble minerals. Pedocals are common in the

Figure 9.15 Approximate boundary between pedocal and pedalfer soils in the United States.

Figure 9.16 Caliche (light zone near hammer) in volcanic soil near Pachuca, Hidalgo, Mexico. *(K. Segerstrom, U. S. Geological Survey)*

western United States (Figure 9.15). Sometimes the calcium carbonate deposition in the upper part of the soil forms a hard crust, commonly referred to as **caliche** (Figure 9.16).

Quartz and clay minerals may be stable end products of weathering in temperate climates, but even they can be broken down further in tropical climates with high rainfall. A wet tropical climate results in such intense chemical weathering that quartz may be dissolved and silica may be removed from the clay minerals. The soluble silica is washed out of the soil by the heavy rainfall, and the residual regolith becomes enriched in the oxides and hydroxides of iron or aluminum. A soil which is enriched in oxides and hydroxides of iron and aluminum is called a **laterite.** A laterite is actually a type of pedalfer. The type of metal concentrated in the laterite is determined by the composition of the parent rock. For example, tropical weathering of a basalt would enrich the regolith in iron, whereas tropical weathering of a granite would enrich it in aluminum. The major source of the world's aluminum is an aluminous laterite called **bauxite** (Figure 9.17).

Biologic activity

Plants and animals play a great role both in initiating soil formation and in aiding its development. We saw earlier in this chapter that simple plants such as lichens can grow on bare rock surfaces, extracting mineral nutrients from the unweathered rock. With time, they form a thin soil which enables higher plants, such as grasses and trees, to become established. The root system of these higher plants, along with the burrows of animals, provide passages through which oxygen and solutions can move. Vegetation holds the soil together, preventing its erosion and retaining moisture around the root systems between rainfalls.

The types of vegetation, such as grasses or trees, also influence the type of soil that will develop. Grasses have numerous small roots which penetrate the soil thoroughly, whereas trees have few and massive roots. Dead grass material is incorporated easily into the soil, resulting in a higher organic content in soils of a grassland region than in soils of a forested region. This is the mechanism by which the productive soils of the American Midwest were produced.

Topography

In a humid region underlain by the same parent material, the soil developed on the steep slopes (uplands) can be quite different from the soil which forms on adjacent areas of gentle slope. The upland soils generally have a lower organic content, are thinner, and are coarser-grained than the soils which form on the adjacent area of gentle slope. The upland soils do not develop thick organic accumulations because organic matter has less opportunity to accumulate before it is removed by erosion. The continuous erosion on the upland slopes may remove soil almost as fast as it forms, resulting in a thinner soil on the steep slopes. Rain falling on steeper slopes has a greater tendency to run off along the slope than to sink in. This increased surface drainage removes the finer particles preferentially, coarsening the upland soils.

Elapsed time

All other factors being equal, soil particle decomposition, soil horizon development, and soil thickness increase with the passing of time. This principle has considerable application in determining the relative ages of surficial de-

Figure 9.17 Deeply weathered boulder of felsic igneous rock in bauxite (Arkansas). The central part of the boulder is less-weathered rock which grades outward into a light-colored layer of kaolinized bauxite which in turn grades out into dark bauxite. *(M. Gordon, Jr., U. S. Geological Survey)*

Weathering on the Moon
A Planetary Perspective

How would rocks and regolith weather if there were no water, oxygen, or biologic activity? The examination of the samples returned during the Apollo lunar missions has provided our first answers to that question.

Weathering on the moon is entirely a process of physical disintegration. Rocks (and regolith) exposed at the lunar surface are broken down constantly and reworked by meteorite impacts. The size of these meteorite impact craters ranges from those large enough to be seen from earth to the micrometeorite impacts made on *individual mineral crystals* (Figure 1). Larger meteorite impacts excavate deeper and larger craters, throwing fresh, coarse-grained rock fragments onto the lunar surface. The smaller meteorite impacts rework the regolith continually, mixing it and breaking down the rock and mineral fragments progressively.

The great heat generated by meteorite impacts is sufficient to fuse the local rocks or

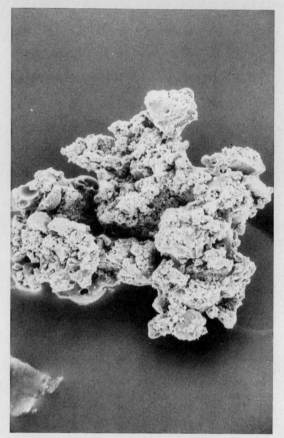

Figure 2 Agglutinate particle. *(NASA)*

Figure 1 Micrometeorite impact on a lunar mineral particle surface. Crater is about 25 micrometers across. *(NASA)*

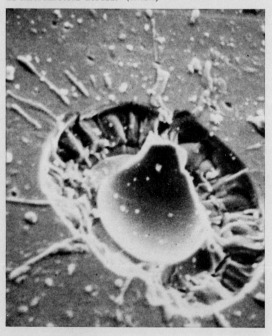

regolith into dark-colored molten glass. This glass may "splash" onto nearby rocks, forming glass-coated boulders. Smaller glass blobs cool into distinct glass droplets. Micrometeorites hitting the regolith fuse particles together into uniquely lunar soil aggregates called **agglutinates** (Figure 2). Thus, the lunar regolith is composed of rock fragments, mineral fragments, glass droplets, agglutinates, and metallic particles.

Lunar soils undergo distinct changes with time. Immature (younger) soils are lighter in color and are composed of a poorly sorted mixture of rock and mineral particles along with a few agglutinates. With the passing of time, continual meteorite impacts break down the rock and

Figure 3 Poorly sorted lunar soil. Note the shiny glass-splattered surfaces on some of the rock particles. *(NASA)*

mineral particles. This forms a finer and better-sorted regolith. As more and more agglutinates form, the soil gets progressively darker. Thus mature (older) lunar soils are darker, finer-grained, and better-sorted. The soil evolutionary sequence we have described is an *ideal* one because deeper meteorite impacts throw fresh, coarse rock onto the lunar surface, mixing it with the finer mature regolith and starting the soil evolutionary process again (Figure 3).

posits in geologic field studies, such as the different glacial deposits in the northern Sierra Nevadas. The degree of decomposition of boulders in several glacial deposits has been used to determine the relative ages of these deposits. The oldest deposits have more deeply weathered rock fragments, whereas progressively younger deposits contain less-weathered particles. Similar studies in different areas have utilized other aspects of soil development, such as relative soil thickness, horizon development, and percentages of easily weatherable minerals, to determine the relative ages of surficial geologic deposits.

Soil Classification

Differences in parent material, climate, biologic activity, topography, and duration of soil formation result in a wide variety of soil types. Prior to 1960, these soil types had been given descriptive names, such as pedocals and pedalfers, which we discussed earlier in this chapter. These different soil types are referred to as the **Great Soil Groups** and are summarized in Table D.1 of Appendix D. They are discussed in detail in Hunt's *Geology of Soils*, which is listed in the Additional Readings at the end of this chapter.

As more was learned about soils, it became apparent to soil scientists that the older classification was inadequate. Many of the terms were imprecise and had little quantitative basis. In addition, many of the terms were subjective because specific origins were implied in the soil names themselves. In 1965, the U.S. Department of Agriculture adopted a totally new soil classification that subsequently has been adopted by many state and foreign soil surveys. The new classification, called the **Seventh Approximation,** is based on observable properties and is quantitative and detailed. Ten major orders of soil are recognized, and these are subdivided into 47 suborders, which can be subdivided further. Detailed descriptions of each of the 10 major soil orders in the new classification, along with their approximate equivalents in the Great Soil Groups, is given in Table C.2 in Appendix C at the end of the book.

SUMMARY

Weathering is a combination of physical and chemical changes that occur in rocks and sediments as they come into contact with the atmosphere, hydrosphere, and biosphere. Physical weathering breaks down rocks into smaller rock fragments with no change in mineralogy. Major physical weathering processes and agents include frost wedging, release of confining pressure, animals and plants, crystal growth, and thermal expansion and contraction. One major effect of physical weathering is the increased surface area of the weathered debris, which increases the effects of chemical weathering.

Chemical weathering decomposes rocks into primary minerals resistant to weathering, secondary minerals produced from the original rock minerals, and soluble ions. The insoluble components of chemical weathering accumulate as residual regolith. In humid areas, soluble ions are removed from the regolith by surface and subsurface waters and accumulate elsewhere, generally in the ocean. In dry areas, soluble salts may accumulate within the regolith. Minerals have different susceptibilities to chemical weathering. In general, minerals formed at higher temperatures are more susceptible to chemical weathering than those formed at lower temperatures. The differential weathering of adjacent rocks is a major agent in the formation of the topography of the earth's surface. Chemical weathering reduces the large number of minerals in igneous, metamorphic, and sedimentary rocks into a residual regolith composed largely of quartz (if present in the parent material), clay minerals, and iron oxides.

One of the major effects of weathering is the production of soils. Soil is a residual surficial material which has been weathered sufficiently to support plant growth. Weathering processes form a series of soil horizons. From the surface down, these horizons are: O (organically rich), A (mineral material leached of soluble components), B (accumulation of material removed from A), C (partially weathered parent material), and R (unweathered parent material). The major components in soil formation are mineralogy of parent material, climate, biologic activity, topography, and elapsed time. The two great soil families are the pedalfers and the pedocals. The pedalfers occur in areas where the rainfall is sufficient to remove most soluble minerals. The pedocals occur in dry areas and are rich in soluble minerals.

Weathering is important in the recycling of elements; the preparation of rock material for transport by wind, water, and ice; the formation of agricultural soils; the development of the topography on the earth's surface; and the concentration of iron and aluminum in economically significant amounts.

QUESTIONS FOR REVIEW AND FURTHER THOUGHT

1. Contrast the end products of physical and chemical weathering.

2. How does physical weathering facilitate chemical weathering?

3. Assuming complete weathering in a warm humid climate, describe the residual regolith which would form on the following rocks:
  a. Granite
  b. Basalt
  c. Quartz sandstone with calcium carbonate cement
  d. Impure limestone

4. Show why water is so important in both physical and chemical weathering.

5. Describe how weathering aids in the formation of landscapes in (a) wet and (b) dry climates.

6. How do each of the following influence the character of soil?
  a. Parent material
  b. Climate
  c. Organic activity
  d. Time
  e. Slope

7. Describe the differences between the O, A, B, C, and R soil horizons.

ADDITIONAL READINGS

Beginning Level

Likens, G. E., and F. H. Bormann, "Acid Rain: A Serious Regional Environmental Problem," *Science*, vol. 184, 1974, pp. 1176–1179.
(A description of the effects of acid rain on the environment)

Lockeretz, W., "The Lessons of the Dust Bowl," *American Scientist*, vol. 66, September-October 1978, pp. 560–569.
(An excellent description of how drought and poor farming methods led to extensive soil erosion. Includes historical development and predictions for the future as well as a detailed list of references for further study)

McNeil, M., "Lateritic Soils," *Scientific American*, vol. 211, no. 5, November 1964, pp. 96–102.
[Origin and distribution of lateritic soils (oxisols). Describes the effects of clearing tropical forests for agriculture]

U.S. Department of Agriculture, *Know the Soil You Build On*, Soil Conservation Service, Agricultural Information Bulletin 320, 1976, 13 pp.
(Well-illustrated and practical treatment of those aspects of soil which are related to construction. Includes a description of soil maps and surveys)

Intermediate Level

Bloom, A. L., "Rock Weathering," in *Geomorphology: A Systematic Analysis of Late Cenozoic Landforms*, Prentice-Hall, Inc., Englewood Cliffs, N.J., chap. 6, pp. 103–135, 510 pp.
(Good overall discussion of weathering and soils. Extensive list of journal articles for further study)

Hunt, C. B., *Geology of Soils: Their Evolution, Classification, and Uses*, W. H. Freeman, San Francisco, 1972, 344 pp.
(Reviews principles of geology and their relation to soil development. Has an annotated index and glossary. A good reference for starting a study of soils)

Advanced Level

Birkeland, P. W., *Pedology, Weathering and Geomorphological Research*, Oxford University Press, New York, 1976, 285 pp.
(A detailed treatment of weathering and soils)

McKay, D. S., R. M. Fruland, and G. H. Heiken, "Grain Size and the Evolution of Lunar Soils," *Proc. 5th Lunar Science Conference*, vol. 1, 1974, pp. 887–906.
(A detailed discussion of the factors governing grain size and sorting in lunar soils. Includes a description of the changes which occur in lunar soils with time as well as an extensive bibliography)

Soil Survey Staff, *Soil Taxonomy: A Basic System of Soil Classification for Making and Interpreting Soil Surveys*, U.S. Department of Agriculture Handbook 436, 1975, 754 pp.
(A detailed description and explanation of the new soil classification)

CHAPTER 10

Mass Wasting

Cliffs along Lake Michigan collapse after a severe storm erodes their bases, the coastal highway south of San Francisco is closed by mud slides, heavy winter rains are followed by flows of mud moving down canyons into the Los Angeles Basin, and steeply inclined sheets of granite slide off a mountain onto a highway in New Hampshire. Events such as these are among the most spectacular geologic processes. Geologists use the term **mass wasting** for the process by which masses of rock or sediment are moved down slopes by the force of gravity.

Mass movements of one kind or another are active in all surface environments and in some parts of the oceans. They are important in moving rock and regolith down slopes and into transportational systems such as coastal currents, streams, and glaciers, which can move sedimentary particles great distances. Mass movements differ from other transportational systems in several ways. Although mass wasting processes move rock and regolith, the material is transported only a short distance. Material is carried not as individual particles, as in water and air, but as *masses* of particles or slabs of rock.

In some mass wasting processes, the material is moved downslope so fast that the initial roar of the masses of rock ripping loose from the slope is followed in a minute or less by the particles cascading down onto the land at the base of the slope. Other types of mass wasting processes are slower, with the regolith deforming into bulging masses which flow downslope as lobes of debris over a period of days, weeks, or months. Still other types of mass wasting are so slow that the only way to tell that downslope movement is occurring is by noting that originally vertical features such as trees, utility poles, and fences have been tilted downslope over a period of years.

Why do the materials on some slopes move downslope rapidly, whereas those on other slopes move so slowly that movement can be documented only over a long period of time? This question may be approached by comparing those factors which tend to aid downslope movement, the **driving forces,** with those which act against downslope movement, the **resisting forces.**

FACTORS IN MASS MOVEMENTS

Some of these factors, primarily gravity, are active driving forces. Others are passive, weakening the material so that active driving forces can move it downslope more easily. In some cases, a factor such as water content can be a driving force under one set of conditions and a resisting force under a different set of conditions. We start our discussion with gravity because it is the major driving force in all mass movements.

Gravity and Its Effects on Materials on Slopes

Rock or regolith on slopes has the potential energy to move downslope. This potential energy is derived from the force of gravity acting on the materials. The way gravity acts can be seen most easily by considering the forces acting on a single rock particle on slopes of different inclinations (Figure 10.1). The force of gravity, G, is directed toward the center of the earth. This gravitational force may be considered as having two components, one acting parallel to the slope, G_t, and one acting perpendicular to the slope, G_p. On gently inclined slopes, such as in Figure 10.1a, G_p is greater than G_t and the particle does not move. At some intermediate inclination, G_t equals G_p and the particle is on the verge of moving (see Figure 10.1b). A slope which has the potential for mass movement is called an **unstable slope**. At any higher inclination, G_t is greater than G_p, and there is a tendency for the particle to move downslope (see Figure 10.1c). Frictional forces (F in Figure 10.1) act upslope and resist the movement of the particle until the gravitational component along the slope is greater than the frictional force resisting the downslope movement. At this point the rock or regolith can break loose and move downslope. The actual movement of slope materials is referred to as **slope failure**. The steeper the slope, the greater the tendency for particles to move down it. As the particles start to move downslope, the gravitative potential energy is transformed into kinetic energy.

Orientation of Layers of Rock and Sediment

The orientation of layers of rock and sediment is also an important factor in the stability of slopes. We mentioned before that steeper slopes tend to be more unstable than gentler ones (Figure 10.1). Actually, many steep slopes can be quite stable so long as the layers of rock or sediment are inclined *away* from the slope. Rock layers inclined toward the slopes decrease the stability of the slopes. Thus, the slope stability on opposite sides of a highway cut may be quite different (Figure 10.2a). Sometimes the rock layers are inclined away from the slope but fractures within the rock are inclined toward the slope, making it quite unstable (Figure 10.2b). Thus, detailed geologic studies are necessary before excavations are made for highways, railroads, or buildings in order to determine the potential for slope failure.

Water

Water is extremely important in many mass movements, but its role is complex. The water may act directly as a driving force by increasing the weight of the sediment or rock, or it may act passively to decrease the strength of the rock or sediment by reducing friction and cohesion between particles. On the other hand, small

Figure 10.1 Effects of gravitational forces on particles resting on slopes of different inclinations.

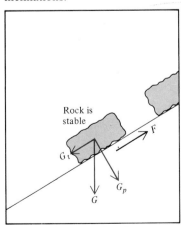

(a) G_t is less than G_p

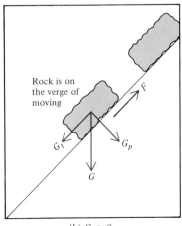

(b) $G_t = G_p$

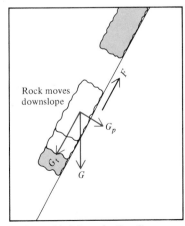

(c) G_t is greater than G_p

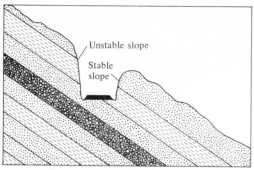
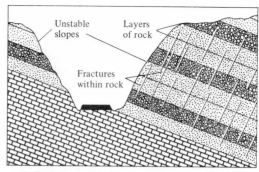

(a) Rock layers are inclined toward (unstable slope) and away from (stable slope) the highway on different sides of the highway cut

(b) Rock layers are inclined away from the highway cut but the fractures within the rocks slope towards the highway cut

Figure 10.2 Slope stability depends on the orientation of rock and sediment layers. Wherever layers or fractures are inclined toward a slope, that slope is unstable and subject to mass movements.

quantities of water in a sediment enable it to resist mass movements.

Loading

Water acts as a driving force by increasing the weight of slope material. A stable mass of dry sand exposed on a slope can have up to about 35 percent of its volume composed of dry pores. These pores may be filled completely with water after a prolonged period of rain. The addition of the water filling the pores increases the weight of the sediment significantly and thus increases the force driving the sediment down the slope. Filling of pores and other openings in rock and sediments by water is one of the major reasons why many mass movements occur during and shortly after intense rainfalls.

Reducing rock strength

Water freezing in rock crevices and pores loosens the rock material by frost wedging (see Chapter 9), which breaks up the rock and enables it to move downslope more easily. Water circulating through the pores or rocks can dissolve cementing materials, reducing cohesion and friction and enabling the grains to move more easily.

Expanding clays

Certain clayey materials called **bentonite** have the ability to absorb great quantities of water, "swelling" to as much as eight times their original volume in the process. Bentonite is com-

posed of fine-grained clay minerals (chiefly of the montmorillonite group) and is formed by the chemical alteration of glassy igneous rocks, such as volcanic ash and tuff, either in weathering or in diagenesis.

The swelling clays may exert pressure on the material above if they absorb water in a confined space. For example, in Austin, Texas, bentonitic clays of the Eagle Ford Formation swell when wet, cracking structures and weakening foundations. During dry periods, the bentonite layers dry out and contract. This forms large vertical cracks in the surface which constitute an additional environmental problem.

If the water-saturated bentonite is on a slope, the potential for slope failure is increased greatly. Swollen, water-saturated clays exposed on slopes have a reduced strength and are more easily subject to mass movements. In addition, they form a "slimy" lubricated surface which reduces friction, facilitating the downslope movement of any layers of rock and sediment above. Expanding clays are also a problem on slopes in cold regions. Bentonite-rich sediments exposed on slopes subject to seasonal freezing and thawing may start to flow when the surface sediments thaw and absorb surface water during the warm seasons.

Liquefaction of clays

Some clays can be transformed quickly from solids to liquids under certain conditions. Clays which can be liquefied quickly are called **quick-**

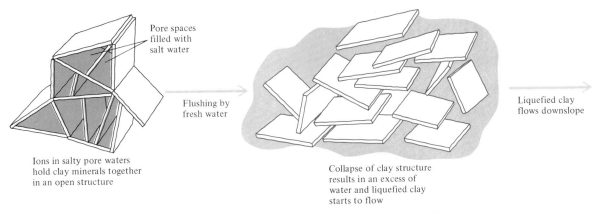

Pore spaces
filled with
salt water

Flushing by
fresh water

Liquefied clay
flows downslope

Ions in salty pore waters
hold clay minerals together
in an open structure

Collapse of clay structure
results in an excess of
water and liquefied clay
starts to flow

Figure 10.3 Formation of quickclay as freshwater flushes out saltwater from the pore spaces between clay minerals.

clay. One type of quickclay was formed origi-nally from clay minerals that accumulated in salt water. The ions in the salty pore waters held clay minerals together as aggregates, forming an open "house-of-cards" structure (Figure 10.3). When such clays are exposed by erosion, they are subject to the subsurface flow of freshwater through their pores. This freshwater flushes out the salt water and the ions that held the open clay structure together. The formerly solid clay becomes unstable. If the slope is subject to any vibrations, the clay structure collapses and the clay is transformed quickly into a viscous fluid, which flows downslope. Quickclay of this type is a serious environmental problem in the St. Lawrence River Valley (Figure 10.4).

Cohesive forces

You might think that the water in the pores between the rock and sediment particles acts as a lubricant which facilitates mass movements. Actually, *so long as the pores are not filled com-pletely,* water films make the particles cohesive. **Cohesion** is the ability of particles to attract and hold each other. The thin films of water lining the pores develop surface tension. **Surface ten-sion** is a force which acts parallel to the water surface and pulls on the water surface (Figure 10.5). Because the water films are also attracted to the particle surfaces, the effect of surface ten-sion is to pull the particles together. This makes the particles cohesive so that they are stable even on steep slopes. The surface tension en-

Figure 10.4 May 1971 landslide at St. Jean-Vianney in the St. Lawrence River Valley of Quebec. The landslide was caused by failure of slopes underlain by Pleistocene clays and resulted in the loss of thirty lives and considerable prop-erty damage. (*Geological Survey of Canada, Ottawa*)

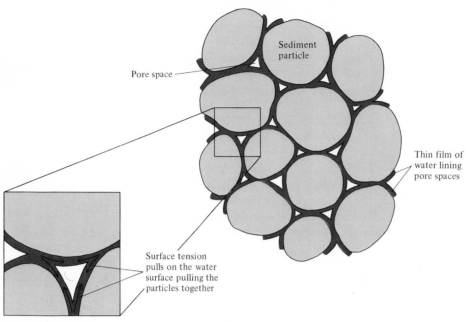

Figure 10.5 Surface tension in the thin films of water lining the pores holds partially saturated sediments together.

ables a castle built of wet sand at the beach to remain standing even though it has vertical walls. As the sand dries, it loses cohesion and the castle begins to crumble. In a similar fashion, if the sand castle is submerged completely in water, the pores become saturated and the castle crumbles into a pile of sand.

Particle Packing

The arrangement of the sand-sized particles in a sediment is called **packing.** There are two extremes of packing. **Cubic** packing (Figure 10.6*a*) is the loosest packing and has the greatest pore space. **Rhombohedral** packing (Figure 10.6*b*) is the tightest packing and has the least pore space. Material which has been dropped or bulldozed into place has more nearly cubic packing. Sedimentary particles which have been vibrated into place or have been deposited by a force acting parallel to the surface of deposition, such as a water or wind current, have more nearly rhombohedral packing.

Change from loose packing to tight packing results in a decrease in volume and a lowering of the surface. You can test this yourself by filling a container partially with any granular material

(coffee, marbles, rice, or sand). Mark the upper surface of the material and shake the container sideways. The particle surface will drop as the grains assume a tighter packing. The surface of loosely packed sediments in nature also will drop if the sediments are shaken by the ground movements associated with highway traffic, construction activities, or earthquakes.

Angle of Repose

When a conveyor belt adds sand continuously to a pile of sand, the sides of the pile collapse periodically as it builds up. A new and higher mound

Figure 10.6 The packing of sand-sized sediment particles affects the stability of the deposit.

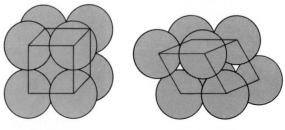

(*a*) Cubic packing   (*b*) Rhombohedral packing

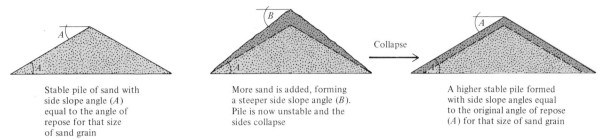

Stable pile of sand with side slope angle (*A*) equal to the angle of repose for that size of sand grain

More sand is added, forming a steeper side slope angle (*B*). Pile is now unstable and the sides collapse

A higher stable pile formed with side slope angles equal to the original angle of repose (*A*) for that size of sand grain

Figure 10.7 Effect or addition of more sand to a pile of sand with a given angle of repose. Slopes greater than the angle or repose for that size sediment are unstable and collapse.

is formed, but its sides have a slope inclined at the same angle as the original mound (Figure 10.7). The maximum angle at which granular material of a given size is stable is referred to as the **angle of repose**. The angle of repose is determined by particle size, sorting, roundness, and the moisture within the sediment. For most sediments, the angle of repose is between 25 and 40°. However, depositional slopes inclined at greater than 40° can occur where larger (boulder-sized) particles are deposited. More angular particles can "interlock" and assume higher angles of repose. Poorly sorted sediments have a higher angle of repose because the smaller particles fit between the larger ones, permitting all the particles to be stable at a greater angle. Partially saturated sediments have a higher angle of repose because of the surface tension of the water films in the pores (Figure 10.5).

Slope failure can occur when construction activities steepen the slope angle. For example, a highway cut made into moist sand can stand at a high angle. However, if there is a long period of drought and the sediment is dried, the slope may fail because the dry sedimentary particles assume a new and lower angle of repose.

TYPES OF MASS MOVEMENTS

Mass wasting processes involve a number of factors such as type of material, type of movement, presence or absence of water, and speed of movement, making the classification of mass movements quite complex. The best overall classification of mass wasting processes is the one given in the article by D. J. Varnes (see Additional Readings at the end of this chapter).

The classification that we will use in this chapter is simplified from that of Varnes, although it is based on the same criteria (type of material, characteristics of the movement, and speed of the movement). Mass movements occur at a variety of speeds. Some occur so slowly that it takes years before the movement is noticeable. Others occur so fast that they become serious geologic hazards. The terms which we will use to describe the speed of mass movements are given in Table 10.1.

Three major types of mass wasting processes can be distinguished based on the type of movement involved. **Falls** involve sediment and rock that move through the air and accumulate at the base of a slope. **Slides** are movements of rock or sediment along planar surfaces. **Flows** are plastic or semiliquid movements of rock and sediment in either air or water. Further subdivisions of these three major types of mass wasting are based on the type of material involved and the speed of the movement. The major mass wasting processes are described and illustrated in Table 10.2.

TABLE 10.1 Relative velocities of mass wasting processes*

Mass wasting	Velocities
	Zero
Extra slow	
	0.06 m/yr
Very slow	
	1.5 m/yr
Slow	
	1.5 m/mo.
Moderate	
	1.5 m/day
Rapid	
	0.3 m/min
Very rapid	
	3.0 m/s
Extremely rapid	

* After Varnes, 1978.

TABLE 10.2   Summary of characteristics of major types of mass wasting processes[*]

Mass wasting type	Character of movement	Subdivision	Speed and type of material	Features
Falls	Particles fall from cliff and accumulate at base	Rockfall	Extremely rapid Develops in rocks	Talus
		Soilfall	Extremely rapid Develops in sediments	
Slides	Masses of rock or sediment slide downslope along planar surfaces	Rockslide	Rapid to very rapid sliding of rock mass along a flat inclined surface	Slide blocks, Inclined rock layer
		Slump	Extremely slow to moderate sliding of sediment or rock mass along a curved surface	Slump blocks, Curved slump plane
Flows	Displaced mass flows as a plastic or viscous fluid	Creep	Extra slow movement of surface regolith and rock	See Figure 10.12
		Solifluction	Very slow to slow movement of water-saturated regolith as lobate flows	See Figure 10.14
		Mudflow	Very slow to rapid movement of fine-grained sediment and rock particles with up to 30% water	Slump, Tonguelike flow, Lobate terminus, Mountain front
		Debris flow	Very rapid flow of coarser debris; commonly starts as a slump in the upslope area	Stream valley, Lobate terminus
		Debris avalanche	Extremely rapid flow; fall and sliding of rock debris	See perspective in Chapter 19

[*] After Varnes, D. J., 1978, Slope movement types and processes in landslides: Analysis and control. Transportation Research Board, National Academy of Sciences, Washington, D. C., Special Report 176.

Falls

The fall of rock particles, through the air or from the face of a cliff, is called **rockfall.** The rock fragments derived from and lying at the base of a cliff or very steep slope are called **talus** (Figure 10.8). Rockfall can be a dry process caused by tree-root wedging, or it can result from frost wedging loosening the particles so that they fall from the cliff face. Rockfall is a very rapid to extremely rapid process (Table 10.2) and thus is a serious geologic hazard, affecting structures near the bases of rock cliffs and in highway road cuts. For example, frost or root wedging along the fractures in the rock exposure shown in Figure 10-2 could result in a rockfall which would bury the highway.

Slides

The movement of rock or sediment along a planar surface is called a **slide.** Sliding movements are differentiated further by the character of the planar surface along which slope failure occurs.

Rockslide

The downslope movement of masses of rock along flat planar surfaces is called a **rockslide** (see Table 10-2). Commonly, the planar surface is along the stratification of sedimentary

Figure 10.8 Rockfall from basaltic cliffs forms talus slopes underlain by coarse-grained and angular debris. Little Canyon tributary to the Snake River, Gooding County, Idaho. *(I. C. Russell, U. S. Geological Survey)*

rocks, but rockslides have developed on exfoliation sheets in granitic rocks (Figure 10.9) or along fractures cutting across layered rocks (Figure 10.2b). Any location where such planar rock surfaces are inclined toward an open space (a road or railway cut or a valley) has the potential for rockslide development (the left-hand slopes in Figure 10.2 *a* and *b*).

Figure 10.9 New talus produced by rock slide. The trees in the foreground were defoliated by the air blast that accompanied the slide. Yosemite National Park, California. *(F. E. Matthes, U. S. Geological Survey)*

Slump

The failure of a mass of material along a curved surface is called a **slump** (see Table 10.2). Slumping is most common in cohesive sediments such as clayey materials, but it also occurs in layered rocks. Slumping commonly occurs because the base of a slope has been eroded, removing support for the material above. This erosion may be due to natural activities such as stream erosion of an adjacent slope or wave attack on a coastal cliff (Figure 10.10). Slumping also can be caused by human activities such as cutting a slope back to expand a housing development or parking lot. Slumping has been an especially serious problem in coastal southern California, where many expensive homes are threatened by the potential failure of the cliffs upon which they are built (Figure 10.11).

Flows

Mass movements in which the material has a plastic or semiliquid behavior resembling that of a viscous fluid are called **flows**. In many cases, mass movements that start as falls, slides, or

Mass Wasting on the Moon and Mars
A Planetary Perspective

Orbital missions around the moon and Mars have shown that mass wasting is not limited to Earth. The Apollo lunar missions documented the fact that meteorite impacts trigger avalanches of rock and regolith on lunar slopes. The range of mass wasting processes on the moon is limited to such dry flows because of the absence of free water. In contrast, the surface of Mars exhibits a wider range of mass wasting processes, many of which date back to times when the Martian

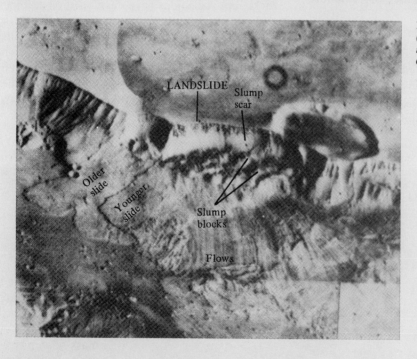

Figure 1 Landslide on Mars. *(Viking photography provided by the National Space Science Data Center)*

slumps are transformed into flows farther down-slope (see "A Planetary Perspective"). Flows exhibit a wide range of characteristics, including both the driest and wettest and the slowest and fastest types of mass movements. We will discuss flows by considering them in order of increasing speed of movement.

Creep

The extremely slow downslope movement of regolith, soil and rock under the influence of gravity is called **creep** (Figure 10.12). Even

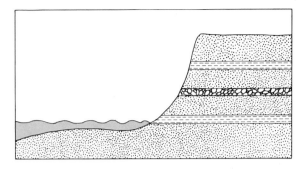

(a) Coastal cliff before a storm

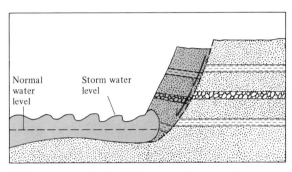

(b) Storm waves erode the cliff base undermining support and causing a slump block to rotate down onto the beach

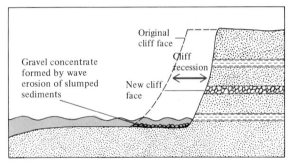

(c) Wave erosion has reworked the slumped material. The cliff face has now receded inland

Figure 10.10 Slumping in coastal cliffs.

surface contained more free water than it does at present.

The Viking spacecraft sent to Mars transmitted clear photographic evidence of slumping on the Martian surface. While slumping on Earth is on a scale of less than a kilometer, the Martian slump was enormous by Earth standards (Figure 1). B. K. Lucchitta (1978) of the U.S. Geological Survey described the slide (see Additional Reading at the end of this chapter). The landslide scar is about 25 km wide and 6 km deep and is concave upwards as in slumps on Earth. There is at least a 700-m difference between the plateau surface and the base of the slump scar. The slide consists of a jumble of slump blocks near the head. Each of the slump blocks shows backward rotation, as in slumps on Earth (Figure 10.10). The slump blocks grade downslope into an apron of flows which extend out across the area at the base of the slope. At least two episodes of slumping and flow are indicated by the superposition of later flow deposits over earlier ones (Figure 1).

The speed of the landslide was calculated as between 27 and 37 m/s (100 to 140 km/h), which is within the range of velocities for large catastrophic landslides on Earth. A number of possible causes of the slide were proposed, including "Mars-quakes," faulting, reduction of friction and cohesion by melting of ground ice, and removal of support by outflow of the lower layers.

though creep is a very slow process, its long-term effects can be seen on the surface through downslope displacements of fences, monuments, telephone poles, and tree trunks. Creep may be the least spectacular of all mass movements, but its continuity of operation and its action over such a wide area make it *the* most important mass wasting process in terms of the total volume of material moved downslope each year.

Figure 10.11 Slumping in cliffs at Pacific Palisades, California. *(E. F. Patterson, U. S. Geological Survey)*

The process of creep is aided by the expansion and contraction of soil by heating and cooling, freezing and thawing, or wetting and drying. For example, repeated cycles of freezing and thawing can move particles downslope in a type of movement called **frost creep.** Each day, water from melting soil ice forms a thin film under the rock particles (Figure 10.13). This thin water film freezes at night, expanding in the process and pushing the particles out at right angles to the slope. When the ice melts the next day, the particle settles to the slope *parallel to the gravitational force.* Each of these cycles displaces the particles farther downslope. Numerous repetitions of this freeze-thaw sequence on large numbers of rock particles move great amounts of material downslope in high latitudes.

Solifluction

The downslope movement of water-saturated regolith is called **solifluction.** Movement rates are faster than in creep and may reach up to a few centimeters per year. Solifluction may occur in any climate in which regolith on slopes becomes saturated with water. However, it is most common in cold climates in which the upper part of the regolith freezes and thaws periodically. Many cold areas are underlain by permanently frozen ground called **permafrost.** During the warmer parts of the year, the uppermost part of the permafrost thaws and releases the water frozen within the sediment. This soggy mass of soil can then flow downslope over the permafrost below. Solifluction creates a topog-

Figure 10.12 Shale layers bent downslope as a result of creep along the surface. Washington County Maryland. *(G. W. Stose, U. S. Geological Survey)*

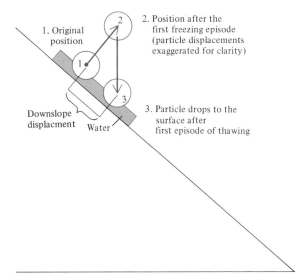

1. Original position

2. Position after the first freezing episode (particle displacements exaggerated for clarity)

3. Particle drops to the surface after first episode of thawing

Downslope displacment

Water

Figure 10-13 Mechanism of frost creep on slopes in high latitudes.

raphy characterized by curved lobate flows on the surface (Figure 10.14).

Mudflows

Flows that contain significant (up to 30 percent) water and consist of a large proportion of fine-grained material are called **mudflows.** Mudflows are common on slopes in semiarid areas where infrequent but very short lived rainstorms convert regolith quickly into a mass of viscous mud and rock which moves downslope. Expanding urban developments in areas marginal to steep mountain fronts, such as those around the Los Angeles Basin, are subject to potentially serious damage from mudflows (Figure 10.15). The problem becomes especially serious when heavy rains follow a period of drought or forest fires which denude a slope's vegetation. In southern California, the severe rains of 1978 triggered mudflows in several areas. Similar storms had occurred in 1952, 1958, 1962, and 1969. Each successive storm did more damage because of the greater property and population that were exposed.

Debris flows

Mass movements in which rock debris and regolith flow very rapidly downslope are called **debris flows** (Table 10.2). Many debris flows start as slumps or slides, which are transformed into flows downslope as the mass breaks up and mixes with air and water.

The Alaska earthquake of March 1964 triggered a massive debris flow from a mountain bordering the Sherman Glacier. The debris flow traveled 5 km from its source and deposited a

Figure 10.14 Solifluction lobes on a slope in the Nome District, Alaska. *(P. S. Smith, U. S. Geological Survey)*

Figure 10.15   Boulders, garage, and car transported by mudflow, December 29, 1965, in Newton Canyon, Los Angeles.   *(R. H. Campbell, U. S. Geological Survey)*

layer of debris about 1.5 m thick as it swept across the glacier without disturbing the fresh snow on the glacial surface (Figure 10.16). The debris which covered the Sherman Glacier showed little sorting with distance, and the lower slope across which it moved was only a few degrees. This raised an interesting question: How do unsorted masses of debris travel so far over gentle slopes? It is believed that at least a few debris flows are able to travel great distances over gentle slopes on a *cushion of air*. The air cushion is formed when tumbling rock debris traps air beneath itself. The presence of an air cushion reduces friction and enables the flow to move over gently sloping land surfaces in much the same way that a "hovercraft" vessel skims over the waves.

Debris avalanches

The general term **avalanche** is used for the most rapid flowing, sliding, and falling mass

Figure 10.16   Following an earthquake-induced land-slide in 1964, ten million cubic meters of slide debris covered 8 km² of the Sherman Glacier in the Copper River Region of Alaska.   *(T. L. Pewe, U. S. Geological Survey)*

The Vaiont Dam Failure

One of the greatest catastrophes of recent times was the 1963 rockslide associated with the building of a massive dam across the Vaiont River Valley in northern Italy. This event is worthy of detailed discussion here because it illustrates many of the principles of mass wasting we have discussed. It also shows the sequence of development and the geologic and human aspects of a large mass movement in a populated area. Our discussion is derived in part from the detailed synthesis of the English and foreign reports of this disaster given by Frank Fletcher (1970) (see Additional Readings).

The Vaiont River flows along the base of a deep, glacially eroded valley which opens out westward into the much broader Piave River Valley opposite the town of Longarone (Figure 1). The valley walls are underlain by a series of limestone formations, ranging from thick to very thin layers and interbedded with shaly limestones (Figure 2). The rock layers on both sides of the valley are inclined toward the center of the valley. A large landslide scar on the northern side of the Vaiont Valley indicates that large-scale mass movements had occurred on the valley walls in the past.

Construction began in 1956 on a 261-m-high concrete arch dam across the Vaiont River Val-

Figure 1 Map of Vaiont Dam area and Piave River Valley showing the surface features mentioned in text. *(Based on research by George A. Kiersch.)*

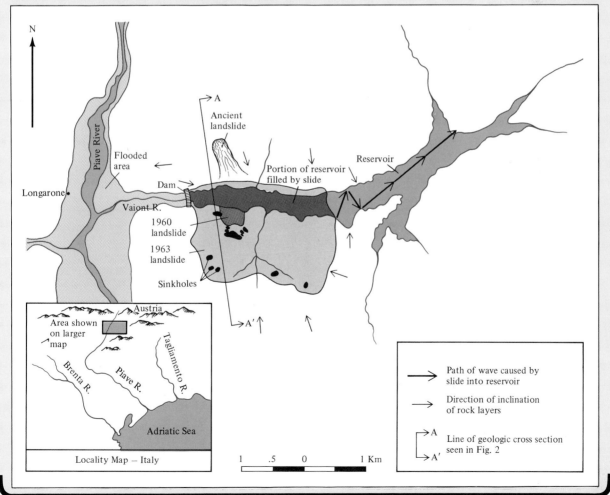

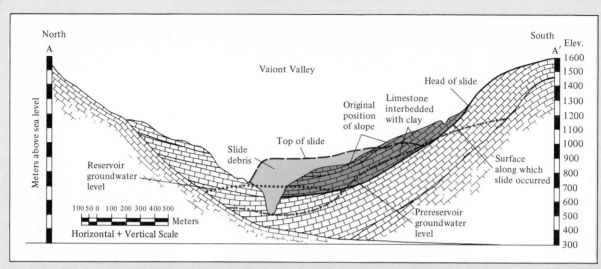

Figure 2 **Section of Vaiont Valley along line AA in Figure 1. All layers except the base are limestones. The mass that slid (dark gray) was limestone interbedded with clay.** *(Based on research by George A. Kiersch.)*

ley. The dam was to create a large lake whose waters would be used for the generation of hydroelectric power. In the summer of 1957, it was noted that concrete being poured for the foundations of the dam was disappearing into the rocks. This suggested that there were fractures or solution cavities or both in the limestone below. Subsequent geologic and geophysical investigations resulted in conflicting reports on the stability of the valley slopes, but work on the dam continued. Instruments were installed to measure any movements in the dam foundation and on the adjacent valley slopes. By 1959, the dam was nearing completion and the reservoir was allowed to fill in several stages. The raising of the reservoir level was to have profound unforeseen effects on the stability of the adjacent valley slopes (see Figure 3).

As the reservoir level increased to 640 m by October 1960, a considerable change in slope stability was noted. On November 4, 1960, a mass of rock and soil with a volume of about 700,000 m³ slid off the south slope into the reservoir, generating a 2-m wave which swept across the lake surface (Figure 1). Studies of engineering models and experimental raising and lowering of reservoir levels suggested to the engineers that the reservoir could be filled safely to near the planned level. Reservoir filling then resumed from a level of 647 m in March 1963 and reached

a level of 710 m by September 4, 1963. Measurements showed that the surface regolith was moving about 3 to 6.5 mm per day, a fairly rapid rate. The creep rate would increase to 12 mm per day by September 15 and nearly 40 mm per day by October 3. By October 7 new cracks had appeared in the south slope, and the road crossing it was closed to traffic. The slope material was also becoming saturated with water because the rainfall for August and September had been three times greater than that recorded for the same 2-month period in any of the last 20 years.

The engineers realized that mass movement was inevitable, but they felt confident that the mass would come down slowly in pieces and blocks because that was the normal behavior of large landslides in that region. Furthermore, scale-model experiments had indicated that there would be no danger from a wave formed in the lake by displacement of water by a landslide.

A number of natural and artificial effects were combining to create a mass movement of gigantic proportions. During the Pleistocene epoch, a glacier had eroded a broad U-shaped valley. Stream erosion by the Vaiont River in postglacial times had eroded a steep-walled inner valley within the broader glacial valley. The erosion of material to form the inner valley removed lateral support from the fractured and steeply inclined rocks within the valley walls.

Great amounts of water had entered the fractures and solution cavities in the limestone beds. This increased the weight of the rocks on the slope. The water in the clayey limestone beds was soaking the clay minerals and reducing their cohesion. The rising reservoir level, plus the great amount of water added to the rock from rainfall, was exerting great pressure on the water within the rock pores. This decreased frictional resistance and in effect made the rock "buoyant," somewhat like a beached boat later picked up in a rising tide.

It was clear to the authorities that mass movement on the south slope was imminent and that people should be evacuated from the surrounding areas. The evacuation order was signed shortly before 10:00 PM on October 9, 1963.

At 10:39 PM, 300 million m³ of fractured and water-laden limestone tore loose from the south slope. The mass slid down the concave slope along a plane of slick clayey limestone beds as *one intact block*. The mass slid into the reservoir, displacing about 50 million m³ of water and forming a wave 200 m high. Hamlets along the lake were wiped out quickly. The well-designed dam took the full force of the wave without failing. The wave reached over the dam and plunged into the narrow river gorge below. A violent destructive wind swept the valley in advance of the flood. The floodwaters surged out of the narrow gorge in a tremendous torrent, which crossed the wide Piave River Valley in a few minutes and totally destroyed the village of Longarone, with a loss of 1450 lives. The total death toll in the area was to be much higher since the floodwaters continued to spread across the Piave Valley and slide-induced waves pounded the villages on the shores of the reservoir (see Figure 4).

The Vaiont Dam rockslide is an excellent example of the complex problems which must be considered in the design and emplacement of large structures. It showed that a very well designed structure can be quite unsafe if it is located in an area of geologic instability. The loss of life and property in this event could have been prevented if the significance of the area's past geologic history and the potential changes that would be brought about by such construction had been realized and acted upon earlier.

The conclusion of Fletcher's description of the Vaiont Dam disaster puts the event in a human perspective.

Figure 3   The Vaiont dam as it looks today. View looking east into the gorge. *(Courtesy of F. W. Fletcher)*

Figure 4   View looking east of upper half of Vaiont dam showing a portion of the landslide mass which filled the reservoir. *(Courtesy of F. W. Fletcher)*

Along the sides of the new road one comes across a crumbling stone wall, all that remains of someone's home, or some simple memorial to the victims—a crude wooden cross in a meadow or, in a shallow rock hollow, a small, carefully placed cluster of fresh, native alpine flowers or, sometimes, just a short list of names, often with the same last names, on a plaque. At the west end of the valley stands the dam, still the highest double-arch dam in the world; but no electricity is produced and exported to Italy's prosperous industries.

Three hundred million cubic meters of rock and soil fill the reservoir and 1899 people are dead.[53]

[53] On November 24, 1968, as the trial began far away in the Abruzzi, the total came to nineteen hundred: Engineer Mario Pancini, his bags packed for the trip to L'Aquila, taped the cracks around the doors and windows of his Venetian room and turned on the jets of his gas range.[*]

[*] F. Fletcher, "A Terrifying Equality: The Story of the Vaiont Dam Disaster," *Susquehanna University Studies*, vol. 8, no. 4, 1970, p. 300.

wasting processes. The roar indicating the failure of the slope above may be followed in only a few seconds or minutes by the avalanching material cascading downslope and spreading across the area beyond.

Very rapid to extremely rapid movements of regolith and rock are referred to as **debris avalanches** (Table 10.2). Many debris avalanches are characterized by amphitheaterlike heads with elongate tongues of debris extending downslope.

SEDIMENT DEPOSITED BY MASS WASTING

The collective term for all sediments deposited by mass wasting is **colluvium.** The short distance of transport of colluvium results in minimal rounding of the particles and a mineralogic composition very similar to that of the source materials. The dense and viscous nature of most mass wasting processes results in little sorting of the sediment particles and the absence of stratification in most cases.

SUMMARY

Mass movements move rock and soil downslope under the influence of gravity. Mass movements are more common where layers of rock or sediment are inclined downslope. Failure of slope materials occurs when the driving forces along the slope exceed the resisting forces.

Water plays a complex role in mass movements, serving both to promote and to resist mass movements. Addition of water to slope materials increases their weight, adding to the driving forces. Water also serves to decrease rock strength by frost wedging, by saturation of swelling clays, by dissolving soluble rock cements such as calcium carbonate, and by promoting the failure of quickclay. A buildup of

excess pore pressure within a rock decreases the frictional resistance and enables the rock to move more readily. However, partially saturated sediments resist mass movement because of the cohesive effects of surface tension.

The most open type of sediment packing is cubic packing. If sediment with cubic packing is subjected to vibration, the grains arrange themselves in tighter rhombohedral packing. A change in packing is accompanied by a drop in the volume of pore space and a subsidence of the surface. This creates a problem if any structures have been built on the surface. The angle of repose determines the maximum angle at which particles of a given size are stable. The angle of repose increases with grain size, angularity, and moisture content. Moist sediments can support slopes steeper than their angle of repose. When the sediment dries, the slope falls, forming a new slope closer to the angle of repose.

There are a wide variety of types of mass movements. They can be divided into three major groups: falls, slides, and flows. Rockfalls (and soilfalls) involve the falling through the air of particles from cliffs and their accumulation at the base of a slope. Rockslides involve the movement of rock along a planar surface. Slumping is a movement along a curved plane. Mass movements in which the material moves as either a dry or a wet flow include creep, solifluction, mudflows, debris flow, and debris avalanches. Creep is a very slow process by which great amounts of rock and soil are moved downslope. Solifluction is the downslope movement of water-saturated sediments. Mudflows are more liquid flows which include a large portion of finer material. Debris flows are relatively dry mass movements involving coarser particles. Some debris flows are thought to be able to move rapidly over low slopes because they travel on a cushion of air trapped below the debris. Debris avalanches are the most rapid type of mass movement. They involve a complex of sliding, tumbling, and flowing rock debris.

Mass movements serve to move weathered material off slopes into transportational systems such as rivers and glaciers and coastal currents. These transporting agents rework the material further and transport it over a wider area. The colluvium deposited by most mass movements tends to be poorly sorted, with angular edges, poor development of stratification, and a mineralogic composition little different from that of the source materials. The catastrophic nature of many mass movements makes it vital that potential mass wasting problems be evaluated in detail prior to the development of an area.

QUESTIONS FOR REVIEW AND FURTHER THOUGHT

1. What is the major driving force in mass wasting? What is the major resisting force?

2. How do mass wasting processes differ from other types of erosional and transportational agents such as glaciers, winds, streams, and ocean currents?

3. Show how water can either inhibit or aid mass wasting processes.

4. Summarize all the characteristics that enable you to distinguish between rockfalls, rockslides, slumps, various types of flows, creep, and solifluction.

5. Describe the sedimentary characteristics of colluvium and explain why they differ from deposits made by streams, wind, and coastal waves.

6. What environmental problems might be associated with the following construction activities?
   a. Building a road through a valley in which the rock layers are inclined into the valley
   b. Expanding a housing development toward a mountainous slope in a semiarid environment
   c. Building a large housing development on a high ocean cliff cut into unconsolidated sediments

ADDITIONAL READINGS

Beginning Level

Briggs, R. P., J. S. Pomeroy, and W. E. Davies, *Landsliding in Allegheny County, Pa.*, U.S. Geological Survey Circular 728, 1975, 18 p.
(Describes different types of mass movements from the standpoint of how they are recognized, what geologic conditions cause different types, and what remedial measures can be taken. Contains numerous photographs and a glossary)

California Division of Mines and Geology, *Southern California Landslides — 1978,* 1979.
(Includes short and well-illustrated articles on (1) slope stability and debris flows, (2) Bluebird Canyon landslide of October 2, 1978, Laguna Beach, California, and (3) mudflow/debris flow damage, February 1978, storm in Los Angeles area) From *California Geology,* vol. 32, no. 1.

Fletcher, F., "A Terrifying Equality: The Story of the Vaiont Dam Disaster," *Susquehanna University Studies,* vol. 8, no. 4, 1970, pp. 271–300.
(A beautifully written description of the Vaiont Dam disaster. Includes geologic background along with numerous human interest stories)

Kerr, P. F., "Quickclay," *Scientific American,* vol. 209, no. 5, 1963, pp. 132–142.
(Detailed description of how clays can be liquefied quickly)

Kiersch, G. A., "The Vaiont Reservoir Disaster," in R. W. Tank (ed.), *Focus on Environmental Geology,* 2d ed., Oxford University Press, New York, 1976, pp. 132–143.
(Detailed description of the geologic and engineering aspects of the Vaiont Dam disaster)

Intermediate Level

Lucchitta, B. K., "A Large Landslide on Mars," *Geological Society of America Bulletin,* vol. 89, 1978, pp. 1601–1609.
(A detailed description of the Martian slide along with a comparison with several similar features on Earth)

Pestrong, R., *Slope Stability,* Council on Education in the Geological Sciences, Publication No. 15 (McGraw-Hill Concepts in Introductory Geology Series), 1974, 65p.
(Describes the factors causing earth movements, recognition of mass movements, and environmental planning in areas subject to mass movements. Includes numerous laboratory exercises for each topic)

Shreve, R. L., *The Blackhawk Landslide,* Geological Society of America Special Paper 108, 1968.
(Detailed description of this ancient landslide and evidence to support the theory that it traveled on a cushion of trapped air)

Varnes, D. J., "Slope Movement Types and Processes," in *Landslides: Analysis and Control,* Transportation Research Board, National Academy of Sciences, National Research Council Special Report 176, 1978, pp. 11–33.
[The best overall reference on mass wasting. Includes a detailed illustrated chart describing mass movements in terms of type of material (rock, debris, earth), type of movement, and speed]

CHAPTER 11

Streams and Stream Sculpture

A mountain glacier in Alaska, the blue-green waters of southern Florida, and Old Faithful geyser in Yellowstone Park all look very different, but they have one thing in common — each represents one of the different forms of water on the earth's surface. Water is present as part of the chemical composition of some minerals; as a solid in glacial ice; as a liquid in streams, lakes, oceans, and underground water; and as a vapor in geysers and clouds. Water in its various forms is essential to many geologic processes and to life itself.

THE HYDROLOGIC CYCLE

The total *amount* of water in, on, and above the earth remains almost constant. However, the *relative percentages* of water in different forms have varied throughout geologic time. The relative percentages of the different forms of water

at the present time are given in Figure 11.1. The greatest percentage of water (97.3 percent) is in the world's oceans, followed by glaciers and larger ice masses (2.14 percent), with all other forms of water making up the remaining 0.56 percent.

Water constantly is transforming into different forms (Figure 11.2). Solar energy evaporates water from the surfaces of oceans, lakes, and rivers. This water vapor condenses as it rises higher into the cooler upper atmosphere and returns to the earth as precipitation in the form of snow or rain. This precipitation is stored temporarily in glaciers, in the bodies of water on the earth's surface, or underground in the pores of sediments and rocks until it is returned to the atmosphere by evaporation. Water absorbed by vegetation is returned to the atmosphere through a process called **evapotranspiration.** The constant circulation of water from the sea, through the atmosphere, and to the land, and

Figure 11.1 Distribution of the earth's water. (*Modified from R. L. Nace, U.S. Geological Survey, Circular 536*)

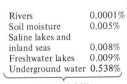

Rivers	0.0001%
Soil moisture	0.005%
Saline lakes and inland seas	0.008%
Freshwater lakes	0.009%
Underground water	0.538%

Other 0.56%

Oceans 97.3%

Glaciers 2.14%

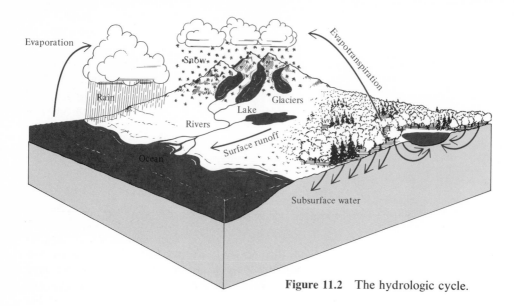

Figure 11.2   The hydrologic cycle.

its eventual return to the atmosphere by way of evaporation from the sea and the land surfaces is called the **hydrologic cycle.**

The time necessary for the return of a water particle to the atmosphere is highly variable and depends on the path taken by the particle. Some raindrops never reach the ground but are evaporated back into the atmosphere as they fall. Some of the rainfall that sinks into the ground may be absorbed by plants and restored to the atmosphere through evapotranspiration within a few hours or days. The rest of this subsurface water moves through the pores of sediment and rocks and may not appear at the surface again for hundreds or even thousands of years.

SURFACE VERSUS SUBSURFACE WATER

What determines whether rainfall will soak into the ground or run along the surface? This is governed by the balance between the rate of precipitation and the rate at which the precipitation can pass through or infiltrate the surface materials. The **infiltration capacity** is the rate (cm/h) at which water may pass through a given surface material. The infiltration capacity starts with an initial value and then decreases steadily as the fine pores of the soil fill with water because clay mineral particles swell when they become wet

and finer particles are washed into larger pore openings, clogging them. Eventually the infiltration capacity reaches a constant value, which is the *effective* infiltration capacity of the soil. For example, consider a soil with an effective infiltration capacity of 3.0 cm/h. If the rainfall is 2.0 cm/h, the rain will infiltrate into the ground at a rate of 2.0 cm/h. If the rainfall were 4.0 cm/h, the infiltration rate would be only 3.0 cm/h (infiltration capacity). When the rainfall rate exceeds the infiltration capacity, as in this example, the excess surface water (1 cm/h) will flow downslope as a sheet of water called a **sheetflood.**

In this chapter we will consider only that portion of the hydrologic cycle in which water moves as streams in channels along the earth's surface. While only one ten-thousandth of the world's water is in streams (Figure 11.1), streams are extremely important in creating landscapes in both wet and dry climates. Our discussion in this chapter will deal only with characteristics of streams in humid areas. Characteristics of streams in dry areas will be discussed in detail in Chapter 14. Many of the concepts of sediment erosion and deposition introduced in Chapter 6 will be developed further in this chapter. We begin our discussion of streams by considering where a stream gets the energy for erosion and for transporting materials.

A STREAM'S ENERGY AND ITS UTILIZATION

Rain dropped high on a mountain has potential energy due to gravity (see Chapter 2). The potential energy is equal to the weight of the water times the difference in elevation between any two points on the stream between which the energy is being determined. As this water begins to flow downslope under the influence of gravity, the potential energy is transformed into kinetic energy (see Chapter 2), enabling the stream to begin erosion and sediment transport.

Some streams flow slowly down gentle slopes, while others rush in torrents down steeper slopes. The relative steepness of a stream channel is called its gradient. The **gradient** of a stream (Figure 11.3) is the vertical drop in elevation over a given horizontal distance. A cross section of a stream from its headwaters to a point downstream is called a **longitudinal profile** (Figure 11.3). The longitudinal profiles of streams are concave upward, with steeper upstream gradients and gentler downstream gradients.

How does a stream utilize its kinetic energy? You might think that most of the stream's energy is used for erosion of material from the stream bottom and for transporting it against the forces of gravity and friction, but that is not the case. Most (as much as 95 to 97 percent) of the energy is used in overcoming frictional forces between the water and its channel, between the water surface and the air, and between the different masses of water within the stream itself. Any conditions that decrease these frictional effects will free relatively more of the stream's energy for erosion and transport. For example, water moving through a smooth rock channel will have relatively more of its energy available for erosion and transport than water moving through an irregular channel.

Streams continue to flow downslope, eroding, transporting, and depositing until they reach a level at which they no longer have the energy to erode down into the land surface. The level at which a stream loses its energy and can no longer cut downward into its bed is called its **base level**. The *ultimate* base level is sea level. However, some streams in the continental interiors never reach the ocean, and they may have base levels which are well above sea level (or even below sea level, such as in Death Valley).

Even though sea level is the ultimate base level for most streams, many streams encounter one or more *local* base levels as they move downslope. We can illustrate the relationship between a stream's energy, local base levels, and behavior by considering the effects on stream behavior when a dam is constructed across the stream (Figure 11.4). The dam impounds river water in the reservoir upstream of the dam. The water surface of the reservoir then serves as a local base level for the stream. As the stream enters the reservoir, it loses energy and can no longer carry sediment. This sediment is deposited in the reservoir, decreasing the stream gradient locally. While the reservoir

Figure 11.3 Cross section along the length of a stream from its headwaters to the ocean, showing the longitudinal profile and the stream gradient. The longitudinal profile of the stream is concave upward, with a steeper upstream gradient and a gentler gradient in the downstream portion.

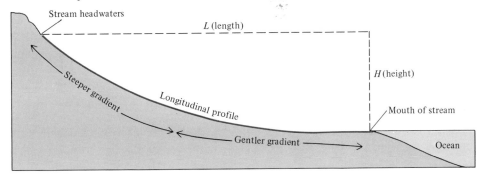

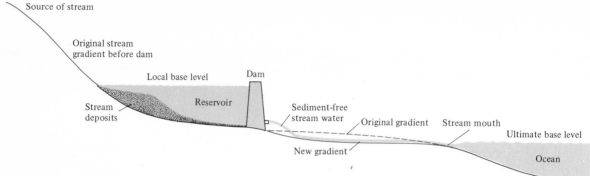

Figure 11.4 Effect of local base levels on streams. The reservoir acts as a local base level for the stream. The stream decreases in velocity as it enters the reservoir. The decrease in velocity results in the stream depositing its sediment, flattening its gradient locally. The sediment-free water exiting from the base of the dam has sea level as a base level and erodes its bed, steepening its stream gradient locally.

water has no kinetic energy, it does have considerable potential energy because the reservoir's surface is well above sea level. When the relatively sediment-free water is released from the dam, its potential energy is converted to kinetic energy. This water carries very little sediment, and, therefore, more of the stream's energy can be used to erode the channel. The accelerated erosion steepens the stream gradient locally. The stream then continues downslope, locally eroding, transporting, and depositing sediment until it reaches sea level.

STREAMFLOW

If you ever have watched a stream over a period of time, you probably have noticed changes both in the characteristics of the flowing water and in the shape of the stream channel itself with the passing of time.

The speed of the stream, called the **velocity,** and the volume of water passing a point in a given time, called the **discharge,** can be quite varied. Variables such as velocity and discharge that describe streamflow are called **hydraulic variables.** There is also a set of **geometric variables** which describe the shape of the stream channel (Figure 11.5). Stream **width** is the horizontal distance measured across the water surface. Stream **depth** is the vertical distance measured between the water surface and the streambed.

Laminar and Turbulent Flow

The paths that individual water particles take in a stream are called **flowlines.** The changes in the flowlines as velocity changes are shown in the stream sections in Figure 11.6. At low velocities, the flowlines are almost parallel and there is little vertical motion in the fluid (Figure 11.6*a*). This type of sheetlike streamflow is called **laminar** flow.

As the stream's velocity increases, the flowlines become more irregular and vertical mixing becomes common. Swirling, whirlpool-like masses of water called **eddies** extend down through the streamflow and move with the current. These eddies die out and form again in another place. Streamflow characterized by vertical mixing and eddies is called **turbulent** flow. In turbulent flows the *net* water movement is downstream. However, local flows can move in all directions (Figure 11.6*b*). The presence of either laminar or turbulent flow is not governed by relative velocity alone but also by the roughness of the stream channel and the depth. A rough stream channel interferes with the laminar flow lines and can initiate turbulent flow at a lower velocity than would be possible in a smoother channel. In a similar fashion, shallower depths result in more turbulent flow for a given velocity and channel roughness. Velocities of streams in nature are usually fast enough so that most flows are turbulent. The multidirectional water movements in turbulent

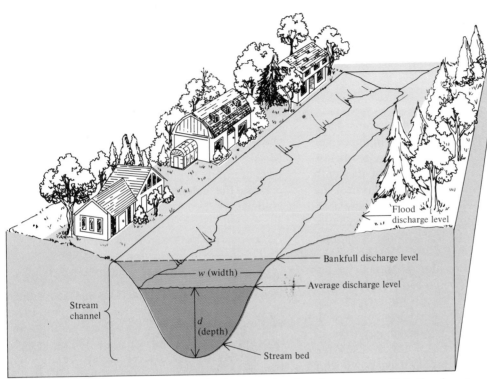

Figure 11.5 Hydraulic and geometric variables used to describe streams. Note the change in stream width *(w)* and depth *(d)* as the stream discharge increases.

Figure 11.6 Types of streamflow. *(a)* Laminar flow: Flowlines are straight and parallel, with no vertical mixing. *(b)* Turbulent flow: Flowlines are highly irregular, with extensive vertical mixing. Local flows are in all directions, but net flow is downstream.

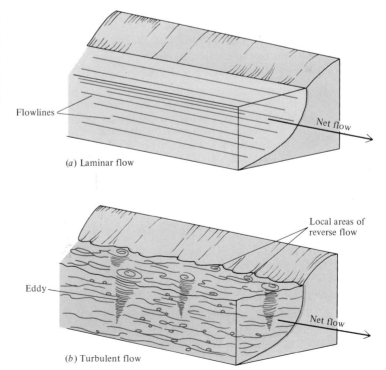

flow aid both in eroding particles from the streambed and in keeping them suspended within the streamflow so that they can be transported downstream (see Chapter 6).

Velocity

Measurements of velocities of streams have shown that maximal velocities occur in the center of the channel and at a depth of from just below the stream's surface to one-fourth of its depth (Figure 11.7). Minimal velocities occur near the bottom and on the channel sides, where frictional effects are greatest. The drop in velocity near the surface reflects the frictional drag of air on the stream's surface.

A number of factors affect a stream's velocity. *All other things being equal,* velocity is higher in streams with steeper gradients, with smoother and straighter channels, and with higher discharge. The last point may not be so obvious as the first two. As stream discharge rises after a rainfall, the stream must flow faster to move the greater discharge through the same channel.

Discharge

The discharge of a stream at any one place can vary a great deal from times of drought to times of flooding. Discharge of a larger stream also increases downstream as more and more tributaries add their waters to the major stream. The Mississippi River illustrates the effects of this downstream increase in discharge. Where do you think a stream's velocity is greater—in the steeper-gradient upstream areas or in the lower-gradient downstream areas? Studies have shown that velocity *increases* downstream in the Mississippi. This is because the increasing discharge of the Mississippi downstream increases the velocity, even though the gradient is lower.

Relationship between Hydraulic and Geometric Variables

U.S. Geological Survey geologists Luna B. Leopold and Thomas Maddock utilized a great number of measurements made in humid areas to derive relationships between hydraulic and geometric variables in streams. They showed that as the discharge increases, there are regular and predictable increases in the width, depth, and velocity of streams.

The discharge of a river may be related to width, depth, and velocity by the following formula:

Discharge (m^3/s) = width (m) × depth (m) × velocity (m/s)

Figure 11.7 Variations in stream velocity with depth. *(a)* Section parallel to streamflow. The maximal stream velocity is located about one-fourth the depth below the stream's surface. The length of each of the arrows is directly proportional to the stream's velocity at that depth. *(b)* Section at right angle to streamflow. The maximal stream velocity is located within the center of the stream just below the stream's surface. The contour lines connect points of equal velocity within the stream.

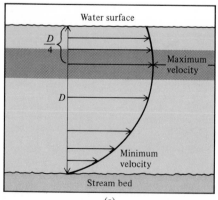

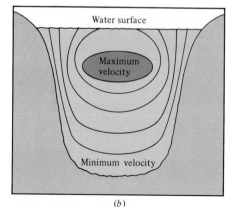

(a)   (b)

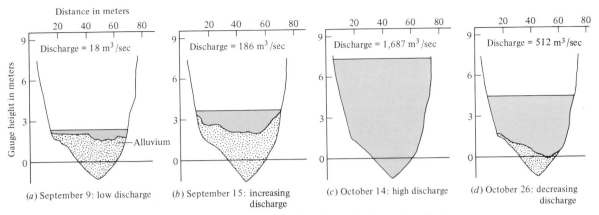

Figure 11.8 Changes in water surface and streambed elevation with increasing discharge in the San Juan River near Bluff, Utah, September 1941 to October 1941. (*a*) During low discharges, the stream flows with a low velocity over thick deposits in the channel. (*b*) A rise in discharge is accompanied by increasing velocity, depth, and width. (*c*) High discharge results in erosion of bed material. The stream reaches maximum cross-sectional area. (*d*) Decreasing discharge results in deposition of stream deposits and lowering of water-surface elevation. (*Modified from* Fluvial Processes in Geomorphology *by Luna B. Leopold, M. Gordon, and John P. Miller. W. H. Freeman and Company,* © 1964)

A system in which changes in one part are balanced by changes in another is said to be in **dynamic equilibrium.** The dynamic equilibrium between hydraulic and geometric variables in streams can be illustrated by the changes that occurred in the channel of the San Juan River near Bluff, Utah, as discharge increased markedly between September and October 1941 (Figure 11.8). The rising discharge was accompanied by higher velocities, a rise in elevation of the water surface (increasing the stream's width), and erosion of the sediment on the streambed (increasing the depth) (Figure 11.8 *a* through *c*). As the stream's discharge began to decrease (Figure 11.8*d*), the velocity decreased and the water level dropped while sediment was deposited, thus building up the streambed. These adjustments brought the stream channel into equilibrium with the lower discharge.

Variations in Discharge with Time

As you might expect, there is a relationship between rainfall and stream discharge over time. In a natural (forest or grass) area in a humid climate, as rainfall continues, a stream's discharge increases slowly. The peak discharge occurs *after* most of the rain has fallen. The time difference between the center of mass of the rainfall and the stream's runoff is called the **lag time.** There is a lag time because it takes a certain amount of time before the rain exceeds the infiltration capacity of the soil. After the ground becomes saturated, the excess rainfall runs along the surface, causing the streams to reach peak discharges some time after the rainfall began.

Flooding

Stream discharges sufficiently great to fill a channel to the top are called **bank-full discharges** (Figure 11.5). Observations of many rivers of different gradients and sizes show that bank-full discharges occur on an average of about once every 1.5 years. When discharge is greater than bank-full discharge, the stream spills out of its channel and floods the adjacent land. The area flooded periodically by a stream is referred to as the **floodplain** (Figure 11.5). Floods are natural events that occur along most streams every few years or so.

Especially damaging floods can occur when local conditions decrease the lag time. Two situations result in potentially dangerous decreased lag times. The first involves sustained and heavy rainfall in steep mountainous areas. The other condition occurs when stream chan-

Flooding in Rapid City

One of the greatest natural disasters in recent United States history was the flooding of Rapid City, South Dakota, on June 9–10, 1972. Rapid City, the second-largest city in South Dakota, lies at the eastern edge of the Black Hills in the southwest corner of the state; it is bisected by Rapid Creek, a (usually) small stream with headwaters in the Black Hills high above the city.

Rainfall began on the afternoon of June 9 when easterly winds forced local air to higher altitudes, where it cooled and the moisture condensed as rainfall. The weak winds at higher altitudes prevented the rainstorm from moving out of the area, and by midnight from 10.16 cm to 30.48 cm of rain had fallen over the area. The "normal" level of Rapid Creek, just above Canyon Lake, was 0.54 m at 7 P.M. During the storm the stream surface rose rapidly to a high level of 4.8 m and to a peak discharge of 883.6 m³/s by 11:15 P.M.

Figure 1   Property damage from the Rapid City flood. *(The Rapid City Journal)*

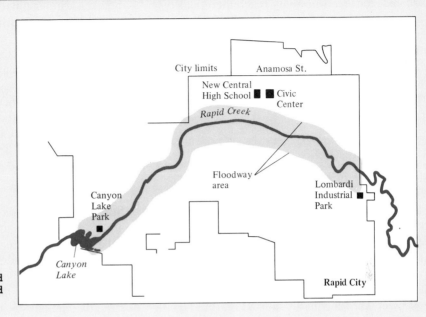

Figure 2 Floodway created in Rapid City after the flood of 1972.

The floodwaters rushed down steep mountainous slopes in stream channels confined mostly to narrow winding canyons. Discharge increased progressively as more and more high-gradient tributaries added their flows to Rapid Creek. A small recreational dam at Canyon Lake was overtopped and partly destroyed as the floodwaters rushed downstream into Rapid City at speeds as high as 7.6 m/s. The discharge greatly exceeded the channel's capacity at Rapid City, and the flood spread out across the city.

The raging torrent killed 237 people and resulted in about $79 million in damage. More than 1335 homes and trailers and 5000 automobiles were destroyed as the rampaging river demolished everything on its floodplain (Figure 1). Flooding lasted for only about 10 h, a relatively short period as floods go but understandable in terms of the speed of the floodwaters. When the floodwaters receded, the people of Rapid City were faced with the choice of rebuilding on the floodplain and taking their chances (as most communities do, unfortunately) or accepting the geologic realities and working around them.

The people of Rapid City chose to create a zoned "floodway" (Figure 2) along the Rapid Creek floodplain. Massive federal aid permitted relocation of almost all the residences and businesses from a 7-mi-wide stretch of the floodplain. Those left behind were protected by floodwalls built by the Army Corps of Engineers. Most of the lower part of the floodplain is now used for community facilities, such as athletic fields and golf courses, which would suffer minimal damage in any future flood. Many other river cities are turning in varying degrees to the floodway concept, which recognizes the inevitability of flooding but attempts to minimize damage to property and loss of life.

nels are modified artificially and formerly vegetated areas are paved over during urbanization. Urbanized streams not only reach their peak discharges earlier (a decrease in lag time), they also have a greater peak discharge than similar streams in natural areas. These changes in discharge can be represented on a type of graph called a **stream hydrograph,** along with the distribution of rainfall with time (Figure 11.9).

STREAM CHANNELS

Streams have three major types of channel patterns (Figure 11.10). A **straight channel** (Figure 11.10a) is the most direct course between two points on the stream. Stream channels composed of a series of sinuous curves are called **meandering channels** (Figure 11.10b). The relative degree to which a stream channel meanders can be determined by calculating the sinuosity of the channel bends (Figure 11.10b). Stream **sinuosity** is the distance measured along the stream valley, D_v, divided by the distance measured along the stream channel, D_c. A perfectly straight stream has a sinuosity of 1.0 (Figure 11.10a). Streams with a sinuosity greater than 1.5 are considered meandering streams. The third type of stream channel is a braided

channel, named for its similarity to the braids in a rope. **Braided channels** are those which break up into numerous smaller channels separated from each other by islands or sandbars (Figure 11.10c).

The same stream may have segments with different channel patterns. In addition, a given stream segment may have a different channel pattern at different times. The ability of a stream to change its channel pattern is another example of the dynamic equilibrium that we discussed earlier in this chapter.

Straight Channels

Straight streambeds have alternating deeper areas called **pools** and shallower barlike areas called **riffles** (Figure 11.11a). These pools and riffles are not located randomly but occur at intervals of about five to seven times the local width of the stream (Figure 11.11b). The flow in straight channels is not parallel to the channel walls but moves from side to side as it heads downstream (Figure 11.11b). The deepest part of a straight stream channel is not necessarily the center of the channel. A line connecting the deepest points along the channel is called the **thalweg.** The thalweg in straight stream channels has a tendency to wind back and forth between the straight channel walls. These observations suggest that even straight stream channels have a tendency to meander. Experimental studies also show a tendency for straight channels to evolve into meandering ones. A straight stream channel cut into sand on an experimental "stream table" evolves slowly into a meandering channel (Figure 11.12). All these observations suggest that straight stream channels develop only when the natural meandering tendency of streams is inhibited locally.

Conditions favoring the development of straight channels include erosion-resistant rocks, rocks with well-developed linear fractures, stream segments with high gradients, and areas of active uplift. Hard rocks inhibit the widening of the channel and the development of a meander. Well-developed linear fractures in rocks provide a path along which erosion is most likely to occur. Thus, the stream aligns itself with the frac-

Figure 11.9 Stream hydrographs of an area before and after urbanization. Urbanization results in a decrease in the lag time and an increase in the peak discharge of the stream.

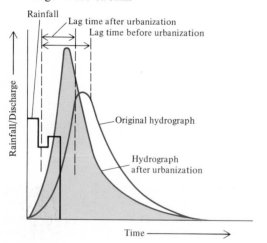

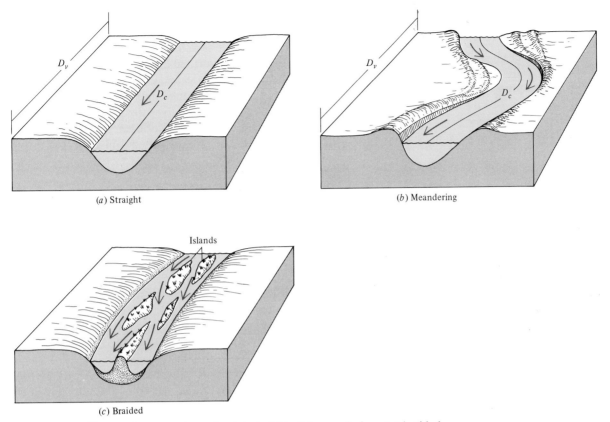

Figure 11.10 Types of stream channels: *(a)* straight; *(b)* meandering; *(c)* braided.

Figure 11.11 Alternating shallows (riffles) and deeps (pools) occur along straight stream channels. *(a)* Side view. *(b)* Top view. Note in *(b)* that the streamflow moves in a curving path from side to side, although the sides of the channel are straight. *(Modified from Luna B. Leopold and W. B. Langbein, "River Meanders," Scientific American, June 1966)*

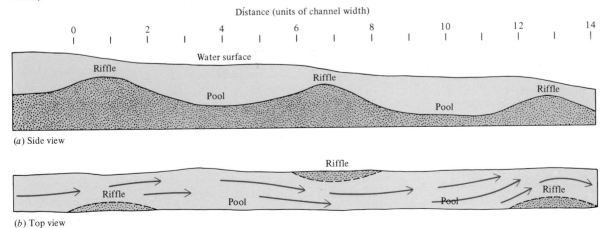

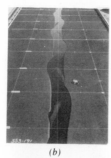

(a)

(b)

(c)

(d)

(e)

(f)

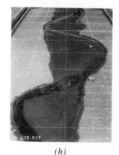

(g)

(h)

(i)

tures.   High gradients result in a greater tendency for the water to move downslope by the most direct path—a straight line. When an area is being uplifted, the vertical distance between the stream channel's altitude and base level (sea level) increases.   This change in base level increases the potential energy of the stream.   As a result, the stream has greater kinetic energy, which serves to erode the stream channel in the most direct path downslope.

Meandering Channels

Meandering channels are the most common type of stream channel.   We mentioned before that flow in straight streams meanders back and forth from one channel side to the other (Figure 11.11). As this meandering flow hits a side of the channel where the bank materials are eroded more easily, it can cut back that side and begin the process of forming a meandering stream pattern.   Erosion of materials on the side of the stream is called **lateral erosion.**   Lateral erosion transforms initially straight stream channels into meandering ones (Figure 11.12).   In order to understand how stream meanders grow and move across the floors of the stream valley, we must look in greater detail at the flow of water in meandering channels.

The idealized streamflow pattern of a meandering stream is shown in Figure 11.13.   Along the straighter sections of the channel (Figure 11.13 *a* and *c*), velocity decreases with depth and all the stream's velocity components are directed downstream.   As the stream flows around the curved parts of the channel, centrifugal force elevates the surface waters along the concave bank. There is a compensating return of water along

Figure 11.12   Sequential evolution of a straight channel into a meandering one on an experimental table. The initial straight channel *(a)* slowly changes into a meandering one *(b)* with the development of point bars and undercut curves *(c, d)*.   The width of the stream valley increases as the meandering channel migrates across the floodplain eroding into the valley walls and leaving behind point-bar deposits on the floodplain *(e–i)*. *(U. S. Army Corps of Engineers Waterways Experiment Station, Vicksburg, Mississippi)*

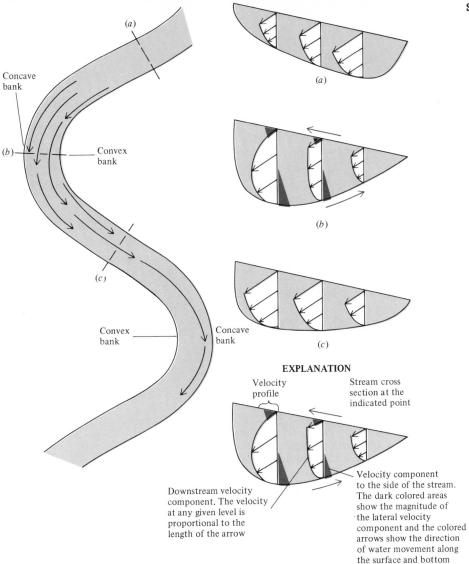

Figure 11.13 Changes in the magnitude and direction of stream velocity along a meandering stream segment. In the straighter stream sections (*a* and *c*) velocity directions are downstream. In the curved channel section there is a velocity component directed toward the concave bank at the top, and toward the convex bank along the bottom. The *net* flow is downstream in all cases. (*Modified from Luna B. Leopold and W. B. Langbein, "River Meanders,"* Scientific American, *June 1966*)

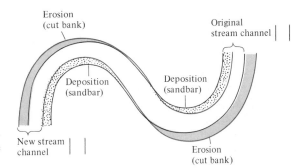

Figure 11.14 River meanders grow laterally by erosion on the concave bends and deposition on the convex bends.

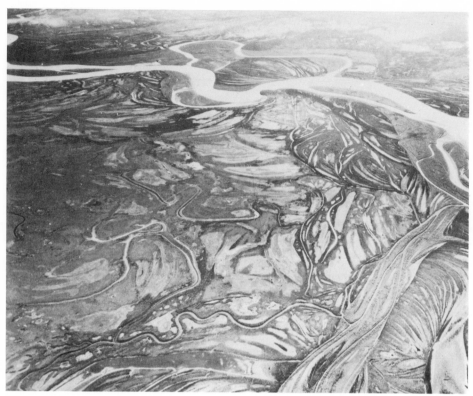

Figure 11.15 Meandering streams and associated channel and floodplain features in the Yukon region, Alaska. The Koyukuk River (dark) joins the silt-laden Yukon River (light) at the right of the photograph. *(T. L. Pewe, U. S. Geological Survey)*

the bottom toward the convex bank (Figure 11.13*b*), and this results in a velocity component directed toward the sides of the stream. The combination of flow directed both downstream and toward the channel walls results in a "cork-screwlike" type of flow called **helical flow.** The helical flow pattern in meandering streams directs the force of the flowing stream along the the concave bank. This results in erosion of that bank (Figure 11.14). Sediment eroded from the channel walls is deposited along the convex bank by the helical flow pattern of the meandering stream. Many floodplains of meandering streams show complex arc-shaped deposits laid down by the stream as it moves laterally along its floodplain (Figure 11.15).

Conditions favoring the development of a meandering channel pattern include easily erodable bank materials and a gentle gradient. Streams with gentler gradients are closer to base

level and have less of a tendency for vertical downcutting and more of a tendency for lateral erosion. Lateral erosion is most marked when the bank materials are unconsolidated sediments such as sands, silts, and clays.

Braided Channels

Braided channels develop where streams are unable to carry part of their sediment load and deposit it temporarily within the channel as bars or islands (Figure 11.16). Braided stream channels are one more example of dynamic equilibrium in streams. Excess sediment is deposited temporarily within the channel until the stream's discharge is sufficient to move it again. As discharge increases, the braided stream channel may become a meandering or even a straight channel.

Braided channels can develop where the stream gradient decreases abruptly, where the stream discharge decreases through evaporation or infiltration into the ground, or where a high-gradient tributary stream adds more sediment to the larger stream than the larger stream has the energy to carry.

Braided channels develop under a wide variety of topographic and climatic conditions. Fast-rushing mountain streams may develop braided patterns at the abrupt decrease in gradient at the base of mountains. This is especially marked in drier climates, where there is not only a decrease in gradient at the mountain-front base but also a greater evaporation rate in the hot and dry air of the desert valleys. Braided channels are also common at the front of melting glaciers. The large volume of sediment released from the melting ice is usually more than the available water can transport from the area.

STREAM EROSION

Streams erode the landscape in several different ways. In some cases, the force of the water alone is sufficient to pry up pieces of rock or erode weakly consolidated sediments in the stream's banks. In other cases, particles carried within the streamflow can erode the loose rock or sediment on the channel floor. Rock fragments carried in turbulent eddies can "drill out" circular cavities in the streambed called **potholes**. Particles of sediment or rock may be "sucked" off the bottom when turbulent eddies sweep through the flow (see Figure 11.6).

The rate of stream erosion increases with both the velocity of the stream and the erodibility of the stream-bank materials. A stream flowing in a granite channel may take thousands of years or more to enlarge the channel appreciably. On the other hand, a stream flowing through unconsolidated sands and gravels may undergo significant channel enlargement in only a few years, or even in one flood. This erosion is most marked in areas with little vegetation, such as deserts, and areas in which vegetation has been removed during construction or killed by pollution.

Some of the mass wasting processes described in Chapter 10 are very important in aiding stream erosion. Most of the erosion of channel banks in rivers takes place when the

Figure 11.16 Braided channels of the Nelchina River, Copper River region, Alaska. *(J. R. Williams, U. S. Geological Survey)*

stream undercuts the channel banks, removing support for the material above. The unsupported materials slump into the stream, and the channel widens as the stream banks are eroded back. The slumped material is reworked by the stream and transported from the area.

So far we have considered only those streams which flow over a uniform type of bed. In reality, streams usually flow over a variety of different types of sediments and rocks. Some rocks, such as siliceous sandstones and granite, are hard and resistant to stream erosion. Other types of rocks, such as shales, are softer and more easily eroded. When a hard rock overlies a soft one, the softer rock is eroded more rapidly, resulting in one type of falls which may occur along a stream (Figure 11.17). The tremendous turbulence generated at the base of the falls results in accelerated erosion of the weaker rock and removal of support for the hard rock above. The unsupported cap rock breaks off and accumulates as a talus at the base of the falls, and the falls recedes upstream a short distance.

Niagara Falls, on the Niagara River between Lake Ontario and Lake Erie, provides an example of retreat of falls. The top of the falls is capped by a resistant dolomitic limestone underlain by more-erodible shales. Undermining of the dolomite by stream erosion of the shale has resulted in a steady upstream retreat of the falls. The average rate of retreat between 1850 and 1950 was a little over 1 m per year. As the falls retreat, they leave behind a steadily lengthening gorge, exposing all the rock layers from the cap rock down to stream level (Figure 11.17). Eventually, the falls will retreat until they connect Lakes Erie and Ontario directly.

Rapids in streams make "white-water" boating exciting. Rapids form under a variety of conditions, including areas where stream discharge increases markedly, where tributaries bring in coarser materials than the main stream can transport, where the stream channel narrows abruptly, and where there are local bodies of resistant rock which stand up above the weaker rocks on either side (Figure 11-17). This

Figure 11.17 Formation of falls and rapids in a stream flowing over rocks of different degrees of erodibility. Turbulent water at the base of the falls erodes the shale and removes support for the sandstone cap rock. The falls retreat upstream with time (dashed lines). Rapids form where resistant rocks stick up above the stream's bottom making the local streamflow more turbulent.

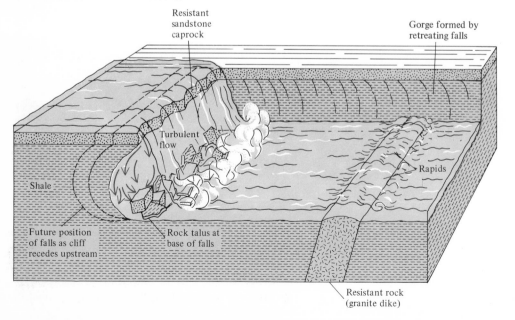

creates shallows which increase the stream's velocity locally and make the water surface more irregular.

STREAM DEPOSITION

Streams transport and deposit a wide variety of sediments ranging in size from boulders to clay through the processes described in Chapter 6. The collective term for all the deposits made by streams is **alluvium.** Alluvium forms several different types of morphologic features. Some of these features occur within the stream channel, some occur on the floodplain, and some occur beyond the floodplain, at the point where a stream enters a lake or the ocean.

Deposits within the Stream Channel

Streams may temporarily deposit a portion of their sediment load within their channels as ridgelike accumulations called **bars.** Even straight streams may have sandbars within the channel (Figure 11.11). Braided stream patterns are formed when a stream branches into a number of smaller channels, each of which flows around the numerous sandbars within the channel (Figure 11.16).

Meandering streams have concave bends in which erosion is dominant and convex bends in which deposition is dominant (Figure 11.14). The arc-shaped bars of sand that accumulate on the convex bends are called **point bars.** Point-bar deposits have good stratification and are composed of coarser sands or gravel at the base, grading upward into finer sands and silts at the top. Such an upward grading sequence is characteristic of point-bar deposits and is used to identify them in the rock record. Successive point bars are deposited as the stream erodes laterally (Figure 11.14). Arc-shaped remnants of older point bars can be seen in the floodplains of many meandering streams (Figure 11.15).

Floodplain Deposits

Floodplain alluvium consists of point-bar sediments deposited as the channel migrates laterally and finer-grained sediments deposited from sus-

pension when floodwaters cover the floodplain. To see how the second mechanism works, we will consider what happens to the sediment in a stream during overbank discharges.

As discharge increases, the stream's velocity and kinetic energy also increase. The increased kinetic energy leads to an increase in the maximum size of sedimentary particle that a stream may carry in suspension (Chapter 6). A stream that carries clay and silt-sized particles in suspension during low discharge may carry sand and silt particles during flood discharge. When discharge exceeds bank-full discharge (Figure 11.5), the stream spills out of its channel and spreads out across its floodplain. The morphologic features and types of alluvium on the floodplain are shown in Figure 11.18.

As soon as the stream spills out of the channel, it slows down and there is an abrupt decrease in kinetic energy. As a result of this abrupt drop in kinetic energy, some of the suspended load and all of the coarse sediment (sand and coarse silt) are deposited near the top of the channel banks. Successive floods build up a low (4- to 5-m) ridge on either side of the channel bank. These low ridges atop the channel banks are called **natural levees** (Figure 11.18). The finer sedimentary particles (fine silt and clay) are deposited across the floodplain as vegetation, and other obstructions decrease the velocity of the floodwaters further. The greater thickness of sediment deposited near the stream in each flood results in a floodplain which slopes away from both sides of the stream channel (Figure 11.18). Breaching of a levee can result in the deposition of coarse stream sediments farther out on the floodplain (Figure 11.18).

In some areas, the natural levees are reinforced artificially and built up with stone and concrete to protect population centers and industrial areas on the floodplain. Nowhere is this more important than in New Orleans, where the Mississippi River is confined between 7.5-m-high levees. The French Quarter bordering the Mississippi River is the highest point in the city of New Orleans, 4.3 m above sea level. The land slopes downward from there, and much of New Orleans is about 1.8 m *below* sea level. Visitors to the French Quarter are frequently startled on hearing boat horns and then looking

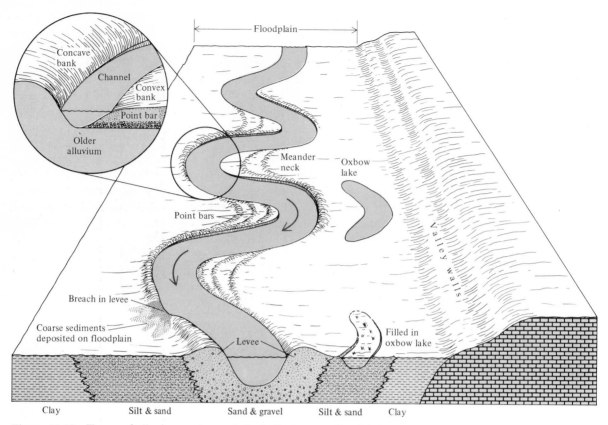

Figure 11.18 Types of alluvium and morphologic features on floodplains.

upward to see the upper portions of large ships passing along the Mississippi River.

As the stream meanders migrate laterally across the floodplain (Figure 11.14), the meanders may become so sinuous (Figure 11.18) that the stream eventually cuts across the narrow neck separating two meander curves. The abandonment of a former meander curve leaves a crescent-shaped lake on the floodplain (Figure 11.18). Arc-shaped lakes formed by cutoff of meander loops are called **oxbow lakes.** Many oxbow lakes subsequently fill with vegetation and fine, suspended flood deposits. Oxbow lakes, filled-in oxbow lakes, and series of point-bar deposits along with sinuous stream channels are the characteristic features on floodplains of meandering streams (Figure 11.15).

Deltaic Deposits

Whenever a stream enters a body of standing water such as a lake, reservoir, or the ocean, its velocity drops and the decrease in kinetic energy results in the deposition of the stream's sediments. A deposit of sediment built by a stream into a body of standing water is called a **delta.** Figure 11.19 shows aerial views of a delta built into the Mississippi River by one of its tributaries. In the 26 years between the first and second aerial photographs, the delta has grown farther into the Mississippi River and the surface of the delta has become stabilized by vegetation. Note that the tributary breaks up into a number of smaller channels that flow across the delta surface into the Mississippi. The smaller channels that branch out across the delta are called **distributaries.** Distributaries commonly are separated by lakes, bays, marshes, or swamps (Figure 11.19).

As the distributaries flow across the delta, their bed load and suspended load (Chapter 6) are deposited on the edge of the delta surface. The coarser bed-load material is deposited

(a)   (b)

Figure 11.19   Growth of the delta of Devil's Creek (Lee County, Iowa) into the Mississippi River over a 26-year period. *(a)* 1930, *(b)* 1956. *(H. P. Guy, U. S. Geological Survey)*

nearest the land, while the finer suspended material is deposited near and beyond the delta edge. This pattern of sedimentation results in a distinctive sequence of layers of sediment (Figure 11.20). The upper surface of the delta is covered with thin beds of distributary sands and the finer organic-rich muds of the inter-distributary areas. The thin and horizontally layered sediments covering the delta top are called **topset beds.** The topset beds grade into the finer, thicker, and inclined **foreset beds** that cover the submerged front face of the delta. Each foreset bed grades in turn into a thin, fine-grained, and horizontally bedded layer called the **bottomset bed.** The bottomset beds extend out from the base of the delta and are the finest and most organically rich deposits of the delta.

The outwardly branching pattern of delta distributaries results in a triangular shape for many deltas. The edges of some deltas protrude very little from the coastline, whereas other deltas, such as the Mississippi, extend outward as a series of fingerlike distributaries (Plate 15). The shape of the delta front is a function of the balance between the supply of sediment from the distributaries and the rate of erosion of deltaic deposits along the shoreline. If more sediment is supplied than can be removed along the coast by current and wave erosion, the delta will extend itself out into the water (Plate 15). On the other hand, any changes in the stream which reduce the supply of sediment can have the opposite effect. The Aswan Dam in Egypt has trapped much of the sediment that formerly reached the Nile Delta on the Mediterranean. This has resulted in increased erosion along the Egyptian shoreline.

STREAM SYSTEMS

Mighty rivers start with raindrops on a slope. The resulting streams evolve into increasingly

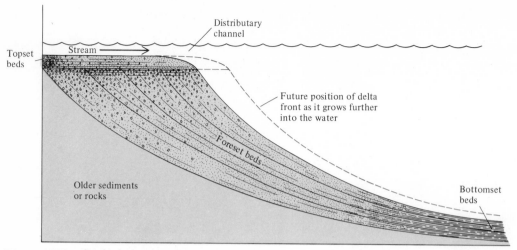

Figure 11.20   Geologic section through a delta growing actively into a body of standing water.

complex and expansive channel systems. Some stream systems, such as the Amazon and the Mississippi, drain areas of subcontinental size. In this section we will discuss how these great stream systems develop and how the patterns they make on the earth's surface provide information about the character of the material over which they flow.

Development of Stream Systems

Rainfall moving down along a smooth inclined slope soon begins to erode shallow and elongate channels into the surface. These channels are localized at places where there are small depressions and at places where the turbulence in the flowing water is stronger. These shallow and temporary channels are called **rills** (Figure 11.21). Once formed, these rills constitute the nucleus of the channel systems which will evolve further in later rainfalls. In subsequent rainfalls, some of these rills are abandoned, whereas others are deepened and joined to each other to form a branching system of stream channels. The area drained by any one of these branching stream systems is called a **drainage basin.** The

Figure 11.21   Sheet and rill erosion begins to form small-scale drainage basins and divides in volcanic ash.   Irazú volcano, Costa Rica.   (*H. H. Waldron, U. S. Geological Survey*)

Figure 11.22 Drainage basins and drainage divides in the San Bernadino Mountains, California. *(J. R. Balsley, U. S. Geological Survey)*

drainage basin of one stream is separated from that of another by a higher intervening area called a **drainage divide** (Figure 11.22).

The growing stream collects water in its headward portion and slowly erodes back into the undissected areas within its drainage basin. Erosion of undissected areas at the upstream ends of stream segments is called **headward erosion.** Sometimes streams with steeper gradients erode headward through drainage divides and intercept the headwaters of streams with more gentle gradients in the adjacent basin. Such an interception and incorporation of another stream's drainage is referred to as **stream capture.** Streams from the smaller drainage basins join with others in an ever-widening branching pattern to form streams with higher discharges. These streams join with other streams to form even larger ones, and so on until master streams, such as the Amazon, Nile, Mississippi, and Congo, are formed.

Stream Networks

Large numbers of interconnected streams are referred to as a **stream network.** The appearance of a stream network on a map or aerial photograph provides the geologist with valuable clues about the characteristics of the surface over which the streams are flowing. As streams erode the landscape, they develop networks with distinctive patterns which reflect the relative erodibility, fracture patterns, and changes in slope in the rocks or sediments being dissected.

A few of the major types of stream network patterns are shown in Figure 11.23. In areas where the surface materials are uniform in character, the streams can erode headward equally in a number of directions. This results in a branch-like stream pattern called a **dendritic** pattern (Figure 11.23*a*). Dendritic patterns constitute the most common type of stream network.

A network pattern in which streams branch out in the *downstream* direction is called a **distributary** pattern. We have already discussed the type of distributary pattern that occurs on deltas as river water fans out into a standing body of water. Another type of distributary pattern occurs on the surfaces of alluvial fans. **Alluvial fans** are triangular-shaped deposits which extend out from the abrupt change in gradient along a mountain front (Figure 11-23*b*).

A stream network composed of elongate

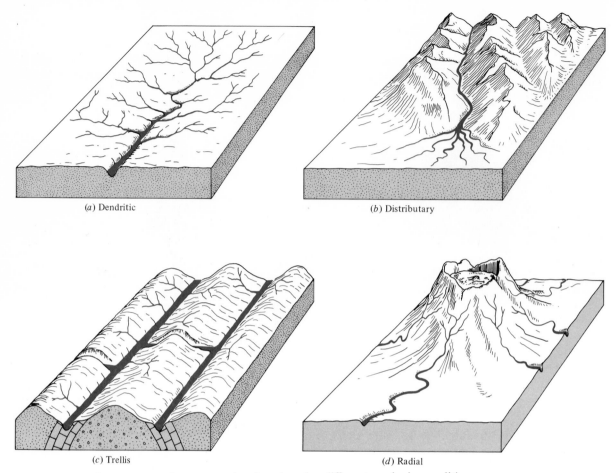

Figure 11.23 Stream network patterns developed under different geologic conditions: *(a)* dendritic; *(b)* distributary; *(c)* trellis; *(d)* radial.

and parallel main channels and short tributaries at high angles is called a **trellis** pattern because of its similarity to a garden trellis. Trellis patterns are common where streams are cutting down through rocks which are folded. The long dimensions of the streams parallel the linear valleys cut into the rocks, while short, high-gradient streams drain the slopes to either side (Figure 11.23c). In areas where the rocks are fractured, the streams follow these fractures, making a series of sharp angular turns in the process. A stream pattern composed of stream segments of roughly equal length making sharp angular junctions with each other is called a **rectangular** pattern. The rectangular pattern differs from the trellis pattern because the main stream segments are not conspicuously elongate

as in the trellis pattern. Streams draining away from all sides of a high area such as a volcano make up a **radial** pattern (Figure 11.23d).

STREAM SCULPTURE

Streams are the most important agents of landscape formation in humid areas and in all but the most arid areas. Stream landscapes consist of both erosional features (valleys) and depositional features (bars, floodplains, and deltas). However, the relative percentages of erosional versus depositional features vary from region to region. We will now look at how streams create landscapes and how such landscapes change with time.

Changes in Stream Landscapes with Time

Fluvial landscapes evolve through a series of stages as the streams erode down to base level and the energy available for stream erosion decreases. Each stage is marked by differences in altitude between the stream channel and the adjacent upland. This measure of relative altitude differences is called the **relief.** Along with the change in relief, each stage also differs in the ratio of stream-channel width to valley width and in the relative importance of lateral erosion versus vertical downcutting by the stream.

To simplify our discussion of stream landscape evolution, we will start with a region of low relief which has been uplifted above base level. The rocks in this area are uniform, and the humid climate and the base level do not change during the evolution of the landscape.

In the beginning stage, the streams in the region are downcutting actively, forming deep valleys whose widths are close to those of the stream channels themselves (Figure 11.24a). There is a great difference in elevation between the actively downcutting streams and the adjacent uplands, and relief in the area reaches a maximum. With the passage of time, the streams erode their channels closer to base level and lateral erosion becomes as important as vertical downcutting.

As time goes on, the streams widen their valleys progressively by lateral erosion. The stream channels then occupy only a portion of the floor of the stream valleys. The lateral erosion of the streams, along with mass wasting, slowly erodes the uplands on either side of the rivers, reducing relief in the area (Figure 11.24b). Continued lateral erosion along with a lesser degree of vertical downcutting leads to the "final" stage of stream erosion. This stage is characterized by streams meandering over a land surface of very low relief near base level. Drainage divides with low relief separate adjacent drainage basins. Little vertical downcutting takes place, and most of the erosion occurs through a combination of lateral erosion and mass wasting (Figure 11.24c). Eventually, stream erosion can reduce the landscape to a featureless plain of very low relief. Such a stream-eroded surface of very low relief is called a **peneplain.**

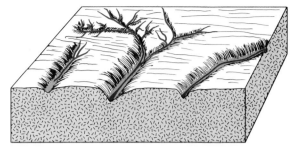

(a) High relief

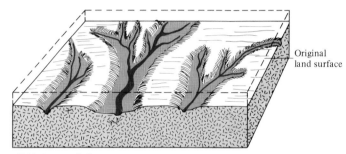

(b) Moderate relief

Original land surface

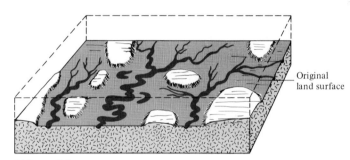

(c) Low relief

Original land surface

Figure 11.24 Stages of stream erosion in an area where the base level has been constant. *(a)* High-relief stage: Vertical downcutting by streams is dominant. The stream channel occupies most of the valley, and regional relief is high. Valley width is close to channel width. *(b)* Moderate-relief stage: Lateral erosion dominates over vertical downcutting. Valley width is significantly greater than stream channel width. Regional relief is moderate. *(c)* Low-relief stage: Streams meander over a wide floodplain. Valley width greatly exceeds stream channel width, and adjacent streams are separated by drainage divides of low relief.

Effects of Lithology, Structure, and Climate on the Evolution of Stream Landscapes

The actual sequence of development of stream landscapes is more complicated than we have illustrated. In our discussion, we assumed that the region was underlain by uniform rocks, that the climate was unchanging, and most important, that the base level remained constant. Actual conditions are usually much more complex. For example, an area underlain by hard unfractured granite may remain in an initial stage of dissection for a far longer period than an area underlain by less-resistant rocks such as limestones and shales. Climatic changes can also affect stream erosion and deposition greatly. A change from an arid to a humid climate is accompanied by a great increase in stream discharge and an acceleration of stream erosion.

Effects of Change in Base Level on the Evolution of Stream Landscapes

The final stage in stream landscapes rarely is achieved because the earth is a *dynamic* body and base levels change more commonly than not. One part of the earth may be uplifted while another part is subsiding. In a similar fashion, the level of the sea may rise or fall considerably over a long period of time. Changes in both land elevation and sea level affect the base level of a stream. When a stream's energy is increased by elevation of the land or by a drop in sea level, the

stream begins to erode its channel actively once again. A stream which begins vertical downcutting again because of a change in base level is called a **rejuvenated** stream (Figure 11.25). On the other hand, a decrease in a stream's energy caused by a drop in elevation of the land surface or a rise in sea level results in accelerated deposition of alluvium. Streams which deposit alluvium actively within their valleys are called **aggradational** streams.

An aggradational-rejuvenational sequence is shown in Figure 11.26. Initially, sea level is low and the stream is downcutting actively and depositing only a thin layer of alluvium across its floodplain (Figure 11.26*a*). A rise in sea level results in stream aggradation and the filling of the valley by a thick deposit of alluvium (Figure 11.26*b*). A subsequent drop in sea level results in stream rejuvenation. The rejuvenated stream cuts downward into its valley fill. In many cases, the excavation of the fill is not complete, and portions of the fill are preserved along the valley walls. These flat erosional remnants preserved above the present level of the stream are called **stream terraces** (Figures 11.26*c* and 11.27).

Figure 11.25 Very sinuous meander in the San Juan River, San Juan County, Utah. Note the narrow neck between the two segments of the river. (*D. Carroll, U. S. Geological Survey*)

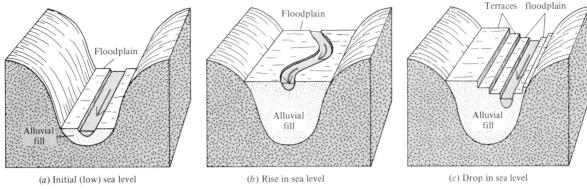

(a) Initial (low) sea level   (b) Rise in sea level   (c) Drop in sea level

Figure 11.26 Stream aggradation and rejuvenation resulting from changes in base level. *(a)* Initial (low) sea level. Stream actively downcutting and depositing a thin alluvial fill as it excavates a valley in older rocks. *(b)* Rise in sea level. Stream deposits alluvium as it adjusts to the higher sea level. A thick alluvial fill now underlies the valley. *(c)* Drop in sea level. Stream erodes into its alluvial fill as it adjusts to the lower sea level. Flat remnants of the fill are left as terraces above the altitude of the present floodplain.

Figure 11.27 Terraces along the north side of the South Fork of the Shoshone River, Park County, Wyoming. Note the slump scars on the face of the lower terrace. *(W. G. Pierce, U. S. Geological Survey)*

SUMMARY

Water exists in solid, liquid, and gaseous states on the earth's surface. The transformation of each of these forms of water into the other forms is called the hydrologic cycle. Streams represent that portion of the hydrologic cycle where surface water is moved in channels under the influence of gravity. The potential energy of the stream is transformed into kinetic energy as the stream flows downslope. Most streams have a longitudinal profile which is concave upward. The profile is composed of a steeper-gradient upstream portion and a gentler-gradient downstream portion. Streams continue to erode the land surface until they reach base level. The base level may be a local one, such as a lake or reservoir, or may be the ultimate base level—sea level.

The geometric and hydraulic variables in a stream are in dynamic equilibrium. Any changes in discharge are balanced by compensating changes in stream width, depth, and velocity. Flooding occurs when discharge exceeds bankfull discharge, causing the stream to overflow its channel and spread over its floodplain. Urbanization has increased both the frequency and severity of flooding by decreasing the lag time and increasing peak discharges. Stream channels may be straight, meandering, or braided depending on the nature of the surface materials and slope, or on the ability of the stream to transport the materials available.

Stream erosion is carried out by the force of the flowing water, by sedimentary particles carried in the water, and by undermining of the stream banks, resulting in slumping of the bank materials into the stream. Falls occur where hard, resistant cap rock overlies weaker material. The falls retreat steadily upstream as erosion of the weak rock undermines the resistant cap rock above.

Streams deposit sediment within their channels, on their floodplains, and in deltas built into standing bodies of water. Alluvium is deposited within the channel as sandbars and on the inside of meander bends as point bars. Levees on top of the channel banks are formed when sediment drops out rapidly as floodwaters spill out of the channel. Floodplain deposits are formed by the lateral deposition of point-bar sediments and by the deposition of suspended sediments from the floodwaters which cover the floodplain periodically. Deltaic deposits built into bodies of water are made up of coarser-grained horizontal topset beds and of finer and inclined foreset beds overlying dark and fine-grained bottomset beds.

Stream landscapes undergo a sequential development with time. This sequence involves a decrease in both regional relief and the ratio of stream width to valley width. With lengths of time and an unchanging base level, the landscape is reduced to a peneplain. Changes in base level resulting from either uplift or subsidence of the land, or rising or falling of sea level, alter the behavior of streams. As base level changes, streams may either aggrade and fill their valleys with alluvium or downcut actively into the valley floor.

QUESTIONS FOR REVIEW AND FURTHER THOUGHT

1. Where does a stream obtain its energy and how is the energy utilized?

2. How does the streamflow formula

Discharge = width × depth × velocity

explain what changes occur in a stream during a flood?

3. What factors favor the formation of the following channel patterns?

   a. Straight

   b. Meandering

   c. Braided

4. How does a stream erode its channel?

5. Describe the differences between stream deposits (*a*) within the channel; (*b*) on the floodplain; and (*c*) in deltas.

6. Describe the evolution of stream landscapes with time. Start with a flat area above base level and assume that this base level and the climate in the area remain constant with time.

7. What effects would the following changes have on a stream?

   a. Elevation of the land surface

   b. Raising of sea level

8. Describe the probable effects of the following modifications on a stream.

   a. Building a dam across a river

   b. Sudden increase in discharge from rapid melting of snow in the headwater area of a stream

ADDITIONAL READINGS

Beginning Level

Belt, C. B., Jr., "The 1973 Flood and Man's Constriction of the Mississippi River," *Science,* vol. 189, 1975, pp. 681–684.
(Describes how human modifications of the Mississippi River have changed its behavior)

Leopold, L. B., "River Meanders," *Scientific American,* vol. 214, no. 6, June 1966, pp. 60–70.
(Describes meanders and the conditions that initiate meandering)

U.S. Geological Survey, *The Channelled Scablands of Eastern Washington — The Geologic Story of the Spokane Floods,* Stock No. 024-001-02507-8,

Catalog No. 19.2 W27/6/974, 1978.
(Detailed and fascinating description of one of the largest floods on the earth's surface)

Intermediate Level

Bloom, A., *Geomorphology: A Systematic Analysis of Cenozoic Landforms,* Prentice-Hall, Englewood Cliffs, N.J., 1978, 510 pp.
(A detailed treatment of stream channels and networks, fluvial deposition, controls on stream erosion, and stream landscape evolution)

Davis, W. M., "The Geographical Cycle," *Geographic Journal,* vol. 14, 1902, pp. 481–504. Reprinted in *Geographical Essays,* Dover, New York, 1954, pp. 249–278.
(Describes the theoretical sequence of landform evolution with time)

Hack, J. R., "Interpretation of Erosional Topography in Humid Temperate Regions," *American Journal of Science,* vol. 258-A, 1960, pp. 80–97.
(Describes the factors working to produce changes in landscapes in humid areas over time)

Leopold, L. B., *Hydrology for Urban Land Planning,* U.S. Geological Survey Circular 559, 1968.
(Describes the effects of urbanization on streams and how proper planning can minimize environmental problems)

Morisawa, M., *Streams: Their Dynamics and Morphology,* McGraw-Hill, New York, 1968, 175 pp.
(A well-written treatment of many aspects of streams, including quantitative analysis of streams and the environmental problems arising from artificial changes made in streams. A good reference to start a more detailed study of streams)

Advanced Level

Schumm, S. A., *The Fluvial System,* Wiley, New York, 1977, 338 pp.
(A detailed modern treatment of many aspects of streamflow and stream erosion)

CHAPTER 12

Groundwater

Water bubbles to the surface in a palm-fringed desert oasis surrounded by sand dunes, streams in humid areas continue to flow for weeks after the last rainfall, water shoots above the surface in wells along a coastline, and limestone is dissolved and redeposited in caverns underground. These different phenomena are all related to subsurface water. Water which accumulates below the ground surface is called **groundwater.** Groundwater rarely reaches the surface naturally, and so it is far less apparent than the surface water we discussed in Chapter 11. However, it is estimated that there is 68 times more groundwater than the surface water in streams and lakes (see Figure 11.1).

Groundwater is one of our most important natural resources. It provides drinking water for many suburban and rural areas. In addition, groundwater supplies enable agricultural and urban centers to exist in dry areas, such as parts of the American West and Southwest, where there is insufficient surface water. To understand how groundwater accumulates and migrates, we must look in more detail at those characteristics of sediments and rocks which enable water to accumulate and flow beneath the earth's surface.

GROUNDWATER ACCUMULATION

Drill a well into sands, sandstones, gravel, or conglomerate (see Chapter 6), and you will probably obtain groundwater. On the other hand, wells drilled into many types of igneous and metamorphic rocks and into fine-grained sedimentary rocks such as shale usually yield little groundwater. Obviously, some types of sediments and rocks have a greater potential for storing groundwater than others. The relative ability of a rock or sediment to store groundwater is called its **groundwater potential.** The groundwater potential is much greater in rocks or sediments with higher porosity (see Chapter 6). Porosity may develop at the time of sediment deposition or sedimentary rock formation. It may also form after the rocks have been buried by younger materials. Some different types of porosity are shown in Figure 12.1. Coarse-grained sediments such as sand and gravel (Figure 12.1a) have pore spaces *between* the sedimentary particles. Such sediments can have porosities as high as 35 percent, and they have great groundwater potential. When sediments are lithified to form sedimentary rocks, the groundwater potential decreases because the addition of cement (or matrix) between the grains reduces the size of the pores (Figure 12.1b). Porosity that develops before or during rock formation is referred to as **initial porosity.**

Many chemical sedimentary rocks, igneous rocks, and metamorphic rocks have no initial porosity because they are composed of tightly interlocking crystals. However, these crystalline rocks may develop porosity *after* they are formed if the rock is fractured subsequently. Porosity that develops after rock formation is referred to as **secondary porosity** (Figure 12.1c). The pores in fractured rocks are not between

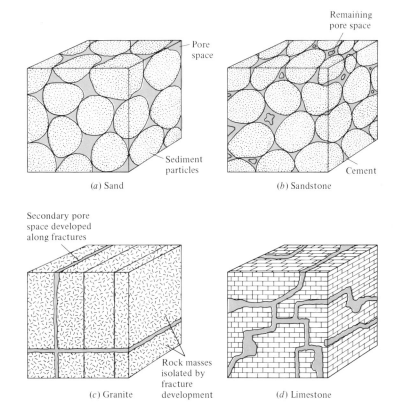

Figure 12.1 Types of porosity developed in sediments and rocks. *(a)* Sand: Pore spaces between grains. *(b)* Sandstone: Pore spaces in areas between grains not filled with cement. *(c)* Granite: Pore spaces develop along fractures within rocks. *(d)* Limestone: Wide pore spaces develop as limestone is dissolved along fracture systems.

the sedimentary particles or crystals but are developed between *masses* of the rock. Groundwater supplies may be obtained from fractured crystalline rocks if the well can be drilled so as to intersect the water-bearing fractures.

A special type of secondary porosity can occur in limestones. Limestone, composed primarily of the mineral calcite, $CaCO_3$, is soluble in acidic groundwater (see Chapter 9). Solution along rock fractures in limestones can create large volumes of secondary pore space (Figure 12.1*d*). We will return to this topic later in this chapter when we discuss the origin of caverns and related features.

GROUNDWATER MOVEMENT

While a stream can flow quite fast, the groundwater in the same area flows much more slowly because it must pass through numerous pores in the sediments or rocks below the land surface. In the following discussion we will look first at

the characteristics that affect groundwater flow and then describe how groundwater flow rates can be determined.

Permeability

In order for groundwater to migrate, the sediment or rock not only must be porous, but the pore space also must be *interconnected*. The percentage of the volume of the sediment or rock that is composed of interconnected pore space is called the **effective porosity**. The effective porosity may be equal to or less than the total porosity. The ability of a fluid to move through a solid is called **permeability.** Well-sorted gravel is permeable, whereas unfractured granite is not. Permeable sediments and rocks must be porous, but porous materials can be impermeable if the pores are not connected. Clay-sized sediments are an example of highly porous material which is almost impermeable. Clay-sized sediments may have total porosities of over 50 percent and hold large volumes of water within their pores when saturated. However, the pores are quite small and few are interconnected, giving clay-sized sediments effective porosities as low as 3 percent. A permeable sediment or rock capable of transmitting groundwater is called an **aquifer.** An impermeable sediment or rock is called an **aquiclude.**

Groundwater moves through sediments and rocks in a number of different ways. Countless minute "threads" of water wind their way slowly through the open pores of sediments (Figure 12.1*a*) and the partially filled pores of sedimentary rocks (Figure 12.1*b*). In contrast, groundwater moves much more rapidly through the larger openings between masses of fractured rock (Figure 12.1*c*). Groundwater moves fastest in solution-enlarged fractures in limestones (Figure 12.1*d*).

Flow Rates

Groundwater moves under the influence of gravity from high to low areas. The rate of groundwater movement first was computed by the French engineer Henri Darcy in 1856. Darcy showed that the discharge of groundwater through a unit cross-sectional area of an aquifer is directly proportional to the permeability of the aquifer and the difference in water level (pressure) between two points, and is inversely proportional to the distance between the two points. This relationship is called **Darcy's Law** and is illustrated in Figure 12.2. Thus, for a given cross-sectional area and length of aquifer, the groundwater discharge increases with both the inclination and permeability of the aquifer.

The well water that you drink may have entered the aquifer long before you were born, perhaps thousands of years ago. Most groundwater flow rates are slow enough that the water moves through the pores in a laminar type of flow (see Chapter 11). The portion of the water closest to the pore walls is slowed down by the molecular attraction between the water and the pore walls, while the water in the center of the pore is affected less and moves faster. In general, the rate of flow of groundwater increases with the size of the pores because a greater proportion of the flow is unaffected by the pore walls. When the pores are very large, as in cavernous limestones, the groundwater may move quite rapidly and in a turbulent type of flow, as in many surface streams (see Chapter 11).

Groundwater flow rates are generally of the magnitude of a few centimeters to less than a meter per day, although higher and lower flow rates occur in special situations. The generally slow rate of groundwater movement has considerable environmental significance. It is this slow flow through interconnected pore spaces that allows oxygenation and soil microorganisms to break down any harmful substances, such as sewage, which may be introduced accidentally into the groundwater. The slow rate of groundwater flow is an important point to consider in light of the great volumes of well water being pumped out each day.

Figure 12.2 The rate of flow of groundwater in an aquifer can be predicted by Darcy's Law.

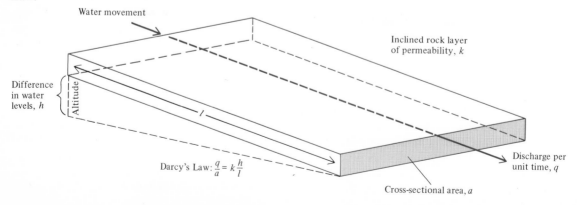

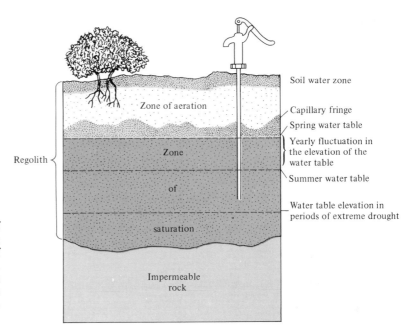

Soil water zone

Capillary fringe

Spring water table

Yearly fluctuation in the elevation of the water table

Summer water table

Water table elevation in periods of extreme drought

Zone of aeration

Zone

of

saturation

Regolith

Impermeable rock

Figure 12.3 Vertical distribution of groundwater. The contact between the saturated zone and the zone of aeration is called the water table. The elevation of the water table varies both seasonally and as excess amounts of groundwater are removed.

UNCONFINED AQUIFERS

Rainfall infiltrates the ground surface and moves downward under gravity flow until it is stopped by an impermeable material. The groundwater then begins to accumulate above the impermeable surface, gradually filling the pore spaces and rising toward the ground surface. If the rainfall is sufficient and spread out over a long period of time, all the pore spaces in the ground may become saturated. However, the permeable rocks and sediments usually are not saturated all the way up to the surface, and the groundwater is distributed in several distinct zones (Figure 12.3).

Vertical Distribution of Groundwater

As the rain infiltrates through the plant cover, small droplets are retained within the root system. This thin belt of partially saturated pores within the root system is called the **soil water zone.** Below the soil water zone is the thicker **aeration zone,** in which the pores are partially filled with water in transit between the surface and the **saturation zone.** The saturated zone is where all the pores are filled with water. The saturation zone extends downward to the im-

permeable materials below. At the very top of the saturated zone is a thin zone, called the **capillary fringe,** where water is held in some of the pores against the force of gravity in much the same way that a dry blotter draws up a drop of water.

The contact between the aeration zone and the saturation zone is a surface called the **water table.** The water table in a given area does not remain at a constant elevation but varies both seasonally and because of the discharge of groundwater in greater volumes than can be replenished by rainfall. The addition of water to the groundwater reservoir is called **recharge.** In spring and fall, recharge is heavy and the water table rises. In summer and winter, recharge is lower and the water table drops in elevation (Figure 12.3). Similarly, if groundwater is pumped out of the ground in greater volumes than can be recharged at that time of the year, the water-table elevation drops.

Water-table elevations also reflect longer-term climatic changes. During a time of extreme drought, the water table can drop below the bottom of local wells, and the wells dry up (Figure 12.3). The prolonged drought in the Northeastern United States from 1963 to 1966

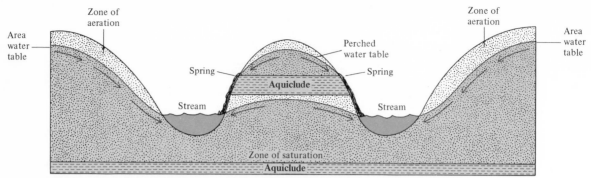

Figure 12.4 Variations in the elevation of the water table with topographic changes. Impermeable beds result in the formation of local perched water tables. Groundwater may migrate across the top of these aquicludes and exit on the sides of the hill as springs.

limited groundwater recharge at a time when many new homes were being built in New York's suburban Nassau and Suffolk Counties on Long Island. Average water-table elevations dropped as much as 2 m during the drought. Builders digging down for basements found no water; when precipitation and groundwater recharge began returning to normal by the early 1970s, water-table elevations rose and basement flooding became a serious problem for many homeowners.

Aquifers such as we have been describing are called unconfined aquifers. An **unconfined aquifer** is one with a free water table which is at atmospheric pressure.

Shape of the Water Table

The water table is a three-dimensional surface that is a subdued "replica" of the land surface above. The water table is highest in elevation under hills and lower in elevation under more gently sloping areas (Figure 12.4). If downward-infiltrating groundwater encounters an aquiclude, it builds up above the aquiclude, forming a local water table called a **perched** water table (Figure 12.4). A more subdued regional water table underlies the local perched water table. If the perched water table is on a hill, the groundwater above may flow along the top of the aquiclude and exit on the side of the hill as a spring. A **spring** is any natural discharge of groundwater that appears at the ground surface or that flows up through surface water bodies such as lakes, rivers, or even ocean. Groundwater is insulated from the effects of surface temperature changes, and, therefore, it maintains a fairly uniform temperature year-round. Consequently, spring water may be cooler in summer and warmer in winter than local surface water.

Effects of Excess Discharge on the Water Table

The elevation of the water table is affected not only by periods of high rainfall or drought but also by the pumping of groundwater. Pumping usually empties the pores of the aquifer at a faster rate than they can be refilled by the slow rate of groundwater flow. This excess withdrawal results in a conically shaped depression called a **cone of depression** in the water table in the area closest to the well (Figure 12.5).

The cone of depression results from the temporary exhaustion of groundwater from the pores nearest the well. As withdrawal of excess groundwater continues, the cone of depression widens. Excess groundwater pumping may result in the local water table being lowered below the level of adjacent wells, drying them up (Figure 12.5). As soon as pumping ceases, the groundwater flow begins to fill in the pores again. If recharge is sufficient, the water table eventually may be restored to the position it had prior to the pumping.

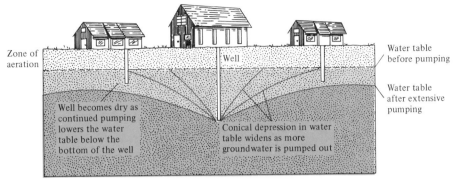

Figure 12.5 Groundwater withdrawal in excess of aquifer recharge forms a temporary conically shaped depression (color lines) in the water table around the well. With continued pumping, the expanding conical depression may draw the water table below the level of adjacent wells, causing them to dry up temporarily.

CONFINED AQUIFERS

Aquifers which are confined above and below by impermeable materials and which contain groundwater under pressure significantly greater than atmospheric pressure are called **confined aquifers.** The conditions necessary for a confined aquifer are shown in Figure 12.6. Water enters the aquifer in a humid recharge area at the base of the distant mountains and flows by the force of gravity downward through the pores of the aquifer. The water in the pores in the deeper parts of the aquifer is under great pressure from the weight of water higher in the aquifer. This pressure cannot be relieved because the aquifer is sealed above and below by aquicludes.

When a well is drilled into the confined aquifer, the pressure is relieved and the water rises upward against the force of gravity toward and even above the ground surface. Wells in which the water flows upward under its own pressure

Figure 12.6 Geologic section showing confined and unconfined aquifers, flowing artesian wells, and the piezometric surface.

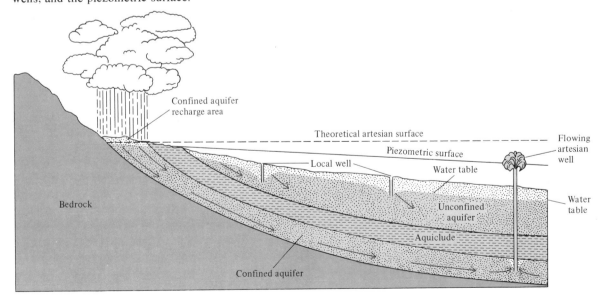

are called **artesian wells.** Theoretically, the water in an artesian well should rise to the same elevation as where it entered the aquifer in the recharge area. Actually, the water rises to a level *below* the theoretical level because of the energy lost to friction as the water moves through the pores of the aquifer. The actual height to which water will rise if a confined aquifer is tapped is called the **piezometric surface.** If the piezometric surface is above the local land surface, the water will flow out of a well when the confined aquifer is tapped. In the early stages of water removal from artesian aquifers, little or no pumping may be required. However, as more and more wells tap the aquifer, the pressure drops and the groundwater eventually must be pumped to the surface.

RELATIONSHIP BETWEEN SURFACE WATER AND GROUNDWATER

There is usually a close relationship between groundwater and surface water. In humid areas, the local water table slopes toward the stream and the groundwater recharges the stream (see Figure 12-4). Streams which receive a portion of their discharge from groundwater are called **effluent** streams (Figure 12.7*a*). It is this year-round groundwater flow that enables streams in humid climates to continue to flow during periods of drought. That portion of the stream discharge which is supplied by groundwater is called the **base flow.**

In dry climates, the local water table slopes *away* from the stream and the stream recharges the groundwater supply. Streams in which the groundwater is supplied by the river are called **influent** streams (Figure 12.7*b*). A good example of an influent stream is the Colorado River. The Colorado rises in the Rocky Mountains and flows southward through the deserts of the American Southwest before it reaches the Gulf of California. How can a river such as the Colorado flow year-round through great distances of desert and not dry up? The explanation is that the Colorado River discharge, plus additional supplies from local springs, exceeds the volume of the water that is lost by evaporation into the dry desert air and by infiltration into the areas bordering the stream. The stream water that infiltrates into the ground supports a narrow belt of vegetation that borders each side of the stream (Figure 12.7*b*).

A stream may be classified on the basis of its year-round flow continuity and its relationship with the local water table. **Perennial streams** flow year-round, supplied largely by base flow (Figure 12-7*a*). **Intermittent streams** flow part of the year, receiving a portion of their water from the local water table when it is high enough.

Figure 12.7 Relationship between surface water and groundwater under different geologic and climatic conditions. *(a)* An effluent stream supplied by base flow from a water-table aquifer is characteristic of continually flowing streams in humid climates. *(b)* An influent stream supplies water to adjacent ground and is characteristic of through-flowing streams in dry regions. Influent flow supports belts of vegetation parallel to the stream banks.

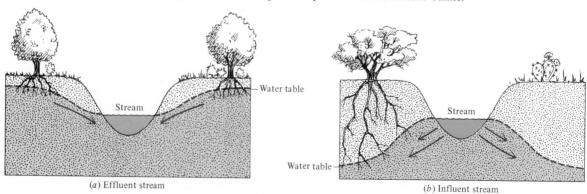

(*a*) Effluent stream   (*b*) Influent stream

Ephemeral streams are dry most of the year and contain water only for a short time after it has rained.

Many streams in the American Southwest are dry for part or most of the year. The local water table is usually below the stream channel and is recharged for only a short period of time after a rainfall. The only vegetation that borders such ephemeral or intermittent streams are plants with very deep roots that can reach the water table below or cactus, which can absorb moisture quickly and retain it through the dry intervals between rainfalls.

Some streams flow in channels cut into impermeable sediments or rocks. In such cases, there may be no interchange between the stream and any aquifers below. Deprived of any base flow, such streams dry up in periods of drought.

GROUNDWATER SYSTEMS

Groundwater may be obtained from a great variety of sediments and rocks. Some of these groundwater sources contain smaller volumes of water that are sufficient to supply small populations or to be used by light industry. Other aquifers extend under whole states or regions and contain immense volumes of groundwater. These regional sources can support large populations and agricultural operations.

Local Aquifer Systems

Most local groundwater resources are derived from unconfined aquifers in permeable surface sediments or rocks. The most commonly tapped aquifers are those which developed in ancient and modern stream deposits. For example, large volumes of groundwater in the Midwest and New England are obtained from sands and gravels originally deposited by streams from melting glaciers. Other communities obtain groundwater from wells in the floodplain and on islands within the channels of present rivers.

Another major type of local groundwater supply occurs in normally impermeable rocks when they are cut by fractures of various kinds. Rainwater infiltrates the surface and percolates downward along the rock fractures. The groundwater builds up above any impermeable rock below. The resulting water table (Figure 12.8) is formed by the interconnected groundwater in the rock fractures. Wells drilled into the intersecting fracture system may tap quantities of water that are sufficient for local use. Groundwater trapped in fractured rock differs from groundwater trapped between the sedimentary particles in sands and gravels in a number of important ways. The water table in fractured rocks is more irregular owing to unequal filling of the rock fractures (Figure 12.8). Groundwater moving through the large pore spaces in fractured

Figure 12.8 Wells in fractured rock.

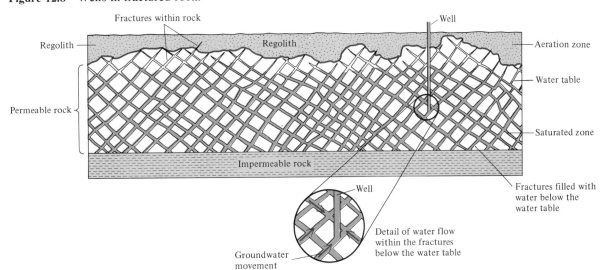

rocks can move much faster than groundwater moving between sedimentary particles. This means that pollutants introduced into an aquifer in fractured rock can travel great distances, polluting other groundwater supplies rapidly.

Regional Aquifer Systems

Water supplies for many large cities in the United States are obtained from artesian aquifers underlying whole states or even regions. Two of the major regional aquifers in the United States underlie the Great Plains, stretching from the Rocky Mountains eastward toward the Mississippi River Valley. These two prominent aquifers are the Cretaceous Dakota sandstone of South Dakota and Nebraska and the Tertiary Ogallala conglomerate of western Texas and New Mexico. The occurrence of groundwater in these two aquifers is described in Figure 12.6.

Other types of regional aquifers occur under coastal plains, islands, areas underlain by vol-

canic rocks, basins between mountain ranges, deserts and semiarid areas, and areas underlain by limestone. Each of these regional aquifers has a wide areal extent, but the characteristics of the groundwater in each are quite different. We will discuss all these aquifer types briefly because each one tells us something about groundwater occurrence, quantity, and quality in different areas.

Coastal aquifers

Many cities on the Atlantic and Gulf Coasts of the United States obtain their water from aquifers which are recharged in inland areas and extend under the coastal plains into the oceans. The structure of these aquifers is similar to the general regional aquifer model shown in Figure 12.6. The aquifers extending under coastal areas differ from the general model in one respect; the pores in the oceanward ends of the aquifers are filled with salt water. Aquifers which are in-

Figure 12.9   Freshwater-saltwater relationships in coastal aquifers. During periods of high rainfall, freshwater displaces salt water in the aquifers, and freshwater springs occur in the bays. During periods of drought, salt water moves up the aquifers (dashed lines) and contaminates coastal wells.

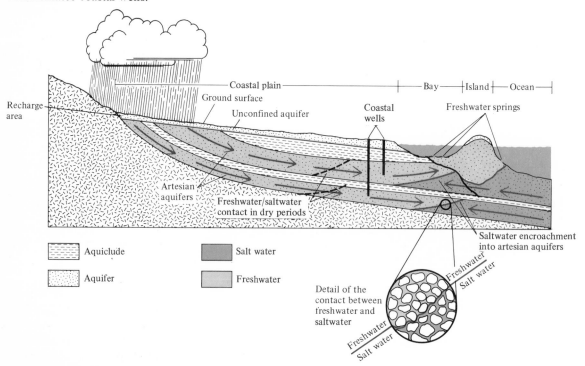

truded partially by salt water and which extend under coastal areas are called **coastal aquifers.** Freshwater and salt water are in delicate balance in a coastal aquifer. The freshwater is lighter (density equals 1.00 g/cm^3) than the salt water (density equals 1.025 g/cm^3), and, therefore, the freshwater tends to "float" on top of the salt water in the pore spaces where the two are in contact (Figure 12.9).

The penetration of salt water into aquifers is called **saltwater encroachment.** Saltwater encroachment in a coastal aquifer increases as the ratio between aquifer recharge and freshwater withdrawal decreases. The saltwater contact shown in Figure 12.9 represents the position of the freshwater-saltwater contact at a time of high rainfall. During this period, the coastal wells would be drawing off freshwater from both the upper and lower artesian aquifers, and freshwater would be leaking upward under pressure as **artesian springs** into the saltwater bay. Freshwater artesian springs are common anywhere that freshwater under pressure is driven upward into an open body of water.

The balance between freshwater and salt water in the coastal aquifers changes markedly in periods of drought. The reduced freshwater recharge into the aquifers allows the saltwater encroachment to reach higher into the coastal aquifers (Figure 12.9). The position of the freshwater-saltwater contact during dry periods is shown by the dashed lines in Figure 12.9. During a period of drought, the coastal wells would be pumping salt water from the aquifers and there would be no freshwater springs into the bay. The island shown in Figure 12.9 preserves its own freshwater supply.

Island aquifers

Have you ever wondered how some islands can have abundant natural supplies of freshwater even though they are surrounded by salt water? The answer lies in the difference in density between freshwater and salt water. So long as there is rainfall and the island surface is permeable, freshwater will infiltrate into the ground. The weight of this freshwater acts like a piston, displacing salt water from the pores

under the island. After a while this mechanism builds up a "lens"-shaped body of freshwater that overlies the salt water in the pores beneath the island (Figure 12.10).

The depth of freshwater at any place on the island can be calculated if the height of the water table above sea level is known. Because of the difference in density between freshwater and salt water, a column of freshwater 41 m high equals the mass of a similar column of salt water only 40 m high. This relationship is called the **Ghyben-Herzberg ratio** and allows us to determine the thickness of the freshwater at any point on the island. For every meter the freshwater table is elevated above sea level, there is 40 m of freshwater below. The thickness of freshwater below any point on the water table may be determined from the following formula:

$$T = 41H$$

where T is the thickness of freshwater below the point and H is the height of the freshwater table above sea level at *that* point (Figure 12.10). For example, if the water table in the center of the island is 1 m above sea level, the total thickness of freshwater below *that* point is

$$T = 41H = 41 \text{ m}$$

Because the water table follows the contour of the land, it has a lower altitude near the ocean, and the thickness of freshwater there is correspondingly less. This relationship gives a lens shape to the freshwater reservoir beneath a permeable island. In dry periods or periods of high groundwater withdrawal, the water-table elevation drops, the size of the freshwater lens below decreases, and saltwater encroachment occurs in wells on the island. Island aquifers are sources of water for people living on barrier islands of the Atlantic and Gulf Coasts of the United States as well as larger islands such as Bermuda, the Bahamas, and the many coral islands of the South Pacific.

On smaller islands with relatively impermeable rocks, precipitation runs off mainly along the surface. On steep islands with impermeable rocks, such as St. Thomas in the Virgin Islands, the fast-moving surface water can also result in dangerous flash floods (Chapter 11). On moun-

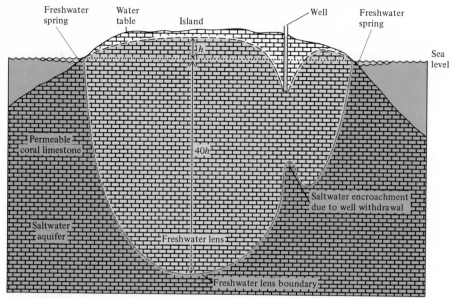

Figure 12.10 Distribution of groundwater under a permeable island. The lens-shaped freshwater body sits atop the salt water below. For every meter the freshwater table is above sea level, there is 40 m of freshwater below that point.

tainous islands, such as St. Thomas, Gibraltar, and Hong Kong, at least a portion of the drinking water is obtained by catchment. **Catchment areas** are large paved-over slopes that empty into catchment (collecting) basins that store the water. Catchment systems are also used to supplement the water supplies on permeable but low-lying islands such as the Bahamas. Because of the low elevations of these islands there is a thinner freshwater lens under them (see Figure 12.10). This thinner lens of freshwater is much more susceptible to saltwater encroachment resulting from low recharge. On these islands, groundwater supplies are augmented by rainwater trapped on roofs and channeled downward into underground storage tanks called **cisterns.**

Volcanic aquifers

Most igneous rocks are relatively impermeable; however, some extrusive igneous rocks, such as basalt, can be quite permeable. Pore spaces in basaltic flows result from shrinkage cracks, lava tubes, gas vesicles, and molds of trees and other objects engulfed by the lava flow (see Chapter 5). Layers of altered volcanic ash and soils developed between basaltic flows act

as aquicludes which contain the water in the pores of the interbedded basaltic flows. The groundwater moves under the influence of gravity and exits at the sides of hills as springs (Figure 12.11).

Large volumes of water are obtained from volcanic aquifers in the Hawaiian Islands. Tunnels are excavated along the areas of contact between volcanic ash layers and permeable basaltic flows. The groundwater accumulates above the aquiclude and is contained within the tunnels by massive gates. One of these tunnels, drilled along an ash layer below a permeable basalt in the Kaui District, Hawaii, had an average daily discharge of 1,862,000 liters (490,000 gallons) per day in 1935.

Alluvial basin aquifers

Basins that are bordered by mountains and filled with stream deposits are called **alluvial basins.** Groundwater is stored in thick aquifers underlying the basins. These aquifers are recharged by rainfall entering the aquifers at the edges of the basins. Groundwater is obtained from alluvial basin aquifers in the numerous basins in Nevada, western Utah, western Colorado, southern Arizona, and southwestern New Mexico.

The most intensive use of basin aquifers is in California. About 40 percent of California is underlain by alluvial basins that contain groundwater. The groundwater resources were especially important in supplying water during the drought of 1976 to 1978, when surface water supplies were minimal. The combined groundwater storage capacity of all of California's groundwater basins has been estimated to be nearly 30 times the total surface water storage capacity of all the reservoirs in the state. However, only about 11 percent of this stored groundwater both is of acceptable quality and can be withdrawn economically. Excessive groundwater withdrawal from basin aquifers in California has created serious environmental problems. We will discuss several of these problems later in this chapter.

Desert aquifers

You may have read about, or seen movies showing, patches of vegetation and standing water called **oases** in the middle of a desert. How can there be abundant water supplies in the absence of streams? A desert oasis results when there is any break in a confined aquifer below, forming artesian springs which rise to the surface (Figure 12.12).

A necessary condition for the formation of an oasis, or for the formation of any desert spring, is the presence of a confined aquifer below with a recharge area in a humid region on the edges of the desert. The confined aquifer also must be fractured in some way to allow the water to reach the surface by artesian flow. Figure 12.12 shows a sandstone aquifer inclined under a desert. The aquifer is broken and offset by a fault which displaces part of an aquiclude so that it stops the aquifer flow. The trapped water is then forced upward by artesian pressure. The water moves up along the fault and exits as flowing artesian springs at the surface. The presence of a local water table in the oasis is indicated by the permanent vegetation and the year-round bodies of freshwater at the surface.

Many deserts do not contain springs or oases, either because the aquifers below are dry or because the saturated aquifers below are not fractured, preventing the groundwater from reaching the surface. Not all desert springs contain drinkable water. Some groundwater dissolves harmful minerals on its way up to the surface.

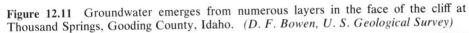

Figure 12.11 Groundwater emerges from numerous layers in the face of the cliff at Thousand Springs, Gooding County, Idaho. *(D. F. Bowen, U. S. Geological Survey)*

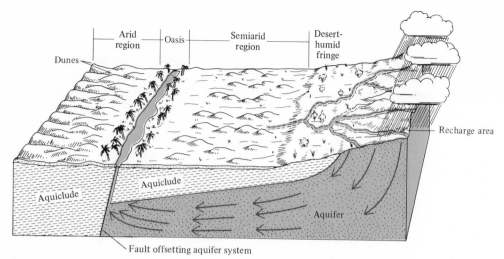

Figure 12.12   Distribution of groundwater beneath a desert. The oasis formed because there was a break in the confined aquifer below, which allowed the water to reach the surface as artesian springs.

Limestone aquifers

Limestone aquifers differ from the other types of aquifers because groundwater not only moves through the limestone but also dissolves the limestone in some places and deposits some of the dissolved calcium carbonate in other places. Groundwater moving through fractures in limestone gradually widens the openings by solution (Figure 12.13), making it possible for even more groundwater to move through the rock.

Local supplies of groundwater are present in many areas of the United States underlain by limestone. The major regional limestone aquifers in the United States are the Roswell Artesian Basin, developed in Permian limestones in eastern New Mexico, and the Florida aquifers developed in Tertiary and Quaternary limestones beneath the peninsula of Florida and in adjacent states.

GROUNDWATER EROSION AND DEPOSITION

We mentioned previously that groundwater moving through limestone is unique in that it has the capability to dissolve the limestone and deposit the calcium carbonate in another area. This solution and redeposition of calcium carbonate results in the formation of unique surface and subsurface features.

Solution of Calcium Carbonate

Limestone is insoluble in pure water but dissolves readily in water made acid by the addition of carbon dioxide. Water infiltrating the ground surface picks up carbon dioxide from decaying plant material in the uppermost part of the soil, forming carbonic acid (see Chapter 9). The solution of the limestone may be represented by the simplified reaction:

$$\underset{\text{Limestone}}{CaCO_3} + \underset{\text{Water}}{H_2O} + \underset{\substack{\text{Carbon}\\\text{dioxide}}}{CO_2} \rightleftharpoons \underset{\substack{\text{Calcium}\\\text{ions}}}{Ca^{2+}} + \underset{\substack{\text{Bicarbonate}\\\text{ions}}}{2HCO_3^-}$$

This reaction is reversible. If the CO_2 in the solution is increased, the reaction moves to the right (more limestone can be dissolved). Conversely, if the CO_2 is lost, the reaction moves to the left and $CaCO_3$ can be deposited from the solution.

The rate of limestone dissolving in solution is increased greatly when the rocks are fractured because there is a greater surface area exposed to solution (see Chapter 9). Not all limestones are pure $CaCO_3$, and most contain small amounts of nonsoluble minerals such as quartz sand and clay minerals. As the limestone is dissolved, the insoluble minerals accumulate to form part of the "cave earth" found in many solution cavities. The presence of leaves and mineral particles which could not have been derived from solution of the enclosing limestone indicates that part of the cave earth is deposited by waters washing regolith through the rock fractures from the surface above.

Solution of limestone is believed to occur most rapidly at and just below the water table. Vertical cutting of solution cavities can be accelerated when the water table is lowered because the groundwater flowing through many solution cavities is connected to the regional drainage system. Any lowering of base level for the regional streams would lead to their incision into the landscape (see Chapter 11). The consequent lowering of the water table results in renewed downward solution of the limestone. Such renewed solutional erosion of cave floors can be caused by a regional uplift of the area or by a drop in sea level (see Chapter 11).

Limestone Solution Features

Progressive solution of the limestone can create large underground cavities which are usually connected to each other by passageways. These features can range in size from openings just wide enough to crawl through to wider and higher **caves** and, under ideal conditions, to massive **caverns** such as Carlsbad Caverns, New Mexico; Mammoth Cave, Kentucky; and Luray Caverns, Virginia. Some caves are systems of long intersecting tunnels, while others consist of much larger passageways opening here and there into high-roofed chambers. In areas of flat-lying limestones, the horizontal part of many caves develops through solution along the bedding of the limestones, while the high chambers develop where there has been accelerated solution of the limestone along vertical fractures. One cavern system developed in this manner is Mammoth Cave in Kentucky (Figure 12.14).

As underground caves and caverns progressively enlarge by solution, the land surface

Figure 12.13 Linear solution cavities in underlying limestone exposed by stripping of the regolith, Oneida County, New York. *(C. D. Walcott, U. S. Geological Survey)*

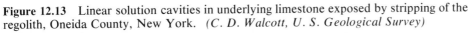

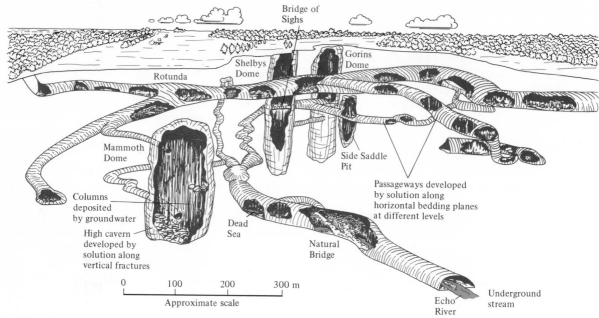

Figure 12.14 Cross section of Mammoth Cave, Kentucky. This cavern system contains a wide variety of limestone erosional and depositional features. The high-roofed "domes" follow vertical rock fractures, and the passageways at different levels follow the horizontal bedding planes of the limestone. Certain passageways have been cut away in the diagram in order to show their internal features. (*Modified from* Geomorphology *by Lobeck, A. K.,* © *1939. Used with permission of McGraw-Hill Book Company*)

above may lose its support and may collapse into the cavern below. A circular depression in the land resulting from solution of underground limestones is called a **sinkhole.** Sinkholes may be dry or filled with water. Dry sinkholes form where the collapsed rock rubble isolates the base of the sinkhole from the local water table. Some sinkholes have lakes in their bases. Sometimes these lakes are surface water accumulations that build up above an impermeable base of collapsed rubble and cave earth. More commonly, the lakes in sinkholes are surface expressions of the water table below. The surface of the sinkhole lake is closest to the land surface in areas where the water table is nearest to the surface.

Karst topography

Areas underlain by limestone contain a distinct set of surface features which result from subsurface solution. They are referred to collectively as **karst topography,** named from an area in Yugoslavia where they are particularly well developed. The formation of these features is shown in the series of block diagrams in Figure 12.15. The area illustrated is "capped" by a resistant layer of sandstone into which perennial streams have cut channels (Figure 12.15a). Subsurface solution of the limestone removes support for the surface, and it collapses, forming sinkholes on the surface. With the exposure of the underlying limestone and its solution cavities, there is a great reduction in surface drainage. Rainfall flows briefly along the surface into sinkholes and fractures and then through underground solution cavities, perhaps emerging on the surface again at a lower elevation. The prevalence of subsurface drainage leaves a series of **dry valleys** at the surface (Figure 12.15b). Sinkholes, the absence of surface drainage, and dry valleys are characteristic features of karst topography. Sinkhole expansion may expose part of the underground stream channels and preserve a portion of the undissected surface as a **natural bridge** (Figure 12.15c). Very little of the original land surface is preserved by this stage, and sinkholes and dry valleys occupy a large part of the land surface. In the latest stage, vertical downcutting and solution cease because an underlying insoluble layer has been reached (Figure 12.15*d*). The same effect could have been attained if the area had been eroded down to regional base level. At this stage, perennial drainage may develop on the surface once again. Residual masses of limestone form isolated **karst towers** above the surrounding lowlands. Karst towers form dramatic landscapes in Puerto Rico, Jamaica, Vietnam, and especially China, where karst towers are traditional features in classic Chinese art (Figure 12.16).

Figure 12.15 Development of karst topography in a humid region, assuming there is no change in the base level of the streams in the region. (*Modified from Lobeck, 1939*). (*a*) Surfical drainage is established on insoluble rocks. Solution of the underlying limestone forms sinkholes into which the surface drainage is diverted. (*b*) Little if any surface drainage as more and more of the limestone is exposed by erosion. Cavern systems extend through the subsurface. (*c*) Sinkhole expansion creates solutional valleys, and cavern system reaches maximum development. (*d*) The area has been reduced to a low relief through extensive solution of the limestone. Surface drainage becomes established once again on insoluble rock. Streams flow between isolated hills of soluble rock. (*Modified from Geomorphology by Lobeck, A. K., © 1939. Used with permission of McGraw-Hill Book Company*)

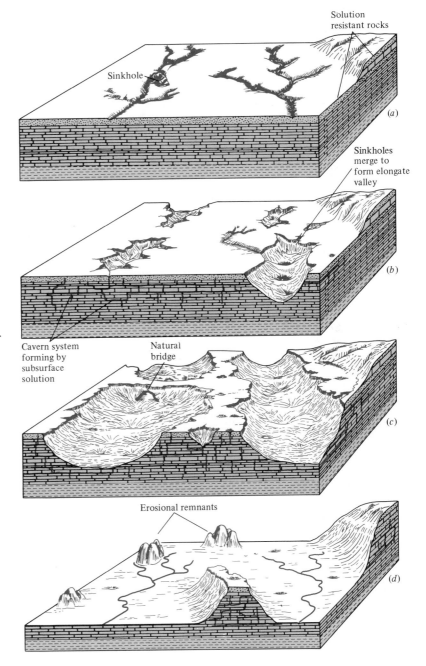

Figure 12.16 Karst towers along Li River near Kweilin in southern China. *(Chang Yunlei, China Pictorial)*

Limestone Depositional Features

Percolating groundwater also may deposit calcium carbonate under the right conditions. Deposition of calcium carbonate from groundwater takes place in the following way. The migrating water contains dissolved carbon dioxide and calcium and bicarbonate ions. As this groundwater migrates through the pore spaces and drips slowly into larger cavities, the excess carbon dioxide diffuses into the cave atmosphere, and calcium carbonate is deposited as a result.

Calcium carbonate is deposited within caves and caverns in a number of different forms (Plate 16). Water dripping from cave ceilings deposits needlelike **stalactites** extending downward from cave ceilings. Water dripping down the stalactites and onto the cave floor below deposits calcium carbonate as broad pinnacle-like **stalagmites.** Broad columnar **pillars** are formed where stalactites and stalagmites meet. Water dripping down cave walls and across cave floors deposits a banded form of calcium carbonate called **flowstone.**

THERMAL SPRINGS AND GEYSERS

So far we have been talking about groundwater springs in which the water generally is cooler than the surface waters. Some springs, called **thermal springs,** are much warmer than the surface waters in their area. Thermal springs are found where magma bodies near the surface heat the groundwater before it reaches the surface. The hot water can contain high concentrations of dissolved ions. As the waters reach the surface, they cool, and many of the ions are precipitated as mineral deposits around the thermal spring.

When groundwater emerges as a column of steam and hot water, it is called a **geyser.** The best-known geyser in the United States is Old Faithful in Yellowstone National Park (Figure 12.17). Geysers form where narrow vertical rock openings fill with water from below. The following is a simplified explanation of geysers; in many cases the "plumbing" of geysers is more complex.

As more and more water accumulates within the rock openings, there is greater and greater pressure on the water below. As the pressure on the water increases, the water boils at a higher and higher temperature. Boiling of the water at the base of the rock fractures starts when either the pressure is reduced or the tem-

Figure 12.17 Old Faithful Geyser, Yellowstone National Park, Wyoming. *(I. J. Witkind, U. S. Geological Survey)*

perature increases. As the basal water begins to boil, steam forms. This steam pushes the water above it upward. This decreases the pressure on the overlying water, causing it to boil in turn. Soon, a jet of steam and hot water shoots above the surface as a geyser. After the geyser erupts, the rock fractures fill with more groundwater, and the cycle begins again.

GROUNDWATER RESOURCES

Groundwater is one of our most important resources. In areas of low rainfall or little surface drainage, such as limestone areas, it may be the only water supply. Increasing development in formerly rural areas has created great demands on our groundwater resources. Unfortunately, the slow rate of groundwater infiltration and movement rarely balances the withdrawal rate. Another problem is that groundwater resources are being polluted with increasing frequency. In this section we will consider how groundwater resources are developed, how they are being recharged, and some of the types of pollution that are reducing groundwater quality worldwide.

Development of Groundwater Resources

As an area undergoes a transition from rural to suburban to urban, its water needs change markedly. The type and quantity of groundwater sufficient at one stage of development may be quite inadequate at another stage.

Rural areas generally obtain their groundwater supplies from shallow wells tapping unconfined aquifers such as permeable glacially deposited sands, floodplain deposits, zones of fractured rock, or limestone aquifers in which the fractures have been widened by solution. This type of groundwater development is adequate so long as withdrawal is modest and wells are far enough from waste disposal sites (cesspools and septic tanks) so that the shallow groundwater supplies are not contaminated.

Suburban areas place a higher demand on groundwater resources. As an area becomes more urbanized, the shallow aquifers become contaminated by runoff from streets and pavements, industrial discharges, and sewage. Such suburban areas must then drill deeper to tap the confined aquifers below. Generally, these deeper confined aquifers are recharged in undeveloped areas far away and are protected from local pollution by aquicludes. Numerous suburban areas, such as Nassau and Suffolk Counties on Long Island, New York, discussed earlier in this chapter, obtain groundwater from deeper confined aquifers.

Groundwater supplies are almost always inadequate for urban areas, especially those with associated industrial complexes. Urban areas obtain most of their water from lakes or reservoirs in distant areas. The surface water is brought into the cities by gravity flow through large underground pipes or surface aqueducts. Among the major cities using such water systems are San Francisco, Los Angeles, Houston, Boston, and New York. The city of Denver is located on the high plains just east of the Rockies. This area is fairly dry, but sufficient water was available when Denver was much smaller. As the city grew, it began to outstrip the local water supply. The city then began to tap surface water on the *western* side of the Rockies and bring it into Denver via tunnels drilled through the Rockies.

Aquifer Recharge

Adequate groundwater supplies require that the aquifers be recharged continually. To preserve the quality of groundwater, it is also necessary to protect the recharge areas from any pollution.

Changes in both the quantity and quality of groundwater in Florida have occurred since the state underwent dramatic increases in both population and industry in the 1950s. The limestone aquifers in Florida are recharged in the high central area of the state, and the groundwater flows away from this area into the Gulf of Mexico on the west and the Atlantic Coast on the east. Prior to 1950 this high central part of Florida was much less developed than it is now, and large volumes of water were able to infiltrate and recharge the aquifers below. With the increasing urbanization of central Florida, more of the surfaces were paved over and aquifer recharge decreased. This decrease in recharge, plus the increased groundwater needs of rapidly growing cities on the Gulf and Atlantic Coasts, has resulted in serious saltwater encroachment into coastal aquifers.

Many areas are minimizing the effects of high rates of groundwater withdrawal by utilizing several methods of local recharge. In suburban Nassau and Suffolk Counties on Long Island, New York, laws require that excavations made for sand and gravel mining be fenced off and maintained to allow surface water to recharge the aquifers below. Culverts transport surface water into the recharge basins, where the water infiltrates slowly into the ground. The recharge basins are cleaned periodically to remove fine sediment and the organic material that reduces permeability at the base of the recharge basin.

In some areas clean surface water and treated waste water are pumped back into the ground to recharge the aquifers below. Wells in which fluids are pumped *down* into the ground under pressure are known as **injection wells.** Saltwater encroachment into the aquifers underlying Los Angeles is being reduced by injection wells located close to the Pacific Coast. Aqueduct water obtained from the Colorado River is pumped down injection wells to displace the salt water in coastal aquifers. The Los Angeles area also recharges the aquifers below by controlled flooding of specially prepared flat areas of permeable ground.

Aquifer Contamination

Many areas of the United States are having problems with the *quality* as well as quantity of groundwater. Local contaminants enter the unconfined surface aquifers first and, with time, filter downward into the deeper aquifers. Hardly a day goes by without documentation of groundwater pollution somewhere in the United States. A wide variety of contaminants affect the quality of groundwater. Rain and snowmelt washing across road surfaces carry oil, vehicle exhaust products, and salt into the ground. Insecticides and nitrate and phosphate fertilizers infiltrate into the ground in agricultural areas. Ruptures of underground storage tanks release radioactive wastes, chemicals, and fuels into the groundwater. Contamination by any of these substances can make local groundwater undrinkable for a long period of time. Furthermore, once contaminants infiltrate into the ground, they are within the migrating groundwater and can affect the quality of groundwater over a wide area.

Groundwater pollution is potentially most serious in a cavernous limestone, in which flow rates are among the highest for any aquifer. Domestic or industrial wastes introduced into limestone aquifers can travel far before they are broken down by soil microorganisms or diluted

Figure 12.18   Contamination of groundwater by sewage wastes in an area underlain by limestone. Sewage wastes from the house on the left leak into the aquifer below and travel by gravity flow under the house on the right, where contaminated water is drawn upward in the well.

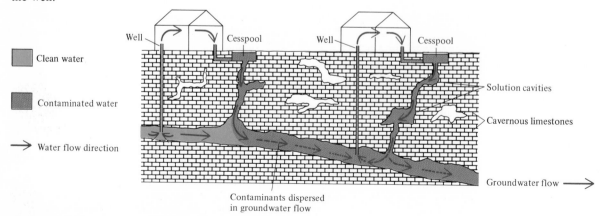

Subsidence due to Groundwater Withdrawal
A Tale of Two Cities

Houston, Texas, and Venice, Italy, seem as totally different as two cities can be. One is relatively young and growing rapidly, and the other is very old and stabilized in terms of growth. Yet, these two seemingly different cities have a lot in common—geology, geography, and environmental problems. Both cities are located near the ocean but are separated from it by a bay and barrier islands. Venice is located on islands within the Venetian Lagoon, separated from the ocean by a barrier island, the Lido (Figure 1a). The Houston area and its suburbs spread out along the northern edge of Galveston Bay, separated from the ocean by the barrier island on which Galveston is located (Figure 1b). Both cities are underlain by unconsolidated sediments from which nearby suburban and industrial areas have withdrawn large volumes of groundwater.

As a result of excessive groundwater withdrawal, there has been extensive subsidence in both cities. This subsidence is especially dangerous because it makes the cities more susceptible to flooding from high tides and from the effects of coastal storms.

Groundwater is being removed from the sediments under the Venetian Lagoon by the rapidly growing industrial suburb of Mestre and the port of Marghera on the mainland (Figure 1a). Estimates place the subsidence in Venice at about 13.7 cm from 1908 to 1961 with an average subsidence rate of about 0.26 cm/per year. Every year it is easier for tidal waters from the Adriatic to sweep into the city of Venice, flooding famous areas such as the Piazza San Marco with water up to 0.75 m deep (Figure 2).

Groundwater is also being removed in large volumes from the suburbs and industrial zones south and southeast of the city of Houston (Figure 3). Major groundwater withdrawal comes from the many industries bordering the Houston ship channel (which connects the city of Houston with Galveston Bay) and from the large industrial complex at Baytown (Figure 3). Many fast-growing suburbs bordering Galveston Bay, such as Clear Lake City and Pasadena, also withdraw large volumes of water from subsurface aquifers. Groundwater withdrawal has depressed the ground surface near Galveston Bay so that it floods more frequently during high tides. As the surface subsides, it may crack, and adjoining

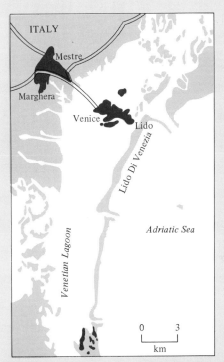

(a) Index map showing features in Venice area.

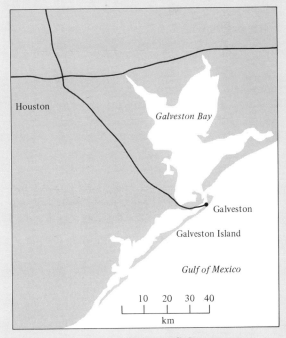

(b) Index map showing Houston–Galveston area.

Figure 2 Flooding in St. Mark's Square, Venice, on December 2, 1966. The water is about 1 m above the usual level. *(United Press International Photo)*

Figure 3 Detailed map (below) of Houston–Galveston Bay area showing localities described in text.

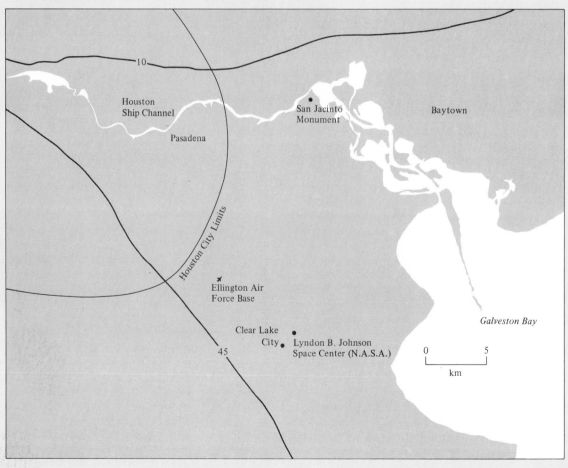

parts of the surface may be offset vertically from each other.

NASA scientists Uel S. Clanton and David L. Amsbury (see Additional Readings at the end of this chapter) have used field and aerial studies to map the subsidence cracks in the land surface southeast of Houston. The small vertical displacements on the subsidence cracks normally are not visible on the ground because erosion by running water evens out the surface quickly. The vertical displacement can be observed only where the surface was protected from erosion during subsidence. Clanton and Amsbury also documented elevation changes and structural damage resulting from subsidence in the area around the Johnson Space Center (Figure 3).

Repeated elevation checks show that subsidence of a little over 0.6 m (2.1 ft) has occurred in two places at the Johnson Space Center between 1964 and 1973. This is a potentially serious problem in an area that has altitudes below 20 ft and is subject to periodic hurricanes. The possibility of serious hurricane damage was especially worrisome to the Johnson Space Center since most of the lunar rock and soil samples are stored there. In 1976, construction began on a new hurricaneproof curatorial facility and storage space for the lunar materials at the Johnson Space Center. The structure, completed in 1979, now protects the collection of lunar samples.

Older buildings in the area bordering the Johnson Space Center show the cumulative effects of subsidence through cracked building walls and pavements and other features resulting from subsidence (Figure 4).

The structures at Ellington Air Force Base (Figure 3) are some of the oldest in the area; they show the cumulative effects of subsidence particularly well. Surface subsidence between 1942 and 1973 was about 1.5 m (4.9 ft). Surface cracks cut across parking lots, paved surfaces, and offset buildings. One of these buildings (Number 732) is cut across by a subsidence crack.

The pier-and-beam construction of this building minimized damage to the building while the surface below was subsiding. Since 1942 the building has protected the segment of the subsidence crack below from erosion. An examination of the crack showed that it had a vertical displacement of about 30 cm and an opening of about 8 to 10 cm (Figure 5).

Figure 4 Subsidence at Ellington Air Force Base, Texas. The crack in the parking lot surface extends under the structure in the distance. Note the sag in the curb and in the building; this is a result of the subsidence. *(Johnson Space Center, NASA)*

Subsidence can be stopped and *some* recovery of elevation may be obtained by pumping water down into the subsurface layers. Subsidence also can be halted by ceasing groundwater withdrawal and by relying on surface water brought into the area by aqueducts or pipelines. The latter course of action is being carried out in the Johnson Space Center–Clear Lake City area at the present time.

Figure 5 Subsidence crack exposed under the floor of the structure shown in Figure 4. There has been a vertical displacement of about 30 cm along the crack since the construction of the overlying building in 1942. *(Johnson Space Center, NASA)*

below dangerous levels. Domestic wastes (sewage) from one home may contaminate the drinking water of a nearby home if the wastes leak into the local aquifer (Figure 12.18).

PROBLEMS ASSOCIATED WITH GROUNDWATER WITHDRAWAL

Collapse

Surface collapse is a potentially serious problem in areas underlain by limestone. Solution by groundwater along rock fractures enlarges the fractures and undermines the surface above. The freshwater filling these cavities provides some support for the surface and may postpone surface collapse. If an area undergoes rapid lowering of the water table by domestic use, industrial use, or pumping to keep rock quarries dry, the surface support is removed and collapse can occur suddenly. Additional factors promoting collapse are the flushing out of residual (insoluble) sediments from the rock crevices and the greater pressure placed on the surface by new structures in areas undergoing development.

One of the largest collapse structures (sinkholes) in the United States formed suddenly on December 2, 1972, in rural Shelby County, Alabama (Figure 12.19). The sinkhole developed in residual clays above deeply weathered dolomite and had the following dimensions: 140 m long, 115 m wide, and 50 m deep. Slump scars (see Chapter 10) were widespread along the sides of the sinkhole. This suggests that the sinkhole will continue to grow as more of the underlying rock is dissolved.

Collapse of the surface also can be more widespread, involving the loss of many lives. R. F. Legget in *Cities and Geology* describes the large-scale collapse in the West Reef area of Johannesburg, South Africa, that occurred after deep pumping of groundwater that started in 1960. The area is underlain by a thick section of Transvaal dolomite and dolomitic limestone which overlies the Witerwatersrand gold-bearing beds. Between 1962 and 1966, eight sinkholes larger than 50 m in diameter and deeper than 30 m had appeared along with many smaller ones. In December 1962, a large sinkhole developed under a rock-crushing plant; the whole plant disappeared and 29 lives were lost. In August 1964, surface collapse caused five deaths as a home dropped 30 m into another suddenly developed sinkhole.

Collapse also can cause contamination of limestone aquifers by introducing surface materials. In May 1978, three sinkholes developed in a West Plains, Missouri, sewage treatment

Figure 12.19 Sinkhole in Shelby County, Alabama. Note the arcuate fractures and slump blocks developed at the edges of the sinkhole. *(Geological Survey of Alabama)*

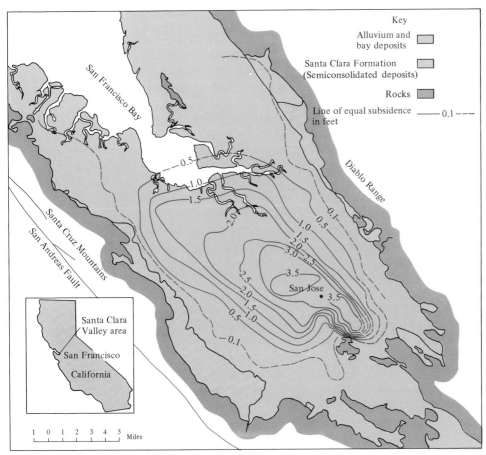

Figure 12.20 Surface subsidence resulting from excess groundwater withdrawal in the Santa Clara alluvial basin in California. The contour lines on the map connect points of equal surface subsidence. *(Modified from Poland, J. F., "Land Subsidence in Western U.S.," 1969, from* Geologic Hazards and Public Problems, *May 27-28, 1969, Conference Proceedings, Olson, R. A. and Wallace, M. W., eds., U. S. Government Printing Office, pp. 77-96. Publication authorized by the Director, U. S. Geological Survey)*

lagoon. Millions of gallons of sewage wastes leaked into subsurface caverns under the area and into the limestone aquifers of Missouri and neighboring Arkansas. At least 700 people became sick, and the National Guard had to truck in supplies of freshwater to the area.

Subsidence

Withdrawal of fluids from pore spaces within unconsolidated sediments causes the grains of sediment to come closer together, resulting in a compaction of the sediment layer. This volume decrease causes a "sagging" in the overlying bed. This gradual downwarping of the surface is called **subsidence.** The lowering of the surface in subsidence may result in cracks in the surface

and in any structures built on it. Structures show a displacement inward toward the area of maximum subsidence.

Marked subsidence resulting from groundwater pumping has occurred in several basins in California. A discussion of the Santa Clara Basin southeast of San Francisco Bay (Figure 12.20) illustrates the subsidence that occurs as a result of excessive groundwater pumping from basin aquifers. Between 1960 and 1967, maximum subsidence in the Santa Clara Valley was 1.18 m (3.9 ft). The greatest subsidence in the area was just south of the city of San Jose (Figure 12.20). The *total* subsidence between 1912 and 1967 has been about 3.96 m (13 ft) in downtown San Jose.

SUMMARY

That portion of the rainfall which infiltrates and accumulates in the subsurface is called groundwater. Groundwater moves by gravity flow through interconnected pore spaces, which may be between individual sedimentary particles or along fractures between large masses of rock. Groundwater fills all the pores in the zone of saturation, while air fills most of the pores in the zone of aeration. The contact between the zones of aeration and saturation is called the water table. The water table follows the contour of the land. Rocks or sediments through which groundwater can move are called aquifers. Impermeable rocks or sediments are called aquicludes. The rate of groundwater flow can be determined from Darcy's Law, which states that the discharge of groundwater through a unit cross-sectional area of an aquifer is directly proportional to the permeability of the aquifer and the difference in water level between the two points, and is inversely proportional to the distance between the two points.

Buried aquifers containing water under pressure are called confined aquifers. There is a close relationship between surface water and groundwater. In humid regions the water table supplies water to streams, whereas in dry areas the stream recharges the water table.

Groundwater is obtained from a wide variety of sources. Local supplies usually are obtained from shallow wells dug into unconfined aquifers such as permeable sediments and zones of fractured rock. Larger volumes of groundwater are obtained in areas of volcanic rocks, from deep basin fills, and from limestone areas. Groundwater may be obtained even in deserts by tapping aquifers below the desert surface. Coastal aquifers contain freshwater in the landward part and salt water in the seaward part. When withdrawal is heavy or recharge is low, salt water moves landward into coastal aquifers. Permeable islands are underlain by lenses of freshwater.

Solution of limestone by groundwater can form caves and caverns. Undermining of the surface by limestone solution may result in collapse and the formation of features such as sinkholes. Groundwater heated by near-surface magma bodies may appear on the surface as thermal springs and geysers.

Excessive groundwater withdrawal may result in collapse and subsidence. In order to ensure continuing groundwater resources for the future, it is necessary to preserve aquifer recharge areas in their natural state and to prevent any contamination of the groundwater supply.

QUESTIONS FOR REVIEW AND FURTHER THOUGHT

1. Describe the various types of porosity that occur in different aquifers.

2. Surface water and groundwater are interrelated. Show the relationship between streams and groundwater in: (a) dry and (b) humid areas.

3. Under which of the following geologic conditions will freshwater accumulate?

   a. In a volcanic area underlain by alternating volcanic ash and basaltic lava flows

   b. On a permeable coral island in a humid area in the South Pacific

4. Describe the origin of each of the following:

   a. A spring

   b. An artesian well

   c. An oasis

   d. A geyser

5. Contrast the environmental effects of the leaking of small amounts of sewage into an aquifer of (a) sandstone and (b) limestone in a humid area.

6. How would you explain the following groundwater phenomena?

   a. Wells dry up in a period of drought.

   b. Narrow belts of vegetation may parallel a stream flowing through a desert.

   c. Urbanization of an area underlain by limestone results in an acceleration of the collapse of the surface.

ADDITIONAL READINGS

Beginning Level

Clanton, U. S., and D. L. Amsbury, "Active Faults in Southeastern Harris County, Texas," *Environmental Geology*, vol. 1, 1975, pp. 149–154.
(Describes the effects of surface subsidence as a result of groundwater withdrawal in the area south of Houston)

Cohen, P. O., O. L. Franke, and B. L. Foxworthy, *An Atlas of Long Island's Water Resources*, N.Y. Water Resources Commission Bulletin 62, 1968.
(A detailed treatment of groundwater resources, usage, and future problems. Covers the water cycle, where the water is stored, the chemical and physical properties of the water, human effects on the system, and management of groundwater resources. Profusely illustrated and a valuable resource for urban planning)

Ford, R. S., "Ground Water—California's Priceless Resource," *California Geology*, vol. 31, no. 2, 1978, pp. 27–32.
(A good overall review of groundwater, with sections describing the chemical quality of groundwater and the different types of wells used to obtain groundwater)

Hack, J. T., and L. H. Durloo, *Geology of Luray Caverns, Va.*, Virginia Division of Mineral Resources Report of Investigations 3, 1962, 43 pp.
(Describes the formation of and features within one of the largest caves in the eastern United States)

Heath, R. C., B. L. Foxworthy, and P. Cohen, *The Changing Patterns of Groundwater Development on Long Island, N.Y.*, U.S. Geological Survey Circular 524, 1966.
(Describes the different stages of development of groundwater resources in an area undergoing rapid development)

Leggett, R. F., "Hydrogeology of Cities," chap. 4 in *Cities and Geology*, McGraw-Hill, New York, 1973, pp. 125–176, 624 pp.
(Describes how a number of different cities obtain water. Includes a discussion of all of the aquifer types discussed in this chapter, along with a discussion of surface water supplies)

Leopold, Luna B., "Surface Water and Ground Water," in *Water: A Primer*, W. H. Freeman, San Francisco, 1974, pp. 8–33, 172 pp.
(A good review of groundwater concepts and the relationship between groundwater and surface water)

Piper, A. M., *Has the United States Enough Water?*, U.S. Geological Survey Water Supply Paper 1797, 1965, 27 pp.
(Describes the different types of water supplies and the demand and availability by geographic regions)

Intermediate Level

Bloom, A. L., "Karst," chap. 7 in *Geomorphology: A Systematic Analysis of Late Cenozoic Landforms*, Prentice-Hall, Englewood Cliffs, N.J., 1978, 510 pp.
(Particularly good discussion of the factors involved in limestone solution and the development of karst features)

Ford, R. S., "Ground Water Use in California," *California Geology*, vol. 31, no. 1, 1978, pp. 247–249. (Describes the sources of groundwater in California and the environmental problems, such as subsidence and saltwater encroachment, that result from excess withdrawal)

Lindorf, D. E., and K. Cartwright, "Ground-water Contamination—Problems and Remedial Actions," *Environmental Geology Notes*, Illinois State Geological Survey No. 81, 1977, 58 pp.
(Discusses a wide variety of contaminants and how they affect the quality of groundwater. Contains numerous case histories)

CHAPTER 13

Glaciers, Glaciation, and Climatic Change

The present extent of glaciers is only a fraction of what it has been in the past 1.5 million years, during the Quaternary Period of geologic time, when thick and widespread masses of glacial ice extended repeatedly from polar regions into more temperate latitudes. The great climatic changes that accompanied the repeated growth and retreat of these thick ice sheets had worldwide effects.

Glaciers stripped soil and loose bedrock and transported it away. Advancing ice sheets blocked preglacial river courses, causing major displacements in regional drainage patterns. Great basins, such as the present Great Lakes, were gouged out by the advancing glaciers and later filled with water from the melting ice. Sea level fell and rose repeatedly as ice was forming and melting on the continents. When sea level was lower than it is at present, the shallower parts of the ocean floor were emergent, creating intercontinental paths for the migration of plants, animals, and humans. Glacial erosion modified the preexisting topographic features, creating some of the most dramatic landscapes on earth (Plate 17). To understand these interesting phenomena, we will start with a discussion of where and how glaciers form.

GLACIERS

Masses of ice which form on land by the compaction and recrystallization of snow and move downslope or outward in all directions under the pressure of their own weight are called **glaciers.**

Where and How Do Glaciers Form?

Glaciers form wherever there is abundant snowfall and the temperatures are sufficiently low to preserve a portion of the snowfall from year to year. The cold temperatures needed to preserve accumulated snow may be found at high altitudes and also in colder latitudes. Wherever winter snowfall exceeds summer melting, patches of year-round snow accumulate as **snowfields.** The lowest altitude of *year-round* snow cover is referred to as the **snow line.** The altitude of the snow line is a function of both the ground temperatures and the precipitation. In temperate latitudes, mountains covered completely with snow in the winter are capped with snow only at colder, higher altitudes during the summer. In polar latitudes, the temperatures may be low enough so that snow cover remains year-round down to sea level. At any given latitude, the snow line is at a lower altitude in the coastal mountains than on mountains in the continental interior farther away from sources of moisture. The greater snowfall in coastal mountains enables snow cover to remain at lower altitudes, despite the warmer temperatures along the coasts.

The transition of snow into glacial ice is similar to the deposition of sedimentary particles and their metamorphism into metamorphic rock (Chapter 7). Therefore, ice may be considered a metamorphic rock because it is formed by the melting and recrystallization of snow particles under increasing pressure. An excavation made on the higher parts of a glacier would show that

Surface ⟶ Increasing depth ⟶

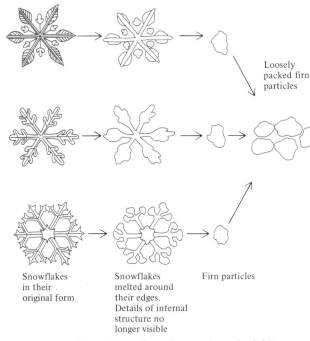

Loosely
packed firn
particles

Closely
packed firn
particles

Glacial ice

Large glacial
ice crystals

Snowflakes
in their
original form

Snowflakes
melted around
their edges.
Details of internal
structure no
longer visible

Firn particles

Figure 13.1   Transformation of snow into glacial ice.

the fresh snow at the surface grades downward into a granular snow with a lower porosity, and eventually into ice.

Freshly fallen snow is made up of a porous (porosity as large as 95 percent) mass of crystalline hexagonal snowflakes (Figure 13.1) which formed as water vapor was cooled below its freezing point. The snow has a great surface area and readily exchanges moisture with the surrounding air. At temperatures below the melting point, this exchange may be by sublimation. **Sublimation** is the direct transformation of snow (or ice) into vapor without intermediate melting. When the temperatures are nearer to the melting point, the exchange may occur by evaporation. Greater moisture losses occur at the thinner projecting points of the crystals rather than across the wider central areas, and the snowflakes become more spherical with time (Figure 13.1).

As the snow is buried by more recently deposited snow, it begins to become more compact, and the porosity decreases. The particles begin to melt at points of contact because the pressure is greater there. This meltwater, along with any meltwater produced by melting of surface snow, moves downward and refreezes around the particles at depth, enlarging them. This melting and refreezing forms a granular form of snow called **firn,** which has a porosity of about 50 percent. The transition of snowflakes to firn may take place within a few weeks at temperatures near the melting point, but it may take many years at the low temperatures of the polar regions.

As the firn becomes buried by greater thicknesses of snow, the increased compaction forces the grains of firn together, steadily eliminating air from pore spaces and forming glacial ice (Figure 13.1). Glacial ice is a solid composed of tightly interlocking ice crystals (Figure 13.1). When sufficient ice has accumulated in the snowfield, the pressure on the ice is sufficient to move it downhill as a tongue extending down a mountain valley or, in the case of glaciers forming at lower altitudes, radially outward from a central area where the snow first accumulated.

Types of Glaciers

Glaciers may be subdivided into two broad categories. **Mountain,** or **alpine, glaciers** have their source on mountain slopes above the snow line. As they move downslope, they take the path of least resistance—commonly a preexisting stream valley. In many cases they are linear in form and are bounded by valley walls. However, they can fan out in an apronlike form as they spread out along the mountain base. **Continental glaciers** have a domelike form in which the ice flows outward from one or more central areas, with little relation to the underlying topography. Continental glaciers are far larger than mountain glaciers, and they cover all but the highest parts of the underlying ground.

Alpine, or mountain, glaciers

Glaciers which form in snowfields above the snow line in mountainous areas are referred to as alpine, or mountain, glaciers. The smallest type of alpine glacier, the **cirque glacier,** forms in snow- and ice-filled rock depressions called **cirques.** When cirque glaciers grow sufficiently so that they extend down the mountain along valleys previously carved by streams, they are called **valley glaciers** (Figure 13.2). One valley glacier may merge with another, flowing down an adjacent valley to form a larger **trunk glacier** (Figure 13.2). This is similar to the way tributary streams join to form larger streams (Chapter 11). However, the ice supplied by each of the valley glaciers is too viscous to mix, and the masses of ice and associated rock debris from each valley glacier flow side by side along the trunk glacier (Figure 13.2).

Trunk glaciers may join together to form a **piedmont glacier,** a type of glacier intermediate between alpine and continental glaciers. Piedmont glaciers are apronlike forms which occupy broad lowlands at the bases of steep mountain slopes and are fed by valley and trunk glaciers. The large Malaspina and Bering Glaciers on the coast of Alaska at latitude 60° N are examples of piedmont glaciers (Figure 13.3).

Continental glaciers

Thick and widespread glaciers covering most of the topographic features in a region are called **ice caps** if they are small, and continental glaciers or **ice sheets** if they are of subcontinental or continental size (Figure 13.4). Continental glaciers differ from valley and trunk glaciers in size, shape, and directions of ice flow. Valley glaciers are small and linear and are confined between valley walls. Continental glaciers are dome-shaped and flow radially outward from a central area. Ice caps today are located in a number of places, including Iceland, Baffin Island in Canada, and Spitsbergen, Norway. However, ice sheets exist in only two places—Greenland and Antarctica. The Greenland ice sheet occupies 80 percent of Greenland (Figure 13.5). It has an area of 1,726,400 km², a volume of

Figure 13.2 Valley glaciers on Mount Fairweather, Gulf of Alaska. *(From D. A. Rahm,* Slides for Geology. *Copyright © 1971 McGraw-Hill, Inc. Used with permission of McGraw-Hill Book Company)*

Figure 13.3 Malaspina Glacier, Alaska, is a piedmont glacier. It is fed by the mountain glaciers seen in the background. *(From D. A. Rahm,* Slides for Geology. *© 1971 McGraw-Hill, Inc.)*

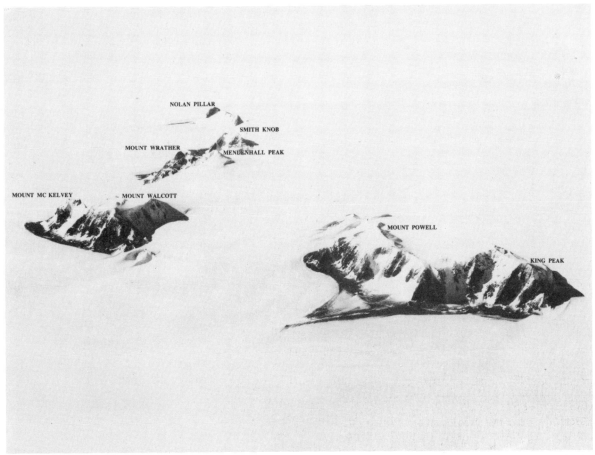

Figure 13.4 The peaks of the Thiel Mountains in Antarctica extend above the Antarctic ice sheet, a continental glacier. The elevation of the snow in the foreground is approximately 1585 m. *(U. S. Geological Survey)*

about 2,600,000 km³, and a maximum measured thickness of about 3.4 km. The ice sheet occupies a rock basin whose floor extends down to −400 m (Figure 13.5). The ice flows radially outward toward the coast from a high central area, streaming through valleys in the coastal mountains and breaking off into the ocean as floating masses of glacial ice called **icebergs.**

The Antarctic ice sheet is far larger than the Greenland ice sheet; it has an area of 12,530,000 km² and a maximum ice thickness of almost 4.3 km. The Antarctic ice sheet is bordered by more than 1,400,000 km² of floating ice. Floating sheets of ice connected to, and nourished by, continental glaciers, snowfall, and the freezing of seawater are called **ice shelves.**

MOVEMENT OF GLACIERS

In our discussion of the scientific method in Chapter 1, we described an experiment by which eighteenth-century scientists proved that valley glaciers move. The experiment also showed that the ice in the central part of the valley glacier moved farther than that at the sides (see Figure 1 in the Perspective in Chapter 1). The explanation of *how* glacial ice moved was to come much later, when geologists began to understand more of the properties of glacial ice.

We will discuss glacial movement by looking first at the mechanisms by which ice moves, then at the factors that influence relative ice velocity within a glacier, and finally at the rea-

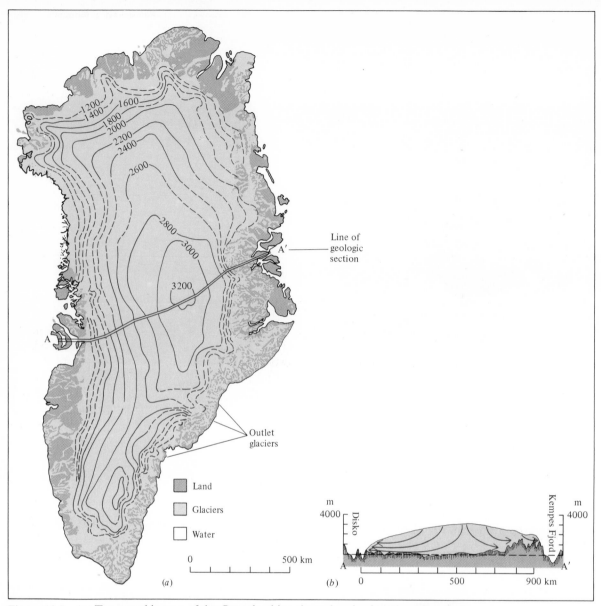

Figure 13.5 *(a)* Topographic map of the Greenland ice sheet showing location of geologic section shown in *(b)* below. Ice streams through the coastal mountains as high-velocity outlet glaciers. The ice reaches altitudes of almost 3.3 km. *(b)* Geologic section through Greenland showing ice sheet and underlying topography. Theoretical direction of ice flow indicated by arrows. Vertical scale exaggerated for clarity. (*Modified from R. F. Flint, Glacial and Quanternary Geology, © 1971, with permission of John Wiley & Sons, Inc.*)

sons why velocity differs from one glacier to another.

How Does Ice Move?

We are familiar with ice as a hard and brittle material which can move only by sliding on its base. The mode of movement of glacial ice can be shown by an experiment in which a hole is drilled vertically through a moving glacier. After a few weeks, the hole is no longer vertical but has been deformed into a curve (Figure 13.6). This experiment shows that only a small part of the glacial flow has occurred by external sliding

Figure 13.6 Motion of ice within a glacier. An originally vertical drill hole through the glacier is deformed into a curve within a few weeks. Only a small portion of the glacial movement is by basal slip along the underlying rock surface. Most of the movement is internal through deformation and recrystallization. The relative speed of the glacier at different levels is indicated by the length of the arrows on the right side of the diagram.

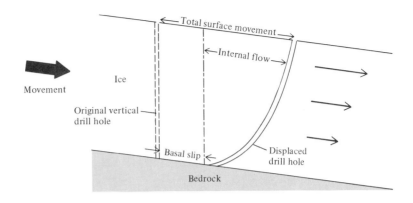

along the base and that most of the movement has been by some type of internal flow.

The internal flow within glaciers is believed to be a result of the deformation and displacement of individual ice crystals. Under the pressure of the overlying ice, firn, and snow, the basal ice can be deformed by gliding along crystallographic planes within the individual crystals. The process by which a crystal deforms by gliding along planes within the crystal is called **intragranular gliding.** This deformation is similar to what happens when a deck of cards is sheared between the thumb and forefinger (Figure 13.7). Some degree of recrystallization accompanies the gliding, and this increases the size of the ice crystals with time (see Figure 13.1). Such ice crystal growth with depth was shown in an ice core drilled into the Antarctic ice sheet at Byrd Station. The average cross-sectional area

of ice crystals was 3.1 mm^2 at 65 m depth and nearly 20 mm^2 at 307 mm depth.

The accumulation of from 30 to 60 m and more of snow, firn, and ice provides sufficient pressure to make the underlying ice flow like a plastic material. Above this basal zone of plastic ice is an upper brittle zone. A moving glacier then is composed of an upper **brittle zone,** which fractures when stressed, and a lower **plastic zone,** which deforms plastically under stress.

Relative Velocity within Glaciers

The relative velocities within a moving glacier are similar to those observed in streams (Figure 11.7). Wherever the glacier is in contact with the ground, frictional effects decrease the velocity of the ice at that point. Maximal velocities within glaciers occur in the portion of the ice

Figure 13.7 (a) Movement of glacial ice by intragranular slippage within the ice crystals. The movement is analagous to the deformation of a deck of cards between the thumb and forefinger (b).

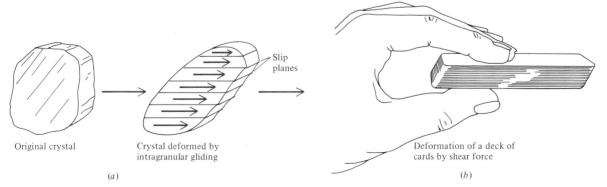

farthest away from the underlying surface and the valley walls (in the case of valley glaciers).

Differences in velocity between adjacent parts of a glacier can be accommodated easily by plastic deformation within the lower part of the glacier but not within the upper, brittle zone. Whenever the upper, brittle zone is stressed, fractures called **crevasses** form which extend down into the top of the plastic zone.

We can illustrate crevasse formation by a few examples from valley glaciers. The ice nearest the valley wall is slowed down by friction relative to the portion nearer the central part of the glacier. This sets up a tensional force, which results in the formation of crevasses in the ice nearest the valley walls. Another type of crevasse occurs when the glacier passes over a rock obstruction or over an abrupt increase in the slope of the valley floor below (Figure 13.8). The ice in the plastic zone adjusts by deforming. However, the upper brittle part cannot deform plastically and is ripped apart, forming crevasses that extend downward toward the plastic zone. After the ice moves over the obstruction on the valley below, the surface crevasses close. Meanwhile, new crevasses form in the brittle ice passing over the obstruction. Crevasses may become covered by thin "bridges" of freshly fallen snow, posing a danger for hikers and skiers.

Velocities of Glaciers

Glaciers show a wide range of flow rates. Some move so slowly that no movement is apparent from one day to the next. Others move so fast that their motion is apparent within a few days.

Why is there such a wide range of glacier flow velocities? Three of the major factors in glacial velocity are the thickness of the ice, the steepness of the slope over which the ice is moving, and the cross-sectional area through which the ice is passing. The thicker the ice, the greater the pressure on its base and the faster the ice can move both by internal flow and by sliding along its base. We saw in Chapters 10 and 11 that the steeper the slope, the greater the component of gravity parallel to the slope (see Figure 10.1). In short, glaciers forming on steeper slopes have a greater tendency to flow faster. For example, valley glaciers can move up to 0.3 to 0.6 m per day, although velocities of 3 to 6 m per day have been observed at "ice falls," where the glaciers move more rapidly over an abrupt steepening in the valley floor below. Velocity of a glacier also increases when the ice is forced to flow through a narrow opening. Velocities of up to 38 m per day have been observed where the Greenland ice sheet streams through narrow valleys in the coastal mountains of eastern Greenland (Figure 13.5). Such fast-

Figure 13.8 Vertical section through valley glacier showing the brittle and plastic zones and the formation of crevasses as the ice moves over elevated parts of the valley floor.

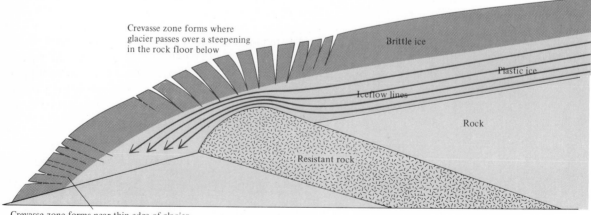

Crevasse zone forms where glacier passes over a steepening in the rock floor below

Brittle ice

Plastic ice

Iceflow lines

Rock

Resistant rock

Crevasse zone forms near thin edge of glacier.
Brittle ice forms crevasses in response to
pressure from ice further up glacier.

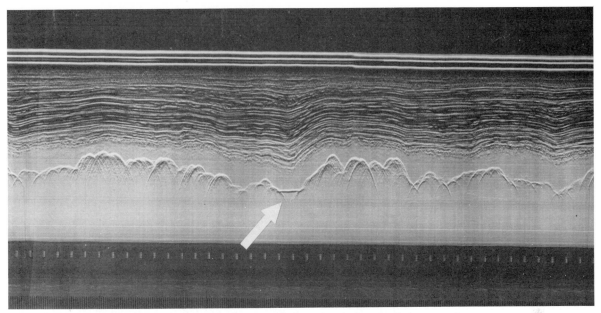

Figure 13.9 Radar probing of Antarctic ice reveals layering in ice, buried mountains, and a layer of water in basin at base of ice (arrow). *(Courtesy of Walter Sullivan, The New York Times)*

moving valley glaciers fed by ice caps or ice sheets are called **outlet glaciers** (see Figure 13.5).

The highest glacial velocities occur in what are called surging glaciers. **Surging glaciers** are those which move with velocities of several kilometers rather than meters per year. The surging is not continuous but is usually short-lived, from a few months to a few years. Exactly why glaciers surge is not understood completely. One possibility is that water builds up rapidly at the base of the glacier, allowing it to slip on its bed. To understand how this can happen, we must look at how pressure affects the freezing point of water. At atmospheric pressure, water freezes at 0°C (32°F). As the pressure on the base of the ice increases, water freezes at slightly lower and lower temperatures. If the pressure is great enough, the ice at the base of a glacier may melt, forming water locally at the same temperature as the surrounding ice. The overlying glacier may move over this water layer, subsequently slowing down once again if the pressure drops and the basal water refreezes. The melting of ice under great pressure and the refreezing of the derived meltwater when the pressure decreases is called **regelation.** Regela-

tion is an important process in glacial erosion, and we shall return to it later in the chapter.

In 1969, a hole drilled 2164 m to the base of the Antarctic ice sheet in central Antarctica encountered water under great pressure. The water rushed up the drill pipe, freezing quickly as the pressure was released. Geophysical studies in 1978 showed that portions of the Antarctic ice sheet sit on a layer of basal water (Figure 13.9). This water is derived partially from pressure melting of the basal ice and partially from the flow of heat from the earth's interior. Such accumulations of basal meltwater can provide the lubricating mechanism for a glacial surge of massive proportions. If part or all of the Antarctic ice sheet were to surge into the surrounding ocean, the large volume of water displaced could have disastrous effects on the world's coastal cities.

Economy of a Glacier

The movement of a glacier is somewhat like a continuously running conveyor belt. So long as there is net accumulation in the snowfields, glacial ice produced above the snow line is delivered continuously to areas below the snow line. In winter, snow may cover the glacier well below the snow line, and the terminus of the glacier may advance. In summer, the snow cover melts and there is a loss of ice below the snow line. In temperate climates much of the loss is by melting, whereas in polar climates the main loss is by evaporation and sublimation. Additional ice losses may occur if the glacial terminus is adjacent to the ocean or any other large body of water; ice is lost when pieces of the glacier break off and float away as icebergs.

The balance between ice accumulation above the snow line and ice loss below the snow line is expressed as the **glacial economy** as follows:

Economy of a glacier	=	accumulation of snow above the snow line	−	loss of snow and ice below the snow line

For purposes of comparison, both the accumulation and loss are converted to units of water. If the economy of a glacier is *positive,* the accumulation is greater than the loss, and the terminus of the glacier advances. If the economy is *negative,* loss is greater than accumulation, and the terminus and ice surface below the snow line melts back, *even though new ice is being supplied continually to the glacial terminus.* If the accumulation balances the loss, the glacial terminus remains stationary. The movement of glacial ice is downward in the zone of accumulation and upward in the zone of loss (see Figure 13.5).

GLACIAL EROSION

The high viscosity of ice and the great pressures that can develop at the base of glaciers along with the rock particles frozen into the ice make glaciers especially effective in erosion of the underlying materials.

Mechanisms of Glacial Erosion

The cold climate in front of a glacier increases the efficiency of glacial erosion. Cold climates in front of a glacier are called **periglacial climates.** Extensive frost wedging (see Chapter 9) in periglacial climates loosens the rock particles at the surface, making them easier to erode

when an advancing mountain or continental glacier overrides them.

Regelation plays an important role in glacial erosion. Glaciers exert great force on particles projecting upward into the ice. The high pressures may result in local melting of the ice on the up-ice side of the rock. The meltwater produced can travel short distances toward the other side of the rock, where reduced pressure causes it to refreeze and attach the rock to the glacier. As the glacier moves past, it carries away part of the rock. Regelation is even more effective when the surface rocks have been fractured by periglacial frost wedging. The process of glacial erosion where particles of rock are removed by the freezing of meltwater along rock crevices, followed by the removal of the rock as the ice advances, is called **plucking.** Considerable glacial erosion is also carried out by the particles which are eroded and carried at the edges and base of the glacier. These particles enable the glacier to act like a piece of sandpaper, wearing away the rock it comes in contact with by the process of **abrasion.** Small particles held within glacial ice moving over a rock cut elongate scratches called **striations** into the rock surfaces. Larger particles of pebble, cobble, or even boulder size carried by the ice can erode wider and deeper elongate **grooves** into the rock surface. The abrasion of the underlying rock by glacially transported particles of all sizes gives the rock a polished, striated, and grooved surface which is characteristic of glacial erosion (Figure 13.10). The grinding of rock particles against each other and the underlying surface produces clay-sized material called **rock flour.**

Glacial erosion also triggers other processes which erode the rocks. For example, erosion of the side of a valley by a glacier may undercut the slope, resulting in mass movements such as rockslides and debris avalanches (see Chapter 10). The material derived from mass wasting may travel on the surface or the sides of the valley glacier or fall into crevasses and be carried within the ice.

Erosional Features

Erosion by valley and continental glaciers modifies the landscape greatly. Some erosional fea-

Figure 13.10 Glacier polish, striae, and grooves in granite in Tulare County, California. The direction of ice movement was diagonally toward the right and away from the camera. *(F. E. Matthes, U. S. Geological Survey)*

tures, such as sculptured rock outcrops and widespread striated and polished rock surfaces, are formed by both types of glacier. However, the most spectacular erosional features are made by valley glaciers in mountain ranges such as the Tetons and the Rocky Mountains in the United States and in Alaska (Figure 13.2).

Cirques and associated features

Snow accumulating in topographically low areas above the snow line is subject to daily and seasonal warming at the top and outer edges of the snow. As the snow melts, it produces water, which penetrates fractures in the rock below. Subsequent freezing of this water fractures the rock thus deepening the basin. By repetition of freezing and thawing cycles, the underlying rock develops into an amphitheater-shaped basin called a **cirque.** The cirque is bounded upslope by a steep **headwall.** If snow accumulation is sufficient, a **cirque glacier** may form and eventually extend down the slope as a valley glacier.

The successive enlargement of the cirques with time produces spectacular scenery in glaciated mountains. The crevasse between a **cirque** glacier and the headwall, called the **bergschrund,** allows any meltwater produced during warm spells to penetrate the rocks in the headwall of the cirque. This water freezes at night or during seasonal cold intervals, bonding the cirque glacier firmly to the headwall. Subsequent movement of the glacier plucks out many rock fragments, causing the headwall to recede and enlarging the cirque. Enlargement of the cirque by headwall erosion produces a variety of features seen in glaciated mountains (Figure 13.11). Two cirques growing side by side produce a narrow ridge called an **arête.** Two cirques

Figure 13.11   Arête in Picket range, northern Cascade Mountains of Washington. The arête was formed by the headward erosion of cirque glaciers on opposite sides of the mountain. *(From D. A. Rahm, Slides for Geology. Copyright © 1971 McGraw-Hill, Inc. Used with permission of McGraw-Hill Book Company)*

eroding toward each other from opposite sides of a mountain cut notches in the divide between their headwalls, forming gaps called **cols.** Cirques eroding headward on all sides of a mountain isolate the peak as a **horn.** The best-known example of this latter feature is the Matterhorn of Switzerland.

Valley glaciation

Glaciers moving out of cirques and down preexisting valleys erode the valleys and tend to straighten them and modify their cross sections.   Stream valleys have a V-shaped cross section, the result of both stream erosion and mass wasting (see Chapter 11). Glaciers fill these valleys far above the level of the former streams. The viscous ice molds itself to the valley, erodes it by plucking and abrasion along its periphery, and forms a **U-shaped valley.**

Even though the upper surfaces of two glaciers meet at the same altitude (Figure 13.2), there may be a considerable difference between the altitudes of their floors. If one valley is underlain by more easily eroded rock or contains a thicker valley glacier, that valley will be deepened preferentially by glacial erosion. After both glaciers have melted back, the tributary

valleys commonly are left elevated as **hanging valleys** above the level of the main valley (see Figure 13.12). Postglacial streams flowing through the tributary valleys plunge down as waterfalls into the main valley below. Several examples of U-shaped valleys, hanging valleys, and falls of this type may be seen in Yosemite National Park in California (Figure 13.12). **Fjords** are deeply eroded glacial valleys that subsequently became inundated by rising postglacial sea level.

Sculptured rock

Glaciers overriding exposures of rock abrade, striate, and polish the side facing the ice and pluck out an irregular surface on the opposite side. The result is an asymmetric rock exposure with the gentle side facing the direction from which the ice moved. Most of these asymmetric rock exposures are only a few meters in height, width, and breadth. However, some are as big as mountains.

GLACIAL DEPOSITION

The greater viscosity of glacial ice relative to that of air or water enables such ice to carry within itself a wide range of particle sizes at the same time. In addition, mass wasting along the valley walls contributes angular blocks of rock, which may be carried on the surface of a valley glacier. This is in marked contrast to the way in which different-sized particles are transported by water and air (see Chapter 6).

The sediment load carried by a glacier may be deposited in two different ways. Some of it is deposited as the ice moves over the land surface. Other types of sediment are deposited as the ice stagnates or as it melts back, forming streams which transport rock debris across the surface in front of the glacier. Recall from Chapter 6 that there will be major differences in the characteristics of these two types of deposits.

The general term for all types of glacial deposits is **drift.** There are two major types of drift. **Unstratified drift** consists of deposits lacking stratification and generally composed of poorly sorted and angular particles (see Figure 6.3). **Stratified drift** consists of layers of sorted and generally rounded particles. We will now consider the formation of these two types of drift in more detail, along with the landforms composed of each type of drift.

Unstratified Drift

Unstratified drift deposited by actively moving glaciers is called **till.** Till is poorly sorted, generally lacks stratification, and contains angular particles, many of which may be striated or polished (see Figure 6.3). It is closest in character to the deposits made by the viscous mudflows discussed in Chapter 10. Till has a particle size range and mineralogic composition that reflects the type of material over which the glacier passed. For example, till derived from the erosion of coarse-grained rocks such as granite and sandstone will have a higher percentage of sand-

Figure 13.12 A glaciated U-shaped valley (center) and a hanging valley and falls (right) in Yosemite National Park, California. *(J. Blair, National Park Service)*

sized particles than till derived from finer-grained rocks such as shales and limestones. The particles of rock in the till may be aligned parallel to the direction of the ice flow and thus provide the geologist with information on the direction of movement of the glacier that deposited them. Particles aligned in a specific direction are said to have a **preferred orientation.** Preferred orientation of rock fragments in till is another example of the paleocurrent indicators first described in Chapter 6.

Till commonly contains rock fragments called **erratics** that have been carried far from their source and deposited on rock of different composition. Among the more interesting erratics are the high-quality diamonds found in glacial drift in Wisconsin, Illinois, Indiana, Ohio, and New York. While no specific data are available, it is believed that these diamonds have a source area in east central Canada. The great viscosity of ice enables it to carry erratics of very large size. For example, the so-called Madison Boulder, a block of granite near Conway, in the White Mountains of New Hampshire, measures 11 by 12 by 27 m and weighs 4700 metric tons.

Some tills contain **indicators,** which are rock fragments or mineral grains with a known and unique source area. Indicators can be used to determine the direction of glacial movement.

For example fragments of pure copper are found in glacial deposits in Missouri, Illinois, and Ohio. These fragments are indicators because they all come from a known source area in the Upper Peninsula of Michigan (Figure 13.13). The approximate direction of glacial movement may be obtained by drawing a straight line on a map between the source area of the indicator and the location where the fragments were deposited (Figure 13.13).

Landforms Composed of Till

Till makes up a variety of landforms deposited by both valley and continental glaciers. The various types of landforms are distinguished by their position relative to the glacial terminus and by their shapes.

Moraines

The general term **moraine** is used to describe many glacial landforms. While most moraines are composed of till, some may be constructed partially of stratified drift deposited at the edges of the glacier. This type of moraine will be discussed later in this chapter. Several different types of moraine are deposited by valley glaciers (Figure 13.14). Material eroded from the valley

Figure 13.13 Native copper as an indicator of glacial movement in the midwestern area of the United States. The native copper deposits are located in the Upper Peninsula of Michigan (×). Lines connecting the points where the copper is found in glacial deposits (●) and the source area (×) approximate the direction of glacial movement between the two points. (*Modified from R. F. Flint*, Glacial and Quaternary Geology, © *1971, with permission of John Wiley & Sons, Inc.*)

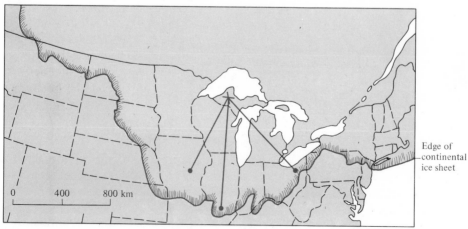

Edge of continental ice sheet

0   400   800 km

Figure 13.14 The moraines in the lower left are steep-sided and 30 to 90 m tall. They were deposited as the glacier progressively melted back up the valley. West slope of Mt. Chamberlin, northern Alaska region. *(G. W. Holmes, U. S. Geological Survey)*

sides forms **lateral moraines** along the valley walls (Figure 13.2). Lateral moraines from two joining glaciers form a **medial moraine** (Figure 13.2). The existence and character of medial moraines are graphic proof that merging glaciers move side by side as separate ice streams for considerable distances without mixing. The intense folding shown in some moraines is a graphic illustration of the plastic deformation that occurs in glacial ice (Figure 13.3).

Three types of moraine occur in association with both alpine and continental glaciers. Sediment deposited as a sheet beneath a moving glacier is called a **ground moraine**. Moraines formed at the terminus of a glacier are called **end moraines**. An end moraine formed at the edge of the glacier at its *maximum* extent is called a **terminal moraine**. The terminus of a glacier may melt back, reach equilibrium, and deposit another moraine. An end moraine deposited by a glacier that melts back from a terminal moraine is called a **recessional moraine**. A glacier has only *one* terminal moraine but may have any number of recessional moraines (Figure 13.14).

Drumlins

Continental glaciers can mold the underlying till into streamlined hills called **drumlins** (Figure 13.15). Some drumlins are composed entirely of till, while others are made up of till plastered on a rock core. In an aerial view, they have a streamlined shape and are elongated parallel to the direction of glacial movement. In a cross section, they commonly have a steeper slope facing the direction from which the ice came. Note that this is the reverse situation compared with the asymmetric sculptured rock forms described earlier in this chapter. Drumlins are generally 1 to 2 km long, 400 to 600 m wide, and 5 to 50 m high. Large numbers of drumlins occur in drumlin **fields** in Michigan, Massachusetts, Wisconsin, and New York State.

Stratified Drift

Stratified drift is deposited by meltwater derived from the melting of glacial ice. Some melting occurs daily or seasonally on many glaciers with positive economies, and this accounts for the formation of the small patches of stratified drift

Figure 13.15 Drumlins near Jameson Lake, Waterville Plateau, Washington. The terminal moraine is visible in the upper part of the photograph. The ice that formed the drumlins moved from the bottom toward the top of the photograph. *(From D. A. Rahm,* Slides for Geology. *Copyright ©️ 1971 McGraw-Hill, Inc. Used with permission of McGraw-Hill Book Company)*

found in some deposits of till. However, most stratified drift deposits are laid down when the glacial economy is at equilibrium, or especially when the glacial economy is negative. When the glacial economy is negative, the terminus melts back faster than it moves forward, and the ice near the terminus thins by melting. The decrease in ice thickness reduces the pressure on the lower portion of the ice. The ice then becomes brittle, and crevasses form. The crevasse system increases the surface area of ice exposed to the air and accelerates melting. Accelerated melting leads to the formation of more fractures, which results in more melting, and so on.

The low viscosity of meltwater makes it a much more effective sorting agent than glacial ice. The wide range of particle sizes originally deposited by the ice are reworked by running water to form beds of differently sized particles. In general, the gravel-sized particles are deposited closer to the glacial terminus, whereas the sand-sized particles can be deposited for a considerable distance in front of the glacial terminus. The finer silt and clay particles are transported in suspension and may be deposited

in lakes at the glacial margin or transported for great distances beyond the glacial margin.

Debris carried under, within, and above glacial ice is reworked by meltwater and deposited within fractures in the stagnating glacier. Stratified drift originally deposited in contact with glacial ice is called **ice-contact stratified drift**. Deposits of stratified drift extending away from the glacial margin are referred to as **outwash**.

Landforms Associated with Stratified Drift

A wide variety of landforms in glaciated areas is composed of stratified drift. Some of these are local features, while others cover wide areas.

Kames

Moundlike hills of ice-contact stratified drift of any size are called **kames**. Some kames are composed of sediment deposited in crevasses and other openings in or on the surface of stagnant glaciers. When the ice melts away, the stratified drift is left in the form of isolated or semiisolated mounds. Another type of kame consists of fans of stratified drift built outward from the edges of a glacier. When the glacier melts back, the masses of stratified drift are left as irregular mounds referred to as a **kame moraine.**

Kame terraces

As a valley glacier melts back, the ice thins and melts away from the valley walls. Meltwater streams deposit stratified drift between the stagnant glaciers and the valley walls, forming a flat-topped **kame terrace** (Figure 13.16).

Eskers

One fascinating glacial landform is the esker. An **esker** is a long, narrow, and often sinuous ridge (Figure 13.17). Some eskers form from sediment moving in open channels on the surface of a stagnating glacier. However, most eskers are believed to be deposited by meltwater streams flowing through tunnels at the base of the stagnating glacier. Sediment deposits are laid down on the tunnel floor, while the ice roof or tunnel walls are widened by melting. When the covering ice melts back, it exposes the tunnel deposits as an esker (Figure 13.17).

Outwash plains

Meltwater streams carry sand and gravel far from the glacial terminus, forming plains underlain by outwash. The outwash from a retreating valley glacier is confined between the valley walls, forming a long, narrow outwash deposit called a **valley train.** In comparison, the outwash from a continental glacier or an ice cap can extend for great distances along the glacial terminus and tens of kilometers away from the terminus. Extensive plains underlain by outwash are called **outwash plains.**

From time to time isolated masses of ice from stagnating glaciers may become buried by outwash. If the climate subsequently warms, these buried masses of ice melt and the surface collapses, forming a basin called a **kettle.** If these basins later fill with water, they are called **kettle lakes** (Figure 13.18).

Proglacial lakes

Lakes fed by meltwater that accumulates in topographically low areas on outwash plains or in stream valleys dammed by glacial deposits are called **proglacial lakes.** Proglacial lakes range in size from small kettle lakes to those which formerly occupied portions of several states. Proglacial Lake Hitchcock filled the Connecticut River Valley during the last retreat of glacial

Figure 13.16 Kame terrace along the Columbia River near Chelan, Washington. The terrace was formed when ice-marginal streams filled the trough between the wasting glacier and its valley wall. *(From D. A. Rahm,* Slides for Geology. *Copyright © 1971 McGraw-Hill, Inc. Used with permission of McGraw-Hill Book Company)*

Figure 13.17 Esker on the Waterville Plateau, Washington. *(From D. A. Rahm,* Slides for Geology. *Copyright © 1971 McGraw-Hill, Inc. Used with permission of McGraw-Hill Book Company)*

Figure 13.18 Kettles and kettle lakes near Pablo, Montana. *(From D. A. Rahm,* Slides for Geology. *Copyright © 1971 McGraw-Hill, Inc.   Used with permission of McGraw-Hill Book Company)*

ice, about 13,000 years ago. Lake Hitchcock was almost 480 km long and up to 16 km wide, had a maximum depth of about 61 m, and extended from Rocky Hill, Connecticut, northward to Lyme, New Hampshire. Meltwater accumulated between the retreating ice front in the north and a stratified drift "dam" at Rocky Hill in the south. Sediment supplied by streams draining into Lake Hitchcock were deposited as rhythmically layered varves, which we described in Chapters 6 and 8.

Among the most famous proglacial lakes are the Finger Lakes of New York State (Figure 13.19). A continental glacier advancing southward from what is presently Lake Ontario moved rapidly through preexisting north-south stream valleys. Glacial erosion deepened and straightened the valleys. As the glacier retreated, it deposited moraines across the lower ends of the valleys. These morainal dams trapped glacial meltwater, forming the Finger Lakes.

The large volume of water impounded in some proglacial lakes can result in catastrophic flooding if it is released suddenly. About 18,000 to 20,000 years ago a glacier dammed the Clark Fork Valley in northeastern Idaho, forming proglacial Lake Missoula, which spread out over western Montana. Lake Missoula covered an area of about 7770 km^2 and contained approximately 2084 km^3 of water. Other glaciers

bordering the lake supplied it with meltwater, and it increased in size and depth until the lake level overtopped the ice "dam." The escaping lake waters cut an ever-deepening channel into the ice plug, thereby releasing even more water. Within a few days the ice dam was destroyed and the lake waters were released in a gigantic flood, which spread westward over Idaho and Washington, eventually reaching the Pacific.

Maximal flood velocities are estimated to have been as high as 72 km/h with a peak discharge of 10,920,000 m^3/s. The peak discharge was about 10 times the combined flow of *all* the rivers in the world (at the present time). The raging torrent cut deep channels into the basalt of the Columbia Plateau, forming a topography referred to as the "channeled scablands." The floodwaters eroded blocks of basalt as large as 10 m across. Material was deposited in streamlined mounds of massive proportions. For example, "ripplelike" ridges 6 to 9 m high and 60 to 90 m apart were formed from gravel and coarser debris. It is believed that similar floods may have occurred in this area several times during the Pleistocene Epoch. However, the last flooding event was the largest and obliterated the evidence of the earlier ones.

DEVELOPMENT OF THE GLACIAL THEORY

The earliest descriptions of glacial deposits attributed them to the "great flood" recorded in the Bible because it was believed that this was the only way that such huge boulders could have been deposited. The development of the so-called glacial theory depended upon getting a scientist to examine both older glacial deposits and present-day glacial environments and to argue persuasively the connection between the two.

In 1829, a Swiss engineer named J. Venetz first postulated that glaciers once had been more extensive than they are today. He believed that the erratic blocks found in northern Europe and Switzerland had been deposited originally by glaciers. These observations inspired Jean de Charpentier to conduct a field study of active glaciers, during which he became convinced that past glacial activity could account for the

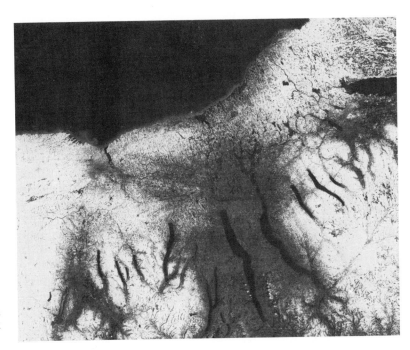

Figure 13.19 Satellite photograph of the Finger Lakes of New York State. *(ERTS)*

features previously attributed to the Biblical flood.

In 1837, the respected Swiss naturalist Louis Agassiz examined the field evidence for glaciation with de Charpentier. That year, Agassiz presented his detailed evidence for glaciation to a meeting of the Swiss Academy of Natural Sciences. His stature as a scientist and his persuasive arguments that glaciers formerly covered large areas of northern Europe and the Alps did much to advance the theory. His 1840 book, *Études sur les Glaciers (Studies of Glaciers)*, contained numerous excellent illustrations which enabled other geologists to understand the deposits of active glaciers and thereby to identify the deposits of former glaciers in their own areas. The first mention of the glacial theory in America was by Edward Hitchcock in a study of the geology of Massachusetts in 1841. Within 5 years, Hitchcock became the first American geologist to endorse the theory. Subsequent work in many areas of the world refined and expanded the glacial theory, and it is accepted universally today. Geologists examining the Pleistocene sedimentary record soon realized that not only was it composed of glacial sediments deposited during former cold periods;

it also was composed of sediments deposited during **interglacials**—warmer periods with climates more similar to today's.

Examination of a great number of Pleistocene sedimentary sequences in North America indicated that there were at least four major glacial episodes separated by at least three major interglacial episodes within the Pleistocene Epoch (Table 13.1). From oldest to youngest, the glacial periods are called the Nebraskan, Kansan, Illinoian, and Wisconsin. From oldest to youngest, the interglacial periods are the Aftonian, Yarmouth, and Sangamon. Evidence which we will discuss later in this chapter indicates that the Pleistocene climatic record was more complex than what is indicated by the incomplete continental sedimentary record. However, these names will subdivide the Pleistocene Epoch adequately for our purposes.

ICE AGES

When we think of ice ages, we naturally tend to think of the most recent ice age—the Pleistocene Epoch. However, there is clear evidence that there were numerous ice ages, some as far back as the Precambrian Era. The details of

TABLE 13.1   Major subdivisions of the Pleistocene Epoch based on the North American continental sedimentary record*

Glacial ages	Interglacial age
	Recent(?)
Wisconsin	
	Sangamon
Illinoian	
	Yarmouth
Kansan	
	Aftonian
Nebraskan	

* The climatic record deduced from fossil animals and plants in ocean sediments indicates that Pleistocene climatic changes were more complex than indicated by the broad subdivisions based on the rather incomplete continental sedimentary record.

these ancient glaciations are minimal because of the incomplete preservation of their sedimentary records. However, if we combine the ages and extents of these ancient glaciations with our detailed knowledge of glacial and interglacial events in the Pleistocene Epoch, we come to an understanding of the events that must be explained by any theory of the origin of ice ages.

Pre-Pleistocene Glaciations

Evidence for ancient glaciations comes largely from rock equivalents of modern glacial types of sediment. The most characteristic glacial rock is poorly sorted and unstratified **tillite** (Figure 13.20). We must be careful in assigning former glacial origins to all such rocks because they may have been formed by other processes, such as mudflows (Chapter 10). However, wherever rocks similar to glacial till are widespread areally *and* overlie a polished and striated surface below, we can reasonably infer that they are tillites. Some tillites grade laterally into rhythmically stratified shales and siltstones which may represent the rock equivalents of varve deposits in proglacial lakes (see Chapter 6).

The evidence for the oldest glaciation comes from rocks of Precambrian age. Rocks of the Gowganda Formation exposed in Ontario, Canada, record an episode of continental glaciation that occurred about 2.2 billion years ago.

Tillites of a younger Precambrian age, about 700 million years, have been found on all continents except Antarctica.

Late Paleozoic to mid-Mesozoic rocks of the Gondwana Formation on all the continents of the southern hemisphere contain distinct tillite sequences in the lower parts. The oldest Gondwana tillites (Devonian) are in South America, while those on the other continents range in age from Pennsylvanian to Permian. The distribution of Gondwana rocks and the ice flow directions deduced from the tillites are shown in Figure 13.21.

These ancient glaciations provide data which we can use to evaluate a theory of ice age origin. Glaciation on a continental scale occurred several times during earth history. The tillites currently are exposed on parts of continents far from the cold of present polar areas. In addition, beds of coal containing fossil warm-climate plants are found in the mountains of Antarctica. These observations suggest that the present location of the continents may be quite different from that of the past. Of the five hypotheses discussed in Chapter 1, only the plate tectonic model explains the distribution of pre-Pleistocene glaciations adequately. In the plate tectonic model, continents can change position on the earth so that landmasses now located near the equator once could have been located

Figure 13.20 Late Paleozoic tillite exposed west of Sao Paolo, Brazil. Note the absence of bedding, the poor sorting, and the large erratic boulder in the tillite. *(Courtesy of Rhodes Fairbridge)*

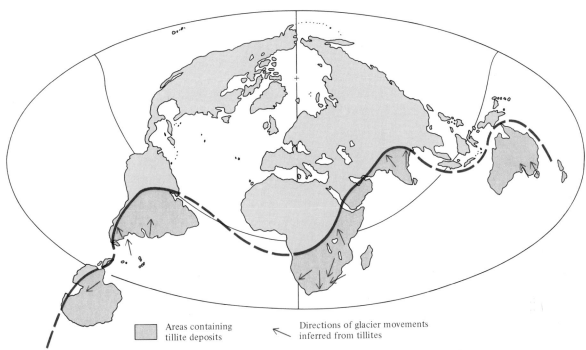

Figure 13.21 Distribution of late Paleozoic tillites in the southern hemisphere. The arrows indicate the direction of glacial movement in each area as the tillites were deposited.

Areas containing tillite deposits

Directions of glacier movements inferred from tillites

near the pole and vice versa. For example, the coal beds exposed in Antarctic mountains could have been formed when Antarctica was located at a warmer latitude.

Pleistocene Glaciations

While the Pleistocene Epoch is commonly thought of as an ice age, a more accurate description of it would be as a *time of great and frequent climatic change*. At any place climates were colder during glacial periods and warmer during interglacial periods than today's climates. The climatic changes were more complex because there were warmer and cooler episodes in both the glacial and interglacial periods.

The last and best-documented glacial advance, the Wisconsin, illustrates the growth and retreat of continental ice sheets through coalescing of glaciers from individual areas of accumulation called **ice centers**. Large-scale Wisconsin continental glaciation started when *average* annual temperatures dropped about 5°C worldwide about 120,000 years ago. This cooling resulted in a lowering of the snow line, allowing

glaciers to expand and occupy areas at lower altitudes year-round. The effects of this can be seen in glacial buildup in North America.

Massive snow accumulations formed ice centers west of the Hudson Bay region and in Labrador. These ice centers were nourished by moist ocean air moving onto the continent from the Atlantic and the Gulf of Mexico. The cold air above the growing glaciers condensed the moisture, causing it to precipitate as snow. Ice flow from these centers was radial in all directions. Coalescence of the ice from the Hudson Bay and Labrador ice centers formed the massive **Laurentide ice sheet** covering most of Canada and eventually extending deep into the northern United States (Figure 13.22). South of the growing Laurentide ice sheet terminus, valley glaciation and local ice caps began forming in highlands such as those in eastern Canada and in the White Mountains of New Hampshire. Ice from these local centers probably was incorporated into the Laurentide ice sheet as it grew southward. Meanwhile, in the Canadian Rockies to the west, valley glaciers grew and coalesced

Figure 13.22 Extent of Pleistocene glaciers in the northern hemisphere (color). Arrows denote local ice flow directions. Dotted line marks the approximate boundary between the Laurentide ice sheet (east, light color) and the Cordilleran ice sheet (west, dark color). Polygonal pattern shows areas of floating ice. Sea level shown is 100 m lower than today. (*Modified from R. F. Flint and B. Skinner,* Physical Geology, *2d ed.,* © *1971, with permission of John Wiley & Sons, Inc.*)

into piedmont glaciers and ice caps, which ultimately formed the **Cordilleran ice sheet.** This ice sheet spread westward toward the Pacific and eastward beyond the Rocky Mountains, merging in places with the Laurentide ice sheet to form a nearly continuous ice sheet extending across Canada and the northern United States (Figure 13.22). The Laurentide and Cordilleran ice sheets are believed to have merged only during the peak of glaciation. At other times there was an ice-free corridor between them. Anthropologists believe that the ice-free corridor probably provided a pathway south for the Asiatic peoples

who populated North America during the late Pleistocene Epoch. The pattern of glaciation in Europe was similar: Local ice caps occurred in the British Isles and on the Alps. The ice sheet growing over Scandinavia merged eastward with an ice sheet developing in the Union of Soviet Socialist Republics to form a massive ice sheet which eventually covered northern Europe and part of northern Asia (Figure 13.22).

These continental ice sheets advanced and retreated with the climatic fluctuations within the 75,000 years or so of the Wisconsin glacial stage. Each glacial advance was accompanied by a drop in worldwide sea level as more and more moisture was used to form continental glaciers. About 15,000 to 18,000 years ago (depending on location), the ice sheets began to thin and melt back from their terminal position. By 12,000 years ago, the Laurentide ice sheet had melted back from the United States to southern Canada. About 7000 years ago, the Laurentide ice sheet ceased to exist and was represented by remnant ice centers in Baffin Bay and Labrador; sea level had risen to within a few meters of its present position and the climate was almost as warm as it is today.

Unglaciated Areas of the Pleistocene

Glaciation is one of the more spectacular aspects of the Pleistocene Epoch. However, only about 30 percent of the earth's land surface was glaciated, and we should consider what was going on in the *unglaciated* areas at and beyond the limits of Pleistocene glaciation. As was mentioned before, the Pleistocene Epoch is described most accurately as a time of considerable, and cyclical, climatic change. We should expect to see the effects of this climatic change in the areas beyond the glacial limits.

Periglacial climates

The cold climate near the immediate margins of former and existing glaciers and ice sheets is called a periglacial climate. We mentioned earlier in this chapter that periglacial frost wedging loosened rocks so that they could be picked up more readily by advancing glaciers and ice sheets. Frost wedging and other periglacial processes formed topographic features far south of the maximal extent of Pleistocene ice sheets. One type of periglacial feature is **block fields**—flat or gently sloping areas covered with a continuous veneer of large (up to 10 m in length) angular and subangular blocks. Block fields usually occur on high, flat-topped mountains or plateaus above the timberline in both temperate and cold latitudes. They formed when intensive frost wedging broke up the underlying bedrock, forming a surficial layer of rock debris. Relict block fields have been reported from the Appalachian Mountains as far south as North Caro-

lina, well beyond the limits of Pleistocene glaciation.

Frost-wedged rocks also form lobe-shaped accumulations of boulder rubble called **rock glaciers.** Active rock glaciers form today near the snow line at cliffs where frost wedging is most active. This accumulation of rock rubble contains snow and ice, and it flows downslope with a form similar to that of a glacier. Melting of the internal ice results in cessation of flow and a deposition of the rock material.

Periglacial climates also affected the regolith. Long-continued freezing of the ground results in permafrost (see Chapter 10). The upper part of the permafrost is subject to thawing in the summer and then refreezes in winter. The freezing and thawing of the upper part of the permafrost results in the formation of surface features such as bands or stripes or polygonal networks of rock debris and polygonally fractured soil (Figure 13.23). These features are known collectively as **patterned ground** and are widespread in northernmost latitudes today.

As surface sediments are frozen, they contract and may develop vertically oriented and downward-closing fractures a few millimeters wide and a meter or more in depth. During the next warm period, meltwater may reach into the crack and refreeze when it hits the colder ground below. As the ground refreezes, the crack forms again. In the next thawing period, another thin layer of ice is added to the progressively widening wedge of ice. After a great number of free-thaw cycles, the wedge-shaped ice mass in the frozen soil may become a meter or more wide at the top. Melting of the ice wedge by postglacial thawing enables the cavity to be filled with washed-in finer sediment, forming an **ice-wedge cast** (see Figure 13.24). These structures are useful to geologists mapping the extent of periglacial belts around former ice-sheet margins.

Loess

Loess is a deposit of predominantly silt-sized particles which were eroded and deposited by winds. Winds blowing across the glacial margins preferentially removed the finer par-

Figure 13.23 Patterned ground in the Barrow district, Alaska. The polygons are 7 to 15 m in diameter. *(T. L. Pewe, U. S. Geological Survey)*

Figure 13.24 Ice-wedge cast on west wall of Wilber Creek, Yukon region, Alaska. *(T. L. Pewe, U. S. Geological Survey)*

ticles from extensive areas of outwash (Figure 13.25). The particles were deposited as sheets of loess extending far away from the glacial margins. Loess sheets may reach thicknesses of 35 m near the sources of glacial sediment. However, the individual layers of loess become thinner and the component particles become relatively finer in a direction away from the glacial source. The fossil shells of former air-breathing snails found in loess deposits are of a type suggesting cooler climates than those of today. Carbon-14 dates (see Chapter 8) obtained from these snail shells can be used to determine the age of the loess deposits. Not all loess is derived from glacial deposits, and extensive loess deposits occur downwind of major deserts, such as the Gobi Desert in China.

Pluvial lakes

During Pleistocene glacial advances, the currently semiarid basins of the western and southwestern United States had cooler and wetter climates than they have today. Such a cooler and wetter climate brought about by advancing glaciers is called a **pluvial climate.** Warmer air moving into the region from the south encountered the cold air associated with the advancing glacier. The moisture in the air condensed as rain, filling up the formerly dry

Figure 13.25 Clouds of silt transported by wind from the Delta River flood plain, Yukon region, Alaska. (*T. L. Pewe, U. S. Geological Survey*)

basins and forming extensive lakes. The increased cloud cover reduced evaporation greatly, enabling the lakes to grow in size during glacial advances. Lakes formed during the heavier rainfall and reduced evaporation of glacial advances are called **pluvial lakes.** One of the greatest pluvial lakes was Lake Bonneville, which covered over 50,000 km² of Utah, Nevada, and Idaho and reached a maximum depth of more than 330 m. A remnant of Lake Bonneville is the Great Salt Lake in Utah. The former shorelines of Lake Bonneville are visible on the sides of the Wasatch Mountains outside of Salt Lake City (Figure 13.26).

Pluvial lakes differ in origin from the proglacial and kettle lakes described earlier in this chapter because meltwater played little part in their expansion. Carbon-14 dating of pluvial-lake sediments has shown that high lake levels occurred at times when glaciers were *advancing*. In a similar fashion, low lake levels coincided with times of glacial retreat. This is opposite to the pattern seen in proglacial lakes, which grew in size as glaciers melted and released water.

Changes in sea level

Glaciers are part of the hydrologic cycle, and any fluctuation in them must affect water distribution on the earth's surface. During a time of active glaciation, moisture evaporated from the oceans falls on the continents as snow, nourishing the expanding glaciers. As more of the earth's water is stored on the continents as glacial ice, there is a reduction in runoff into the ocean, and therefore a progressive lowering of sea level occurs. It is estimated that glacial sea levels may have been as much as 100 m lower than today's. At a glacial maximum, the Atlantic

Figure 13.26 Former shoreline positions of pluvial Lake Bonneville along the Wasatch Mountain front in Utah. (*R. B. Morrison, U. S. Geological Survey*)

Figure 13.27 Suffolk Scarp at Chuckatuck, Virginia. The scarp marks a former interglacial sea level 15 m above the present one which can be traced from northern Virginia
into Florida. The house on the right is built on fossil dune sands which accumulated
above and landward (west) of the shoreline. The mounds are peanut plants staked for
drying in the sun. *(Nicholas K. Coch)*

shoreline lay 150 km east of the present site of
New York City, with the intervening coastal
plain forested by spruce and pine, and its population including mammoths, mastodons, and
other extinct mammals. The drop in sea level
during glacial advances would have changed the
stream base level so that streams would accel-
erate their downward erosion, excavating their
valleys (see Chapter 11). In the same manner,
interglacial sea levels should have been higher
than at present because the large volumes of
water from melting glaciers would have raised
sea levels worldwide. As sea level rose, base
level was raised and streams began to deposit
sediments within their valleys. Evidence of sea
levels higher than at present occurs on the Atlantic and Gulf Coastal Plains in a series of emerged
coastal features extending up to an altitude of
over 40 m above present sea level (Figure 13.27).

CAUSES OF ICE AGES

Determination of the origin of ice ages is challenging because not only must we have a mechanism for cooling the earth's surface so that continental glaciation may begin, we must also
account for the numerous cyclical climatic
changes indicated by the Pleistocene sedimentary
and fossil record. In the following sections we
will review briefly some of the mechanisms that
have been proposed to account for the climatic
changes documented by the Pleistocene sedimentary and fossil record.

Cooling of the Earth

There was an overall decrease in temperature
at midlatitudes between the late Cretaceous
Period (85 million years ago) and the onset of
Pleistocene continental glaciation (about 1

million years ago). Evidence for this temperature change comes from studies of fossil land plants and land organisms and from oxygen-isotope analysis of the shells of marine organisms. The rate of temperature drop increased markedly by the start of the Oligocene Epoch (35 million years ago), falling 8°C between then and the onset of Pleistocene continental glaciation. A number of hypotheses have been proposed to account for this cooling.

Mountain building

Late Cenozoic mountain building had increased the earth's relief significantly relative to that in the early Cenozoic Era. These rising landmasses could lower temperatures significantly in the continental interiors by acting as barriers to the moderating ocean winds and possibly by interfering with the poleward transfer of heat from the equator. The cooler elevated highlands would be new sites for valley glaciation. However, it is believed that such mountain building could account for only about 3°C of the observed temperature drop. In addition, most continental ice sheets accumulated on lowland areas far from mountains.

Volcanic dust

It has been proposed that increasing amounts of volcanic dust during the Cenozoic Era partially blocked the influx of solar energy, thus cooling the earth's surface. In addition, the particles could have served as nuclei for the condensation of atmospheric water vapor. However, sporadic periods of volcanism rarely coincide in time with major glaciations. For example, periods of extensive volcanism, such as during the Mesozoic Era and Tertiary Period in western North America, did not cause accelerated glacial activity.

Decrease in solar radiation

Small variations in solar energy have been noted over historic time. The period from 1645 to 1715 was one of minimal sunspot activity and apparently decreased total solar radiational output. Temperatures fell worldwide, and the snow-line altitudes dropped significantly. The time period is known as the Little Ice Age.

While there is other evidence of small-scale reduction in solar energy in historic time, there is no evidence that there were any significant variations in the past or that variations of so small a magnitude would be sufficient in themselves to trigger glacial episodes.

Latitudinal changes in continents

Throughout geologic time, polar areas were cooler than lower latitudes because the concentration of incoming solar radiation is much lower at the poles than at the equator. However, there may not have been ice at the poles always. Whenever continental masses were located in polar areas with access to oceanic moisture, the potential for extensive glaciation increased.

The plate tectonic model (see Chapter 1) provides the mechanism for moving continents (continental drift) into colder latitudes. By the early Tertiary Period, continental drifting had placed Antarctica over the South Pole and Europe, Asia, and North America around the North Pole. Thus, continental areas were situated in cold latitudes with access to oceanic moisture. These continental movements provided the conditions necessary for widespread glaciation. Geologic evidence indicates that major glaciers existed in west Antarctica by the late Miocene or early Pliocene, in east Antarctica by the Pliocene, and in the coastal mountains of southern Alaska by the late Miocene.

The increasing masses of ice in polar areas could have cooled the earth by increasing its reflectivity, or **albedo.** More solar radiation is reflected back into the atmosphere from high-albedo, light-colored surfaces such as snow and ice. The loss in solar energy could have contributed to a continuing decrease in temperature, ultimately triggering continental glaciation into temperate latitudes by the Pleistocene Epoch.

Cyclical Fluctuations in Pleistocene Climates

One or more of the hypotheses we have just discussed may have *some* relevance in explaining the *triggering* of glaciation, but we must look elsewhere for an explanation of the cyclical changes in Pleistocene climates. One of the most promising theories to explain cyclic climatic changes is known as the **astronomic,** or **Milankovitch, theory** after its strongest advocate.

The Milankovitch theory attributes cyclical climatic changes to the periodic differences in the position of the earth relative to the sun. According to the Milankovitch theory, three factors govern the relative position of the earth and sun. The first factor is eccentricity, or change in shape of the earth's orbit. The earth's orbit around the sun changes from circular to elliptic and back to circular about every 92,000 years. As a result, the nearest and farthest distances from the earth to the sun may differ by as much as 3 million miles over the 92,000-year cycle. The second factor is the periodic changes in the inclination of the earth's rotational axis relative to its plane of rotation around the sun. The inclination may vary from 21.8 to 24.4° over a cycle of about 40,000 years. The third factor is the precession of the equinoxes, the periodic shifting of the earth's axis as a result of gravitational attraction by the sun and moon (period is about 21,000 years). According to the Milankovitch theory, extensive glaciations began when the combinations of the three periodic factors we have described produced colder summers in the northern hemisphere. When the summers were cold enough so that only a portion of the previous winter's snows melted, snow would accumulate from year to year and the snow cover would eventually be thick enough to form glaciers. Presumably, the Milankovitch effect has acted over all of geologic time, although glaciations did not occur in the past at the regular intervals predicted by the Milankovitch theory. This suggests that the Milankovitch effect does not *cause* ice ages but can trigger cyclic glacial and interglacial activity once other factors have cooled the earth sufficiently for extensive glaciation to occur.

ORIGIN OF ICE AGES

The origin and cyclicity of glacial ages is a controversial topic. However, enough data are now available to present a working hypothesis to explain both the causes of ice ages and the cyclical glacial-interglacial climatic changes that occur within them. We will speak here specifically of the late Cenozoic glacial ages, although the same mechanism also could explain the earlier glacial ages recorded in the rock record.

Separation of the continent of Antarctica from Australia and its movement over the South Pole by the Miocene would have provided both the cold temperatures and the source of moisture required for the growth of continental glaciers on Antarctica. The growing Antarctic ice sheet increased the earth's albedo, further cooling the climate and thus favoring the growth of even more glaciers in a "feedback" mechanism.

By the Pleistocene Epoch, the earth had been cooled sufficiently so that continental glaciers began to move from cold latitudes into temperate latitudes. At this stage, the Milankovitch effect took over and produced the cyclical climatic changes that occurred within the late Cenozoic glacial age.

FUTURE CLIMATE

How will our climate change in the future? Extrapolations based on past climatic trends are unclear as to future climate. Do we have some time before we reach the warmest temperature of an interglacial period, or have we now passed that temperature high and are headed toward a progressively colder climate? Some scientists feel that the increasing amounts of particulate matter in the air that result from manufacturing and the burning of fossil fuels will reflect incoming solar radiation back into space, thus cooling the earth.

Other scientists feel that our increasing use of fossil fuels is increasing steadily the amount of carbon dioxide, CO_2, in the atmosphere, which will lead to steadily warmer climates. The increasing atmospheric CO_2 is penetrated easily by solar radiation. However, the heat generated on the earth's surface is blocked from escaping back into space by the CO_2 layer. This buildup of heat could increase atmospheric temperatures by 3°C by the year 2050. This would create conditions for a very different climatic distribution than at present. It is ironic that the human race, which has evolved amid the great temperature changes of the late Cenozoic, now may have the ability to determine the course of climatic change by widespread and accelerated pollution.

SUMMARY

Glaciers form wherever there is a net accumulation of snow from year to year. As the snow is buried under thick accumulations of younger snow, it is transformed into firn and finally into ice. Glaciers may be subdivided into two broad categories—alpine, or mountain, glaciers and continental glaciers. Glaciers move through a combination of plastic internal flow and slippage along their bases. The upper part of a glacier is brittle, while the lower part is plastic. The economy of a glacier is the balance between accumulation above the snow line and loss from all causes below the snow line. If accumulation is greater than loss, the glacier front advances. If loss is greater than accumulation, the glacier front retreats by melting and by breaking up as the ice thins.

Glaciers erode by plucking and abrasion. Mountain glaciers form cirques, arêtes, cols, horns, and U-shaped valleys. Erosion of rock surfaces by both mountain and continental glaciers forms striations, groves, polished rock surfaces, and sculptured rocks. Unstratified drift is deposited directly by the glacier as ground and end moraines. Glacial meltwater deposits stratified drift as kames, kame terraces, eskers, and outwash plains.

The Pleistocene Epoch was a time of great and repeated climatic change and was composed of a number of glacial and interglacial ages. The sedimentary record shows that extensive continental glaciation occurred earlier in the earth's history, in the Precambrian and Paleozoic Eras.

The latest episode of continental glaciation began as a result of late Cenozoic cooling. Glaciation started first in Arctic and Antarctic areas by the Miocene Epoch. Continued cooling resulted in a further lowering of the snow line by the Pleistocene Epoch, enabling ice sheets to form and to expand into lower latitudes. These continental glaciers expanded and retreated in response to cyclical climatic changes in the Pleistocene Epoch.

Many climate-related changes were also occurring in the unglaciated 70 percent of the earth's surface during the Pleistocene Epoch. Great pluvial lakes developed in currently semiarid basins. Sea level fluctuated many meters both above and below present levels. Great amounts of wind-blown silt, derived from outwash plains, were deposited across continental interiors. Cold climates characterized by extensive frost wedging and frozen-ground phenomena extended for great distances in front of the ice. Finally, the human race was evolving rapidly and adjusting actively to the changing climatic conditions, developing a rich culture in the process.

QUESTIONS FOR REVIEW AND FURTHER THOUGHT

1. Describe the formation of glacial ice.

2. What evidence suggests that glaciers consist of both brittle and plastic ice?

3. Describe as many differences as you can between alpine, or mountain, glaciers and continental glaciers.

4. What factors determine the velocity of glacial movement?

5. Describe how unstratified and stratified drift are deposited. How does this account for the great differences in these two types of sediment?

6. How do each of the following erosional features form?

   a. Hanging valley

   b. Sculptured rock

   c. Cirques

   d. Striations and polished surface

   e. U-shaped valleys

   f. Cols

7. How can each of the following be used to determine the direction of glacial movement?

   a. Striations

   b. Drumlins

   c. Indicators

8. Describe where and how each of the following depositional features is formed:

   a. Kame terraces

   b. Eskers

   c. Drumlins

   d. Ground moraines

   e. End moraines

9. How would you distinguish between glacial and interglacial sediments?

10. How did glacial advances and retreats affect the unglaciated areas beyond the terminal moraines?

ADDITIONAL READINGS

Beginning Level

Calder, Nigel, "Head South with All Deliberate Speed—Ice May Return in a Few Thousand Years," *Smithsonian,* vol. 8, no. 10, 1978, pp. 32–40.
(A well-written and well-illustrated explanation of the Milankovitch theory of cyclical climatic change)

Canby, T. Y., "Search for the First Americans," *National Geographic,* vol. 156, no. 3, September 1979, pp. 330–363.
(Describes the migration of humans from Asia into the Americas. Includes excellent maps showing the distribution of continental glaciers and human habitation sites at 18,000 and 11,000 years before the present)

Matsch, C. A., *North America and the Great Ice Age,* McGraw-Hill, New York, 1976, 123 pp.
(A detailed and well-illustrated discussion of many aspects of Pleistocene glaciation. The best place to start a more detailed study of glacial and Pleistocene geology)

Matthews, S. W., "What's Happening to Our Climate," *National Geographic,* vol. 150, no. 5, 1976, pp. 575–615.
(Describes the techniques used in analyzing climates and the possible climatic changes in the future. Many excellent color photographs, maps, and figures supplement a fascinating discussion)

Robin, G. de Q., "The Ice of the Antarctic," *Scientific American,* vol. 207, no. 3, September 1962, pp. 132–142.
(Describes the character and structure of the Antarctic ice. Contains informative photos of snow and ice crystals)

U.S. Geological Survey, *Glaciers: A Water Resource,* 211-345/76, 1976, 23 pp.
(Describes the relationship between glaciers and streamflow. Includes numerous excellent aerial photographs of different glaciers)

Intermediate Level

Beaty, C. B., "The Causes of Glaciation," *American Scientist,* vol. 66, 1978, pp. 452–459.
(Shows how a number of different factors working together can trigger continental glaciation)

Dansgaard, W., et al., "One Thousand Centuries of Climate Record from Camp Century on the Greenland Ice Sheet," *Science,* vol. 166, 1969, pp. 377–381.
(Describes how the record of past climates can be obtained from oxygen-isotope analysis of ice in deep cores, thus revealing the complexities of late Pleistocene temperature changes)

Fairbridge, R. W., "Glacial Grooves and Periglacial Features in the Saharan Ordovician," in D. R. Coates (ed.), *Glacial Geomorphology,* Publications in Geomorphology, Binghamton, N.Y., 1974, pp. 315–327, 398 pp.
(Describes the features associated with an ancient glaciation)

Sharp, R. P., *Glaciers,* Oregon State System of Higher Education Publication, Eugene, Ore., 1960, 75 pp.
(Discusses glaciers in detail but in a very readable manner. A good reference to start a detailed study of glaciers)

Advanced Level

Flint, R. F., *Glacial and Quaternary Geology,* Wiley, New York, 1971, 892 pp.
(Covers every detail about glacial and Quaternary geology worldwide. An excellent overall reference)

Ruddiman, W. F., and A. McIntyre, *Northeast Atlantic Paleoclimatic Changes over the Past 600,000 Years,* Geological Society of America Memoir 145, 1976, pp. 111–146.
(Describes how ancient climates can be deduced from studies of marine microfossils)

CHAPTER 14

Dry Regions and Wind Activity

DESERTS

When you mention the word *desert,* most people's minds conjure up pictures of windy, waterless, unvegetated, hot areas of rolling sand dunes, perhaps even with Foreign Legionnaires or mounted Arab horsemen. There certainly are such sandy deserts, but as we shall see, most of the world's desert areas are far different in appearance and characteristics.

What Are Deserts?

Deserts are more appropriately thought of as areas of low precipitation and high evaporation. The evaporation rate greatly exceeds precipitation in most cases. Many geologists define **deserts** as land areas receiving less than 25 cm of precipitation a year. Deserts usually are bordered by **semiarid areas** with precipitation of less than 50 cm per year. Desert landscapes include extensive areas of sand dunes, broad rocky expanses, areas of sparse vegetation, and at certain times of the year lush vegetation in full bloom. Plate 18 illustrates some of these landscapes.

The dry stream channels found in many deserts give the impression that stream erosion and deposition are unimportant. However, at certain times of the year, many deserts experience short periods of heavy rainfall during which stream erosion and deposition are extensive. Another striking feature of desert landscapes is the sharp outlines of many of the rock formations. Low precipitation in deserts results in sparse development of vegetation and reduced rates of chemical weathering. This in turn results in the absence of the thick soils which smooth out the topography in more humid climates.

Another misconception about deserts is that they are always hot. If we define deserts on the basis of yearly precipitation, we technically must include the *cold* deserts which form in polar areas because of low precipitation. These **polar deserts** have stark, rugged landscapes similar to those of warm deserts (Figure 14.1). In addition, wind plays a major part in the erosion and transport of material in cold deserts just as it does in warm deserts. Henceforth, when we use the term desert we will be referring to warm desert areas only.

Most of the geologic processes active in humid areas, such as weathering, streams, and wind, also are important in deserts. However, the low precipitation and reduced vegetation in deserts change the relative rates for each of these processes compared with the rates in humid climates. You will find it helpful to refer to our discussions of weathering (Chapter 9), mass movements (Chapter 10), streams (Chapter 11), and groundwater (Chapter 12) in humid areas for comparisons as we describe the processes of erosion and deposition in desert areas.

Types of Deserts

Deserts are of three different types. **Climatic deserts** are controlled by moisture distribution

Figure 14.1   An Antarctic dry valley (polar desert).   *(Polar Information Service, National Science Foundation)*

related to global atmospheric patterns. **Topographic deserts** are caused by mountainous masses which block the flow of moisture-laden oceanic air into the continental interiors. **Coastal deserts** are caused by adjacent cooler oceanic currents that reduce the moisture which falls on the coastal zone.

Climatic deserts

A map of the world's climatic zones shows that most deserts exist in a tropical latitudinal belt from 10 to 30° north and south of the equator (Figure 14.2). The largest deserts on the earth, such as the Sahara in Northern Africa and the Arabian desert in the Middle East, are located in these latitudes.

To explain the formation of these huge climatic deserts let us consider the generalized atmospheric circulation patterns on the earth's surface. Atmospheric circulation from equator to pole is divided into more localized circulation "cells" which govern the prevailing wind patterns in corresponding latitudinal belts. Examination of those circulation cells between the equator and latitudes 30° north and south will

help explain why many of the earth's great deserts occur there (Figure 14.3).

The greatest concentration of solar energy per unit area of the earth's surface occurs in the equatorial region. The surface air there is heated and rises, cooling as it reaches higher and higher altitudes. As the air cools, it releases moisture, which falls as torrential rains in the equatorial regions. The rising air is much drier as it continues to move northward and southward toward the poles. This warm, dry air descends to the surface around latitude 30° north and south. The descending warmer air is capable of holding additional moisture and is unlikely to release any of the moisture it contains as precipitation. These warm, dry winds sweeping across the surface provide the conditions necessary for the formation of the great climatic deserts found at latitudes of about 30° north and 30° south of the equator (Figure 14.2). The air then moves toward the equator to complete the circulation cell.

The air between latitude 30° N and the equator does not move back to the equator in a straight north-to-south direction but rather in a northeast-to-southwest direction. This deflec-

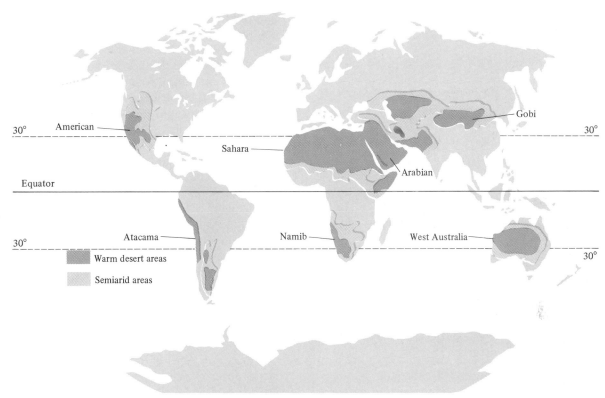

Figure 14.2   Geographic location of warm deserts and semiarid areas.

tion is a result of the earth's rotation and is known as the **Coriolis effect.**   As a result of the Coriolis effect, a moving object in the northern hemisphere (flowing air, for example) moves to *its* right.   Since winds are named by the direction *from* which they blow, the prevailing winds in this latitudinal belt are called the northeast trade winds (Figure 14.4).   In a similar fashion, the winds moving equatorward from latitude 30° S veer to *their* left.   This generates the southeast trade winds, which are the prevailing winds in this southern latitude belt.

Figure 14.3   Air circulation patterns resulting in the formation of climatic deserts at latitudes 30°N and 30°S.   Only the northern circulation cell is shown here.

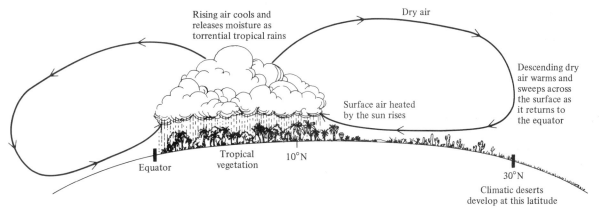

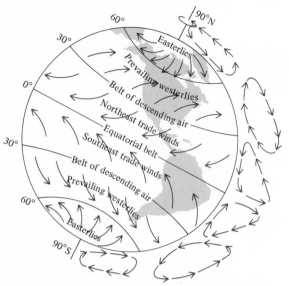

Figure 14.4 Generalized atmospheric circulation patterns and major wind systems on the earth's surface.

Topographic deserts

Topographic deserts are formed in areas where intervening mountainous terrain bars the transfer of moisture from oceanic areas. Good examples of topographic deserts are those which have formed in the western and southwestern United States (Figure 14.2). At these latitudes, the dominant winds are from the west and are called the prevailing westerlies. Moisture-laden oceanic air driven inland by the prevailing westerly winds rises over the coastal ranges of

California (Figure 14.5). As the air mass rises over the coastal areas, it cools and releases excess moisture. Additional moisture is lost as the air mass moves farther inland and up over the Sierra Nevada Mountains (Figure 14.5). The air mass then descends into the area to the east of the Sierra Nevada Mountains, warming as it loses altitude. This warm, dry air dehydrates the land, thus forming the deserts in the American Southwest (Figure 14.5). The Gobi Desert is the largest one in Asia (Figure 14.2) and is located in the belt of prevailing westerly winds (Figure 14.4) in the Mongolian Republic and the Peoples Republic of China's Inner Mongolian Autonomous Region (Figure 14.4). Very little moisture reaches the Gobi because it is deep within the continental interior and separated from sources of moisture by intervening mountains.

Coastal deserts

The third type of desert is the elongate coastal desert, such as those found along the western sides of continents in the southern hemisphere. At first it might seem unusual to find a desert adjacent to an ocean, but it is the significant difference in temperature between the coastal lands and the adjacent oceanic currents that brings about the dry conditions under which deserts form. The ocean waters in the southern part of the southern hemisphere are driven eastward by the prevailing westerly winds in an

Figure 14.5 Conditions favoring the development of topographic deserts in the American West and Southwest.

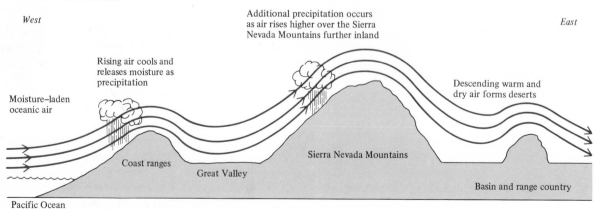

West

Additional precipitation occurs
as air rises higher over the Sierra
Nevada Mountains further inland

East

Rising air cools and
releases moisture as
precipitation

Descending warm and
dry air forms deserts

Moisture–laden
oceanic air

Coast ranges

Great Valley

Sierra Nevada Mountains

Basin and range country

Pacific Ocean

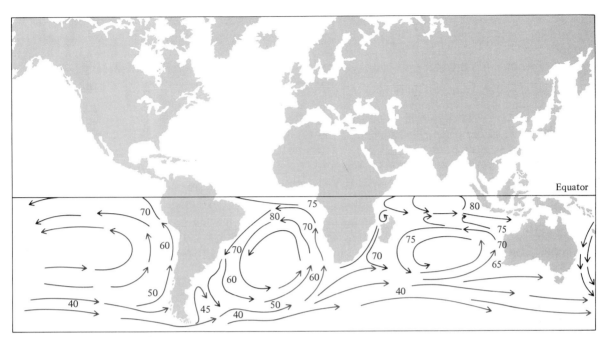

Figure 14.6 Major oceanic surface currents and temperatures. The color arrows show cold currents moving up the western sides of continents in the southern hemisphere, thus providing the cool coastal air masses that form coastal deserts such as the Atacama (South America), Namib (South Africa), and the western Australian. The numbers on the arrows are temperatures in °F. (*Modified with permission from H. F. Garner*, The Origin of Landscapes: A Synthesis of Geomorphology, © *1974 by Oxford University Press.*)

oceanic current which circles the Antarctic continent (Figure 14.6). This cold current is deflected to its left (northward) by the Coriolis effect; when it reaches South America, Africa, and Australia, it moves up the western coasts of these continents. These cold currents, such as the Peru, Benguela, and west Australian (Figure 14.6), result in cool air masses, even when the adjacent continent is warmer. The cool air moving inland from these cold currents is moisture-deficient and cannot release what moisture it has unless coastal mountains force it up to higher altitudes where it can cool further. If the coast bordering the cool current is low-lying and hot, conditions for the formation of a desert are present. As this cooler ocean air passes inland over the hot, low-lying coast, it is warmed. The air is then capable of holding more moisture, and thus it is less likely to release any of its moisture as precipitation. The Atacama of Chile and the Namib of Southwestern Africa (Figure 14.2) are examples of coastal deserts.

Characteristics of Deserts

Deserts can be characterized by unique temperatures, precipitation, vegetation, weathering and soils, and stream activity.

Temperatures

Deserts may be very hot in the daytime but quite cool at night. Why are there such extremes of temperature in "warm" deserts? Lack of cloud cover in most dry regions allows the ground surface to receive a high proportion of incoming solar radiation, and surface temperatures can reach very high values by the middle of the day. One of the highest temperatures recorded anywhere was 58°C (136.4°F) *in the shade* at El Azizia in the Libyan portion of the Sahara Desert. Exposed rock and sediment surfaces can reach much higher temperatures than the air. After the sun has set, heat from solar radiation is dissipated rapidly into space, resulting in very low nighttime temperatures. H. F.

Garner (1974) provides a fascinating account of the extremes of temperature which can be encountered in desert areas (see Additional Readings at the end of this chapter).

> The writer will not soon forget arising at 3 A.M. in the Algerian Sahara in early September. The bucket of water for washing presented a thin film of ice, and a heavy wool sweater and leather jacket were comfortable while riding in an open jeep. By 9:30 A.M., the air temperature was nearing 85°F; by 11:30 A.M., a pocket thermometer registered over 105°F, and at 200 P.M. the same thermometer registered 127°F. What passed for a local pub in a nearby oasis cooled its beer by wrapping the bottles in wet sacking and laying them in the sun. Evaporation occurs at an almost unbelievable rate under such circumstances, and effective ground moisture levels are fantastically low. The foregoing account amounts to a record of a daily temperature variation of at least 95°F. The beer was delicious.*

Precipitation

High daytime temperatures in many deserts affect the amount of precipitation that can reach the ground. Precipitation formed by condensation of moisture in the air at higher altitudes may be evaporated back into the atmosphere as it falls into lower, warmer air. The small amount of water vapor in desert air may condense as dew on rock and sediment surfaces during cool nights. This condensation of moisture occurs in all but the most arid deserts; where present, it aids in chemical weathering.

Deserts have a wide range of average annual rainfall. The average yearly rainfall within any one desert may vary considerably from place to place, even for adjacent areas. For example, the average yearly precipitation at Furnace Creek station, on the floor of California's Death Valley, is about 4 cm. However, the precipitation in the adjacent mountains is higher. The

greater precipitation in desert highlands results in greater chemical weathering and erosion. The weathered material is carried down into the adjacent lowlands by mountain streams, which generally flow for only a short period after a rainfall.

Vegetation

Deserts exhibit a wide range of vegetation densities and types, ranging from sparse vegetation to almost total absence of vegetation in the most arid deserts and moving sand "seas" in the Sahara and Arabian Deserts (Plate 18). Trees, common in humid climates, are rare in deserts. Those present are located along channels of intermittent streams or places where the groundwater table is near the surface.

Desert vegetation shows remarkable evolutionary adaptations for survival in a dry climate. The sparse vegetation in deserts greatly affects the processes of weathering and erosion. In humid climates, organic material is important in physical and chemical weathering of rocks (see Chapter 9). Plant roots penetrate the soil and make it easier for water to infiltrate around the root openings, while the root system anchors the soil and makes it resistant to erosion by water and wind. A greater proportion of the rainfall in humid areas is able to infiltrate the soil rather than run off along the surface. This soil water keeps the soil moist between rains, aiding the weathering processes but making the soil resistant to erosion by wind as well as recharging the water table (see Chapter 12). In contrast, limited vegetation in deserts reduces the efficiency of chemical weathering and increases the amount of material that can be removed by wind or water erosion.

Weathering and soils

Physical weathering is more visible and probably more important than chemical weathering in dry regions. Salt crystal growth, diurnal temperature differences, and frost wedging are especially important (see Chapter 9). For example, shattered rock fragments litter many desert floors, and in places these fragments can be reassembled into larger cobbles and boulders. The rate of chemical weathering is reduced greatly in deserts because of the low rainfall and scarcity of vegetation. Some of this chemical weathering is carried out during wet seasons, while a portion may be carried out by the dew that condenses out of the desert air at night. Greater soil moisture retention in shaded areas and around patches of vegetation may increase chemical decomposition in those areas.

One indication of chemical weathering is the thin dark red, brown, or black coatings which form on rock particles and surfaces after long exposure to desert air. These coatings are composed of oxides of iron with traces of manganese oxide and silica and are referred to as **desert varnish** (Figure 14.7). There are two major hypotheses for the origin of desert varnish. Some believe that it is formed by solution of iron and manganese within the rock particles; the material is deposited on the outside of the particles as the water evaporates. Others believe that the iron and manganese coatings are derived from an external source which currently is not identifiable. The well-developed desert varnish on quartzite and other types of rock which do not contain any iron or manganese suggests that this mechanism may account for the formation of at least some desert varnish.

The soils developed in deserts usually are thin and patchy, a result of the decreased chemical weathering and the great amount of soil erosion by water and wind. They commonly contain soluble minerals, are lacking in organic accumulations, and contain generally angular rock fragments derived from physical weathering. Soil horizons usually are not present or are weakly developed.

Drainage

The streams in some deserts never reach the sea because the desert basins are ringed by mountains. This type of drainage, called **internal drainage,** results in the physical and chemical stream load being deposited within the desert.

Most of the drainage within desert areas is by ephemeral streams (see Chapter 12), which carry water for only a short time during and after a rainfall. Precipitation which falls on the steep and largely bare desert slopes usually exceeds the ground's ability to absorb it. This

Figure 14.7 Desert varnish. The dark coating is especially visible on the fragments nearest the shovel blade. *(Courtesy of Donald Doehring)*

results in more surface runoff than infiltration into the ground. If the rainfall is heavy and relatively short in duration, water can quickly fill and spill out of intermittent stream channels, flowing along the surface of the ground as a shallow and fast-moving sheetflow (see Chapter 11). Some of this surface flow infiltrates into the ground and recharges the groundwater reservoir below. If the zone of saturation is recharged sufficiently, intermittent streams can maintain their flow between rainfalls. These intermittent streams will dry up if the water table drops below the elevation of their channels (see Chapter 12).

If stream discharge is great enough to overcome losses by infiltration into the ground and evaporation into the air, excess water can accumulate in low areas on the desert floors as temporary **playa lakes.** These shallow and wide-

Figure 14.8 Layer of salt covering the surface of the Racetrack Playa northwest of Death Valley, California. *(Courtesy of John S. Shelton)*

spread lakes may form within hours and persist for days or weeks after the rainfall. When the lake dries up, the dissolved ions are deposited as a bed of crystallized salts (Figure 14.8).

Some deserts have **external drainage** and are traversed by through-flowing streams, which derive their discharge from areas with higher precipitation outside the desert. Examples of such through-flowing streams are the Nile River in Africa and the Colorado River in the southwestern United States. These streams are able to flow through and irrigate the desert areas near the channel because their discharge is greater than the combined water loss by infiltration into the ground and evaporation into the dry desert air. External drainage usually results in most of the physical and chemical stream load being deposited outside the desert. However, human modifications on the Colorado and Nile have resulted in changes in normal depositional patterns. Most of the physical load of the Nile is now trapped upstream because of the Aswan Dam. Dams on the Colorado River also have trapped most of the sediment load. In addition, increasing withdrawal of river water for agricultural and domestic uses has increased the concentration of salt in the lower Colorado.

Stream Activity

When you visit a desert, the work of the wind is apparent in moving sand, dunes, or even a dust storm, and it may seem that wind is the most important factor in shaping the landscapes. Look more carefully and you may see alluvial fans extending out from the mountains, and steep-sided channels that are floored with sand and gravel and cut across the desert floor. Both these features provide evidence of stream activity. We now know that deserts experience infrequent but intense stream activity and that stream erosion is far more important (although less apparent) than wind activity in forming desert landscapes.

We should also remember that desert stream activity in the *past* was significantly greater than it is at present. In Chapter 13 we described pluvial climates, during which many present-day deserts were more humid than they are today. A major portion of a desert landscape may have been formed during the increased chemical weathering and stream activity of pluvial climates. This relict landscape is being *modified* today only by winds and infrequent stream activity.

Stream erosion

Even though desert precipitation is infrequent, the special conditions in deserts maximize the ability of the ephemeral and intermittent streams to erode and deposit material. The ability of streams to erode is increased by both the high slopes and the sparse vegetation. In the absence of a cover of vegetation, the impact of falling rain results in considerable soil erosion. The water flowing as sheetflows over the surface moves faster because there is little vegetation blocking its path. Consequently, desert rainfalls result in high stream discharges and sheetflows which can incorporate large quantities of loose sediment and rock material into the flow. Some desert streamflows have relatively low sediment loads and are similar to streams in more humid regions. Other flows have a higher proportion of sediment to water and are more like mudflows (see Chapter 10). Mudflows form when water is lost through evaporation and infiltration while there still is a large amount of sediment to carry. Desert streams and mud-flows carve out steep-sided canyons called "arroyos" or "dry washes" in the United States and "wadis" in the Middle East (Figure 14.9).

Another desert erosion feature is the pediment. **Pediments** are gently sloping (less than 5°) bedrock surfaces of low relief that slope away from the bases of mountains (Figure 14.10). The origin of pediments is quite controversial. Most geologists who have studied the problem agree that pediments are erosional features developed on bedrock and that the formation of the pediment is associated with the erosional retreat of the mountain front, extending the pediment mountainward with time. The disagreement is over *how* the erosion is carried out. Some attribute pediments to lateral erosion by streams in channels or by sheetflows. Unfortunately, it is hard to explain how stream-channel migration or sheetflooding can be effective enough to erode such expanses of rock. Geologist John Moss, working in the granite pediments in the Usery Mountains east of Phoenix, found that the pediments there were

Figure 14.9 The walls of this arroyo in Wayne County, Idaho, were cut laterally 2 m toward the cook tent during a desert flood on July 10, 1936. *(C. B. Hunt, U. S. Geological Survey)*

Figure 14.10 Pediment surface at Sheephole Pass Pediment, Mojave Desert, California. *(Courtesy of Donald Doehring)*

underlain by crumbly granite extending downward to an irregular granite surface below. From this evidence, he suggests a two-stage mechanism for pediment formation, starting with deep weathering in a formerly wetter climate in the Tertiary Period and a removal of the weathered debris in the Quaternary Period. This hypothesis has the advantage of explaining how present-day stream erosion can form pediments because the streams need remove only weathered rock debris.

When sedimentary rocks are topped by resistant, nearly horizontal rocks (commonly lava flows), topographic features called mesas are formed. A **mesa** is a broad, flat-topped erosional remnant bounded on the sides by steep slopes or cliffs. A **butte** is similar to a mesa, but it has a less extensive surface area (Figure 14.11).

Stream deposition

The bases of mountain fronts in deserts are mantled with sediment supplied by the ephemeral and intermittent streams draining the mountains. Extending basinward from the mouth of many of these stream channels are alluvial fans (Figure 14.12). Alluvial fans are found at the bases of some steep slopes in humid regions, but

Figure 14.11 Buttes and mesas near Canyon de Chelly in northern Arizona. *(Courtesy of Donald Doehring)*

they are much more common and larger in deserts. The size of the fans is quite variable, from hundreds of meters to tens of kilometers in length. The fan surface is concave upward, with slopes of 5 to 10° near the apex of the fans at the mountain front to less than 1° at the basin edge of the fan.

Streams and mudflows traveling down mountain slopes drop much of their sediment as they lose kinetic energy at the base of the mountains and as water is lost through percolation into older sediments and evaporation into the hotter and drier air of the desert basin. The merging of adjacent alluvial fans built out from the mountain front forms a basinward-sloping depositional surface called a **bajada** (see Figure 14.12).

Alluvial fan sediments in desert areas grade from complexly bedded gravels and sands near the mountain front to sands and silts in the basin portion. Finer alluvial fan sediments interfinger with soluble salt deposits derived from desiccation of playa lakes occupying the lowest portions of the desert floors (Figure 14.13). Such basin sediments can reach thicknesses of thousands of meters when the adjacent mountains are being lifted up or the basins are subsiding.

The Erosional Cycle in Deserts

Assuming no changes in base level, the topography in deserts undergoes a sequential series of evolutionary changes similar to those in humid regions (see Chapter 11). Stream erosion and pediment formation on several sides of a mountain mass form breaks in the mountains and isolate erosional remnants as **inselbergs** (Figure 14.14). If the drainage is internal, the physical and chemical loads of the streams are deposited within the basins. In the last stages of erosion,

Figure 14.12 Alluvial fans along the north slope of the Avawatz Mountains in the Mojave Desert of California merge to form a basinward-sloping bajada (center). *(J. R. Balsley, U.S. Geological Survey)*

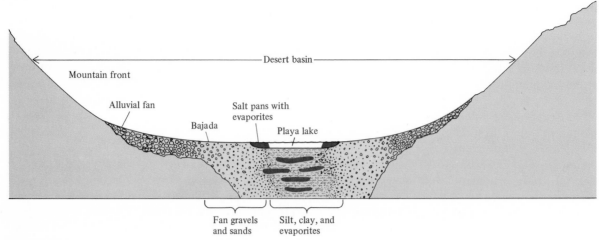

Figure 14.13 Geologic section showing sediment facies deposited in a desert basin with internal drainage. Alluvial fan gravels and sands grade basinward into silts and clays. The center of the basin is underlain by alluvial silts and clays interlayered with evaporites derived from the desiccation of playa lakes.

the low relief of the desert surface is broken occasionally by isolated highlands (inselbergs) surrounded by pediments and thick basin fills composed of alluvial fan and playa sediments.

The end stage of erosion in dry climates is similar to the low-relief surface that is the end stage of stream erosion in humid climates. The resistant rock masses that stand up above extensive erosional surfaces in humid areas have their parallels in the inselbergs of dry areas. While these two end-stage erosion surfaces are similar in appearance, they are very different in their mode of formation and underlying materials. In humid areas, through-flowing streams have removed most of the material eroded from the adjacent highlands, and the largely erosional land surface is underlain by a relatively thick soil rich in chemically resistant minerals such as quartz, clay minerals, and oxides of iron and aluminum. However, the low-relief surface that results from desert erosion and local deposition in the absence of through-going streams is quite different. The desert surface is underlain by very poorly developed soils containing soluble minerals which can overlie great thicknesses of sediment. The sediments are more angular and generally contain chemically unstable minerals, reflecting the decreased effect of chemical weathering in dry areas.

WIND ACTIVITY

Some appreciation of the work of wind in deserts can come from observing what happens in a desert windstorm. The first gusts of wind pick up loose plant debris and sand grains, moving them across the desert floor. The sand grains move by saltation, bouncing along the surface and dislodging other sand grains in a "chain reaction" effect until there is a 1- to 2-m-thick "carpet" of sand moving along the desert floor, bouncing off rocks and other obstacles. If wind velocity and turbulence (see Chapter 6) are sufficient, the finer silt and clay-sized particles may be eroded and carried in suspension high above the carpet of sand in saltation. A full-fledged desert dust storm reduces visibility to zero. As the wind velocity drops, the saltating sand particles are deposited first, accumulating in streamlined piles behind obstacles and between rock particles on the desert floor. The finer sedimentary particles may be carried in dust clouds for great distances beyond the desert.

The abundance of wind erosional and depositional features attests to the importance of wind activity in deserts. Although it is less effective in humid areas, wind action is still important. While wind activity is not the *dominant* force in shaping desert landscapes, wind cer-

(a) Mountains and basins formed by faulting. Uplifted blocks begin to be eroded and alluvial fans build out from mountain front. Mountain masses are progressively dissected. Temporary playa lakes form in low points in basin after rainfall.

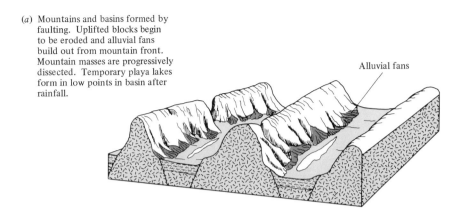

Alluvial fans

(b) Relief in the area decreases as mountains are eroded and alluvial fans join to form a bajada along the mountain front. Pediments become more prominent. Thickness of the basin fill has increased.

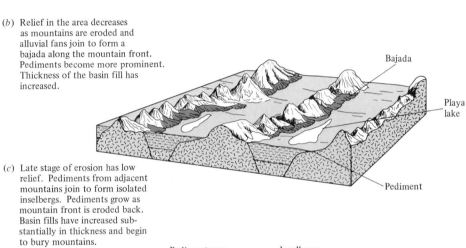

Bajada

Playa lake

Pediment

(c) Late stage of erosion has low relief. Pediments from adjacent mountains join to form isolated inselbergs. Pediments grow as mountain front is eroded back. Basin fills have increased substantially in thickness and begin to bury mountains.

Pediment

Pediment pass (rock floored valley)

Inselbergs (remnant mountains)

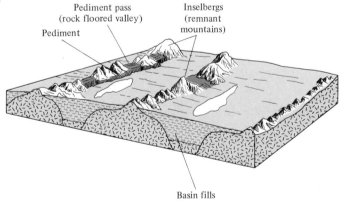

Basin fills

Figure 14.14 Evolution of landscapes in a desert. In this area the basins and mountains originally were formed by faulting (a). With increasing time, the mountains are eroded down to a low relief and the thickness of the basin fill increases (b, c). This sequence occurs when mountains and basins are stable and the drainage is internal.

tainly reaches its maximum *effectiveness* in areas of dry soil, sparse vegetation, and frequent winds.

Much of what we know about wind activity and dune formation started with the work of

R. A. Bagnold, who carried out exhaustive studies in the deserts of North Africa as a British officer before and during World War II. His book, *The Physics of Blown Sand and Desert*

Dunes, was published in 1941 (see Additional Readings). Bagnold's observations have been referred to time and time again, and they form the basis of our discussion of the subject. We will describe the work of the wind by first looking at how wind erodes and transports material and then at how depositional features such as dunes are formed.

Erosion and Transport by Wind

Wind can readily erode sand and silt and clay-sized particles. Once eroded, the coarser and finer particles are transported in different ways.

Sand-sized particles

The wind blowing across the desert may be moving quite fast a meter or so above the ground surface. However, as with flowing water (see Chapter 11), the velocity decreases toward the ground surface, where the effects of frictional drag are greatest. The air closest to the ground, a zone equal in thickness to about one-thirtieth of the predominant particle size in the deposit, is relatively motionless. Only those grains which protrude above this thin zone of "dead" air can be eroded. As the speed of the wind increases, a critical, or **threshold, velocity** is reached in which some of the sand grains start to move. They begin rolling and climbing over other grains and obstacles, while some begin saltating in parabolic leaps. Laboratory experiments have shown that once some of the grains start moving at the threshold velocity, grains will continue to move, even if the velocity drops below the velocity needed to *start* grain movement. This is because once grains start saltating, they set other grains into motion by colliding with them in a chain reaction. Sand particles are transported by wind in a number of ways. Some particles hit hard objects such as rocks and ricochet back into the air. The force of impact of particles upon a sandy surface displaces some of the surface particles, throwing them into the air. As some grains embed themselves in the sandy surface, they can push other grains laterally along the surface in a type of grain movement called **surface creep.** Before long there is a zone, generally less than 2 m thick, above the ground surface in which there is extensive saltation with each sand grain impact moving others in a chain reaction.

Silt and clay-sized particles

Wind of a given velocity may start movement of sand-sized particles before it can start eroding silt and clay particles. The finer-grained silt and clay particles are harder to erode for a number of reasons. Their fine size makes them more cohesive, and they form a surface with a low degree of roughness, making it harder for the wind to move them. They also are so small that they cannot extend up through the layer of motionless air near the surface to be affected by the force of the wind.

How are these fine particles set into motion? This occurs when the surface is disrupted by natural or human action. For example, a farmer's plow, truck, or cattle, or a motorcycle, "dune buggy," or any other off-road vehicle (ORV), moving along a fine-grained surface is trailed by clouds of dust as the ground surface is broken and the wind begins to remove the fine material. In the absence of such human activity, the surface can be broken by the impact of saltating sand grains or the rolling of gravel, which throws the finer-grained material up into the higher-velocity air above the ground. The low weight of these fine particles enables them to be carried in suspension for great distances, even out of the desert area. Such dust storms may travel in both the lower and upper parts of the atmosphere. Low-altitude suspensions travel a few tens of kilometers, becoming both thinner and finer away from the source. If dust particles can be carried into the jet stream, they may travel great distances, with little apparent settling or deposition of particles en route, until they reach a downdraft and are deposited.

Some recorded instances of dust transport point out how far these particles can travel. Fine dust blown from the "Dust Bowl" in the southwestern United States in the 1930s fell on snow in New England and most likely far out into the Atlantic Ocean. Examination of fine dust collected on the Caribbean island of Barbados* by

* A. C. Delany, A. C. Delany and D. W. Parkins, Airborne Dust Collected at Barbados; *Geochimicta and Cosmochimica Acta,* vol. 31, 1967, pp. 885–909. Specific ref. to pp. 907–908.

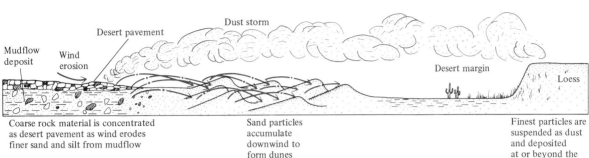

Wind direction

Dust storm

Desert pavement

Mudflow
deposit

Wind
erosion

Desert margin

Loess

Coarse rock material is concentrated
as desert pavement as wind erodes
finer sand and silt from mudflow

Sand particles
accumulate
downwind to
form dunes

Finest particles are
suspended as dust
and deposited
at or beyond the
margins of the
desert.

Figure 14.15 Deflation of a poorly sorted desert mudflow deposit results in the formation of three separate sediment deposits (desert pavement, dune sand, and loess), each of which has a different average size but has better sorting than the mudflow deposit.

A. C. Delany and others in 1967 showed that it contained mineral and biologic components that were picked up in Europe and Africa, thousands of kilometers across the Atlantic Ocean.

Wind erosional features

Wind action can result in a number of distinctive erosional features as well as being a very effective sediment-sorting agent. The erosion of loose sedimentary particles by wind is called **deflation.** Air has only one one-thousandth of the density of water. Therefore, wind can transport only sand and smaller-sized particles. To illustrate the sorting effect of wind, consider what happens in the deflation of a desert mudflow deposit composed of particle sizes from cobbles to clay (Figure 14.15). Sand particles are eroded first and travel downwind in a layer of saltating particles less than 2 m thick. These sand grains may be deposited downwind as dunes. Finer particles are lifted higher and may travel as dust storms for great distances before being deposited beyond the edges of the desert as loess (see Chapter 13). As the mudflow deposit is deflated progressively, only coarser particles remain. Eventually, a lag deposit of coarse rock particles, called a **desert pavement,** covers the surface and prevents further deflation. Stream erosion also may contribute to the formation of desert pavement. Sheetflows may wash out some or all of the finer sedimentary particles. Subsequent wind erosion can remove

the remaining particles, producing a surface concentration of rock fragments which act like "armor," protecting the underlying sediments from further erosion. Desert pavements are a common surface in many deserts (Figure 14.16).

Figure 14.16 Desert pavement of coarse angular rock particles. Note that many are covered with desert varnish. Death Valley National Monument. *(C. B. Hunt, U. S. Geological Survey)*

Figure 14.17 Ventifacts from Sweetwater County, Wyoming. (*M. R. Campbell, U. S. Geological Survey*)

Numerous sand grains in saltation along the ground surface have considerable erosive force. Exposed rock surfaces may be smoothed and polished by grain impacts to form faceted and polished rock clasts called **ventifacts** (Figure 14.17). The abrasive nature of the particles in saltation within 2 m of the ground requires that structures such as wooden telephone poles be shielded with metal collars at the base to prevent them from being weakened by sandblasting. Sand grains in saltation also can erode the bottoms of rock outcrops preferentially, forming pedestal-shaped rocks (Figure 14.18).

Wind Deposition: Dunes

Materials eroded by the wind are deposited in a number of ways. Fine materials may be deposited as loess far from the site of original deflation, whereas sand-sized particles are deposited as sheets and mounds of sand attached to obstacles and as streamlined masses of sand known as **dunes** closer to where they originally were eroded.

Dune formation

The only things needed to form dunes are dry granular material and wind. In most cases,

this material is quartz sand, but dunes can be built of other granular materials, such as gypsum in dunes of the White Sands National Monument in New Mexico or shell fragments in the fossil Pleistocene dunes on Bermuda. In fact, Bagnold describes the formation of small dunes composed of very cold ice crystals.

Dunes begin to form when sand particles moving in saltation encounter an obstruction. The presence of an obstacle (Figure 14.19) forces the airflow lines to deflect around the obstacle, creating zones of quiet air, or **wind shadows**, both immediately in front (windward) of the object and at the rear (leeward) of the object. Sediment carried over the obstacle by wind settles in these wind shadows, building up a deposit of sand. Bagnold showed that sand can travel rapidly over a bare surface but that it tends to accumulate on a sand-covered surface. Consequently, so long as there is a constant wind direction and a source of sand, this accumulating sand pile will continue to grow upward, becoming slightly higher at the downwind end. This growing pile of sand becomes an even larger obstacle to the airflow, causing even more sand to accumulate. The pile grows upward with a steadily increasing angle on the leeward edge. When a minimum height of about 30 cm is reached and the angle of the leeward face reaches

Figure 14.18 Granite outcrop etched and pitted by the abrasive action of windblown sand. Atacama Province, Chile. *(K. Segerstrom, U. S. Geological Survey)*

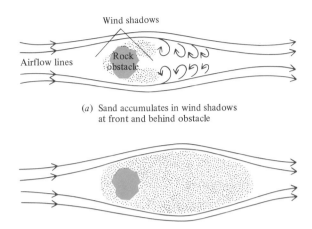

(a) Sand accumulates in wind shadows at front and behind obstacle

(b) Sand drifts behind obstacle have coalesced into a dune

Figure 14.19 Aerial views showing the accumulation of sand in wind shadows around an obstacle. The accumulations form a mound which may develop into a dune.

34°, the angle of repose for dry sand (see Chapter 10), a streamlined sand body called a dune is formed.

Dune migration

As the dune continues to build up, it also starts to migrate downwind (Figure 14.20a). Saltating sand grains landing on the windward slope kick other sand grains into the air and also push sand grains up the slope by surface creep.

Particles move up slowly toward the crest of the dune by surface creep, building it up to a steeper angle as sand grains are trapped in the zone of reverse flow leeward of the dune crest and accumulate over the upper part of the leeward slope (Figure 14.20b). When the crest of the dune is built up steeper than the angle of repose for dry sand (about 34°), the slope collapses and the sand grains slide down the steep leeward slope, or **slip face.** Each of these collapse events generates a foreset bed (see Chapter 11) along

Figure 14.20 Side view of the formation and migration of dunes. See text for details.

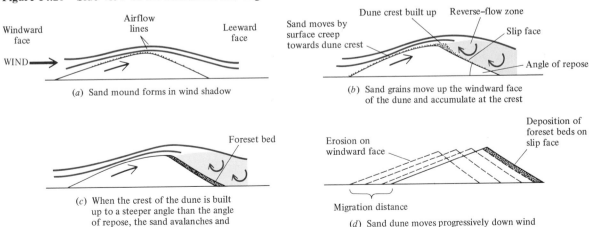

(a) Sand mound forms in wind shadow

(b) Sand grains move up the windward face of the dune and accumulate at the crest

(c) When the crest of the dune is built up to a steeper angle than the angle of repose, the sand avalanches and forms a foreset bed along the slip face of the dune

(d) Sand dune moves progressively down wind

the slip face (Figure 14.20c). The dune advances downwind by successively depositing a series of foreset beds on the slip face at the same time that the wind is eroding the material on the windward slope. The end result of this process is the formation of a streamlined dune with a gently sloping windward face and a steeper-sloping slip face which is inclined downwind (Figure 14.20d).

Dune types

There are a number of different types of sand dune. Each type has a distinct shape and occurs under different conditions of vegetation, sand supply, wind velocity, and variability of the wind direction and the surface over which the dune is moving.

Barchan dunes One of the most aesthetically pleasing dunes is the crescent-shaped **barchan**, which has a slip face that is curved concavely downwind (Figure 14.21). Barchans are unusual in that they can move as isolated dunes across bare rock or gravel surfaces. The sand moves up the convex windward slope and cascades down the concave slip face and out along the tips of the barchan, moving the dune progressively downwind. Barchans are best developed when there is a limited supply of sand and a constant wind direction.

Longitudinal dunes Massive dunes whose long axes parallel the major wind direction are called **longitudinal dunes**. They form in areas with a large supply of sand and high wind velocity. They are common in the center of great sandy deserts such as the Sahara, where they are referred to as "seif" dunes. These dunes can reach heights of 100 to 200 m and may range in length from 400 m to more than 100 km. The crests of these dunes are sinuous, with curved slip faces on one of the sides (Figure 14.22). Bagnold believed that these dunes were formed by steady winds blowing from one direction, which alternated with brief storm winds at a small angle to this direction.

Transverse dunes Dunes that form at right angles to the dominant wind direction in areas of large sediment supply and moderate winds are called **transverse dunes**. One type is the elongate sinuous desert dunes found in areas with a large sand supply (Figure 14.23). These transverse dunes may grade into barchan dunes on the edges of the dune field, where there is less sand. Another type of transverse dune is the coastal dune found along oceanic coasts and the shorelines of major lakes. The abundant sand made available by coastal erosion and transport is reworked readily by onshore winds into irregular and elongate transverse dunes parallel-

Figure 14.21 Individual (top) and compound (bottom) barchan dunes west of Salton Sea, California. The wind that formed the dunes blew in a direction from the upper left to the lower right of the photograph. (*Courtesy of John S. Shelton*)

Figure 14.22 Longitudinal dunes in the Sahara Desert. The ripple marks in the foreground are also the result of wind action. *(Richard Harrington, Phototrends)*

ing the coast. Such dunes may form in equilibrium with coastal vegetation; as the dunes grow upward, the shrubs and grass also grow upward, keeping pace with the dune's growth.

Parabolic dunes Another type of vegetated dune is the parabolic dune. **Parabolic dunes** resemble the crescent shape of barchan dunes in aerial view (Figure 14.24). However, the tips of parabolic dunes point upwind. One type of

Figure 14.23 Transverse dunes in Florence, Oregon, were formed by a wind that blew in a direction from the upper left to the lower right of the photograph. *(From D. A. Rahm,* Slides for Geology. *Copyright © 1971 McGraw-Hill, Inc. Used with permission of McGraw-Hill Book Company)*

parabolic dune forms in coastal areas with strong onshore winds and at least a partial vegetation cover. This type of parabolic dune starts to form from transverse coastal dunes in local "pockets" where the dunes are less vegetated or where the vegetation has been removed by natural or human activity (Figure 14.24). The local absence of vegetation allows the dunes to be eroded by onshore winds, forming a deflation basin which becomes elongated parallel to the direction of the wind. Material eroded by deflation piles higher and higher on the convex downwind dune crest. The sand avalanches down slip faces which are inclined landward (downwind). At the same time that the central part of the dune is being extended, the ends of the dune are kept intact by any vegetation on the sides. The advancing slip faces of such parabolic dunes can create environmental problems when they start to bury inland forests, structures, and roads (Figure 14.24).

Characteristics of dune deposits

Sand particles transported by wind typically are fine-grained, well sorted, and well rounded and may have a "frosted" grain surface. The dune sands are fine-grained and well sorted because the low density and low viscosity of air severely limit the particle sizes which can be transported. Saltating particles frequently collide with each other during wind transport. These impacts are all the stronger because the

Wind Activity on Mars

Photographs made by the orbiting Mariner 9 spacecraft in late 1971 showed landforms on the Martian surface that are very similar to dune fields on Earth. The features were noted in a dark-floored crater in the Hellespontus region of Mars (47.5° S, 331° W). The area measured 60 km by 30 km and showed sinuous and coalescing ridges and individual forms (Figure 1). Earlier observations had shown that there are seasonal changes in color and tone on the Martian surface. These observations made it seem possible that, in spite of the much thinner Martian atmosphere, there may be considerable wind transport across the Martian surface.

The pictures made by the cameras on the Viking I lander in July 1976 and the Viking II lander in September 1976 showed a landscape that was very similar to desert areas on Earth (Figure 2). The surface was strewn with angular-faceted particles, similar to ventifacts on Earth, and there were streamlined sedimentary accumulations on one side of the rock particles. Larger, dunelike accumulations with slip faces and exposed stratification were noted in the distance.

These observations confirmed that at least some part of the Martian surface has a great similarity to deserts on Earth (Figure 3). In fact, sediment actually was observed moving across the foot pad of the Viking lander in a sequential series of photographs. In 1977, Carl Sagan and others (see Additional Readings) utilized low-pressure wind-tunnel experiments plus Bagnold's equations on wind transport to determine the wind velocity needed to start sedimentary particles saltating (saltation threshold) on the Martian surface. Monitoring of the surface wind velocity by Viking lander instruments showed

Figure 1   The landform is about 130 km long and 64 km wide.   Dune-like landforms within a Martian crater. *(NASA)*

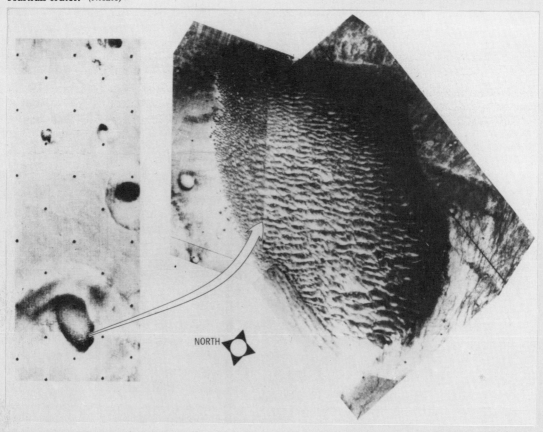

NORTH

Figure 2   Martian surface features at the Viking 1 lander site. The photograph shows dunes, dune crossbedding, and sand shadows extending like tails from rock fragments.   These features all suggest winds that blew across the surface from left to right. *(NASA)*

that the saltation threshold velocity was exceeded more than once over the time period in which the wind velocities were measured. Sagan and others believe that the high wind velocities needed for particle saltation on the Martian surface are likely to be provided by the steep latitudinal temperature gradient and

by the major differences in elevation on that planet. The seasonal changes in color and tone on the Martian surface may be due in part to the burial of older features by windblown materials or the erosion of the materials covering them by the strong winds that periodically sweep across the Martian surface.

Figure 3   Surface features in the coastal desert near Nazca, Peru. The landscape is very similar to that at the Viking 1 landing site (compare with the right side of Figure 2). *(Courtesy of William K. Hartmann)*

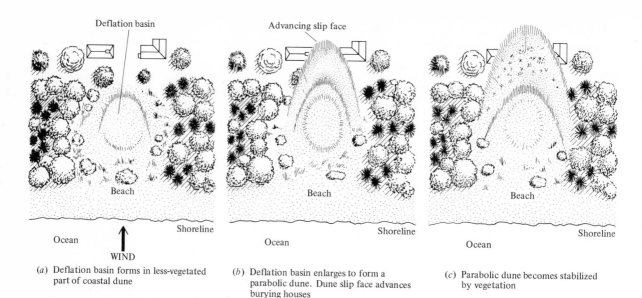

(a) Deflation basin forms in less-vegetated part of coastal dune

(b) Deflation basin enlarges to form a parabolic dune. Dune slip face advances burying houses

(c) Parabolic dune becomes stabilized by vegetation

Figure 14.24 Formation of a parabolic dune from the deflation of a transverse coastal dune. Growth of these dunes can threaten structures such as homes and roads.

intervening air provides less of a "cushioning" effect than water does. The innumerable impacts among sand particles in dunes wears away the rough edges and rounds the particles. The surfaces of the grains have a great number of microscopic impact scars which collectively make the surface appear frosted. While frosted grain surfaces may be formed in the "sandblasting" of wind transport, not all frosted grain surfaces are of this origin. Similar surface features can be produced by postdepositional chemical etching of particle surfaces. The beds of cross-stratified sand in dunes tend to be much thicker than those in water-laid sediments because these dunes generally are higher than those which are formed during water transport (Figure 6.15).

Determining Paleowind Directions

Ancient wind deposits represented in sedimentary rocks usually show well-defined foreset bedding. The geologist can determine the effective wind direction, or **paleowind,** during the accumulation of the foreset beds shown in the exposure because the foreset beds always are inclined downwind (Figures 6.14 and 6.15). Analysis of paleowind directions in rocks in many places in the world indicate wind directions that are not compatible with the *present* location of those places. This suggests that the continents were located at a different latitudinal position when the windblown sands originally were deposited. Of the five hypotheses discussed in Chapter 1, only the plate tectonic model adequately explains the discrepancy between paleowind directions recorded in rocks and the present prevailing winds at those latitudes. In the plate tectonic model, continents can change position on the earth so that a wind deposit originally formed in the belt of northeast trade winds could be moved to a latitude where the predominant wind is the prevailing westerlies (Figure 14.4).

SUMMARY

Deserts are land areas with less than 25 cm yearly precipitation and evaporation rates which greatly exceed precipitation rates in many cases. Deserts often are bordered by semiarid areas with less than 50 cm per year of precipitation. The largest deserts on the earth (Sahara and Arabian) form in latitudes 10 to 30° north and south of the equator, where global atmospheric circulation patterns result in warm, dry air moving equatorward along the surface. Other deserts form in the center of continents far from oceanic sources of moisture (Gobi), where mountains bar the inland passage of moisture-laden ocean air (American Southwest), and in coastal areas with cold currents flowing offshore and mountainous areas barring moisture-laden winds from reaching the area (Atacama, Namib).

Deserts are characterized by wide daily temperature fluctuations and low annual precipitation. Precipitation generally occurs in a few widely spaced rainfalls separated by dry periods. Vegetation is sparse and consists of specialized types which have the abilities to obtain and store available water. The sparse vegetation and the large exposed area increase the efficiency of both water and wind erosion. Physical weathering is more important than chemical weathering, and the desert soils are thin and discontinuous.

Desert stream activity is particularly effective because the sparse vegetation cannot break the force of the rain or anchor the soil. Large quantities of sediment are transported from highlands down to desert floors by ephemeral streams and sheetflows. Some of the stream discharge or groundwater flow or both reaches the lowest part of the desert basin, forming temporary shallow playa lakes from which soluble salts are deposited upon evaporation. Desert stream activity erodes the highlands slowly, depositing the material on alluvial fans and on the desert floor. In many cases, desert drainage does not reach beyond the desert basin, and the bordering mountains are buried slowly in the sediments derived from their erosion.

Wind activity is particularly effective in the dry periods between rainfalls. The sand, along with dry silt and clay, is deflated by the wind, leaving behind a layer of rock particles which form a protective pavement on the surface and inhibit further deflation. The silt and clay can be carried in suspension far from the source and form deposits of loess at the edges of the desert and beyond. The sand moves along the surface by saltation and surface creep, and it can be molded into streamlined forms called dunes. Different types of dunes are formed under different conditions of sand supply, wind strength, constancy of wind direction, and amount of vegetation. Wind erosion and deposition also are important in more humid areas, especially along sandy shorelines.

QUESTIONS FOR REVIEW AND FURTHER THOUGHT

1. Describe the conditions which result in different types of deserts.

2. Describe how the abundance and distribution of vegetation in deserts affect weathering and erosion.

3. Compare stream activity in dry and humid climates.

4. Continued erosion in both humid and arid climate eventually results in a land surface with low relief; however, these surfaces are very different with respect to the underlying materials and the processes which formed them. Explain.

5. What makes wind so effective as an agent of erosion in deserts?

6. Describe how a dune forms and how it migrates.

7. *a.* Name four different kinds of dunes.
   b. Describe how each of these dunes differs in shape from the the others.
   c. What factors favor the formation of each of the four different kinds of dunes?

8. Why is wind erosion such an efficient sediment-sorting agent?

ADDITIONAL READINGS

Beginning Level

Gore, R., "The Desert: An Age Old Challenge Grows," *National Geographic*, vol. 156, no. 5, November 1979, pp. 586–639.
(A well-illustrated account of deserts, desertification, and attempts to retard the advance of deserts)

Hartman, W. K., "Searching for Mars on Earth," *Astronomy*, vol. 4, no. 10, October 1976, pp. 20–26.
(A review of weathering and wind activity in deserts. Excellent color photos of various deserts and a discussion of their similarities with Martian deserts)

Intermediate Level

Bagnold, R. A., *The Physics of Blown Sand and Desert Dunes*, Methuen, London, 1941, 265 pp.
(The best overall reference on wind activity and dune formation)

Bloom, A. L., *Geomorphology*, Prentice-Hall, Englewood Cliffs, N.J., 1978, 510 pp., chap. 14, "Eolian Processes and Landforms."
(A detailed description of arid and semiarid environments and morphology)

Cutts, J. A., and R. S. U. Smith, "Eolian Deposits and Dunes on Mars," *Journal of Geophysical Research*, vol. 78, no. 20, July 10, 1973, pp. 4139–4154.
(A description and analysis of wind features on Mars)

Doehring, D. O. (ed.), *Geomorphology in Arid Regions*, Publications in Geomorphology, SUNY, Binghamton, N.Y., 1977, 272 pp.
("The Formation of Pediments: Scarp-Back Wearing or Surface Downwasting?" by John Moss, pp. 51–78, provides a good review of the pediment problem. Other papers deal with additional aspects of erosion, deposition, and landscape development in dry climates)

Garner, H. F., *The Origin of Landscapes*, Oxford University Press, New York, 1974, 734 pp., chap. 6, "Arid Geomorphic Systems and Landforms."
(A good discussion of the effects of alternating arid and humid climates on landscape development)

McGinnies, W. G., B. J. Goldman, and P. Paylore, (eds.), *Deserts of the World: An Appraisal of Research into Their Physical and Biological Environments*, University of Arizona Press, Tuscon, 1968, 788 pp.
(An excellent overall reference on both geologic and biologic aspects of deserts. A good place to look first for data on a specific desert)

McKee, E. D. (ed.), *A Study of Global Sand Seas*, U.S. Geological Survey Professional Paper 1052, 1979, 429 pp.
(An up-to-date and lavishly illustrated collection of articles about sandy deserts, dunes, and ancient dune deposits)

Solbrig, O. T., and G. H. Orians, "The Adaptive Characteristics of Desert Plants," *American Scientist*, vol. 65, no. 4, July–August 1977, pp. 412–421.
(A review of climate in warm deserts and a good account of how different types of vegetation adapt to the harsh climates in warm deserts)

Advanced Level

Glennie, K. W., "Desert Sedimentary Environments," *Developments in Sedimentology*, Elsevier, New York, vol. 14, 1970, 222 pp.
(Discusses many geologic aspects of deserts)

Mabbutt, J. A., *Desert Landforms*, M.I.T. Press, Cambridge, Mass., 1977, 340 pp.
(Discusses many geologic aspects of deserts)

Sagan, C., D. Pieri, P. Fox, R. E. Arvidson, and E. A. Guiness, "Particle Motion on Mars Inferred from the Viking Lander Cameras," *Journal of Geophysical Research*, vol. 82, no. 28, September 30, 1977, pp. 4436–4438.
(An interesting application of Bagnold's principles to analyzing Martian wind features)

CHAPTER 15

Coastal Zones

Coastal zones, where land meets sea, are areas of great natural beauty and diversity, ranging from the high-relief cliffed shorelines along much of the Pacific Coast and part of the New England coast, to the low-relief barrier island shorelines of much of the Atlantic and Gulf Coasts, to the colorful coral reef shorelines of southern Florida. Although we normally think of coasts as land areas bordering oceans, most of the same processes and features may be seen along the shores of any large lake.

Only a small area of the continents is composed of coastal zones, but the population density in coastal areas is far greater than anywhere else. This creates a potentially serious problem because coastal zones represent one of the most dynamic areas on the earth's surface. During a coastal storm tremendous energy is released, and rapid and extensive geologic changes can occur along the coast. The recent increases in population and building in coastal areas, combined with artificial changes in shorelines, have created serious problems of erosion on beaches and dunes. These factors have made coastal areas even more vulnerable to the effects of storms.

COASTAL PROCESSES

A wide variety of processes are active in coastal zones. Most of these processes are related to wave activity, but currents generated by the rise and fall of the tides also are important in erosion and deposition within the coastal zone.

Waves

Erosion and deposition within coastal zones are largely a result of either direct or indirect wave activity. The waves usually are formed where winds move along the water surface, distorting it into a series of undulations. On a clear and windless day at the coast, large waves may break onshore. Offshore, fishing boats may rock back and forth with the passing waves. Such wave activity in the absence of local winds seems puzzling until we realize that some of these waves are related not to local conditions but to storm centers far away.

The turbulent and gusty winds in a storm center exert a frictional drag on the water surface. This transfer of wind energy to the water forms ripples, which grow upward into waves with time. The waves produced by winds in the storm center are irregular and appear to move in several directions close to that of the predominant wind. As the waves move out from the direct influence of the wind, they become longer and lower waves referred to as **swells.**

The symmetrical deep-water swells are described by the terms shown in Figure 15.1. The highest part of the wave is the **crest,** and the lowest part is the **trough;** the vertical distance between them is the **wave height.** The horizontal distance between any two similar points on two waves is the **wavelength,** represented by the Greek letter lambda (λ). The wave crest lines are the **wave fronts,** and the direction the wave is moving in is shown by the **wave normal.** The wave normal is perpendicular to the wave fronts

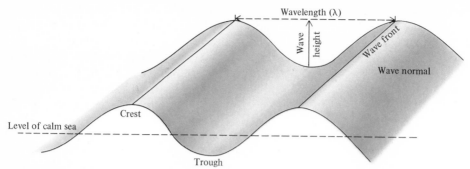

Figure 15.1 Terms used to describe waves.

at any point.   The time (in seconds) required for one wavelength to pass a given point is called the **wave period.**

The height and period of waves depend on the speed and duration of the winds generating them and the **fetch,** or length, over which the

Figure 15.2 Motion of a ball within a wave. As the wave forms move from left to right, the ball makes a circular orbit, returning to its original position after one wavelength has passed. The curved arrow shows the motion of the ball, and the straight arrow shows the direction of movement of the wave forms.

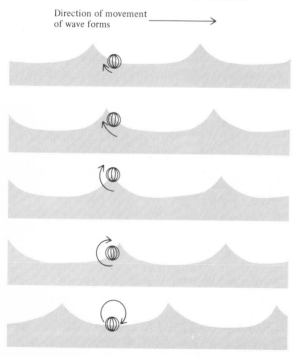

wind acts.   For example, if wind velocity and duration are the same, waves will be higher on large lakes than on the smaller ones because the fetch on the larger lakes is greater.   Swell moving away from storm centers can travel thousands of kilometers across deep water before the energy within the wave is released as it crashes onto a distant coast.   During the migration of swell, the wave heights may drop as much as 50 percent in the first 1600 km, but the loss of energy is negligible.

Waves are a mechanism by which energy is transferred along the water surface.   While the wave *form* moves across the surface, the water particles within the wave have an orbital motion.   You can experience this yourself if you float at the surface somewhere seaward of the breaking waves along a coast.   *If there is no local current or wind,* you merely will bob up and down in a circular motion (Figure 15.2) as the waves pass through.   After one wavelength has passed through, a floating object returns to its initial position.   The diameter of the circular orbit at the surface is equal to the wave height.   Now dive down a few feet and you will continue to move in an orbital path as waves pass overhead, but the diameter of the orbit will be less than it was at the surface.   If you dive down farther, the orbital motion will be even less.

Experimental studies and field observations show that the water's orbital motion decreases with depth and ceases to exist at a depth equal to about one-half the wavelength of the waves above (Figure 15.3).   This downward limit of orbital water movement is referred to as the

wave base. Above the wave base is a zone of agitated water, while below the wave base the water is not influenced by wave motion. The water depth for the waves shown in Figure 15.3 is greater than the wave base for those waves, and thus the bottom is undisturbed by the passing of the waves above. However, if waves with a larger wavelength move across the area, the wave base may intersect the bottom, and the bottom materials may be reworked by the agitated water. Such changes in wave base are important in the erosion and redeposition of nearshore bottom sediments. For example, waves from distant storms may have wavelengths as great as 600 m, corresponding to a wave base of 300 m. Waves of this magnitude can stir up bottom sediments far seaward of the coast.

So far, we have been talking about symmetrical deep-water waves in which the wave base is less than the water depth. When a wave reaches shallow water where the depth equals one-half of the wavelength, the wave begins to "feel the bottom." The upper part of the wave continues to move landward, while the lower part is retarded by frictional drag along the bottom. From this point landward, the character of the waves changes markedly. Wavelength decreases, velocity increases, and the waves get higher and more asymmetrical (Figure 15.4). The zone in which these changes occur is called the **breaker zone.** Eventually, the oversteepened waves topple over within the **surf zone,** where there *is* an actual forward movement of the water with the wave. Within the surf zone the energy stored in the wave is utilized for erosion and deposition along the shoreline.

Wave Refraction

The wave fronts approaching a coast are not slowed down uniformly but start to slow down first at places where the offshore bottom is shallower, such as off of projecting points or headlands

Figure 15.3 Water motion within a wave. The orbital motion of the water particles becomes minimal at a depth equal to one-half the wavelength (wave base).

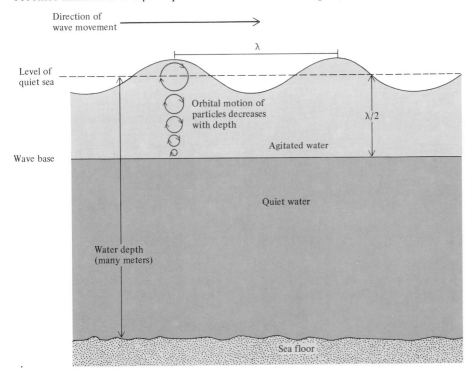

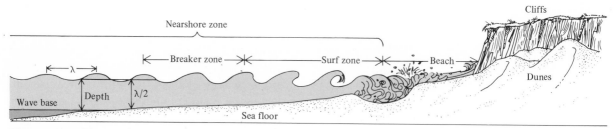

Figure 15.4 Changes in waves as they approach a coast. When the water depth equals half the wavelength, the waves begin to break. The wave height increases, the wavelength decreases, and the wave becomes asymmetrical. The waves eventually tumble over in the surf zone.

(Figure 15.5). The portions of the wave that still are in deeper water continue to move landward without slowing down. This differential wave movement makes the wave fronts bend, or **refract,** as the waves advance toward the shore.

Wave refraction results in a concentration of wave energy on protruding headlands. Equal segments of the deep-water waves in Figure 15.6 (A-B, B-C, C-D, and so forth) have equal amounts of energy. However, as the waves refract, more of this energy is focused on the headlands while wave energy is dissipated across the intervening bays, as shown by the patterns of the wave normals in Figure 15.6. The net result of continued wave refraction is a tendency for the waves to erode headlands and smooth out the shoreline.

Longshore Drift

Waves breaking at an angle to the shoreline result in the movement of sediment along the coast. We can see how this process works by isolating one sand grain and observing its movement in response to waves approaching the shoreline at an angle (Figure 15.7). Onrushing waves move the sand grain up the beach face parallel to the wave normal (A-A'). The retreating water moves the grain down the steepest slope of the beach face (A'-B). Through this pair of movements, the grain has moved a small distance (A-B) along the shoreline. The grain at B is picked up by another onrushing wave, repeating the process and moving the grain even farther along the shoreline. Material that is carried

Figure 15.5 Wave refraction at Point Reyes, California. The wave fronts converge on the headlands but diverge across the embayments. *(From D. A. Rahm,* Slides for Geology. *Copyright © 1971 McGraw-Hill, Inc. Used with permission of McGraw-Hill Book Company.)*

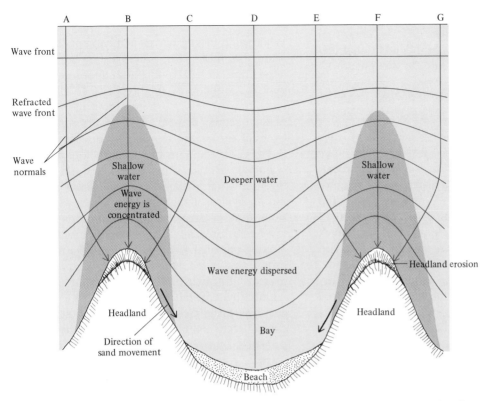

Figure 15.6 Aerial view of wave refraction along an irregular shoreline.   Wave refraction concentrates the wave's energy on the headlands and disperses it over the bays.   As the wave fronts enter shallower water, they are refracted progressively so that they become more parallel to the bottom contours.

Figure 15.7 Wave fronts advancing at an angle to the coast move particles of sediment along the shoreline.

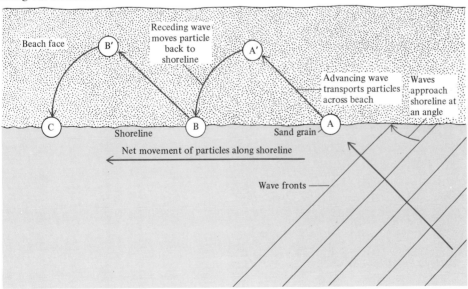

along the shore by waves and currents is called **longshore drift.** Longshore drift is important in the formation of the coastal depositional features that we will discuss later in this chapter.

Longshore and Rip Currents

Waves approaching a coast at an angle (Figure 15.7) generate a **longshore current** which flows parallel to the shoreline. The longshore current, along with the wave mechanism shown in Figure 15.7, transports sedimentary particles along the shoreline and is important in the formation of coastal depositional features. Wherever waves approach a shoreline at an angle, they tend to "pile up" water within the nearshore area. There is a tendency for the water to escape seaward. Masses of water moving seaward from the shoreline, commonly referred to as "undertow," are called **rip currents.** In many cases, the rip currents carry suspended sediment seaward and can be recognized as "plumes" of turbid water extending out beyond the surf zone.

Tidal Currents

While wind-driven waves and wave-induced currents are the major processes in erosion and deposition along stretches of exposed shoreline, currents induced by tidal forces (see Chapter 2) become relatively more important in other parts of an oceanic coastal zone. Currents induced by tidal forces are relatively more important in the bays and lagoons behind the shoreline and at openings in the shoreline, called **tidal inlets,** through which oceanic and bay waters exchange during cycles of high and low tides. Tidal currents moving into the coastal zone as the tidal level rises are called **flood currents,** while those moving out of the coastal zone as the tidal level drops are called **ebb currents.** These moving masses of water erode and transport sediment within the coastal zone (Figure 15.8). Flood currents bring fresh oxygenated water into the coastal areas behind the shoreline. Ebb currents transport nutrients and organic particles seaward, providing the materials necessary for oceanic life (Figure 15.8).

Suffolk scarp Barrier Island

Estuary

Bay

Cape Hatteras

Turbid water moving seaward

Estuary

Inlet

Spit

Cape Lookout Plumes of turbid water moving out of inlets on ebb tide

Figure 15.8 Satellite view of the Cape Hatteras area, Virginia and North Carolina, showing ebb tidal currents moving turbid water out of the estuaries and through inlets into the Atlantic Ocean. The long, light-colored ridge in the upper left corner is the Suffolk Scarp, a Pleistocene shoreline (see Figure 13.27). The other features labeled here will be discussed later in this chapter. (*NASA.*)

COASTAL EROSION

Both erosion and deposition occur on almost all shorelines. However, the relative role of each in forming shoreline features varies from coast to coast. Wherever wave erosion removes more material than is being deposited, *net* erosion occurs and the coast is cut back. In a similar fashion, wherever more sediment is being deposited than eroded, net deposition occurs and the shore builds out seaward. The dominant erosional force along shorelines is the tremendous energy released by the waves crashing against the coast.

Wave erosion is most effective along low-lying sandy coasts. The noncohesive nature of the shore materials, along with the low altitudes along the coast, enables storm waves to remove vast quantities of sand quickly. Removal is by direct wave attack along the shore and by overwashing of the shore materials landward through the coastal areas.

Some coasts are composed of high bluffs of sediments. Wave erosion at the base of a cliff undermines the cliff and causes slumping of the front of the cliff into the surf zone (Figure 10.10). Waves wash out the finer particles of sediment from the slumped mass and remove them either along the shore or seaward. The coarse particles of sediment and rock remain behind, forming a submarine **wave-built terrace** which slopes gently toward the sea from the base of the cliff.

Wave erosion also is effective along rocky coasts, although the rate of erosion is much slower than it is along other types of coasts. The rocks in coastal cliffs are weakened and prepared for erosion by a number of the physical and chemical weathering processes described in Chapter 9. For example, water freezing within rock crevices exerts great pressure, loosening rocks in the process. Waves breaking against rock cliffs push water under great pressure into rock crevices. This wedges out blocks of rock, which fall into the surf. The rock debris is picked up by the waves and thrown repeatedly against the cliff, eroding a **wave-cut notch** at the base of the cliff. This may undermine the cliff

sufficiently to cause a fall of debris and an accumulation of talus at the cliff base (see Chapter 10). The talus acts as armor for the cliff, protecting it temporarily from further erosion. Smaller particles eroded from the talus are moved back and forth by the surf, eroding a flat, gently sloping **wave-cut bench** in the rock layers beyond the base of the cliff. The wave-cut bench may be exposed at extreme low tide (Figure 15.9).

So far we have been talking about coasts composed of rocks with uniform resistance to wave erosion. In actuality, most coastal cliffs contain rocks that differ in resistance to wave erosion either because of their mineralogy or because of fractures cutting across them. Wave erosion erodes the weaker rocks preferentially, cutting through cliff headlands on opposite sides to form **sea arches** and isolating other masses of rocks as **sea stacks** (Figure 15.10).

As the cliffs recede and the wave-built terrace or wave-cut bench extends farther landward, an equilibrium tends to develop. At some point the growing terrace or bench tends to dissipate the incoming wave energy, resulting in a reduction in the wave energy available for erosion at the shoreline. This lower energy is sufficient to move sediment but not to erode the cliff. A temporary equilibrium is reached under which sediment can accumulate as beaches against the cliffs. A subsequent storm may produce waves high enough to strip the beach and erode the cliffs again. After a number of such storms, a more permanent equilibrium can be reached in which cliff erosion is reduced and beaches are preserved for longer periods of time.

COASTAL DEPOSITION

Coastal sediments come from three major sources: rivers, reworking of nearshore sediments, and erosion of material landward of the shoreline. Large volumes of sediment are supplied to coastal zones, where major rivers, such as the Mississippi, form deltas (see Chapter 11). However, postglacial rise in sea level (see Chapter 13) "drowned" many of the rivers in the

Figure 15.9 Wave-cut bench exposed seaward of cliffs at Bolinas Point, California. *(Courtesy of John S. Shelton.)*

Figure 15.10 Sea arch, Pacific City, Oregon. *(From D. A. Rahm, Slides for Geology. Copyright © 1971 McGraw-Hill, Inc. Used with permission of McGraw-Hill Book Company.)*

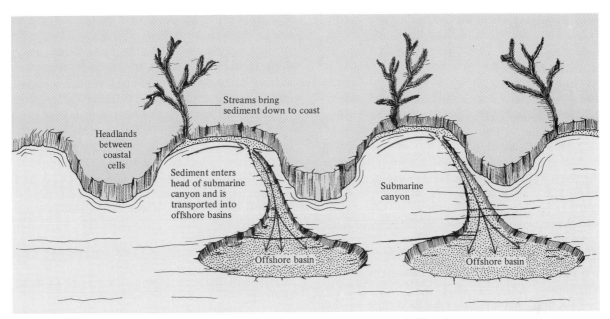

Figure 15.11   Localized sand movement cells along the coast of southern California.

United States, forming branches of the sea called **estuaries.** These estuaries trap most of the coarser river sediment and allow only the finer suspended particles to reach the ocean coast (Figure 15.8). In coastal areas bordered by mountains, such as much of California, steep-gradient streams can supply large volumes of sediment to the coast. However, water impoundment dams to prevent flash flooding in more heavily populated areas (see Chapter 11) have trapped large quantities of sediment, progressively "starving" some of the California beaches.

The offshore sediments are also a source of at least some of the coastal sands. During the Pleistocene Epoch, when sea level was much lower than at present, streams flowed out along much of the present offshore area and deposited vast volumes of sediment (see Chapter 13). There is no question that some of this material is moved into the coastal zone by storms, but at the present time we do not know how much material is moved.

Most coastal sediments are derived from erosion of materials exposed along shorelines. Along some coasts, the sediments are derived from wave erosion of rock and sediment from cliffs. Erosion of older coastal plain deposits and recently deposited beach and dune sand provides most of the material along low-lying sandy coasts. Along some tropical shorelines, such as those in southern Florida and the Bahamas, most of the coastal sediments are composed of shells, shell fragments, and skeletal debris derived from marine plants and animals and from erosion of older fossiliferous rocks exposed along the coast.

Once sediment reaches the shoreline, it can be moved along the shore by waves and longshore currents. Sediment can be moved for great distances along some sandy Atlantic and Gulf coastlines. However, this is not possible along parts of the southern California coastline because submarine canyons there have their heads in the nearshore area (Figure 15.11).

Sediment brought to the coast by streams in California is moved along the shore until it intersects the heads of one of these canyons. Sediment entering the canyon heads is transported seaward and deposited in deep basins offshore. Five such cells of local coastal circulation occur along the California coast between Point Conception and just south of San Diego. Each cell begins with a stretch of rocky coast with limited sand supply. The beaches down-drift of this point become progressively wider. Each cell is terminated by a submarine canyon which captures the longshore drift, preventing it from extending any farther downdrift. The next cell begins with a rocky coast without beaches.

Coastal sediments are deposited in a number of ways. Newly deposited beach material exposed by a falling tide may dry out sufficiently so that it can be blown by the wind into dunes many meters above sea level. Particles moving along the shoreline may be deposited where there is an increase in depth or a decrease in current velocity. For example, particles of sediment may be swept into tidal inlets by flood currents and deposited in the relatively quieter waters of bays and lagoons landward of the shoreline.

Biologic activity also plays a role in the deposition of coastal sediments. Grasses and other plants serve as "baffles," slowing down tidal currents and resulting in deposition of their bed load and a portion of their suspended load (see Chapter 6). Other types of organisms, such as mussels, live on organic material suspended within the water. They draw in water, feed on the organic particles, and excrete the associated sedimentary particles as fecal pellets.

COASTAL ZONES

The coastal zone includes the area from the maximum extent of tidal action inland to the nearshore area seaward of the surf zone. Within the coastal zone are three major environments (estuaries and bays, tidal wetlands, and barrier islands and beaches) which grade into one another (Figs. 15.8 and 15.12).

Estuaries and Bays

All rivers draining coastal zones become brackish-water, tidally influenced estuaries near the coast (dark color in Figure 15.12). Estuaries are influenced by tides and contain brackish and saline water, which flows seaward on ebb tide and landward on flood tide.

Estuaries have a number of different origins. Most estuaries are former river valleys that were "drowned" by the worldwide postglacial rise in sea level (New York Bay, Chesapeake Bay, San Francisco Bay, and Puget Sound). Some estuaries in higher latitudes are glacially deepened valleys called fjords that were drowned by postglacial rise in sea level (see Chapter 13). Fjords differ from other types of estuaries in that they have sheer rock walls which often extend far below the water level. Estuaries are the sites of some of the major American cities (Boston, New York, Norfolk, Charleston, San Francisco, and Seattle) and are important resources for food (shellfish and fish), transport, cooling water, and waste disposal.

Tidal Wetlands

Portions of the coastal zone which are covered partially or completely by tidal waters (light color in Figure 15.12) are called **tidal wetlands.** Tidal wetlands are composed of tidal flats and tidal marshes. **Tidal flats** are unvegetated areas which are exposed fully during low tide. **Tidal marshes** are partially or completely vegetated areas which may be covered to different degrees during the tidal cycle. Wetlands are sites of accumulation of finer-grained sediments and great amounts of organic material. Extensive wetlands occur around the periphery of estuaries, bays, and sounds (Figure 15.8). Many major American cities (Boston, New York, Norfolk, and San Francisco) have extended their land areas by filling in former wetlands. Such elimination of wetlands causes severe environmental problems. Wetlands provide sanctuary for juvenile stages of marine organisms, enabling them to develop before moving into the oceans. Organic particles and nutrients swept out of wetlands into the oceans provide nourishment for oceanic life (see Figure 15.8). Wetlands also reduce the flooding effects of high stream discharges by absorbing large volumes of water and then releasing them slowly.

Beaches and Barrier Islands

Accumulations of coarse sediment which form along portions of the coastal zone exposed to wave action are called **beaches.** The beach extends from the low-water line landward to where

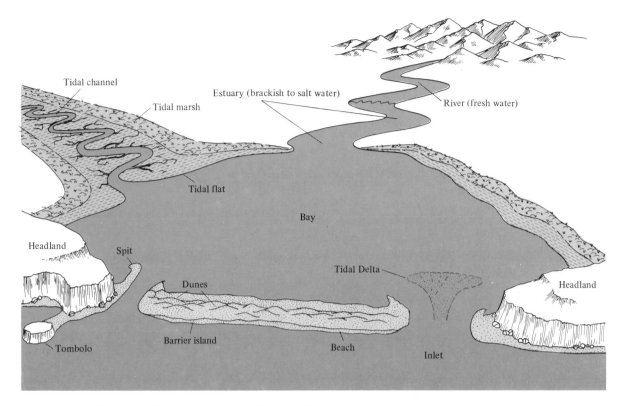

Figure 15-12   Features of coastal zones mentioned in the text.

there is a change in morphology, type of sediment, or vegetative cover. In general, beaches terminate landward at rock or sediment cliffs, dunes, or the seaward limit of permanent vegetation (Figure 15.4).

Beaches occur in a variety of forms (gray areas in Figure 15.12). Small beaches are formed along concave portions of cliffed shorelines. More extensive beaches are formed along **spits,** which are elongate bars of sand built by sediment derived from headlands or sandy islands. Beaches also form on **tombolos,** which are bars of coarse sediment that connect offshore islands with the mainland or with each other.

The widest and most extensive beaches are found along **barrier islands** (also gray in Figure 15.12)—elongate bodies of sand separated from the mainland by a bay or sound. This type of shoreline makes up most of the Atlantic and Gulf coastlines of the United States (Figure 15.8). Several major American cities are located on barrier islands, including Galveston, Texas; Miami Beach, Florida; and Atlantic City, New Jersey.

Barrier islands are composed of a wide variety of sedimentary deposits. Very large changes in grain size occur between adjacent areas, reflecting changes in both intensity of transport and agent of transport. For example, coarse sand, gravel, and broken shells may be deposited in the surf zone at the same time that wind is depositing fine sand in dunes as little as a few tens of meters from the surf zone.

Beaches are divided into a number of zones with different characteristics. The **nearshore zone** includes the breaker and surf zones. The **foreshore zone** is the sloping front face of the beach which is exposed at low tide. The **backshore zone** is the area between the foreshore and the cliff (in cliffed rocky beaches) or dunes (in barrier island beaches). On the landward side of the barrier island are the wetlands and bay.

Carbonate Shorelines

Most of the American coast fits within one of the coastal types already described. However, the shorelines in southern Florida and in many other tropical areas are considerably different. These shorelines have many of the same features described so far in this chapter (Figure 15.12). However, the shoreline features there are made up largely of the deposits of calcium carbonate–secreting organisms such as calcareous algae and solitary and colonial corals. These calcium carbonate–secreting organisms make up massive wave-resistant structures called **reefs.**

Conditions favoring the formation of coral reefs include warm (between 25 and 30°C), shallow, clear, and agitated waters. The colonial corals and coralline algae making up the framework of most reefs grow fastest in the direction of open water or in the direction facing the prevailing winds or both because these conditions provide the oxygenated waters and food required for coral growth. Clear water is necessary because particles of sediment suspended in cloudy water cover the corals and kill them. In several areas of the world today, coral reefs are being threatened not only by pollution but by the turbid water resulting from blasting through coral reefs for navigation channels and by dredging to create harbor facilities and housing developments.

Reef-building corals are animals, but they live symbiotically with algae. It is believed that algal photosynthesis influences the deposition of the calcium carbonate which constitutes the stony structure in which the corals live. Because of this, corals require sunlight for growth, and coral growth is limited to the upper 40 m of water. Wherever all of these favorable conditions occur, coral reefs can continue to grow outward and upward, even if sea level rises. Tropical storms can cause considerable damage to coral reefs. However, shortly after the storm, new coral growth begins again, and soon a layer of live coral covers the reef once more. The skeletal debris eroded from coral reefs is transported by tidal and wind-driven coastal currents and is built into the same sort of features, such as spits, tidal deltas, and beaches, that are found along shorelines in higher latitudes (Figure 15-12).

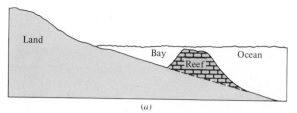

Figure 15.13 Types of coral reefs. *(a)* Barrier reef. *(b)* Barrier reef, Seychelle Islands. *(Michael Friedel, Woodfin Camp.) (c)* Fringing reef, Maldive Islands. *(Michael Friedel, Woodfin Camp.) (d)* Wake Island, an atoll in the South Pacific. *(U. S. Navy.)*

There are three basic types of coral reef. A **barrier reef** is separated from the mainland by a bay or lagoon (Figure 15.13*a* and *b*). As Figure 15.13*b* shows, channels through the reef permit the exchange of waters between the bays and the ocean. The longest reefs in the world are the barrier reefs along the east coast of Australia and along the east coast of Yucatan and British Honduras (Belize).

Fringing reefs are built right against the coast and have their greatest growth rates where the surf is strongest. This type of reef is more common along shorelines with steeper offshore slopes (Figure 15.13*c*). Fringing reefs are common around the edges of volcanic islands in the Caribbean and South Pacific.

The third type of reef is the circular **atoll,** which is common in the South Pacific (Figure 15.13*d* and Plate 19). In 1842, Charles Darwin first proposed an origin for these structures based on his field observations. Other theories for formation of atolls have been proposed, but Darwin's theory still fits most of the facts. Darwin proposed that atolls form from the subsidence of volcanic islands and their associated fringing reefs (Figure 15.14). The fringing reef continues to grow upward as the volcano subsides. When most of the island has subsided, a barrier reef is formed, encircling the island (Figure 15.14*b*). After all of the original volcanic island has subsided, a lagoon takes its place as the circular reef pattern of the atoll forms. Storms breaking on the circular atoll reefs pile up material above sea level as islands on the reef (Figure 15.14*c*). While many of the smaller islands are uninhabit-

(b)

(c)

able, many of the larger ones are more favorable for habitation. If Darwin's subsidence theory is correct, there should be volcanic rock under the carbonate rocks of an atoll. Support for Darwin's theory came after World War II, when deep drilling on the Eniwetok and Bikini atolls revealed volcanic rock at depth.

COASTAL STORMS

Tremendous changes are made in the coastal zone with the passing of a storm. The rain associated with most storms swells the rivers, pouring large volumes of sediment-laden freshwater into the estuarine environments, resulting in major disruptions for organisms living there. The greatest effects of a storm are felt on low-lying sandy coasts. The generally low topog-

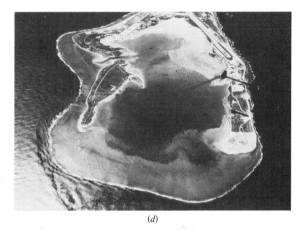

(d)

raphy along these coasts and the tremendous force of storm-driven waves and currents combine to result in maximal erosion of coastal structures. The effects of coastal storms are

Figure 15.14 Formation of an atoll according to Charles Darwin.

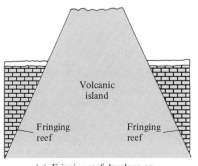

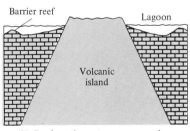

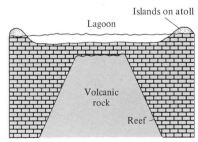

(a) Fringing reef develops on edges of subsiding volcanic island

(b) Reef continues to grow upward keeping pace with the island subsidence. Fringing reef evolves into barrier reef with development of lagoon

(c) After volcanic island subsides below sea level, a reef is built up on the remnants. Barrier reef forms the circular atoll around a central lagoon

greatest in concave sections of the shoreline such as the Texas coast, the Mississippi-Alabama-western Florida coast, or any embayment in which storm waters can build up height, such as Florida Bay, Chesapeake Bay, and Delaware Bay.

Factors Governing Severity of Damage

As a cyclonic (low-pressure) storm approaches a coast, the low pressure associated with the storm causes a rise in the neighboring ocean surface. At the same time, strong winds associated with the storm raise the ocean surface in front of the storm. The increased elevation of the water surface resulting from low pressure and high, persistent winds is called a **storm surge.** Storm surges can increase the level of the sea several meters along a wide stretch of coast in front of the storm. The actual level of the sea surface at the coast will be equal to the tidal level at that time plus any storm surge. If there is a high storm surge coinciding with a spring tide, the results can be disastrous. Willard Bascom, in *Waves and Beaches,* describes the effects of the most deadly storm surge in American history.

> **Probably the most famous example of a storm surge was the Galveston, Texas, "flood" of 1900. On that occasion a hurricane with winds of one hundred and twenty miles per hour raised the water level along the shore of the Gulf of Mexico fifteen feet above the usual two-foot tidal range. The storm waves, probably another 25 feet high, rode in atop the storm and demolished the city; some five thousand people drowned.***

The factors that govern the severity of a given storm on a given coast are summarized in Table 15.1.

Geologic Changes Resulting from Coastal Storms

Recent research in the coastal zone has shown that the changes which occur in the short periods

* Excerpt from *Waves and Beaches* by Willard Bascom. Copyright © 1980 by Willard Bascom. Copyright © 1964 by Educational Services Incorporated. Reprinted by permission of Doubleday & Company, Inc.

TABLE 15.1   Factors determining the severity of coastal storms

Factor	Effect
Height of storm surge	The higher the storm surge, the more extensive the damage
Tidal stage	All effects are greater at high tide and even greater at spring tides
Coastal shape	Concave shorelines have the highest storm surges for a given wind
Storm winds	The greater the wind velocity, the greater the damage
Storm center movement	The slower the storm moves over the coast, the greater the damage; the worst possible condition is a storm that "stalls" near the coast
Nature of coast	The higher the coast and the more resistant the material, the less the damage; low-lying mainland beaches and barrier islands sustain the greatest damage
Previous storm damage	A coast weakened by a previous storm will be subject to proportionally greater damage in a subsequent storm

of storms may represent a much larger proportion of the *total* yearly coastal change than had been thought previously. Beaches along rocky headlands and cliffed coasts commonly are stripped of loose material, but damage is nowhere near as great as on low-lying barrier island shorelines.

You may have noticed that beaches look quite different in the summer and winter (Figure 15.15). In winter, beaches are narrower and steeper and are composed of coarser-grained particles because larger storm waves remove sand faster than it can be replenished by longshore drift. This sand is returned to the beaches during the calmer period from late spring to fall. Sequential studies of beach changes at Carmel, California, show that the beach there widens by over 60 m (200 ft) from spring to fall. These seasonal changes are modest compared with those which result from major storms.

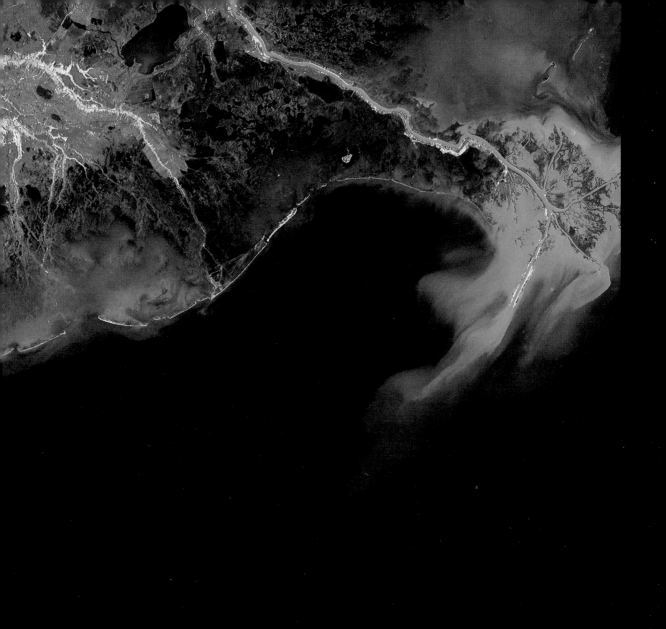

Plate 15   Satellite view of the Mississippi River delta showing the cloud of silt flowing into the Gulf
f Mexico. The colors are a result of the filters used. (ERTS)

Plate 16   Cave interior, Carlsbad Caverns, New Mexico. (*Fred E. Mang, Jr., National Park Service*)

Plate 17   Mountain glaciers in the Alps near the Italian-Swiss border. The near glacier is the Pres de Bar Glacier as seen from the Grand Col de Ferret. Several alpine glacial features are visible. (*Courtesy of Jerome Wyckoff*)

A

Plate 18 Deserts. (a) Sandy desert, Death Valley, California. (Cecil W. Stoughton, National Parks Service) (b) Sparsely vegetated rocky desert, Granite Mountains, Mojave Desert, California. (Courtesy of Donald Doehring) (c) Desert vegetation in bloom in April, Boyce Thompson Arboretum, west of Superior, Arizona. (Courtesy of Jerome Wyckoff)

B

C

Plate 19   Aerial view of Moorea in the Society Islands. Note the volcano, the lagoon, and the barrier reef. (Courtesy of Stanton A. Waterman)

Plate 20   Satellite view of North and South America and eastern part of Pacific Ocean taken
January 31, 1968 at an altitude of 6800 m. (NASA)

Plate 21   Hot, mineral-rich waters emerging from submarine vents in the Galapagos Rift Zone. *(Courtesy of Dudley Foster, Woods Hole Oceanographic Institution)*

Plate 22   Metal crystals deposited by hot mineral-rich waters emerging from submarine vents in the Galapagos Rift Zone. *(Courtesy of Emory Kristoff)*

Plate 23   Images of the four Galilean satellites of Jupiter obtained by the Voyager spacecraft in 1979. The satellites Io (upper left), Europa (upper right), Ganymede (lower left), and Callisto (lower right) are shown in their proper relative size, color, and brightness. Each satellite has different surface features and a different history. *(Jet Propulsion Laboratory)*

Plate 24   Mosaic image of Saturn obtained by the Voyager spacecraft in 1980. The soft velvety appearance is due to a haze layer around the planet. The width of the rings at the center of the disk is estimated to be 10,000 km. *(Jet Propulsion Laboratory)*

Plate 25   Image of Jupiter obtained by the Voyager spacecraft in 1979. The image illustrates cloud activity in Jupiter's atmosphere. The large, roundish object in the lower, central part of this face of Jupiter is the Great Red Spot, an atmospheric storm that has been observed for the past 300 years. Most of these observations of the Great Red Spot were obtained with telescopes. *(Jet Propulsion Laboratory)*

(a)

(b)

Figure 15.15 Seasonal changes in beach particle sizes along Boomer Beach, La Jolla, California. *(a)* Winter (February 1980): beach composed entirely of rock fragments. *(b)* Summer (September 1980): beach covered with sand. *(Courtesy of John S. Shelton.)*

One of the most severe storms to affect the Atlantic Coast in recent years occurred on March 7–8, 1962. The Army Corps of Engineers has documented the changes that occurred at Virginia Beach, Virginia (Figure 15.16). The gray area in Figure 15.16 shows the normal variation in elevation from fall 1956 to spring 1957; very little of the dune surface changed during this time interval. The light color region shows the dune profile right before the storm, and the heavy line shows the surface of the beach after the extreme dune and foreshore erosion caused by the storm. Approximately 30 percent of the "normal" foreshore and dune was eroded during the storm. Coastal dunes provide the first line of defense against storms. The dunes at Virginia Beach lost more than 50 percent of their volume, lowering the dune crest from 4.9 to 3.4 m, enabling future storm surges to rise over the dunes and flood inland areas.

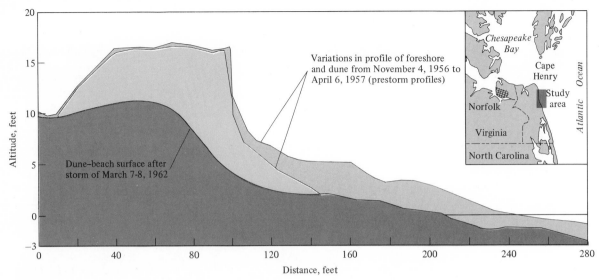

Figure 15.16 Beach and dune erosion at Virginia Beach, Virginia, after the March 7–8, 1962, coastal storm. *(Modified from Miscellaneous Paper 6-64, U. S. Army Corps of Engineers.)*

Storm waters not only erode the face of the dunes but also sweep through lower portions of the dunes, depositing sediment across the back area of the barrier island. Deposition of a series of such overwash deposits by successive storms increases the width of the barrier island gradually. Especially severe storms may cut through the barrier island itself, forming new tidal inlets and depositing sediment in the bays (see Figure 15.17). Inlets formed during storms normally are closed after a few days by longshore drift. When the storm waters recede, the deposits of sediment are exposed and may become vegetated, forming the islands found in many bays.

Coastal Recovery After Storms

What happens to the sand removed from beaches during storms? Part of it moves along the shoreline through wave action and longshore currents, part is transferred across the beach by overwash, part moves into bays via storm-cut inlets, and the remainder is moved offshore, to be returned to the beach during calmer weather.

Figure 15.17 Effects of the storm of March 7-8, 1962, at Long Beach, New Jersey. The ship was washed ashore during the storm. Sand was washed from the beach across the barrier island as a series of washover lobes. *(U. S. Army Corps of Engineers.)*

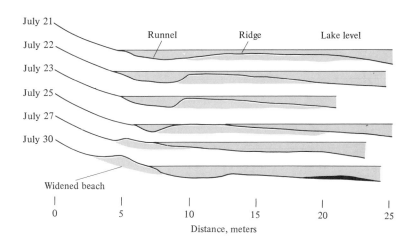

Figure 15.18 Natural beach replenishment after a storm on Lake Michigan. An offshore ridge migrates gradually landward and eventually joins the beach, increasing its width. During its landward migration, the ridge is separated from the beach face by a water-filled runnel. *(Modified with permission from Davis et al., "Natural Beach Restoration on Lake Michigan,"* Journal of Sedimentary Petrology, *vol. 42, p. 416, Fig. 4. Society of Economic Paleontologists and Minerologists.)*

How are beaches restored naturally after storms? Detailed studies of Lake Michigan beaches after storms have shown that sand is returned to the beaches through formation of a submarine ridge and its landward migration (Figure 15.18). During its landward migration the ridge is separated from the beach by a water-filled trough referred to as a runnel. In little over a week, the ridge shown in Figure 15.18 had "welded" itself to the storm beach, increasing the width of the beach considerably.

CHANGING SEA LEVEL

We have been referring to sea level as relatively constant, moving small distances up and down in tides and rising higher in short periods of storm surge. However, barely measurable but long-term changes in sea level are occurring at present because of a number of factors. Such long-term changes in sea level have occurred through most of geologic time, as shown by the changing depositional environments preserved in the earth's sedimentary rock record (see Chapter 6).

A small but steady rise in sea level is occurring today as a result of continued melting of glacial ice. Variations in the level of the ocean surface resulting from increasing or decreasing volumes of ocean are called **eustatic** changes. Eustatic changes in sea level are felt everywhere. Rising postglacial sea level (see Chapter 13) is resulting in a slow landward

migration of the shoreline and an increase in shoreline erosion. The beach restoration described earlier in this chapter (Figure 15.18) represents only an *apparent* building out of the beaches. Detailed studies of Atlantic and Gulf Coast barrier islands have shown that *net* long-term erosion is occurring along these shorelines. The long-term effect of this landward shoreline advance can be seen by projecting the present rate of rise in sea level into the future (Figure 15.19). Such projections indicate that many of the great coastal population centers could be submerged eventually.

Along with eustatic changes in sea level we must consider the changes resulting from vertical movement of the land itself. Tectonic forces may result in either elevation or submergence of coastal areas. Portions of the earth's surface that were covered by great thicknesses of ice during the recent Pleistocene Epoch (see Chapter 13) were pushed down by the weight of the ice. As the glaciers withdrew, the land surface began to "rebound," tending to restore the original elevation. This is a slow process that is still going on in many previously glaciated areas. In many of these areas, beaches formed along Pleistocene shorelines were uplifted subsequently by isostatic rebound. In some cases, these areas show a number of former beaches in a "stair-step" arrangement (Figure 15.20).

Another type of tectonic change is caused by earthquake movements. During the 1964

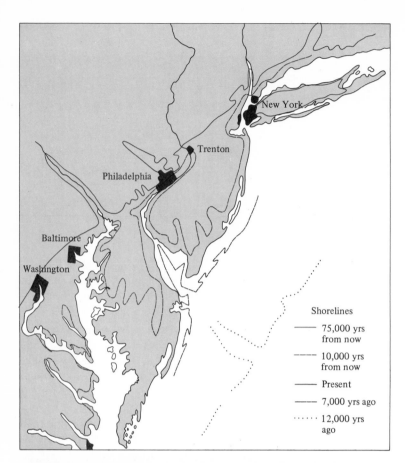

Figure 15.19 Location of past and future shoreline positions on the Central Atlantic Coast. Projection of present processes of coastal change and rates of rise in sea level indicates that major population centers on the coastal plain may be submerged in as little as 10.000 years. Modified with permission from J. C. Kraft (1973). *(Redrawn with permission of Donald R. Coates.)*

Shorelines

⎯⎯⎯ 75,000 yrs from now

⎯ ⎯ ⎯ 10,000 yrs from now

⎯⎯⎯ Present

⎯⎯⎯ 7,000 yrs ago

· · · · · 12,000 yrs ago

Figure 15.20 Elevated beach ridges in the Hudson Bay region, Canada. These elevated beach ridges demonstrate that the land has risen since the disappearance of the most recent Pleistocene ice sheet. (Department of Energy, Mines, and Resources, Canada.)

earthquake in Alaska, portions of the coast either were warped upward or subsided beneath the sea. The combined effect of tectonic and eustatic changes in sea level makes it difficult to determine the magnitude of change in sea level in many areas.

The actual rate of change in sea level varies from coast to coast because the rate at any one place reflects the combination of both eustatic and tectonic changes. While all coasts are affected by eustatic changes, only some coasts are affected by tectonic changes. For example, the rate of rise in sea level along a tectonically stable coast will be less than the rate along a coast which is subsiding.

COASTAL RESOURCES

The coastal zone is the ultimate repository for most of the minerals and organic material brought down to the sea by streams. Several important minerals are reworked by coastal waves and currents into economically minable placer deposits of titanium, zircon, diamonds, tin, monazite, and iron. In Namibia (formerly South-West Africa), for example, diamonds, weathered out of intrusive igneous rock called kimberlite, were transported to the coast by ancient streams and concentrated by coastal waves and currents in beach gravels which subsequently were covered by thick layers of beach sand. Giant earth-moving machines are now being used to recover vast quantities of gem-quality diamonds from these sands. For example, in 1977 alone, 2,001,217 carats of diamonds were recovered.

Large amounts of petroleum and natural gas were formed from the deposition of organic material in ancient coastal environments. Phosphate rock, essential for fertilizers, is believed to have been formed in ancient nearshore coastal environments and is now being mined from deposits in Florida and North Carolina. Deposits of sand and gravel concentrated by ancient surf action are mined extensively for use in highway fill, concrete, and high-quality optical glass.

HUMAN EFFECTS ON COASTAL ZONES

Geologists have long realized the dynamic nature of the coastal zone and are not surprised at the tremendous erosion and shoreline changes which can occur over a long period of time. These changes are due largely to storms, but some are a result of the slow but steady postglacial rise of sea level. These two "normal" changes caused relatively little structural damage and loss of life until the late 1940s because the coastal zone was relatively sparsely settled. Major changes in the coastal zone since then include (1) major expansion of coastal cities, (2) conversion of summer homes in resort areas to year-round use, and (3) human modifications along shorelines to prevent erosion, minimize storm damage, and provide easier access between bays and the ocean. These changes not only have increased the structural damage and loss of life from storms, they also have accelerated geologic changes both during storms and in good weather.

The increasing number and severity of problems in coastal zones in recent years is due in large part to the fact that too little attention is paid to the interrelated nature of the different aspects of the coastal zone; changes made in the dynamic balance along a coast at one place frequently cause additional problems somewhere else. Geologists use the terms "updrift" and "downdrift" to refer to geographic positions along a shoreline in relation to the direction of transport of sediment along that shoreline (Figure 15.21). This usage is similar to the usage of the terms "upstream" and "downstream" in studies of streams.

Structures built into the nearshore coastal zone alter the dynamic equilibrium along coasts by cutting off the sand supply needed to nourish nearby beaches. Rock, concrete, or stone structures called **groins** commonly are built into the surf to trap sand and build out beaches. Figure 15.21 illustrates the consequences. The beaches updrift of the groin grow by net deposition. However the beach downdrift of the groin suffers net erosion because the groin prevents the downdrift beaches from being replenished by sand.

Sediment transport along a coast also is disturbed when tidal inlets are stabilized by the construction of rock or concrete **jetties** into the surf zone on either side of the inlet to prevent it from closing through sand deposition. These

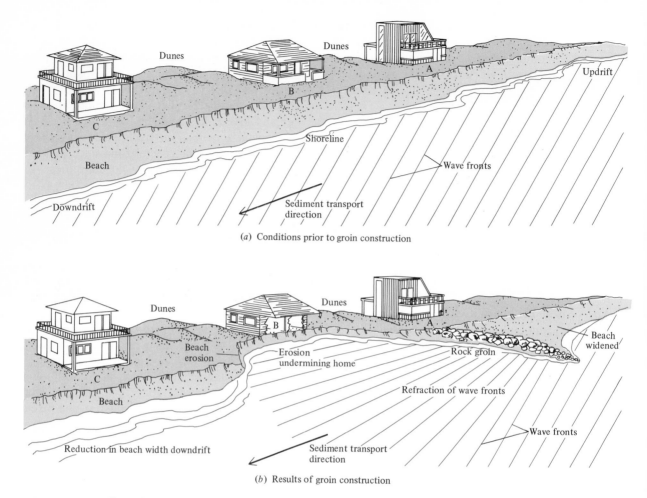

Dunes

Dunes

A

Updrift

Shoreline

Wave fronts

Beach

C

Downdrift

Sediment transport
direction

(*a*) Conditions prior to groin construction

Dunes

Dunes

B

A

Beach
widened

Beach
erosion

Erosion
undermining home

Rock groin

C

Refraction of wave fronts

Beach

Reduction in beach width downdrift

Sediment transport
direction

Wave fronts

(*b*) Results of groin construction

Figure 15.21 Effect of groin-construction on coastal erosion and deposition. Homeowner A decides to widen the beach by building a stone groin at the edge of the property. The beach widens at A by sand deposition updrift of the groin. However, the groin has prevented the beach downdrift from being replenished, and net erosion is occurring there, undermining the home at B.

stabilized inlets provide convenient access to the ocean for boats, but they also cause severe problems of erosion downdrift. Sand moving along the coast is blocked partially by the updrift jetty and accumulates there, widening the beach. This is similar in effect to the situation depicted in Figure 15.21. Most of the sand particles that are able to pass the updrift jetty do not make it across to the downdrift side because they are swept into, or out of, the inlets by tidal currents and are deposited either in the ocean or in bays. The beaches on the downdrift side of the inlet therefore are cut off from the longshore drift

needed to balance the effects of wave erosion. Consequently, severe beach erosion is common downdrift of such stabilized inlets. Some ingenious methods have been devised to minimize the effects of stabilized inlets on the shoreline. At Lake Worth, Florida, sand is pumped from a "trap" on the updrift side of the inlet and deposited on the beach downdrift of the inlet. Other methods of artificial beach restoration, such as rebuilding dunes with sand dredged out of bays, have had mixed results. Many have negative environmental effects, and most provide only temporary and costly relief.

SUMMARY

Coastal zones are areas of rapid geologic change located at the junction of land, water, and air. Wind flowing across the water surface generates waves which move away from the storm center as swell. When the waves reach water depths equal to one-half their wavelengths, they decrease in wavelength, increase in height, increase in velocity, and lose their symmetry, eventually collapsing in the surf zone and rushing up the foreshore. Wave fronts approaching the coast at an angle move sand grains up and down the beach face and gradually move large volumes of sand along the shoreline as longshore drift.

Waves approaching an irregular coastline are refracted as they enter shallow waters. Wave refraction concentrates the wave energy on the headlands and disperses it over the intervening bays. The concentration of wave energy on headlands erodes them preferentially, and the shoreline tends to become straighter with the passage of time. Wave erosion along rocky coasts is accomplished by the force of the waves driving water under great pressure into rock crevasses. Rock debris carried by the waves erodes wave-cut notches and wave-cut benches. Differential erosion forms sea caves, arches and stacks. Cliffs of unconsolidated sediments are undercut by the waves, resulting in a slumping of a section of the cliff onto the beach. The coarsest particles eroded from the cliff materials may be deposited within the nearshore area as a wave-built terrace.

A number of currents move sediment in the coastal zone. Particles of sediment are carried along the shore by the longshore current. Rip currents move sediment from the beaches out into the nearshore waters. Tidal currents move sediments in and out of tidal inlets and erode and deposit materials in the tidal channels, flats, and marshes of the estuarine portion of the coastal zone.

Coastal zones are composed of estuaries and bays; wetlands made up of tidal marshes, tidal flats, tidal channels; and beaches and barrier islands. Storms can make great changes in the coastal zone. The intensity of a storm is a function of the height of the storm surge, the tidal stage, the shape of the coast, the wind velocity, the speed of movement of the storm, the nature of the coast, and previous storm damage. Coastal storms result in massive erosion and deposition in the coastal zone. Beaches tend to restore themselves after storms with sand moved in from deeper water as a series of landward-migrating ridges.

Tropical low-lying coastlines commonly are composed of calcium carbonate secreted by organisms and broken apart by wave action into calcareous sand and mud. Colonial coral and coralline algae build wave-resistant reefs. Reefs may be fringing, barrier, or atolls depending on the offshore slope and the subsidence history of the area.

A slow eustatic rise in sea level has been caused largely by the postglacial melting of glacial ice. Tectonic changes in sea level result from upward and downward movements of coastal areas. The rising postglacial sea level moves the shoreline landward, resulting in long-term net erosion along many parts of the United States coastline. Diamonds, tin, titanium, and iron minerals were concentrated by past wave erosion in placer deposits and are being mined today. Phosphate, limestone, petroleum, and natural gas are resources which are being obtained from ancient coastal deposits today. Human activities have caused serious problems in the coastal zone. Structures such as groins, jetties, and stabilized inlets have blocked longshore drift, causing erosion downdrift of the structures. The increase in population in coastal zones has brought more people and structures into an area of potential geologic change.

QUESTIONS FOR REVIEW AND FURTHER THOUGHT

1. How can you get an idea of the offshore slope by seeing where waves begin to break off a beach?

2. An irregular rocky shoreline is being submerged by postglacial rise in sea level. What changes in shoreline shape should occur with

time? Why?

3. *a.* What is wave base?

b. How does wave base change with different types of waves?

c. What important geologic work is done as wave base changes continually along a coast?

4. The tidal range along a coast is 1 m. A large house sits on a concrete slab atop the dune. The concrete slab is 3 m above high tide. Is the house safe in a storm? Why?

5. Describe each of the following currents and show how they contribute to erosion and deposition within the coastal zone.

a. Longshore currents

b. Rip currents

c. Tidal currents

6. An east-west trending shoreline faces the ocean to the south. During good weather the winds come from the southwest, but during storms the winds come from the east.

a. Which way is sediment moving along the shoreline on a good swimming day?

b. Which way is sediment moving along the shoreline on a stormy day?

c. If 2 units of sand move by a point on a good swimming day and 10 move by on a stormy one, what is the net amount moved and in which direction is it transported?

7. *a.* What are the effects of building groins and jetties in the coastal zone?

b. What happens when a tidal inlet is stabilized by building jetties into the surf zone on either side of the inlet?

8. What conditions are necessary for the formation of a coral reef?

9. What geologic conditions favor the formation of the following structures in an area in which corals can thrive?

a. Fringing reef

b. Barrier reef

c. Atoll

10. Name the factors that govern the severity of a storm on a given coast. Show how each of these factors works.

ADDITIONAL READINGS

Beginning Level

Bascom, W., "Beaches," *Scientific American,* Reprint 845, W. H. Freeman, San Francisco, 1960, 11 pp.
(Description of many aspects of coasts in a very readable manner)

Dolan, R., B. Hayden, and H. Lins, "Barrier Islands," *American Scientist,* vol. 68, January-February 1980, pp. 16–25.
(Describes the formation and characteristics of barrier islands. Includes a discussion of the environmental aspects of development in coastal zones)

Griggs, G. B., and E. J. Rogers, "Coastline Erosion in Santa Cruz County, California," *California Geology,* vol. 32, no. 4, April 1979, pp. 67–76.
(Describes the factors that determine rates of coastal erosion in both rocks and sediments)

Howard, A. D., and I. Remson, *Geology and Environmental Planning,* McGraw-Hill, N.Y., 1978, 478 pp., chap. 4, "Coastal Environments."
(A very good treatment of environmental problems and planning in the coastal zone)

McGowen, J. H., C. G. Groat, L. F. Brown, W. F. Fisher, and A. J. Scott, *Effects of Hurricane Celia: A Focus on Environmental Geologic Problems of the Texas Coastal Zone,* Texas Bureau of Economic Geology, Geol. Circular 70-3, 1970, 34 pp.
(Reviews the environmental damage caused by coastal storms. Well illustrated with maps and photos of storm damage. Also discusses environmental planning in coastal zones)

Schuberth, C. J., "Barrier Beaches of Eastern America," *Natural History,* vol. 79, no. 6, June-July 1970, pp. 46–55.
(Well-written and illustrated discussion of the natural and artificial changes that occur in barrier islands)

Shepard, F. P., and H. R. Wanless, *Our Changing Coastlines,* McGraw-Hill, New York, 1971, 579 pp.
(An extremely well-illustrated discussion of the coastal regions of the United States. The best place to look for details on a particular coastal area)

U.S. Army Corps of Engineers, *Land against the Sea*, U.S. Army Coastal Engineering Research Center Misc. Paper 4-64, 1964, 43 pp.
(Reviews the origin of coastal features. Detailed discussion of the artificial structures used to maintain access to the coast by boats and to minimize coastal erosion)

Intermediate Level

Bloom, A. L., *Geomorphology*, Prentice-Hall, Englewood Cliffs, N.J., 1978, 510 pp., chap. 19, "Shore-Zone Processes and Landforms," and chap. 20, "Explanatory Description of Coasts."
(Particularly good coverage of rocky shorelines, carbonate shorelines, and coastal classification)

Inman, D. L., and B. M. Brush, "The Coastal Challenge," *Science*, vol. 181, no. 4094, 1973, pp. 20–32.
(Good overall quantitative review of the energy and sediment budgets along shorelines and the effects of human intervention on coastal processes and water quality)

Leatherman, S. P. (ed.), *Barrier Islands from the Gulf of St. Lawrence to the Gulf of Mexico*, Academic Press, New York, 1979, 325 pp.
(A number of individual articles on the origin, evolution, and characteristics of barrier islands)

Shepard, F. P., *Submarine Geology*, 3d ed., Harper & Row, New York, 1973.
(Well-illustrated and detailed treatment of coastal morphology and sediments)

Advanced Level

Davis, R. A., *Coastal Sedimentary Environments*, Springer-Verlag, 1978, 420 pp.
(A series of articles by experts on different coastal sedimentary environments)

Komar, P. D., *Beach Processes and Sedimentation*, Prentice-Hall, Englewood Cliffs, N.J., 1976, 429 pp.
(A detailed quantitative study of coastal dynamics and sedimentation)

CHAPTER 16

Oceans

Only 30 years ago we knew very little about the sea floor and only a bit more about the water that covers it in spite of the fact that oceans cover 72 percent of the earth's surface (Plate 20). Research in the past 30 years has resulted in a drastic reappraisal in our thinking about the oceans and ocean basins. For example, scientists once thought that the ocean basins were just submerged, relatively flat parts of the continents. Today we know that far from being flat, the oceans contain the greatest mountain ranges on the planet and that many aspects of the oceans are fundamentally different from those of the continents.

Oceanography is the study of the physical, chemical, biologic, and geologic aspects of the oceans. In this chapter, we will concentrate only on those aspects of oceanography which enable us to understand the shape of the sea floors, the composition and circulation of ocean water, and the nature of sediment deposition in the oceans. In addition, most of the evidence now used in support of the plate tectonic model has come from recent studies of the sea floors. We will see how adequately each of the five hypotheses outlined in Chapter 1 explains the features now recognized in the ocean basins.

EXPLORING THE OCEANS

Modern oceanographic studies began on December 21, 1872, when a British research ship, the H.M.S. *Challenger,* left Portsmouth, En-

gland. Three and a half years and 68,890 miles later, the *Challenger* returned to England after conducting oceanographic measurements in all the oceans except the Arctic. The data on water depths; current directions; physical, chemical, and biologic aspects of the water; and nature of the bottom sediments and life were to be published eventually in 50 volumes which changed forever our ideas about the ocean.

Remarkable advances in scientific theory and technology have been made in the century since the *Challenger*'s voyage. Modern ocean research vessels can study the oceans and ocean basins in ways that nineteenth-century scientists could only dream of. Changes in the techniques of measuring water depths and collecting samples of sediments illustrate the extent of these advances.

Measuring the Depth of the Oceans

Scientists on the H.M.S. *Challenger* obtained the first accurate depth readings. They lowered a weighted line and recorded the depth as the length of line let out before the line went slack as the weight rested on the bottom. The process was long and tedious. In deep water, it took as much as an hour and a half to reach the bottom.

Today water depths can be obtained rapidly day or night and under adverse weather conditions by using an electronic device called a **precision depth recorder** (PDR). A research vessel generates a sound source, and the depth is calculated according to the time it takes for

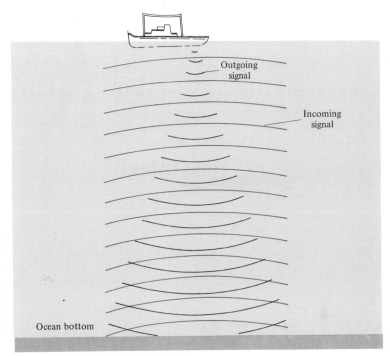

Outgoing
signal

Incoming
signal

Ocean bottom

Figure 16.1 Precision depth recording (PDR). A sound impulse generated by the research vessel is reflected off the bottom and recorded by receivers on board. The depth of water is calculated by knowing the velocity of sound in water and the delay time between impulse generation and return of the signal.

the sound signal to reach the sea floor and be reflected back to the ship (Figure 16.1). The results of the PDR analyses can be viewed as a continuous profile of the bottom topography, as shown in Figure 16.2. In order to detect very small relief features, PDR records utilize a greater vertical scale (water depth) than a horizontal scale (distance). This gives a **vertical exaggeration** to the bottom, making it appear more "rugged" than it really is (Figure 16.2).

Collecting Samples of the Sea Floor

When weighted lines were used to measure water depth, the cup at the bottom of the weight was often coated with sticky tar or wax so that it would bring up a small sample of the sediment from the bottom. Larger samples were collected by dredging—lowering a device to scoop up whatever lay on the bottom. Dredge samples still are collected because large samples can be obtained simply and relatively inexpensively. However, dredge samples are not useful for detailed studies of layers of sediment because the procedure tends to mix the surface layers together. Cylindrical coring tubes driven by weights can recover samples up to 30 m long. These core samples reveal the sucessive sediment layers which were deposited in that area in the past (Figure 16.3).

Today scientists can descend even into the deepest parts of the ocean basins in small research submarines. With these vessels they can observe submarine processes directly and use mechanical "arms" (Figure 16.4) to select, examine, and even take samples of sediments and rocks.

The Deep-Sea Drilling Project (DSDP)

The Deep-Sea Drilling Project is an attempt to determine the structure and geologic history of the ocean basins by drilling through the bottom sediments into the underlying rock. The ability to obtain sediment and rock core samples by drilling in water as deep as 6500 m is a great technological achievement. A specially designed ship, the *Glomar Challenger* (Figure 16.5), is used in the drilling.

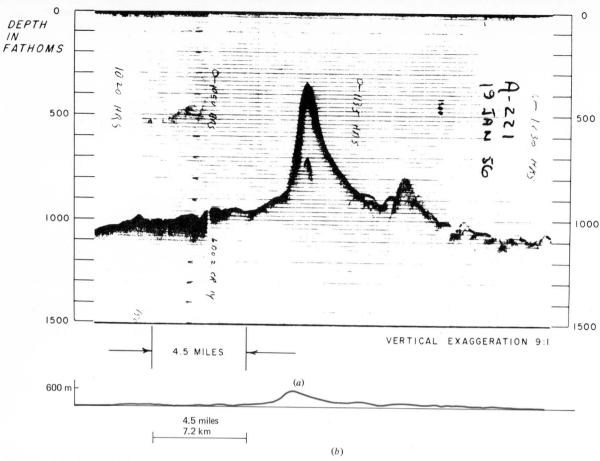

Figure 16.2 *(a)* Precision depth record (PDR) of a seamount. The vertical axis is exaggerated 9 times compared with the horizon-axis to clarify details. *(Woods Hole Oceanographic Institution.)* *(b)* A redrawing of the same profile without the vertical exaggeration.

Computer-controlled thrusters at the front and rear keep the ship in place over the drill site without anchoring. Sonar positioning enables the drillers to remove cores and replace pipe during the drilling. By 1975, over 514 holes had been drilled at 352 different sites. Just as the H.M.S. *Challenger* changed our ideas about the characteristics of ocean water, life, and bottom topography, its twentieth-century counterpart, the *Glomar Challenger,* has revolutionized our understanding of the origin and evolution of ocean basins. We will examine these data and ideas later in this chapter.

SEA-FLOOR TOPOGRAPHY

PDR studies of the ocean basins show them to be quite rugged in appearance in some areas but so flat in others that it is necessary to use a vertical exaggeration to show many of the features (see Figure 16.2). Oceanic islands and midocean ridges rise high above the ocean floors, while oceanic trenches extend far below the average depth of the seas. The maximum relief on the ocean floor—the distance between its highest and lowest points—is greater than the relief on the continents. Indeed, if Mt. Everest (altitude 8848 m), the highest mountain on the continents, were placed in the deepest oceanic trench (depth 11,515 m), its top would not reach sea level.

PDR data have been combined to make three-dimensional maps of the ocean bottom so that for the first time, geologists have had an accurate map of the entire surface of the earth beneath both its atmosphere and its hydro-

Figure 16.3 Layering in piston cores reveals changes in ocean sedimentation at any one place with time. *(Courtesy of Charles D. Hollister, Woods Hole Oceanographic Institution.)*

Figure 16.4 The research submarine *ALVIN*. *(Woods Hole Oceanographic Institution.)*

sphere. Figure 16.6 is a map of the North Atlantic Ocean Basin showing the major topographic features. In our discussion of ocean-floor topography, we will refer frequently to both Figure 16.6 and an idealized profile with a high vertical exaggeration showing a topography similar to that off the east coast of North America (Figure 16.7).

Continental Shelves

Every continent is surrounded by a **continental shelf**—a gently sloping surface extending seaward from the shoreline (Figures 16.6 and 16.7). Continental shelves have very gentle slopes, typically with about 1 m vertical drop over a horizontal distance of 500 m. This slope may also be expressed as a gradient of 1:500. The continental shelf terminates at the shelf break, which is marked by a sharp increase in gradient.

The continental shelves are the shallowest parts of the oceans, with water depths at the shelf breaks ranging from about 35 m to about 240 m. The widths of the shelves also are highly variable. Extremely narrow shelves border major mountain ranges like the Andes, but wide expanses are typical along the east coast of South America. In tectonically active zones such as the area off southern California, the normally flat continental shelves may be broken into a series of elongate basins by faults.

Because the continental shelves are the shallowest parts of the ocean basins, they were affected most severely by the large-scale changes in sea level associated with Pleistocene glaciation which we discussed in Chapter 13. When continental glaciers reached their maximum extent during the Pleistocene, sea level may have been more than 100 m lower than it is today. The shoreline would have been far out on most continental shelves, and much of what is now beneath water would have been exposed to the air. In colder latitudes, continental glaciers moved across the areas of the exposed shelf, leaving an irregular glacial topography similar to that seen on the continents. In warmer latitudes, sediments were deposited far out on the shelves by streams that flowed across wide coastal plains. We can trace the courses of these ancient streams by PDR data that outline

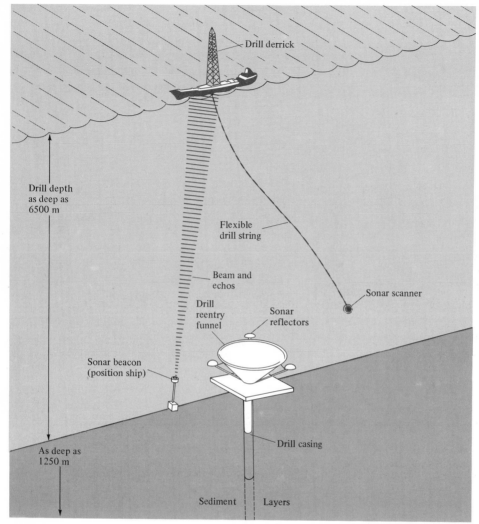

Figure 16.5   The *Glomar Challenger* is specially equipped to take long cores of the ocean bottom sediments and underlying rocks in great depths of water.   The vessel is positioned by sonar over the drill site and is maintained in that position by powerful thrusters on the ship.   The drill string may be removed from the hole and replaced by using a sonar scanner.

elongate depressions called **shelf valleys** cut into the shelves.   Many shelf valleys, such as the Hudson shelf valley (Figure 16.8), connect with the mouths of our modern rivers.   This indicates that the Hudson River at one time flowed across what is now the continental shelf.

The topography and sediments found on the continental shelves thus are largely the product of subaerial processes of stream and glacial deposition and erosion and have been modified only recently by oceanic processes.   The continental shelves are the only part of the ocean basins for which this can be said, and they are, therefore, the only part that can be thought of as submerged portions of the continents.

Continental Slopes and Submarine Canyons

The **shelf break** at the outer edge of the continental shelf marks the beginning of a more

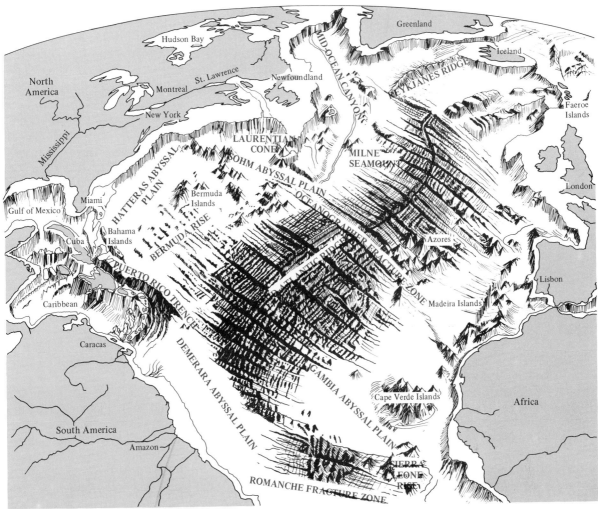

Figure 16.6   Topographic features on the bottom of the North Atlantic Ocean.   *(Based on oceanographic map used by courtesy of the Aluminum Company of America.)*

Figure 16.7   A generalized profile showing the submarine topography across the western part of the North Atlantic Ocean.

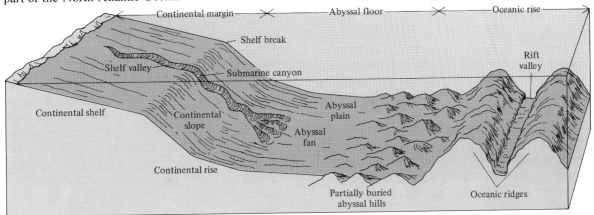

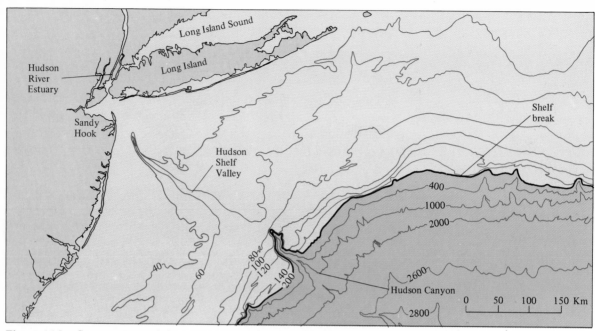

Figure 16.8 Contour map of the continental shelf and slope south of Long Island, New York. Note the shelf valley that cuts across the shelf between the Hudson River Estuary and the Hudson Submarine Canyon.

steeply sloping region called the **continental slope** (Figures 16.6 and 16.7). Gradients on the continental slopes range from 1:2 to 1:40, and even the gentlest slope gradient is different enough from the gradients of the shelves to permit easy identification. One consequence of the steeper gradient of the continental slope is that sediments deposited on the slope tend to be unstable and move downward into deeper parts of the ocean if they are disturbed.

The continental slopes are incised deeply in places by steep V-shaped valleys called **submarine canyons** that extend across the slope and onto the deep sea floor below. Major submarine canyons sometimes have tributary canyons joining the main valley, as with stream drainage networks on land. Some submarine canyons, such as those off southern California, have their heads almost in the surf zone (see Figure 15.11), but others begin only at the shelf break. Some submarine canyons, such as the Hudson submarine canyon, obviously were linked at one time to modern streams. The Hudson Canyon connects with a shelf valley

that in turn is aligned with the mouth of the present Hudson River Estuary (Figure 16.8). However, other canyons seem to have no counterparts among modern rivers. They may have been related to older rivers whose channels subsequently were buried by continental shelf sediments.

The upper parts of submarine canyons could have been cut by subaerial stream erosion at the same time as the shelf valleys, during low stands of sea level in the Pleistocene Epoch. At such times the Hudson River would have flowed across the continental shelf and deposited its sediment near the upper edge of the continental slope. However, the lower parts of the canyons could not have been carved by streams because they are found at depths far lower than any exposed by glacial lowering of sea level. They must have been formed by some type of submarine process of erosion. Most oceanographers believe that the deeper parts of the submarine canyons were eroded by dense sediment-laden currents that moved down the continental slope from their initial sites of deposition. We will examine this topic more closely later in this chapter.

Continental Rises

At the base of the continental slope there is an area of gentle gradient (average 1:150) called the **continental rise** that marks the transition from continental slope to the deep sea floor (Figures 16.6 and 16.7). The smooth surfaces of the continental rises are broken by channels that appear to be extensions of submarine canyons. Recent geologic studies have shown that the continental rise is a depositional feature underlain by sediments that have been transported down the continental slopes. Presumably, the sediment-laden currents that carve the submarine canyons deposit much of their sediment at the base of the continental slope. This is a process similar to the formation of an alluvial fan at the base of a mountain range on land (see Chapter 14).

Abyssal Plains and Abyssal Hills

The continental rises gradually grade into the flattest parts of the ocean basins, regions called **abyssal plains,** where gradients are typically as low as 1:1000. The abyssal plains are covered by sediment from two sources: continent-derived clastic sediment transported into the depths by turbidity currents and surface and deep oceanic currents, along with biogenic and chemically precipitated sediment that forms within the ocean.

The flatness of the abyssal plains is broken by **abyssal hills**—topographic highs rising a few meters to hundreds of meters above the plains and ranging up to 10 km in diameter. Deep sea drilling has shown that the abyssal hills are extinct basaltic volcanoes capped by a thin layer of sediment. The hills are less abundant near continental margins, possibly because those near the continental rises have been buried completely by continental sediment.

Figure 16.9   The interconnections between the worldwide network of ocean ridge systems. The ridge systems are offset by fracture zones.

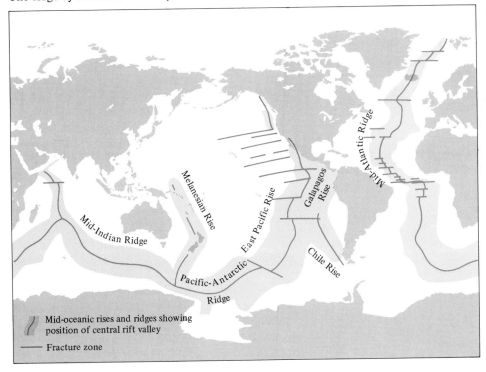

Ocean Ridges and Fracture Systems

Large linear mountain ranges called **ocean ridges** rise out of the depths of every ocean basin and are interconnected in a worldwide network (Figure 16.9). The ridges are subdivided and given geographic names for purposes of identification. Some of the ridges are very long—the Mid-Atlantic Ridge, for example, extends for about 21,000 km—but others, such as the Galapagos Rise, are short. Some ridges lie near the center of their ocean basins and have been called midocean ridges, but others, such as the East Pacific Rise, are located far from the ocean center, near a continental margin.

All segments of ocean ridges, regardless of where they are located in their ocean basins,

display similar features. The ocean floor gradually rises above the abyssal plain, and sea-floor relief becomes more rugged toward the central part of the ridge. At the very crest of the ridge is a deep valley called a **central rift valley** that is flanked by steep rift mountains (Figure 16.10). The ridge crest is not a continuous line but is offset in several places by fractures that cut across the axes of the rift valleys at high angles. For example, look back at Figure 16.6 and notice how the Reykjanes Ridge is offset from the Mid-Atlantic Ridge. These fractures extend across the width of the ridges and most or all of the adjacent abyssal plain. In places they come close to the continental slopes, and some geologists believe that certain fractures can be traced onto the continents. Individual ocean

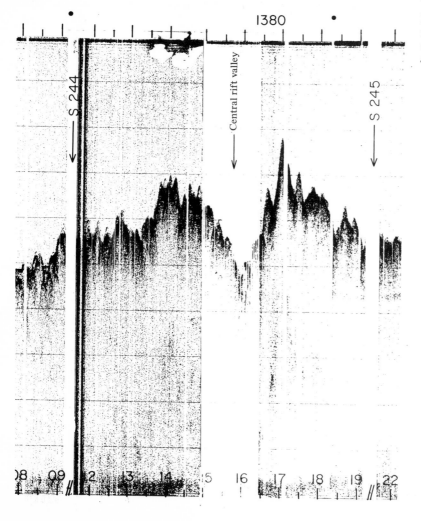

Figure 16.10 PDR profile across rift valley in South Atlantic Ocean. *(Lamont Doherty Geological Observatory.)*

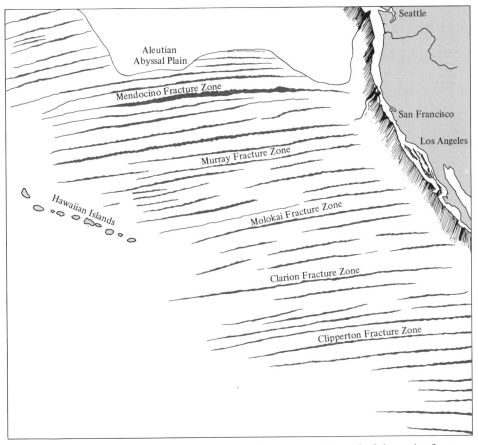

Figure 16.11   Fracture zones in the eastern Pacific Ocean.   Several of the major fracture zones extend all the way from the Hawaiian Islands to the west coast of North America.

fractures can be traced for thousands of kilometers, and many of the larger ones have been named (Figure 16.11).

Although the ridges rise thousands of meters above the abyssal plains, their tops generally are still well below sea level. A few peaks on an ocean ridge may stand above sea level as islands such as the Azores. A large segment of the Mid-Atlantic Ridge, including the central valley, is exposed in Iceland. From the rocks on such islands and the results of DSDP cores, we have learned quite a bit about the composition of the ridges.

Ocean ridges are composed of basaltic lava and are sites of frequent volcanic and earthquake activity. Radiometric age dating of basaltic rocks collected from various parts of the worldwide ridge system shows that the basalts nearest the ridge axes are the youngest and that ages increase systematically away from the axes on both sides. The ridge crests are covered by a thin veneer of sediments, but the sediment blanket thickens away from the axes and is thickest over the oldest basalts. The ocean ridges also are regions of abnormal earth magnetism, and we shall examine this property in detail in Chapter 20.

Ocean Trenches, Island Arcs, and Marginal Seas

The deepest parts of the world's oceans are found near the ocean margins in narrow arc-shaped deeps called **oceanic trenches** (Figure 16.12). The greatest water depth ever recorded — 10,915 m — was measured in the Marianas Trench in the South Pacific. The trenches have

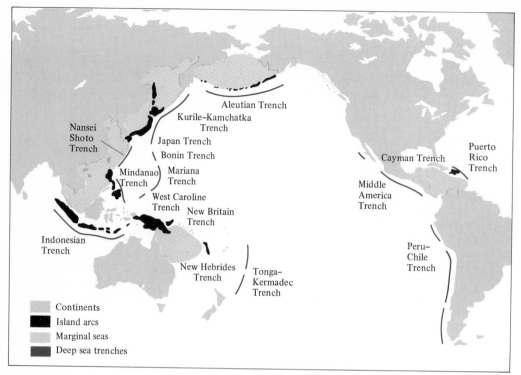

Figure 16.12 Oceanic trenches (dark color), island arcs (black), and marginal seas (light color) in the Pacific.

a systematic geometric orientation. They are consistently oriented so that their convex sides face the deep ocean while their concave sides face the nearest continent (see Figure 16.12).

On the continent or concave side of most trenches is an **island arc**—an area of active volcanism and frequent earthquakes. For example, the Aleutian, Japanese, and Philippine island systems are island arcs bordered by trenches. Volcanism on the island arcs is more varied in composition than volcanism along the ocean ridges, consisting primarily of andesite with lesser amounts of rhyolite and basalt. In one instance in the modern oceans, a continent borders an ocean trench. The Peru-Chile trench lies just west of the coast of South America; rather than an island arc, the Andes Mountains lie near the trench.

Behind (on the concave side of) most island arcs are relatively shallow basins called **marginal seas,** or back-arc basins. These basins separate the island arc-trench system from the nearest continent. The Sea of Japan and the Philippine Sea are two of the larger marginal seas.

Island arc-trench systems and their associated marginal seas are more abundant around the margins of the Pacific than in any other ocean. In the Atlantic, for example, only the Puerto Rico Trench-Greater Antilles Arc-Caribbean Sea system is present (see Figure 16.6). The shallow marginal seas serve as sediment traps. They accumulate the debris shed from the continents and prevent this sediment from reaching the abyssal plains beyond the island arcs. Where arc-trench systems nearly surround an ocean basin such as the Pacific, this severely restricts the amount of continent-derived sediments supplied to the basin and leads to a lower rate of sedimentation within the ocean.

Seamounts and Guyots

Features resembling submerged volcanoes that rise at least 1 km above the sea floor are called

seamounts. Seamounts are particularly abundant along the ocean fractures that offset the ocean ridges (see Figure 16.11). They are extinct basaltic volcanoes that apparently never managed to build their cones high enough to reach sea level.

Submarine hills resembling seamounts in size and shape, but having flat tops rather than jagged peaks, are called **guyots** (pronounced gee-yō). Shallow-water fossils collected from the sediment cover on guyots and the similarity of their flat tops to wave-cut terraces (see Chapter 15) suggest that guyots are extinct volcanoes that were eroded by wave action, planed off at sea level, and then slowly subsided beneath the surface of the sea. Atolls are formed where coral reefs can maintain their growth while oceanic islands subside (see Figure 15.14).

Basaltic Island Chains

Linear groups of oceanic islands composed of basaltic shield volcanoes are called "Hawaiian-type" islands after their most famous example. Although linear like the ocean ridges and island arcs, Hawaiian-type islands are associated with neither central valleys nor trenches. Another unique feature is the age of the volcanoes in island chains such as the Hawaiian Islands (Figure 16.13). Only the island of Hawaii, the southeasternmost island of the chain, contains active volcanoes. The dormant and extinct volcanoes which make up the rest of the chain increase in age progressively toward the west and northwest, as shown in Figure 16.13. The ages of basalts continue to increase steadily in the adjacent atolls and islands of the Midway Group and in the Emperor Seamount Chain still farther north.

The relative ages of the Hawaiian Islands, Midway Group, and Emperor Seamount Chain suggest that they had been formed in the southeast and carried steadily northwest by a mechanism similar to that of a conveyor belt. We will return to this topic when we discuss the origin of oceanic features later in this chapter.

Figure 16.13 Ages of volcanism in the Hawaiian Islands, Midway Islands, and the Emperor Seamount Chain (in millions of years before the present). Ages in black are those of Hawaiian volcanoes that are above sea level. Ages in color are from submerged basaltic volcanoes that underlie atolls and coral islands (Midway Island Chain) or seamounts (Emperor Seamount Chain). *(Ages of volcanic rocks from I. McDougall, "Potassium-Argon Ages from Lavas of the Hawaiian Islands,"* Geological Society of America Bulletin, *vol. 75, pp. 107–128, 1964.)*

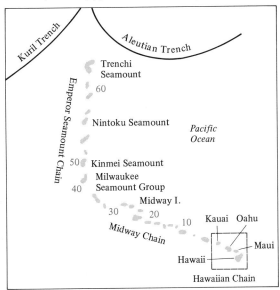

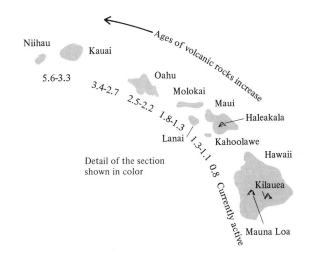

SEAWATER

Although you might not determine much variation in local seawater characteristics, there are large differences in seawater characteristics over broader areas. Along with changes in biologic content, seawater varies in temperature, composition, salinity, dissolved gases, density, and concentration of suspended particles. Large volumes of ocean waters with distinctive characteristics are called **water masses.** Water masses move through the ocean basins much like the hot and cold air masses which move across the earth's surface and influence the weather from place to place by their interaction.

Composition

The most notable characteristic of seawater is its **salinity**—the measure of the concentration of dissolved solids. It is this dissolved load that gives seawater a salty taste and makes it unfit for human consumption. All the gases found in the atmosphere are dissolved in the water of the oceans also, but we will consider gases separately from dissolved solids.

Salinity

Ions dissolved during chemical weathering on the continents are carried away by streams, and most are carried eventually to the oceans. Roughly half of the naturally occurring elements have been detected in solution in seawater, including only slightly soluble, rare elements such as gold and silver. However, the most abundant dissolved ions are the most soluble ones, and seven of these account for more than 99 percent of the ocean's dissolved load, as shown in Table 16.1. It is easy to see from this table why minerals such as halite, NaCl, and gypsum, $CaSO_4 \cdot 2H_2O$, are common evaporite minerals.

Salinity values generally are expressed in parts per thousand (ppt) of dissolved ions in water. The average salinity of surface ocean water is 35 ppt, but salinity varies widely in different areas. Low salinities occur in tropical areas where high rainfall dilutes surface salinity and in polar regions where melting ice from glaciers has the same effect. Higher salinities occur in subtropical areas where high evapora-

TABLE 16.1   The most abundant ions dissolved in seawater

Ion	Formula	% of all dissolved ions
Chlorine	Cl^-	55.04
Sodium	Na^+	30.61
Sulfate	$(SO_4)^{2-}$	7.68
Magnesium	Mg^{2+}	3.69
Calcium	Ca^{2+}	1.16
Potassium	K^+	1.10
Bicarbonate	$(HCO_3)^-$	0.41
Total		99.69

tion rates exceed rainfall and dissolved ions are concentrated to a larger degree. The highest salinities occur in very dry areas such as the Persian Gulf and Red Sea. In the shallower waters in these very dry areas, the waters may be **hypersaline,** containing more than 40 ppt of dissolved ions. Coastal waters in temperate climates may be brackish as a result of the mixing of freshwater and salt water (see Chapter 15).

Dissolved gases

All the gases of the atmosphere, including the inert ones, are found in ocean waters, but oxygen and carbon dioxide are probably the most important to living creatures and in ocean sedimentation. Both these gases are present in large concentrations in the upper few meters of the water column, where exchange and mixing with the air are greatest, but their relative concentrations change with depth.

In the **photic zone**—the upper few tens of meters in which plant life flourishes because of the penetration of sunlight—carbon dioxide is depleted and oxygen given off during the process of photosynthesis. However, below the photic zone, carbon dioxide is not used by organisms, but dissolved oxygen is extracted from seawater by fish and other creatures. Some oxygen also is consumed by oxidation of organic debris falling toward the sea floor. As a result, dissolved oxygen tends to decrease with depth. If it were not for oceanic circulation, which will be discussed later in this chapter, most of the bottom waters would be devoid of oxygen and hence of most life forms.

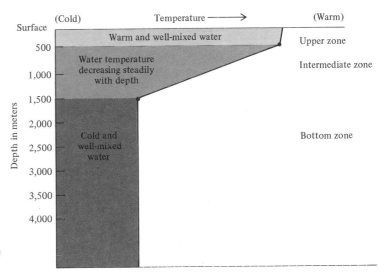

Figure 16.14 Generalized temperature variations in ocean waters.

Temperature

Ocean water temperatures also vary with latitude and depth. Surface water temperatures are highest, as would be expected, in equatorial regions and decrease steadily toward the poles. A typical profile through the ocean at a middle latitude can be divided into three zones on the basis of water temperature (Figure 16.14).

The uppermost zone, extending to depths of about 500 m, is the warmest and also is characterized by a relatively homogeneous temperature. The warmth comes from the absorption of heat from the sun, and the homogeneity is a result of waves and surface currents that act to mix the water thoroughly (see Chapter 15). The bottom layer is homogenized by deep currents but is extremely cold, often just barely above 0°C, the freezing point of water. In the intermediate zone, temperatures decrease steadily with depth, reflecting an attempt at heat balance between the upper and lower zones.

Density

The density (mass per unit volume) of seawater depends on several factors, including temperature, pressure, salinity, and amount of suspended sediment. Increases in salinity and amount of suspended sediment increase the mass of the water and thus increase its density. Hypersaline waters thus are significantly denser than brackish waters. Water expands (increases volume) when it is heated, and so warm tropical water is less dense than cold Arctic or Antarctic water.

Density is one of the controlling factors in ocean circulation because the movement of some water masses is initiated when dense water sinks beneath less-dense water. These movements will be examined later in this chapter.

OCEANIC CIRCULATION

The waters of the oceans are in a state of constant movement brought about by several different agents. We have seen already how winds produce waves and longshore currents (Chapter 15) and how the sun and moon help generate daily tidal movements (Chapter 2). In addition to these, there are well-established global oceanic circulation patterns in which water masses move both along the surface and at depth. These water movements—called surface currents and deep currents—play important roles in sedimentation, climate modification, and the production of marine food resources.

Surface Currents

In Chapter 14 we saw that earth's atmosphere has a well-established circulation pattern (see Figure 14.4). The prevailing wind system in an area exerts a force on the surface waters in the oceans that sets them in motion to produce the **surface currents** (see Figure 14.6). Wind is the basic driving force, but the actual direction of the currents is affected by the Coriolis effect (discussed for air movement in Chapter 14) and by deflection around large landmasses. The major surface currents are shown in Figure 16.15 and indicate that water masses can move great distances. We might expect surface current motions to be restricted to the shallower depths of the oceans, but there is evidence that some "surface" currents may extend downward to depths of 2000 or 3000 m.

The importance of these currents to humans is illustrated by one of the best known—the Gulf Stream. Surface waters in the North Atlantic are driven southward into equatorial regions by the northeast trade winds but then are deflected northward to move along the east coast of North America as the Gulf Stream. Warm equatorial waters thus move into the middle latitudes, and this warmth moderates what otherwise would be extremely severe climates in such places as Iceland and the British Isles.

Deep Currents

While surface currents basically are horizontal movements of water, there are currents in the oceans that involve significant vertical movements as well. These are called **deep currents** and are generated by differences in the density of water masses rather than by surface winds. The factors controlling water density—temperature, salinity, and turbidity—produce different forms of deep ocean density currents.

Hot and cold currents

Cold water at the poles sinks because of its density and flows along the bottom toward the warmer equatorial regions. In the section of the Atlantic Ocean shown in Figure 16.16, these movements are of the North Atlantic Deep

Figure 16.15 Map of the world's major surface currents. The currents in the northern hemisphere curve clockwise, while those in the southern hemisphere turn counterclockwise as a result of the Coriolis effect.

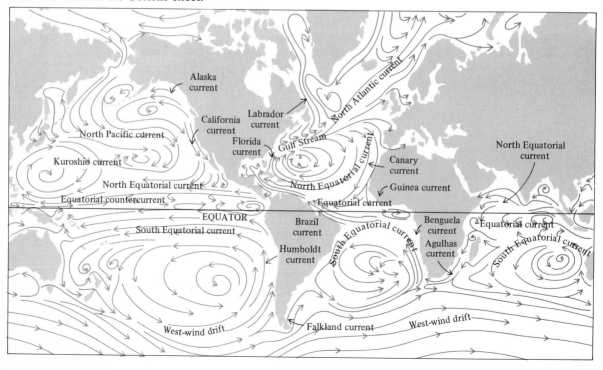

Figure 16.16 Vertical north-south section through the Atlantic Ocean showing deep current circulation resulting from density differences between the water masses.
  ABW = Antarctic Bottom Water
  AIW = Antarctic Intermediate Water
NADW = North Atlantic Deep Water
  MW = Mediterranean Water (saline)
(Modified from G. Neumann and W. J. Pierson, Jr., Principles of Physical Oceanography, *© 1966, p. 466, with permission of Prentice-Hall, Inc.)*

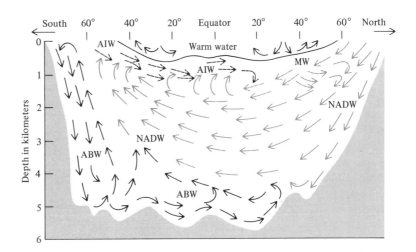

Water and Antarctic Bottom Water masses. This circulation ensures a mixing of ocean water and a homogeneously cold region near the sea floor as described earlier (see Figure 16.16).

Saline currents

The difference in density between highly saline and "normal" marine waters is responsible for the density currents that control water circulation in the Mediterranean Sea (Figure 16.17). The Mediterranean Sea is situated in a semiarid region (see Chapter 14), and evaporation has produced waters that are more saline (hence denser) than those of the Atlantic Ocean. At the Strait of Gibraltar, where the two bodies are connected, relatively light (normal salinity)

Atlantic Ocean water flows into the Mediterranean while denser saline Mediterranean water flows beneath it out into the Atlantic. The Mediterranean water sinks to about 1000 m in the Atlantic before being dissipated gradually (see Figure 16.17). This circulation pattern effectively brings about a complete change in the waters of the Mediterranean Sea in approximately 75 years.

Turbidity currents

Sediments deposited near the edges of the continental shelf become unstable when they are piled up more steeply than the angle of repose (see Chapter 10) or when they are shaken by an earthquake. They then move down the

Figure 16.17 Density current movement between the Mediterranean Sea and Atlantic Ocean. Relatively cool Atlantic water flows along the surface from the Atlantic into the Mediterranean. The water is warmed, evaporation increases, and the water becomes more saline and sinks. The warm saline Mediterranean water flows back into the Atlantic as a distinct density current which sinks to about 1000 m in the Atlantic.

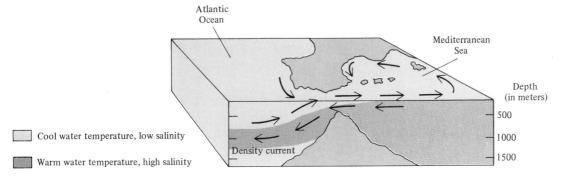

continental slopes as a **turbidity current,** a cloudy mass of water and suspended sediment that sinks because it is denser than its surroundings (Figure 16.18). Turbidity currents are important mechanisms for carrying sediment from the continental margins into the deep oceans. They are the source of most of the sediment that has built up the continental rises, covered some of the abyssal hills, and spread out onto the abyssal plains (see Figure 16.7). In addition, abrasion of the sea floor by particles suspended in the turbidity currents may bring about the carving of submarine canyons and their extensions across the continental rises.

THE FLOORS OF THE OCEANS

Modern oceanographic techniques permit scientists to study not only the submarine topography and water characteristics of the ocean basins but also the nature of the materials underlying the sea floors. In all oceans there is a simple layered structure, which is shown schematically in Figure 16.19. Beneath a relatively thin surface veneer of unconsolidated sediment and lithified sedimentary rock, there is an intermediate zone of interlayered sedimentary

rock and basaltic lava flows. Beneath this lies a uniformly basaltic basement in which the basalts are very similar to those now found at the ocean ridges. The sediment and sedimentary rock that cover the basalt are not nearly so uniform in composition as the basalt because of the wide variety of factors governing ocean sedimentation.

Ocean Sedimentation

The samplers and piston cores described earlier in this chapter have provided data on the uppermost portion of the ocean bottom at a great number of sites. Information on deeper deposits has been obtained at a few hundred sites by deep drilling (DSDP). Additional information on deeper deposits is obtained from **seismic** techniques, in which research ships set off much more powerful energy sources than those used for PDR work. This energy not only is reflected from the sea floor but also penetrates below it and is reflected by different layers up to a few hundred meters below the ocean floor. Continuous seismic studies can produce a "cross section" showing both the stratification and the structure of the sediments and rocks beneath the ocean floors (Figure 16.20).

Figure 16.18 Turbidity currents and the deposition of turbidites. Slumping of sediment on the upper continental slope results in the formation of a dense suspension of sediment and water which flows down the continental slope and across the abyssal floor. Particles settling out of the turbidity current produce a turbidite layer—a graded bed in which the average size of the particles of sediment decreases from the base to the top. The turbidite overlies a layer of pelagic ooze deposited under quiet water conditions in the area.

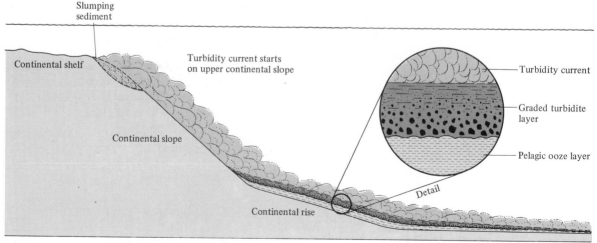

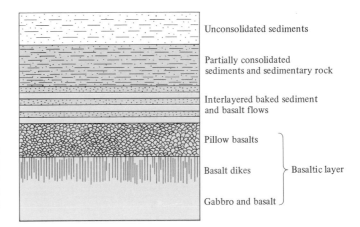

Figure 16.19 A generalized geologic cross section showing the layered structure of the materials below the ocean floor. The thickness of the sediments and the sedimentary rock layers in any area depends on sediment type and the rate of sedimentation.

Oceanographers recognize three major classes of ocean sediments: terrigenous, biogenic, and authigenic. **Terrigenous sediments** are clastic sediments of several types that are derived from the continents or oceanic islands. Biogenic sediments are formed by the accumulation of skeletons of marine organisms on the sea floor, and authigenic sediments form in the sea by chemical precipitation (see Chapter 6).

Terrigenous sediments

Most terrigenous sediments are concentrated on the continental shelves flanking the large landmasses and on the sea floor surrounding oceanic islands. Rivers bring sediment to the oceans, where it is redistributed by waves and currents. Terrigenous sediments derived from tropical environments generally are fine-grained because of the effectiveness of chemical weathering in reducing grain size, whereas those from colder latitudes generally are coarser-grained because of the incomplete nature of chemical weathering and the input of coarse glacial debris. Volcanic islands supply large quantities of tephra to the adjacent sea floors. The tephra deposits are thickest on the downwind and down-current sides of the islands. Many volcanic islands in tropical areas are surrounded by coral reefs, as described in Chapter 15, and wave erosion supplies broken reef debris to the adjacent sea floor. Complexly interlayered lava flows, tephra, and reef debris thus surround many islands.

Relatively coarse-grained terrigenous sediment also can be transported to the deeper parts of the ocean basins. In high latitudes, massive sediment-laden icebergs break off the seaward edges of glaciers and float out to sea, where they eventually melt. As they melt, the poorly sorted sediment they carry is deposited in the oceans. The deposition of sediment from melting icebergs is called **ice-rafting.** Terrigenous sediments also may be transported down the continental slopes and onto abyssal plains by turbidity currents, as described earlier. The deposits of turbidity currents are known as **turbidites.**

Winds may blow fine-grained terrigenous particles far out across the oceans. Surface and deep ocean currents can carry these and other fine-grained particles great distances from where they first entered the ocean. Eventually these fine particles settle through the water and are deposited on the ocean bottom. Deposits which form from organic and inorganic particles and shells of microorganisms which settled through the water are called **pelagic** deposits. Marine geologists also refer to these deposits as **pelagic oozes** in reference to the consistency of the sediment when it is extruded from cores. Oxidation of the iron in pelagic deposits produces a reddish-brown color, and as a result the pelagic oozes are also referred to as **brown clays.**

Biogenic sediments

Biologic activity provides a significant portion of the sediment in many areas of the oceans,

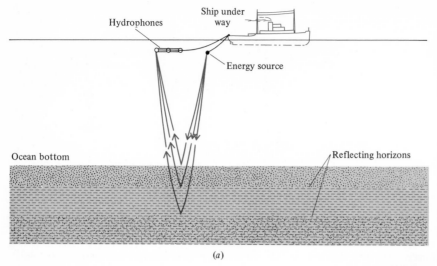

Hydrophones

Ship under way

Energy source

Ocean bottom

Reflecting horizons

(a)

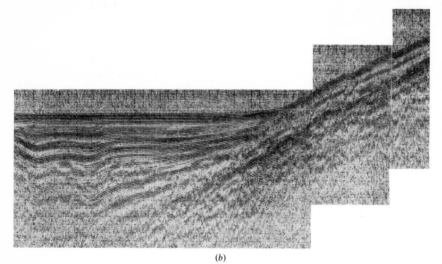

(b)

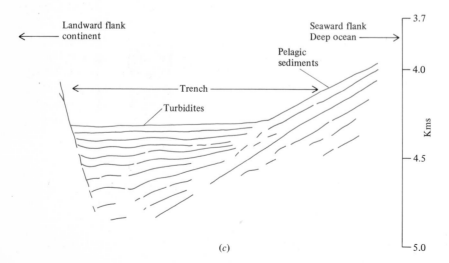

Landward flank
continent

Seaward flank
Deep ocean

Pelagic
sediments

Trench

Turbidites

Kms

3.7

4.0

4.5

5.0

(c)

Figure 16.20 Seismic reflection studies of the structure of the ocean floor. *(a)* Seismic reflection study in progress. Survey ship emits sound waves powerful enough to penetrate the bottom and be reflected back from the subsurface layers. The reflected waves are picked up by a receiver towed behind the ship, analyzed by computer, and shown visually as a continuous seismic reflection profile. *(b)* Seismic reflection profile of an abyssal plain area in the bottom of a deep sea trench. The subsurface horizons are delineated in a tracing of the profile. *(c)* This interpretation of the seismic reflection profile in *(b)* delineates the various subsurface horizons. The turbidity current deposits from the upper part of the continental slope (left) are prevented from reaching the ocean basin (right) by the trench. *(Redrawn with permission of David A. Ross, Woods Hole Oceanographic Institution.)*

particularly in areas where organic activity is high and the input of terrigenous sediment is low. The biogenic sediment consists of the skeletons of small organisms that are composed of either calcareous, $CaCO_3$, or siliceous, SiO_2, material.

Calcareous oozes are extremely fine grained sediments composed of the external skeletons and skeletal parts of calcareous organisms (Figure 16.21). Calcareous oozes are abundant in the South Pacific, Indian, and Atlantic Oceans except for the colder-latitude areas. This does not mean that these are the only places where calcareous-shelled organisms live, because the abundance of calcareous forms in surface waters is not necessarily reflected in the bottom sediments.

The calcium carbonate from which these organisms build their skeletons becomes more soluble as pressures increase and water temperatures decrease. Both increased pressure and colder water are characteristic of deep ocean regions (Figure 16.14) so that when an organism dies and its skeleton sinks, the calcium carbonate is dissolved gradually. At a depth known as the **carbonate compensation depth,** pressures and temperatures are such that almost all the calcareous material is dissolved before it can reach the sea floor. The compensation depth varies somewhat from ocean to ocean but is usually between 3.5 and 4.2 km. In areas of deeper water, then, even if there is a large community of calcareous-shelled organisms living at the surface, calcareous biogenic sediments may be completely absent from the sedimentary record.

Siliceous oozes are fine-grained biogenic sediments composed of siliceous organisms (see Figure 16.21). Siliceous oozes occur in great volumes in the colder waters of high latitudes, partly because a large amount of nutrients is contained in this water.

Authigenic sediments

Perhaps the most intriguing ocean sediments are the authigenic sediments (see Chapter 6) that form on the sea floor through chemical reactions between ions dissolved in seawater and materials on the bottom. Authigenic sediments are more abundant in areas where terrigenous sedimentation rates are relatively low. The most common

Figure 16.21 Siliceous and calcareous microfossils found in ocean sediments. *(Scripps Institution of Oceanography.)*

authigenic deposits are **manganese nodules,** which are rounded masses rich in manganese that cover wide areas of the sea floor (Figure 16.22). Similar manganese-rich materials form coatings on rocks exposed at the ocean bottom. In addition to the manganese, significant concentrations of iron, nickel, cobalt, and copper are found both in the nodules and in the coatings.

Figure 16.22 Manganese nodules on the floor of the South Central Pacific Ocean at a depth of 5288 m. *(Smithsonian Oceanographic Sorting Center for the National Science Foundation.)*

Apparently manganese nodules form by precipitation from solution, with preferential deposition on hard objects already on the sea bottom. The manganese is deposited in several stages, as indicated by the series of concentric layers that is observed when the nodules are sliced open. Growth rates appear to be highly variable and are reported by researchers to range from less than 1 mm per 1 million years to more than 100 mm per 1 million years.

Phosphorite nodules, another type of authigenic deposit, contain up to 30 percent P_2O_5. These nodules form closer to the continental margin than the manganese nodules but also occur in areas of relatively low rates of terrigenous sedimentation.

Oceanic Sediment Facies

The type of sediment at any one location on the ocean floor reflects the relative input of terrigenous, biogenic, and authigenic sediments. The geographic distribution of the major oceanic sediment facies is shown in Figure 16.23. As we have come to learn more about the origin of modern oceanic sediment facies, we have been able to use uniformitarian principles to reconstruct the paleogeography of ancient oceans from their sedimentary rock record.

THE AGE OF THE OCEANS

The oldest fossils *ever* collected from DSDP cores are of only Jurassic age (roughly 180 million years old). Radiometric dating of the basaltic basement of the oceans has produced similar results. The basalts at the ocean ridges and on the Hawaiian-type islands where eruptions are continuing today are understandably young, but the oldest basalts ever recovered from the basement of the sea floor are also only Jurassic.

These results hardly could be more surprising. It seems that the features that cover more than 70 percent of the earth's surface have been here for only 180 million of the 4.6 billion years of the planet's history. Serious questions about the origin of the ocean basins and the evolution of the earth are raised, but one problem finally is solved. The static view of the earth (see Chapter 1) hardly can be considered valid if such major planetary features are so newly formed.

ORIGIN OF THE OCEANS

If the oceans are indeed so young, how did they form so late in the history of the earth? This involves two separate questions: Where did the water in the oceans come from? How did the present ocean basins form?

There is general agreement among earth scientists as to the origin of oceanic waters. During the early stages of earth history, volcanic eruptions brought water into the atmosphere just as they do now. When the surface of the planet cooled sufficiently, most of this water accumulated on the surface to form the hydrosphere. Over the billions of years that the hydrosphere has existed, soluble ions freed by chemical weathering have been carried to the oceans, converting them to the saline waters that we have today.

There is less agreement as to the origin of the ocean basins, although most oceanographers now favor the plate tectonic model over the others proposed in Chapter 1. Indeed, the plate tectonic model is a development of the past 20 years and owes much of its support to data obtained during exploration of the oceans. As a result, plate tectonics is the only model that deals specifically with features of the ocean basins as well as those of the continents. It is the first really *worldwide* tectonic model. Theories of an expanding, shrinking, or pulsating earth were proposed to explain features on the continents before we had our present understanding of the features of the ocean basins. Nevertheless, we should review and evaluate the four remaining theories.

Four Models for the Origin of the Ocean Basins

According to the expanding earth model, ocean basins form when the earth's surface stretches and then breaks, forming topographically low areas (oceans) and high areas (continents and mountains) (refer to Figure 1.5). Ocean ridges would represent the tension cracks through which basaltic magma could move upward easily. Continental margins would have resulted from the ripping apart of large continents during expansion.

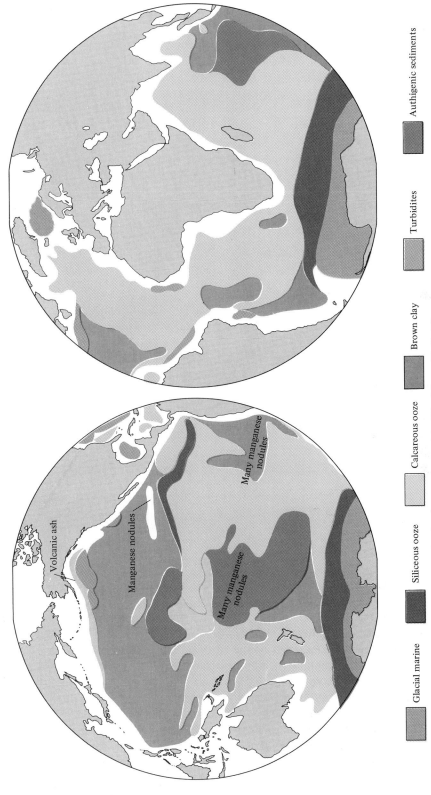

Figure 16.23   Distribution of sediment facies in the oceans.   *(Modified from F. P. Shepard, Submarine Geology, 3d ed., Fig. 14.3, © 1973 by F. P. Shepard, with permission of Harper & Row, Publishers, Inc.)*

The Galapagos Rift
Window on the Earth's Interior

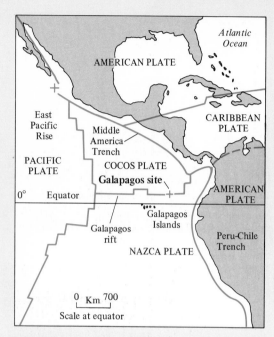

Figure 1. Index map of East Central Pacific region showing location of Galapagos rift study area.

A recent oceanographic expedition, Galapagos II, has provided a fascinating insight into the biologic, chemical, and geologic processes along ocean ridges. The researchers utilized the research submarine *ALVIN* (Figure 16.4) to study a portion of the Galapagos Rift, part of the ocean ridge system in the East Central Pacific (Figure 1). Scientists in the *ALVIN* found that conditions on the ridge were quite different from what they had expected. The sea floor was not barren but was covered with great numbers of animals.

How could such concentrations of life survive in such cold, deep (2.5 km), and dark waters? The answer lies in the relationship between the bottom water and the rock below the sea floor. Measurements and direct observations indicated that cold bottom waters are sinking into rock fractures on the sea floor. As the waters descend, they become heated and dissolve mineral matter from the hot rocks below the surface. The hot waters eventually return to the surface, rich in minerals and gases such as hydrogen sulfide, and emerge through fractures and vents on the sea floor.

This heated gas-rich water makes the area productive. Some of the bacteria which normally are present in the ocean can utilize the hydrogen sulfide in metabolism, and they multiply in great numbers in areas where this warm and gas-rich water emerges from the sea floor. These bacteria-rich areas apparently attract larvae of larger organisms drifting by in the current, and soon the bottom is covered with unusual varieties of worms, crabs, mussels, and clams. For example, huge clams, 10 by 30 cm, grow in

The shrinking earth model envisages ocean basins as large downbuckled parts of the earth's outer surface. Topographically high features such as ocean ridges, island arcs, and Hawaiian-type island chains, as well as low areas such as ocean trenches, could represent differential crumpling within the basins. The pulsating earth model explains the different physiographic features by a combination of expansion and con-

traction. The basins themselves may form during expansion, along with the ocean ridges, but other features such as island arcs probably formed during compressional (shrinking) phases.

According to the plate tectonic model (Figure 16.24), the ocean basins are constantly in a state of simultaneous growth and destruction. Ocean ridges are places where new basaltic ocean floor basement is brought to the surface

great numbers around the warm-water vents (Figure 2). The Galapagos II scientists had discovered a whole new ecosystem, one based on chemical syntheses by bacteria. This showed that sunlight is not always the main source of energy for life.

Perhaps the most spectacular accomplishment of the study occurred when scientists observed metals being deposited in mounds around the vents as the emerging hot (350°C) mineral-rich waters cooled rapidly by mixing with the cold (2°C) bottom waters (Plate 21). Examination of a piece of the metallic crust showed that it had a dull exterior composed of zinc sulfide and a bright golden interior composed of iron sulfide (Plate 22). In some areas the metallic crusts were interlayered with pelagic sediments, suggesting that the metal precipitation is an intermittent process.

The researchers suggest that the crust at the Galapagos Rise breaks apart and is the site of extensive basaltic extrusion about every 10,000 years. As the new sea floor cools, cracks develop through which seawater moves into the hot rocks below, eventually reappearing at the surface as hot, mineral-rich waters. Bacteria which thrive on the hydrogen sulfide in the waters feed a variety of organisms which populate the sea floor. As the hot solutions cool rapidly from contact with the cold bottom waters, layers of metallic minerals are deposited around the vents. The newly formed ocean crust moves away slowly from the ridge crest and is buried progressively by pelagic sediments. After a period of time, basaltic eruptions create new ocean crust at the Galapagos Rift, and the cycle starts again.

Figure 2 Crab and clam populations around active hot-water vents on Galapagos Rise. *(Woods Hole Oceanographic Institution.)*

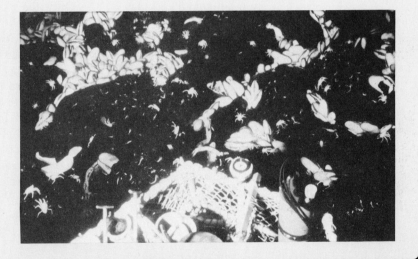

from the mantle. Forces in the mantle spread the ocean basins wider and wider while new basalt rises to make new ocean floor. In contrast, island arcs and trenches are places where the ocean floor is returned to the mantle by the process of subduction. The entire process is much like a gigantic natural recycling mechanism involving the ocean floor.

As ocean floors (plates of oceanic crust) pass over fixed "hot spots" (sources of heat from the mantle), there is temporary local melting, forming a volcanic island. When the moving plate isolates the volcano from the hot spot, the volcano becomes extinct while a new volcano is forming on the part of the ocean bottom which currently is over the hot spot. This mechanism eventually can form a basaltic island chain, such as the Hawaiian Islands, in which

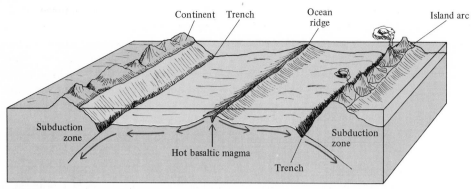

Figure 16.24   The plate tectonic model for the origin of the oceans. Sea-floor spreading involves rifting and extrusion of basalt at the ocean ridges to produce new ocean floor materials. The cool older ocean floor is subducted into the asthenosphere at ocean trenches.

the ages of the islands become greater away from the island currently over the hot spot (see Figure 16.13). The extinct volcanoes in the basaltic island chain are reduced by weathering and coastal erosion at the same time that the cooling volcanic rock is contracting and subsiding beneath the sea. If coral reefs can grow on the sides of the subsiding islands, it is possible to form atolls eventually (Figure 15.14).

Evaluating the Models

The expanding, shrinking, and pulsating earth models have severe drawbacks in their explanations of the topographic features of the ocean basins. For example, an expanding earth does not explain features like seamounts, guyots, and Hawaiian-type islands satisfactorily because these features require subsidence beneath the sea during their evolution. On an expanding earth, a finite amount of seawater would have to fill an ever-increasing volume of ocean basins, and there should be a general lowering of sea level as a result, not a drowning of formerly high features.

A shrinking earth would require significant folding of sediments and basement rocks near the margins of the oceans, but seismic data suggest that most of the sediments there are relatively undeformed. Furthermore, the sharp shelf break and the nature of the continental margins suggest that they were formed by tension rather than compression.

A pulsating earth would produce ocean ridges during expanding phases and island arcs during contracting phases. However, earthquake and volcanic activity indicate that both arcs and ridges are active today.

Only the plate tectonic model, built as it was on data obtained during oceanographic research, seems to explain all the features of the ocean basins satisfactorily. But it need not be totally correct, and we must collect data constantly to revise the theory. To check, we must look at the state of deformation in the oceans to see whether it fits that predicted by the plate tectonic model and how it compares with deformation on the continents. Only then can we see whether a single model can explain the topographic and geologic features of both continents and oceans. We will examine deformation in Chapter 17 and analyze mountain-building and ocean-forming processes in Chapters 18 and 21.

OCEAN RESOURCES

With our new technology and knowledge of the oceans, we are beginning to turn more and more toward the oceans as sources of the things we need to survive: food, water, raw materials, and energy.

Oceanic food resources, such as fish, shellfish, crustaceans, and certain varieties of marine plants called algae, have been utilized since the times of the earliest humans. Because the concentration of nutrients from land sources is greatest on the continental shelves, they provide over 90 percent of our oceanic food resources, even though they occupy only about 8 percent of the surface of the ocean basins. In comparison, the deep oceans are relatively barren, except for certain parts of the ocean ridge system (see the Perspective on the Galapagos Rift in this chapter).

Unfortunately, this 8 percent of the oceans is receiving nearly 100 percent of oceanic pollution. Oil spills and the dumping of waste materials near the continents provide the most dramatic examples of pollution, but far more pollutants are supplied to the oceans as dissolved ions and particles suspended in streams. Continued degradation of ocean waters poses a serious threat to all levels of the marine food chain and thus to an important source of food for us.

Desalinization of salt water to produce drinking water technically is possible but is very expensive because energy is needed to heat the seawater until it evaporates, producing freshwater vapor and a residue of dissolved materials. Thus this process is practical only where there are limited natural supplies of freshwater and a plentiful supply of energy, as in the Middle East.

The oceans are *potentially* a source of incredibly large amounts of raw materials if we can learn to extract the elements dissolved in seawater and mine the manganese and phosphorite nodules on the sea floor. Only a tiny percentage of the substances found in seawater (common salt, for example) currently are extracted from it because of the enormous expense involved. The low concentrations of most elements in seawater require that great quantities of water be processed at an enormous expenditure of energy (and money). This comes at a time when our energy resources already are strained. Prior to the 1960s, for example, large amounts of magnesium and bromine were extracted from seawater, but today they are recovered less expensively by drilling into subsurface pools of brine on the continents. In addition, the jurisdiction over these ocean resources is currently a matter of great controversy.

We have also turned to the ocean in the search for new energy resources. Oil and gas from beneath the continental shelves already have made a contribution to our society, having reached our homes and automobiles from the North Sea, the Gulf of Mexico, and the north slope of Alaska.

SUMMARY

Modern methods of oceanographic research have given us a picture of the oceans and ocean basins that is quite different from what was anticipated by early researchers. Continents are surrounded by gently sloping surfaces covered by shallow water (the continental shelves) that pass seaward into more steeply sloping regions called continental slopes. The deep, flat portions of the oceans (the abyssal plains) contain small abyssal hills that appear to be extinct volcanoes and large linear mountain ridges called ocean ridges. The ridges are offset but not completely truncated by cross-cutting oceanic fracture systems. The deepest parts of the oceans, the oceanic trenches, are found close to the continental margins and are associated with either volcanic island arcs or continental mountains such as the Andes. Island arc-trench pairs often are separated from the continents by shallow seas called marginal seas. Basaltic islands may occur singly as seamounts, in submerged flat-topped form as guyots, or in large clusters of shield volcanoes as Hawaiian-type islands.

The seawater that fills the oceans varies with respect to several properties, including temperature, composition, and density. Average seawater contains 35 ppt of dissolved solids in the form of ions, of which sodium and chlorine are by far the most abundant. All the gases in the atmosphere also are found in solution in the oceans, particularly oxygen and carbon dioxide. The density of seawater increases as salinity and suspended sediments increase and as temperature decreases. Water masses may rise or sink past others, depending on their relative densities.

Ocean water is in a constant state of circulation that involves both surface and deep waters. Surface water movement is driven by the world's prevailing winds, while the deep currents are a result of density adjustments. Cold, highly saline, and highly turbid waters initiate oceanic circulation in some areas by sinking beneath lighter water of more normal salinity.

The ocean floors are underlain by a layered section composed from top to bottom of unconsolidated sediment, sedimentary rocks, interlayered sedimentary and volcanic rocks, and basaltic lava flows. The sediment on the sea floor comes from a continental source (terrigenous), from biologic activity (biogenic), or from chemical precipitation (authigenic). Coarse terrigenous sediments reach their ocean deposition sites by stream transport, ice-rafting, and turbidity currents. Finer particles, some blown into the ocean by winds, are carried in suspension by surface and deep currents for long distances. Eventually, these fine particles settle out of suspension and are deposited as pelagic ooze. Biogenic sediments consist of fine-grained calcareous and siliceous skeletal remains of near-surface organisms that die and sink to the bottom, accumulating as calcareous and siliceous oozes. Calcareous organisms tend to dissolve in the deep cold waters of the oceans at the carbonate compensation depth. Therefore, fossil calcareous shells are not found as abundantly in the ocean sediment record as their present abundance in surface waters would suggest. Authigenic sediments such as manganese and phosphate nodules form by precipitation in areas of the ocean basins where the rate of sedimentation is known to be relatively low.

Dating of sediment layers by fossils and lava flows by radiometric means shows that the ocean basins are young—no older than Jurassic. Of the five hypotheses postulated in Chapter 1, only the plate tectonic model explains the age and physiography of the oceans adequately.

Oceans are potentially a vast reservoir of food, mineral, and energy resources if we can develop the technology necessary to extract what we need cheaply and without damaging the fragile ocean environment.

QUESTIONS FOR REVIEW AND FURTHER THOUGHT

1. Construct a profile of a typical continental margin. Which features in this profile are the result of continental processes and which result from marine processes? Which are depositional and which are erosional?

2. How would the profile constructed in Question 1 have been different if drawn at a different latitude?

3. Several physiographic features of the ocean basins stand up in strong positive relief: ocean ridges, island arcs, seamounts, and Hawaiian-type islands. By what mechanisms do these features form?

4. Describe the properties that permit distinction between different water masses. What accounts for the global variation in these properties?

5. Even though shelf valleys, submarine canyons, and valleys that cut through the continental rises all may be aligned, they probably were not all formed by the same process. Explain.

6. During the Pleistocene Epoch, sea level was lowered more than 100 m below present sea level. How would this change have affected both the quantity and areal distribution of different types of ocean sediments in the Pleistocene?

7. Discuss the formation of each of the following types of ocean sediments:
   a. Manganese nodules
   b. Brown clay
   c. Ice-rafted sediments
   d. Calcareous ooze

8. What factors control the abundance of each of the sediments described in Question 7 at any one place on the ocean floor?

9. There is much more gold dissolved in the oceans than there is stored in solid form at Fort Knox. Why then is most of our supply of gold still coming from mines on the continents?

ADDITIONAL READINGS

Beginning Level

Ballard, R. D., and F. L. Grassle, "Return to the Oaes of the Deep," *National Geographic,* vol. 156, no. 5, November 1979, pp. 689–705.
(Contains a description of the dive by the submarine *ALVIN* on the Galapagos Rift in the Pacific. Contains spectacular underwater color photographs of the unusual features observed on this ocean ridge)

Davis, R. A., Jr., *Principles of Oceanography,* 2d ed., Addison-Wesley, Reading, Mass., 1977, 505 pp.
(A well-balanced treatment of biological, chemical, physical, and geological oceanography. Includes particularly good detailed discussions of processes, life forms, and features in the nearshore portions of the oceans)

Heezen, B. C., and C. D. Hollister, *The Face of the Deep,* Oxford University Press, New York, 1971, 659 pp.
(A collection of outstanding bottom photographs of animals and surface features on the ocean floor. The accompanying text and diagrams describe the origin of the features)

Menard, H. W. (ed.), "Ocean Science," reprints from *Scientific American,* W. H. Freeman, San Francisco, 1977, 307 pp.
(A collection of articles dealing with the history of oceanography, marine geology, the sea and its motions [physical oceanography], and marine life and resources)

Nierenberg, W. A., "The Deep Sea Drilling Project after Ten Years," *American Scientist,* vol. 66, no. 1, January–February 1978, pp. 20–29.
(Describes the techniques used to obtain long cores of ocean sediments and rocks and what these data and other types of exploration techniques have told us about the structure and evolution of the ocean basins)

Ross, D. A., *Introduction to Oceanography,* 2d ed., Prentice-Hall, Englewood Cliffs, N.J., 1977, 438 pp.
(A well-balanced treatment of all aspects of oceanography, along with a discussion of marine resources, law, and pollution. Reference list contains many articles dealing with more advanced treatments of the topics covered in the book)

Advanced Level

Shepard, F. P., *Submarine Geology,* 3d ed., Harper & Row, New York, 1973, 517 pp.
(Detailed treatment of geological oceanography. The best place to look first for the geology of a particular ocean)

CHAPTER 17

Deformation of Rocks

In the preceding chapters we have seen that the earth is a dynamic planet whose surface is in a state of constant change brought about by weathering and erosion. The earth's interior is as dynamic as its surface. Internal changes bring about the eruption of volcanoes, the uplift of new mountain ranges, and the opening of new ocean basins. All hypotheses dealing with the origin of mountains and oceans recognize that these internal changes must involve intense deformation of rocks in the crust and mantle.

Fortunately, rocks preserve a record of their deformational history in what are called **geologic structures**. The most important of these structures are **folds** (Figure 17.1), flexures that form when rocks bend and buckle in response to internal earth forces, and **faults** (Figure 17.2), fractures that form when rocks break and the materials on opposite sides of the fracture are displaced from their original positions. In this chapter we first will learn how rocks deform and how to interpret their deformational histories and then use this knowledge to evaluate the different mountain-building hypotheses.

HOW DO ROCKS DEFORM?

In our experience, rocks are solid, hard, and brittle. When struck with a hammer, they break and splinter. We might expect that natural forces would produce similar breaking; this does occur, as shown in Figure 17.2, but the folded rocks shown in Figure 17.1 have not behaved brittlely

at all. Instead, they have behaved ductilely by flowing like children's modeling clay while being bent. How can solid rock behave in a ductile manner if it is brittle? Can rock behave brittlely in some circumstances and ductilely in others? There are many instances in which this appears to be the case. What is it that determines whether a rock will behave in a brittle or ductile manner, whether it will be faulted or folded?

The Importance of Time

Everyday household events give clues to the behavior of rock during deformation and show that *time* is an important factor. A straight clothes pole or flat bookshelf becomes warped permanently if it is left heavily loaded for a long time. Glass panes become thicker at the bottoms of old windows because gravity causes the ions in the glass to move downward, and over a long enough period of time the change in shape is noticeable to the naked eye. We saw in Chapter 13 that solid ice in a glacier flows under the influence of gravity if enough time is allowed for the process to proceed.

When forces act on wood, steel, glass, and ice over long enough periods of time, these materials behave enough like fluids to change their shape. Their constituent ions change position relative to one another, causing a slow flow of material which results in the new, deformed shape. The amount of time needed for the movement of ions ranges from a few weeks for wood to a few hundred years for glass and centuries or

Figure 17.1 Large folds north of Borah Peak, Idaho. View is looking northwest. *(Courtesy of John S. Shelton.)*

thousands of years for ice. Given still longer periods of time and more intense forces, rock material also can flow in the solid state and thus deform in a ductile manner.

Several different processes, or mechanisms, are involved in the movement of rock material during ductile deformation (Figure 17.3). One

process, called **intragranular slip,** involves minute displacements along atomic planes within a single mineral crystal or grain (Figure 17.3*a*) and is the same process discussed in Chapter 13 by which glacial motion is accomplished partially. Individual displacements are far too small to be seen, but, when multiplied billions of times for all

Figure 17.2 Small-scale faults cutting sandstone beds exposed in a roadcut north of Mt. Carmel Junction, Utah. Amount of offset is indicated by the separation of the dark-colored beds in the two fault blocks at the left. *(Courtesy of John S. Shelton.)*

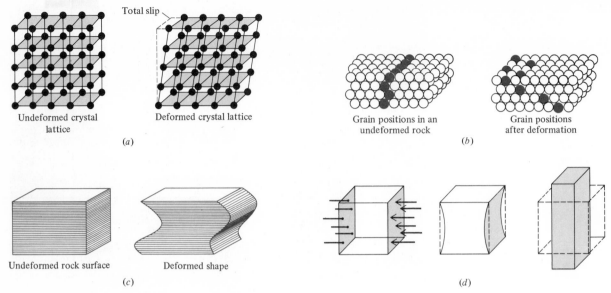

Figure 17.3 Mechanics of rock deformation. *(a)* Intragranular slip: Very small dislocations along atomic planes (color) bring about a change in the shape of a single grain. *(b)* Intergranular slip: Mineral grains shift positions as forces are applied. Colored grains are shown for reference. *(c)* Small-scale movement along parallel planes in a rock, similar to slip within a deck of cards. *(d)* Solution and precipitation caused by pressure. As pressure is applied to an undeformed mineral (left), the mineral dissolves in the zone of greatest pressure (middle, color) and then precipitates in the areas of least pressure to produce changes in shape.

the affected atomic planes in all the affected mineral grains in a rock, they produce visible results.

Another process, called **intergranular slip,** involves movement of entire grains past one another in rock (Figure 17.3*b*). Slippage in this process is on a much larger scale than displacements within mineral structures, and it takes advantage of weaknesses in the cohesive properties of a rock. Displacement on an even larger scale takes place in some rocks along closely spaced parallel planes (Figure 17.3*c*). This form of slippage is similar to the sliding of cards past one another in a deck, and it ranges in scale from microscopic displacements to those clearly visible with the naked eye.

Movement of material also can be brought about by a fourth process, known as **pressure solution** (Figure 17.3*d*). Pressure applied to a rock can cause minerals to dissolve partially along grain boundaries where the pressures are felt most intensely and to reprecipitate in spaces where the pressures are felt the least. The entire

rock does not dissolve, but enough material can be transferred in this manner to cause changes in the rock's shape.

During folding, any combination of these mechanisms may be active in a sequence of rocks. Slippage, if present, is minor, and the general coherence of a rock is not changed. During brittle deformation, a rock loses cohesiveness and breaks along planes of weakness. These planes are called **fractures** if oriented randomly, **joints** if oriented systematically, or **faults** if the rocks on opposite sides have moved relative to one another. This kind of deformation occurs when deforming forces overcome grain-to-grain cohesion in a rock or cause breakage within minerals. The columnar joints shown in Figure 4.11 are produced by the shrinking of a cooling igneous rock, and there is essentially no displacement of material. Faults such as the San Andreas Fault of California are produced by far more intense forces and may involve displacement of hundreds of kilometers of rock.

Figure 17.4 Experimental deformation of marble. A cylinder of marble (left) is deformed in a laboratory by squeezing it in a press. At low confining pressure, the marble is shortened by brittle fracturing (center). At higher pressures, the marble behaves ductilely and flows while undergoing exactly the same shortening (right). *(Courtesy of M. S. Paterson, Australian National University.)*

Undeformed   Pressure = 200 atm   Pressure = 445 atm

Behavior of Rocks

We have seen something of *how* rocks actually deform. Let us turn to a study of the behavior of rocks during deformation to see what controls the type of behavior (ductile or brittle) that can occur. Recently, geologists have developed experiments designed to reveal the factors that control the way in which rock deforms (Figure 17.4). Rock samples are subjected to forces that compress them, as in Figure 17.4, or stretch them, and the results are studied carefully. The experiments are made on many different types of rock and are repeated under a variety of temperature and pressure conditions to help geologists understand the structures found in rocks.

The laboratory experiments indicate that the factors determining the behavior of rock during deformation are rock composition and texture, presence or absence of water, nature and intensity of the deforming forces, and temperature and pressure of the environment in which the deforming forces act. There are many factors to be considered, but two concepts summarize all the variables and greatly simplify the study of rock behavior during deformation: stress and strain.

Stress

Forces acting to deform a rock are opposed by the internal strength of the minerals present and by the cohesive forces that hold mineral grains or crystals together in the rock (Figure 17.5). We may think of a rock as exerting internal forces against the external deforming forces; wherever the opposing forces interact, the rock is said to be in a state of **stress**. The

Figure 17.5 A rock is in a state of stress when the internal cohesiveness of the rock (black arrows) resists external forces (color arrows).

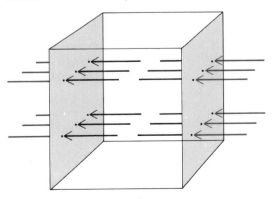

magnitude of the stress indicates the intensity of the deforming forces and is measured in units of force per unit area over which the force is applied: kilograms per square centimeter, or pounds per square inch. If the external forces overcome the internal cohesiveness, deformation takes place.

We shall consider three types of stress, classified according to directional properties: compressional, tensional, and shear stress (Figure 17.6). Compressional and tensional forces act at right angles to the deformed body. Opposing external forces directed toward one another produce **compressional** stresses (Figure 17.6*a*), while those directed away from one another result in **tensional** stresses (Figure 17.6*b*). **Shear** stresses are produced in a manner similar to what happens when the two blades of a pair of scissors glide past one another (Figure 17.6*c*). The opposing deforming forces produce stress in a plane that is *parallel* to their direction rather than perpendicular to it as in tensional and compressional stresses.

Strain

Rocks respond to stress by changes in volume or shape. The amount of change in shape is called **strain** and usually is measured in terms of the amount of change compared with the original shape. Stress thus measures the strength of the deforming forces as they interact with rocks, while strain measures the response of the rock to those forces.

Some rocks are deformed much more easily than others. Rocks which change shape easily are called **incompetent** rocks, and those which resist and change only under great stresses are called **competent** rocks. Whether a rock is competent or incompetent sometimes depends on the type of stress affecting it. Some rocks may resist changes under great tensional or compressional stresses but deform easily when subjected to small shear stresses. However, some rocks are incompetent under all types of stress and deform readily.

Behavior of rocks during deformation

Rocks may behave in very different ways during deformation, but experiments show that all behavior of rocks may be classified into one of three major types: elastic, ductile, or brittle.

Elastic behavior   **Elastic behavior** is similar to what happens when a weight is suspended from a spring or rubber band (Figure 17.7). If a small weight is hung from a spring, the spring stretches. If twice the weight is hung from the spring, twice the elongation will occur; three times the weight will produce three times the elongation. The strain (essentially the elongation) is exactly proportional to the amount of stress applied (the weight attached to the spring). This is one important characteristic of elastic behavior. Another is that once the weights are removed, the spring returns to its original length, and there is no evidence that deformation had once taken place. Elastic deformation thus is temporary and disappears once the stresses are relaxed. While Figure 17.7 shows elastic behavior induced by tensional stresses, the same type of behavior can be caused by compression.

Figure 17.6   Types of stress. Color arrows represent deforming forces. (*a*) and (*b*) Compressional and tensional stresses are perpendicular to the shaded surfaces. For simplicity, only the external forces are shown. (*c*) Shear stress is tangential to the shaded plane rather than at right angles to the external forces.

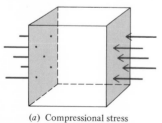

(*a*) Compressional stress

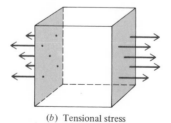

(*b*) Tensional stress

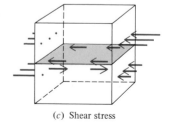

(*c*) Shear stress

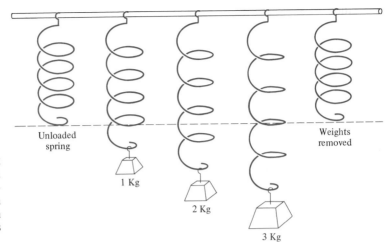

Figure 17.7 Elastic behavior during tension. During elastic behavior, each additional increment of weight produces a proportional amount of strain (lengthening of the spring). When weights are removed, the spring returns to its original position.

In that case the springs respond by becoming shorter.

Ductile behavior Ductile behavior is quite different from elastic behavior: The resulting deformation is permanent and can be observed even after stresses are relaxed. During ductile deformation, stress and strain are not proportional as they are in elastic deformation (Figure 17.8). A small stress may produce a given response; twice that stress may yield five times the strain; three times that stress, eight times the deformation, and so on. Ductile deformation is like the deformation of modeling clay. The sheets bend easily under deforming forces and remain bent after the forces are removed.

Figure 17.8 Ductile behavior. At first the spring behaves elastically, as in Figure 17.7. With additional weight, it deforms ductilely; when the weights are removed, the strain caused during ductile deformation is retained permanently.

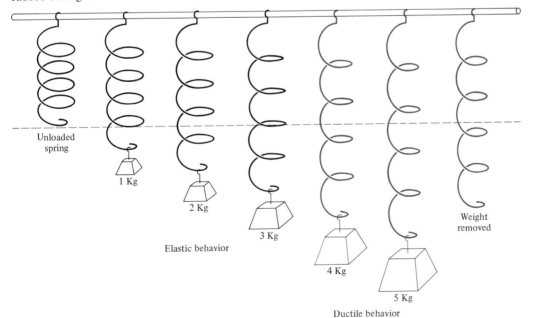

Brittle behavior   **Brittle behavior** differs from elastic and plastic behavior in that the rocks actually break as the spring does in Figure 17.9. All rocks have a maximum strain beyond which their internal cohesive forces can no longer hold them together. When a rock experiences more strain than this limit, it ruptures (breaks).

Experiments show that most rocks undergo all three types of behavior, depending on the amount of applied stress and the temperature and pressure conditions of the environment in which the stresses are applied. Most rocks behave elastically at conditions of low stress, but as stresses increase, they experience ductile behavior and eventually break. As weights were added to the springs shown in Figures 17.7, 17.8, and 17.9, they deformed elastically, then deformed plastically, and finally broke. Sandstones, shales, granites, and schists behave in the same manner.

In general, high temperatures and pressures promote ductile behavior and make brittle deformation less likely to occur. High pressures reinforce the internal cohesiveness of rock so that brittle deformation is unlikely. At high temperatures there is more heat energy available to the ions in mineral structures, enabling them to move freely in their structural positions. This makes the structures more flexible, promoting ductile behavior and also making it less likely for a rock to break. Rocks deformed at great depths, where temperatures and pressures are high, thus tend to be folded intensely rather than faulted. Rocks deformed close to the surface, where pressures and temperatures are lower, break more easily and commonly are faulted as well as folded.

A rock's texture and composition also play major roles in controlling which type of behavior will take place. Rocks composed of silicate minerals generally behave more competently than nonsilicate rocks such as limestone, marble, and evaporite minerals. Granites, gabbros, and gneisses act competently, resist deforming forces until high strains are built up, and as a result generally undergo little ductile deformation before breaking. Beds of halite, in contrast, require very little stress before they begin to deform ductilely. In some areas the weight of overlying rocks is enough to make halite behave extremely ductilely. The halite flows like a highly viscous fluid and actually can be injected into more competent surrounding rocks much in the way that magma is injected into host rock (Figure 17.10).

EXAMINATION OF DEFORMED ROCKS IN THE FIELD

Experiments such as those shown in Figure 17.4 help us understand how rock deforms and under what conditions different types of deformation can be expected to occur in different types of rock. Comparisons of naturally deformed rock with rock deformed in the laboratory can tell us about conditions of stress and strain in the earth, just as laboratory studies of mineral stability help us understand the temperatures and pressures of metamorphic reactions. If we can understand the nature of deformation and date its occurrence in the earth, we can begin to understand the major internal forces which cause the deformation.

What Can Be Seen?

Ductile and brittle deformation are permanent in rocks, and folding and faulting are records of

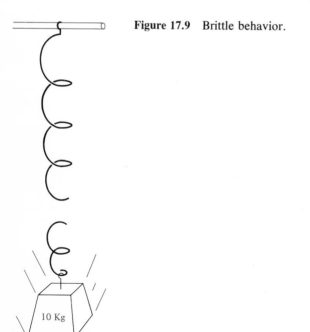

Figure 17.9   Brittle behavior.

10 Kg

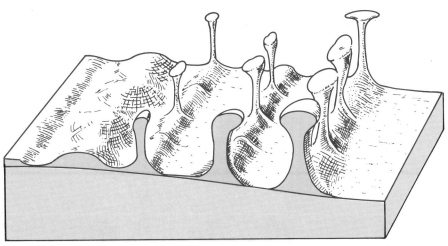

Figure 17.10 Salt domes. Schematic diagram showing shapes of salt dome structures. (The overlying rock has been left out for clarity.) With increased pressures from the overlying rock (from left to right in diagram), the ductile behavior of halite increases and the salt domes increase in size.

these types of behavior. However, we saw that deformation during elastic behavior disappears once stresses are relaxed so that rocks hardly would be expected to record any history of elastic deformation. However, on a large scale the earth does display elastic behavior.

The melting of vast continental glaciers some 10,000 years ago removed a large mass from the continents, comparable to the removal of weights from a spring. Just as the spring returns to its relaxed position, the continents now are rising by what is termed **elastic rebound** (Figure 17.11). The return to the unloaded or unstressed state takes fractions of a second for a spring but thousands of years for continent-sized masses. The actual amount of uplift is shown in Figure 17.11 by the color contour lines, each of which represents a specific amount of rise. The greatest rebound in North America is centered near Hudson Bay in Canada; presumably this is where the greatest amount of ice was concentrated. (See Figure 15.20.)

Evidence of ductile and brittle deformation is abundant and covers extensive periods of the far distant geologic past. We will concentrate on interpreting these folding and faulting events.

Measuring the Orientation of Deformed Rocks

In our study of deformed rocks, we will focus on sedimentary rocks because the degree to which their bedding has been changed from its original horizontal position is a measure of the amount of deformation to which they have been subjected (Figure 17.12). This does not mean that igneous and metamorphic rocks are never deformed but rather that sedimentary rocks possess a built-in indicator of deformation that is generally absent from igneous and metamorphic rocks.

Strike and dip

To describe the deformation of a sequence of sedimentary rocks, we must be able to describe the position of the bedding planes in space. Geologists use two terms to describe the orientation, or attitude, of a planar feature such as a bedding plane: strike and dip (Figure 17.13).

The **strike** of a planar geologic feature is the orientation, measured with a compass, of a horizontal line drawn on that plane. In Figure 17.13, the surface of the lake is horizontal so that the intersection of the water and the bedding plane is a horizontal line—the strike line. Strike di-

Figure 17.11 Rebound of eastern North America following glacial retreat. The Laurentian ice sheet that advanced over the northeastern and north-central United States during Pleistocene times was centered in the vicinity of Hudson's Bay. As it melted, a large mass was removed from the crust, and the earth is now rising upward in response to the unloading. Each contour line indicates the amount of uplift in feet above sea level since the area was deglaciated. *(Modified with permission from the National Atlas of Canada, 1972, Canadian Department of Energy, Mines, and Resources, MCR 1128.)*

rections are generally recorded as compass bearings relative to north or south. Thus, the strike of N53°E (read north 53 east) means that the horizontal line drawn on the bedding plane trends 53° east of north, as shown in the diagram.

The **dip** of a planar feature is a measure in degrees of how much the plane has been changed from a horizontal position. The amount of inclination of a bedding plane is measured downward from the horizontal with a clinometer and is recorded as the dip angle. Dip is always measured at right angles to the strike direction, but for the bed striking at N53°E there are two directions in which this measurement could be

made: Dips toward the northwest and southeast are possible. In the illustrated example, the bed is dipping to the southeast. A complete description of the attitude of a bed must thus include its strike, dip angle, and general direction of dip. The attitude of the bed shown in Figure 17.13 is thus N53°E 60°SE. In a qualitative way, beds may be described as gently, moderately, or steeply dipping.

The symbols shown in Figure 17.13 commonly are used on maps to indicate the strike and dip of planar features. The long bar points in the direction of strike, the short bar in the dip direction, with the amount of the dip angle in-

Figure 17.12 Bedding as an index of deformation. *(a)* Undeformed sequence of sedimentary rock. *(b)* Bending of bedding planes (a fold) indicates ductile deformation. *(c)* Offset of bedding planes indicates brittle behavior (a fault).

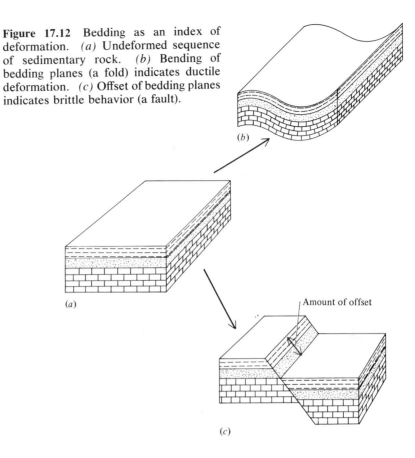

(b)

(a)

Amount of offset

(c)

Figure 17.13 Strike and dip.

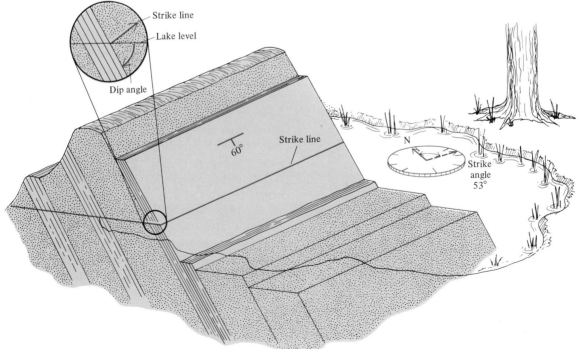

dicated. The orientation of any planar feature may be recorded by using strike and dip, and a geologist studying structures in deformed rocks records the strike and dip of fault planes, fractures, foliations, and joints as well as bedding planes.

Folds

Folds are a permanent record of ductile deformation and range in scale from those visible only with a microscope to those several miles across (look back at Figure 17.1). Folds are valuable to geologists because they preserve a record of deformation and deforming forces. A close look at the anatomy of a fold and at the different types of folds enables the structural geologist to reconstruct the stresses at the time of deformation.

Anatomy of a fold

Every fold consists of two sides, called **limbs,** which join at the area of maximum curvature (Figure 17.14). Imagine a sequence of sedimentary layers that have been folded as in Figure 17.14. A line can be drawn on every bedding plane that traces the area of maximum curvature and essentially separates the fold limbs. This line is known as the **fold axis.** A fold axis can be drawn on every bedding plane, and a plane called

an **axial plane** may be constructed that passes through all the axes, as shown in Figure 17.14.

Types of folds

There are basically two kinds of fold. Those folds in which the limbs dip toward one another and thus toward the fold axis are called **synclines,** and those in which the limbs dip away from one another and away from the fold axis are called **anticlines** (Figure 17.14). Several varieties of these two basic types are shown in Figure 17.15. Anticlines and synclines are classified further by the attitudes of their axial planes and axes, by the dips of their limbs, and by the angles between their limbs.

For example, if *axial planes* are vertical, the folds are said to be **upright.** If axial planes are horizontal, the folds are **recumbent.** Anything in between is an **inclined** fold. Folds whose *axes* are horizontal are said to be **nonplunging,** in contrast to those folds in which the axis is inclined at some angle to the horizontal (**plunging** folds). Note that upright and inclined folds may be either plunging or nonplunging.

Features observed in folded rocks

It is not difficult to study the deformation of an area such as that shown in Figure 17.1 because the nearly continuous exposures of bedrock give a clear picture of what has happened to the rocks. However, in other areas, it is not so clear. Where bedrock exposures are small and isolated because of sediment cover, geologists often must try to unravel complex deformational histories with very little data. Consider the area shown in Figure 17.16a. Bedding in the scattered outcrops maintains constant strike and dip directions, but this can be accomplished in several ways. The simplest explanation is shown in Figure 17.16b) where there is no folding at all but rather a simple tilting of the beds. A somewhat more complex history is postulated in Figure 17.16c, where tight isoclinal folding is indicated. Fortunately, intense folding produces small-scale features in the deformed rocks that enable geologists to recognize that something more than gentle tilting has occurred.

Figure 17.14   Anatomy of a fold.

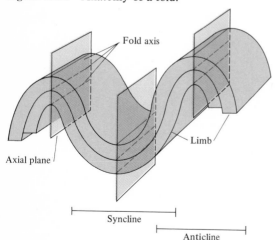

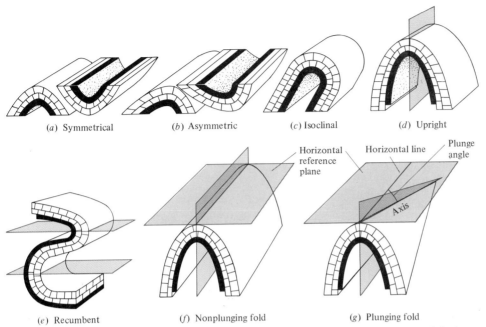

Figure 17.15 Geometric classification of folds based on *(a* through *c)* dip of limbs, *(d* and *e)* attitude of axial planes, and *(f* and *g)* attitude of fold axis. *(a)* Symmetrical: Limbs dip equally but in opposite directions. *(b)* Asymmetric: Limbs have markedly different dips. *(c)* Isoclinal: Limbs dip equally and in the same direction. *(d)* Upright: Axial planes are nearly vertical. *(e)* Recumbent: Axial planes are horizontal. *(f)* Nonplunging: Axis is horizontal. *(g)* Plunging: Axis is inclined; the angle of inclination is called the angle of plunge.

Cleavage In many folds, closely spaced planar fractures develop more or less parallel to the axial planes (Figure 17.17). These fractures are called **rock cleavage** and should not be confused with the cleavage of individual mineral grains. Rock cleavage affects entire beds or groups of beds, not just a single mineral crystal or grain. Rock cleavage planes develop on the limbs as well as at the axial planes of folds, and they commonly cut across the bedding planes, in-

Figure 17.16 Interpretation of geologic structure from scattered outcrops. *(a)* Aerial view showing scattered outcrops of sandstone (white) and shale (gray). Color lines are inferred contacts between rock types. *(b)* Interpretation 1: Sequence of interbedded sandstones and shales has been tilted gently. *(c)* Interpretation 2: Sequence of interbedded sandstones and shales has been folded tightly.

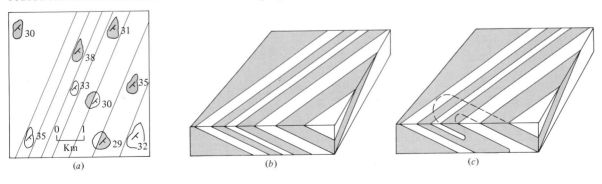

(a)

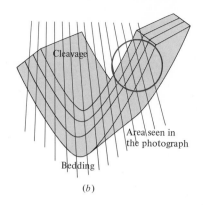

(b)

Figure 17.17 Cleavage in folded rocks. *(a)* Bedding and cleavage in alternating beds of sandstone and shale. Cleavage is well developed in the dark, shale layers, and less well developed in the lighter sandstones. *(G. K. Gilbert, U. S. Geological Survey.)* *(b)* The sketch shows that cleavage is associated with a large-scale fold.

dicating that they formed after the bedding. If cleavage had been observed in the outcrops shown in Figure 17.16a, the hypothesis shown in Figure 17.16c would have been preferred by a geologist mapping in the region.

Cleavage generally forms perpendicular to the direction of maximal compression, but the mechanism by which it forms is unknown. Some cleavage planes appear to mark the planes of slip in folds that have formed by the "deck-of-cards" mechanism shown in Figure 17.3c. Some cleavage seems to mark planes of pressure solution in rocks, but this mechanism does not appear to be possible in all instances. Perhaps more than one mechanism is involved.

Foliation We saw in Chapter 7 that regionally metamorphosed rocks often develop a foliation of platy minerals. When metamorphism accompanies deformation, this foliation commonly is parallel to the axial planes of the major regional folds because, as with cleavage, it forms at right angles to the dominant compressional stresses.

Foliation differs from cleavage in that it is a parallel alignment of minerals rather than a plane of fracture.

Boudins Interbedding of competent and incompetent types of rock commonly results in contrasting behavior during folding. Incompetent rocks flow easily, whereas the interbedded competent types cannot and rupture instead. The incompetent rock materials then flow into the breaks, producing features called **boudins,** from the French word for *sausage* (because of the resemblance of this feature to links of sausage) (Figure 17.18).

Outcrop patterns of folded rock When folded sedimentary rocks are subjected to erosion, individual beds trace out patterns on the land surface that help geologists identify the presence of folds (Figure 17.19). The effect of erosion on horizontal and tilted beds is shown in Figure 17.19a and b, on folded beds in Figure 19.17c and d. Rock layers that are resistant to erosion form ridges, and nonresistant

Figure 17.18 Boudin of chert (dark-colored rock just above hammer handle) in Inwood marble, Isham Park, New York City. Note the ductile flow of the marble around the boudin. *(Courtesy of Peter Mattson.)*

Figure 17.19 Outcrop patterns caused by erosion of deformed and undeformed rock.

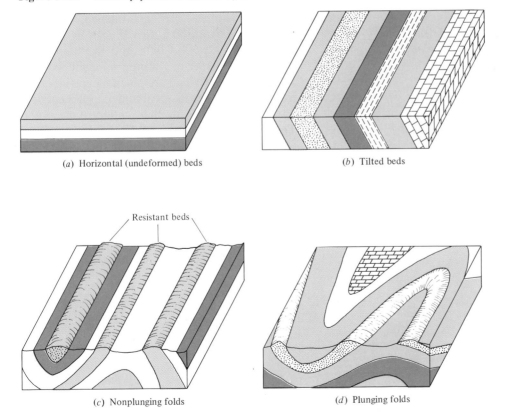

(*a*) Horizontal (undeformed) beds

(*b*) Tilted beds

Resistant beds

(*c*) Nonplunging folds

(*d*) Plunging folds

layers form valleys that can be followed as topographic markers. Tilted beds and beds deformed into nonplunging folds yield outcrop patterns composed of parallel bands of rock units. They can be distinguished by noting what material makes up the individual bands. No individual bed will appear more than once in a sequence of tilted rocks, but folding causes repetition of individual beds several times in an area. The outcrop pattern of plunging folds is unique, consisting of nonparallel bands of rock that converge in areas that are called **fold noses.** Anticlines plunge in the direction indicated by the convergence of their limbs, whereas synclines plunge in the opposite direction, away from their fold noses.

Faults

When stresses cause more strain than a rock can absorb by ductile flow, rupture occurs. When the resulting fractures are planar and when the blocks of rock on opposite sides of the fracture are offset relative to one another, the fracture planes are called **faults** (look back at Figure 17.2). Faulting can be induced by any type of stress, but most rocks rupture more easily under shear or tensional stress than under compressional stress.

Consider a rock undergoing deformation (Figure 17.20). As stresses build up, the rock undergoes first elastic deformation, then it undergoes ductile deformation, and finally it ruptures. At the rupture point, the strain is relieved. If the deforming forces continue to operate after the opposite sides of the fault have

stopped moving, the strain once again may be built up in the rock until it is released during a new episode of faulting. When we look at a fault today and measure its offset, we must remember that the displacement probably did not take place in a single catastrophic movement but rather in a series of small movements. For example, the San Andreas Fault of California exhibits displacement of hundreds of kilometers; but the fault has been active for approximately 60 million years, and this offset could have been accomplished by many small movements over that time period.

Faults, like folds, can be observed at several scales. For example, each slippage plane in Figure 17.3c can be considered to be a fault since there is displacement of rock material on opposite sides of the fracture planes. Faults may be seen within the confines of a single outcrop, as in Figure 17.2, or may be concentrated in large, regional-scale fault zones.

Anatomy of faults

Faults may be thought of as consisting of the three simple elements labeled in Figure 17.21. These are the two blocks of rock on opposite sides of the fault and the fault plane itself. The blocks of rock on opposite sides of an inclined fault plane are given names inherited from the early British coal miners, who encountered faults often in their underground work. The **hanging wall** rests on (or hangs upon) the fault plane, whereas the **footwall** forms the foundation (or footing) upon which the fault plane and the hanging wall rest.

Figure 17.20   Stepwise development of large-scale fault displacement.

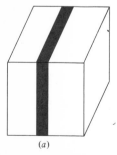

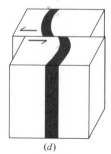

(a)   (b)   (c)   (d)

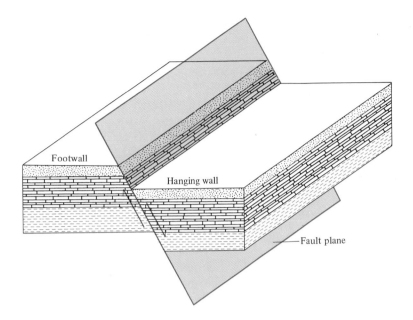

Figure 17.21 Anatomy of a fault. The arrows show the relative sense of displacement.

Types of faults

Faults are classified according to the nature of movement of the blocks (Figure 17.22). However, it is nearly impossible to determine the *absolute* direction of movement of the blocks. They may well move in opposite directions as shown in Figure 17.22a, but both sides can equally well move upward (or downward) by different amounts to produce the apparent movement shown in the diagram. All we can be sure of is the apparent or *relative* sense of motion. This is what the arrows indicate and what the classification scheme is based on.

Some faults are called **dip-slip faults** because the displacement along the fault planes has been up or down the dip of the fault, with little if any lateral displacement. Others are **strike-slip faults**, with horizontal displacement parallel to the strike of the fault plane and with little or no vertical offset. Complex fault motion is found in oblique-slip and rotational faults (see Figure 17.22).

Dip-slip faults Dip-slip faults in which the footwall has moved upward relative to the hanging wall are called **normal faults**. There is nothing abnormal about other kinds; this type of fault formed the vast majority of those in the coal

Figure 17.22 Types of fault. *(a)* Normal fault: Hanging wall moves down relative to footwall; displacement is up or down the dip of the fault (dip-slip displacement). *(b)* Reverse fault: Hanging wall moves up relative to footwall; displacement is up or down the dip of the fault (dip-slip displacement). *(c)* Strike-slip fault movement is parallel to strike of the fault plane; no displacement up or down dip. *(d)* Oblique slip: Combination of strike-slip and dip-slip displacement. *(e)* Rotational fault.

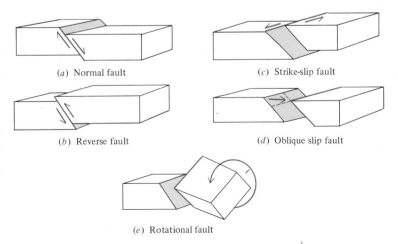

(a) Normal fault

(c) Strike-slip fault

(b) Reverse fault

(d) Oblique slip fault

(e) Rotational fault

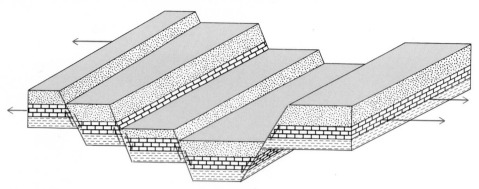

Figure 17.23 Normal faults. The hanging wall has moved downward relative to the foot-wall in each of these faults. These faults are the result of tension.

fields which gave the faults their nomenclature and thus was considered to be the "normal" type. Normal faults form where the earth's crust is subjected to tension (Figure 17.23). Complex systems composed of several normal faults that dip alternately toward and away from one an-other result in alternatingly downthrown and upthrown blocks such as those shown in Figure 17.24a. In the American Southwest, the alternating downdropped and uplifted blocks are called **basins** and **ranges,** respectively. In Europe, similar structures are referred to as **horst**

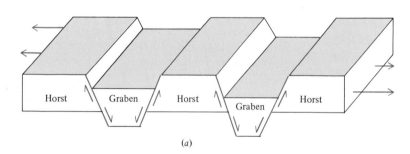

(a)

Figure 17.24 (a) Horst and graben structure. (b) Wild Rose graben near Ballarat, California. The depression is bounded by fault scarps on both left and right. (Courtesy of Jerome Wyckoff.)

(b)

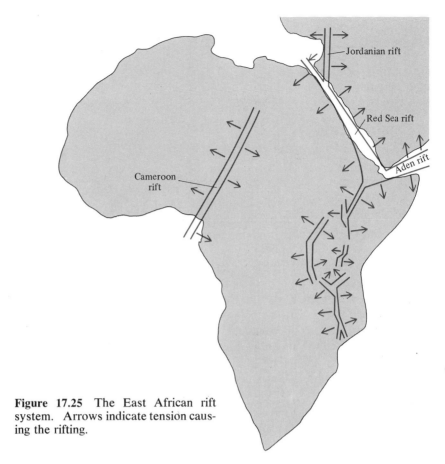

Figure 17.25 The East African rift system. Arrows indicate tension causing the rifting.

(uplifted) and **graben** (downdropped) structures.

Large, linear regions of tension produce long graben-like structures bounded by uplifted blocks. East Africa today is experiencing tensional stresses, as indicated by the extensive East African rift system, a series of several graben (Figure 17.25). One graben in this system is the low area occupied by the Red Sea. Another is responsible for Olduvai Gorge, in which so many important fossil remains of ancient humans have been found. The topography of the crest areas of the ocean ridge systems is very similar in form to the African graben, and geologists believe that this points to a state of tensional stress in every ocean basin.

Dip-slip faults in which the footwall has moved downward relative to the hanging wall are called **reverse** faults because they have the opposite sense of motion from normal faults. Reverse faults are caused by compression (Figure 17.26) and are found in association with strongly folded rocks, showing that folding too

generally is a compressional process. **Thrust faults** are a class of reverse faults characterized by extremely gentle dips, often only a few degrees. Many thrust faults are responsible for the transportation of large masses of old rock upward and eventually over younger rocks, thus reversing normal sequences predicted by the principle of superposition. However, the opposite sense of displacement is also possible in cases in which young rocks may be transported long distances and thrust over older rocks. Because thrust faults dip gently, the upward movement is mostly horizontal, and rocks may be moved horizontally several kilometers from their original deposition sites (Figure 17.27).

We might expect a large mass of rock to be crumpled and broken as it is thrust for great distances over other solid rock. This happens in some thrust faults, with friction between the two blocks causing the breaking, but is strangely absent in others. It is thought that in some thrust faults, the pressure of water trapped in

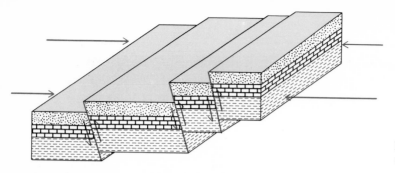

Figure 17.26 Reverse faults are a result of compression (deforming forces).

pores counteracts the weight of the overthrust block and thus reduces the frictional drag enough so that the entire transported block is moved intact. (See Chapter 10 Perspective.) In other thrusts, movement is along layers of highly incompetent rock such as gypsum and halite that flow so easily that they act like a lubricant for the overthrust block. In such faults, it is this lubricant that reduces friction and permits the overthrust block to move without breaking up. Thrust faults are found in many mountain ranges, including the Alps, Rockies, and Appalachians.

Strike-slip faults Faults characterized by nearly horizontal displacement parallel to the strike of the fault plane are called strike-slip faults (see Figure 17.22c). From the viewpoint of a person standing on one of the two fault blocks, the movement of the opposite block appears to be to either the left or the right. This provides the nomenclature for strike-slip faults:

right-lateral or **left-lateral.** Shear stresses are responsible for strike-slip faults, and these stresses may be generated during major episodes of regional tension or compression. Strike-slip faults thus may accompany normal or reverse faults in structurally complex areas, and they are found throughout the world. As a matter of fact, strike-slip faults are among the best-known geologic features in the world after volcanoes. The San Andreas Fault is a right-lateral strike-slip fault, while the Great Glen Fault of the Scottish Highlands is a left-lateral strike-slip fault.

Transform faults Many large-scale strike-slip faults now are believed to be a special class of fault called **transform faults.** These faults coincide with the oceanic fracture zones and, shown in Figure 16.9, appear to offset active ocean ridges. Transform faults are not restricted to the oceans, and many geologists believe that the San Andreas Fault is a transform fault that

Figure 17.27 Thrust faults in the Appalachian Mountains. Cross sections of the Valley and Ridge province of the Tennessee Appalachians showing thrusting of sedimentary rocks toward the west. Individual thrust faults are outlined in color. *(Modified with permission from P. B. King, "Tectonic Framework of Southeastern United States," 1950, American Association of Petroleum Geologists, vol. 34, pp. 635-671.)*

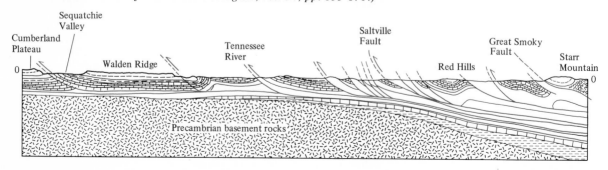

has offset segments of the East Pacific Rise (Figure 17.28). In the plate tectonic hypothesis, transform faults are boundaries between lithosphere plates that are sliding past one another. Transform faults may separate ocean ridge segments, join island arc segments, or connect two other transform faults.

Transform faults may appear to be simple strike-slip faults, but detailed study of movement in modern transform faults shows that they are far more complex than typical strike-slip faults. The evidence for movement of opposite blocks in transform faulting is based on analysis of earthquakes generated during the faulting and will be discussed in detail in Chapter 19. The reason for the complexity of the fault motion is that the two blocks separated by the transform fault are themselves undergoing some kind of motion, as shown in Figure 17.29. When a transform fault separates two ridge segments, each segment itself is undergoing rifting and essentially is being split apart. Where transform faults separate island arcs, according to plate tectonic models, each arc segment is undergoing intense compression and subduction.

To understand the complexities that distinguish transform faults from typical strike-slip faults, let us examine the evolution of a transform fault that cuts a midocean ridge (Figure 17.29). At the point of initial rupture, the ridge segments are separated by some distance along the fault, giving an impression of

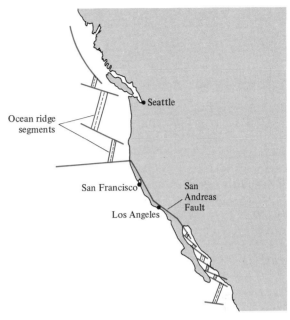

Figure 17.28 The San Andreas Fault interpreted as a transform fault. Segments of the East Pacific Rise are offset by several transform faults, of which the largest is the San Andreas Fault.

left-lateral offset (Figure 17.29a). However, each block is undergoing tension, and new crustal material forms at both segments of the ridge crest as rifting opens the ridge wider (Figure 17.29b). With continued rifting, a conveyor-belt-like mechanism is established that carries newly created crustal material outward

Figure 17.29 Evolution of a transform fault. (a) Sea-floor spreading begins along a discontinuous ocean ridge. (b) Spreading continues. Old basaltic crust is carried away from the ridge crest, and new basalt (gray) forms in its place. *The distance between the ridge crest segments remains the same as in (a).* (c) Sea-floor spreading continues, but the ridge crests do not change position relative to each other. Arrows show that between ridge segments, fault movement is the opposite of that inferred in (a).

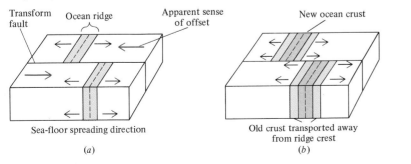

(a)   (b)   (c)

from the ridge crest segments on both sides of the fault. Along the portion of the transform fault that lies between the two ridge segments, the *actual* motion is precisely the opposite of the left-lateral motion expected! Indeed, along the fault everywhere but between the two ridge segments, the opposing blocks appear to be moving in the same direction.

If our ideas about transform faults are correct, the distance between two offset ridge crest segments need never change. Materials erupted at the ridge crests may be transported thousands of kilometers apart, but the spreading centers themselves can remain in their original positions, as shown in the sequence of illustrations in Figure 17.29. This is certainly not the case for

typical strike-slip or dip-slip faults, and so transform faults are indeed different from the other types of faults that we have studied. We shall return to them in greater detail in Chapter 21.

Recognition of faults

Some faults, such as the one in Figure 17.2, can be recognized easily because all three elements—the fault plane and both fault blocks—are clearly visible. Others, including some regional faults, are not quite so obvious, and geologists must examine the rocks carefully for signs of brittle deformation. For example, recently active faults may offset topographic features such as streams or even human-built

Figure 17.30 Strike-slip fault offsetting rows of plants in a garlic field in California. *(Alan Pitcairn from Grant Heilman.)*

structures (Figure 17.30) and are easy to recognize.

Ancient faults also offset geologic features, but millions of years of erosion may mask some of the evidence. As in the case of folds, outcrop patterns of deformed rock help point toward the presence of a fault (Figure 17.31). The abrupt offset, repetition, or sudden absence of a well-defined rock sequence or an individual rock unit is evidence for faulting.

Other clues lead to recognition of faults in places where these kinds of evidence may not be obvious. Rocks in fault zones commonly are broken and crushed by the grinding that takes place when two blocks of rock move past one another. The two blocks abrade one another as they move, and angular fragments of the rocks are broken off along the fault plane. Fault motion grinds some of these into extremely fine grained materials but leaves a few larger pieces. Such crushed and broken rocks are called **fault breccias,** or **fault gouge** (Figure 17.32). With continued grinding, the larger fragments also are pulverized, the smaller grains are elongated, and a foliation is produced. The result is the cataclastically textured rock *my-*

lonite described in Chapter 7.

The finely pulverized material produced in fault zones commonly smoothes and polishes the plane of the fault in much the same way that a jeweler's abrasive powder polishes a gemstone. Slightly larger, more-resistant grains are dragged across the polished surface and produce fine grooves (Figure 17.33). These grooved, polished surfaces are called **slickensides** and are useful in showing the movement of the fault blocks. Each groove marks the motion of a particle during faulting so that displacement can be interpreted as having been parallel to the groove. *Which* way parallel to the groove may not be so easily determined, but the slickensides permit identification of the strike-slip, dip-slip, or oblique-slip nature of the fault.

Ancient faults may have topographic expression. Gouge and mylonite generally are less resistant to weathering and erosion than the unbroken rock materials outside the fault zone. Differential erosion then produces topographic lows along the strike of the fault plane. These lows may consist of aligned ponds, lakes, or aligned valleys separated by small divides (Figure 17.34).

Figure 17.31   *(a)* Omission and *(b)* repetition of strata due to faulting.

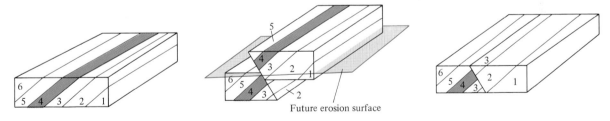

(a) Omission of a bed from a well-defined sequence by fault uplift and subsequent erosion

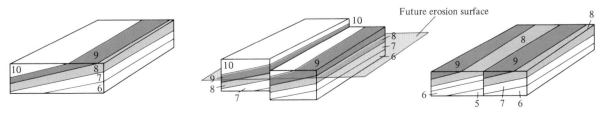

(b) Repetition of a sequence by fault uplift and subsequent erosion

Figure 17.32 Fault zone near Hell's Gate, Death Valley, California, showing crushed zone containing fault gouge. (*Courtesy of Jerome Wyckoff.*)

DEFORMATION AND THE HYPOTHESES OF MOUNTAIN BUILDING

The first geologic hypotheses concerning the origin of mountains came from studies of deformed rock. When geologic mapping expanded across Europe and North America in the 1800s, geologists discovered that the mountain systems of the Alps, Pyrenees, Urals, Apennines, Appalachians, and Rockies contain some of the most intensely deformed rocks on the continents. The expanding, shrinking, and pulsating earth models were proposed in order to explain the structures found in these mountain ranges.

Figure 17.33 Slickensides. The grooves on this fault-polished rock parallel the movement of fault blocks. (*Earthquake Information Bulletin, U. S. Geological Survey.*)

Two important advances have been made during the twentieth century in our study of deformation, and these have revolutionized interpretations of earth history completely. First, we have extended the study of deformed rocks to those of the ocean basins and thus can speak for the first time of truly worldwide studies. Second, we now are able to measure strain that is building up in rocks as we observe them in the field, and thus we can study the current state of deformation throughout the world. Therefore, we now can apply uniformitarian principles to deformed rocks—that is, we can use the modern state of deformation to interpret ancient folded and faulted rocks. To a great extent, the theory of plate tectonics has emerged from this approach.

Modern Deformation

Two kinds of information vital to an understanding of mountain building have emerged from studies of modern deformation. One deals with the nature of the faulting process, the other with the extent and direction of global-scale deforming forces.

Several types of instruments have been devised that enable us to measure the buildup of strain in rocks of active fault zones. Strain

Figure 17.34 Small sag ponds (aligned depressions) exposed along the trace of the San Andreas Fault zone, approximately 24 miles southeast of Hollister, California. *(Courtesy of John S. Shelton)*

gauges attached to rocks in the field monitor the accumulation of strain and tell something of how faulting occurs: how rapidly strains build to the rupture point, and how often the buildup-rupture cycle occurs. Extremely precise surveying instruments, some using laser beams to measure distances to within fractions of a millimeter, measure the amounts of displacement in episodes of faulting. We thus learn about the mechanics and rates of fault movement and can apply this information to ancient faults.

Detailed studies of earthquakes have enabled geologists to determine the dips of large-scale fault planes and the directions of movement of fault blocks. From these data we can identify the type of fault (normal, reverse, strike-slip, or transform) and the type of stress involved (tension, compression, or shear). The results show that the earth's crust and outer mantle are in a complex state of stress, as shown in Figure 17.35. Tensional stresses are operating today at the ocean ridges, compression at the ocean trenches and beneath the island arcs, and shear in the ocean fractures. Compare Figure

17.35 with the diagram showing plate motion in Chapter 1 (Figure 1.7). It was from data such as these that plate motions were postulated originally.

If uniformitarian principles can be applied and the earth can be shown to have experienced similar stress patterns in the geologic past, three of the remaining hypotheses of mountain building can be discarded. Expanding and shrinking earth models require worldwide tensional and compressional stresses, respectively, and the pulsating earth model entails alternating periods of global tension and compression. If the ancient earth exhibited simultaneous tension and compresssion, as the present earth does, these models would be invalidated, leaving the plate tectonic hypothesis as the only viable one.

Interpreting Ancient Stresses

Geologists can also use data from folded rock to determine the nature of ancient stresses. Most folds are compressional, and upright folds generally have their axial planes oriented at right

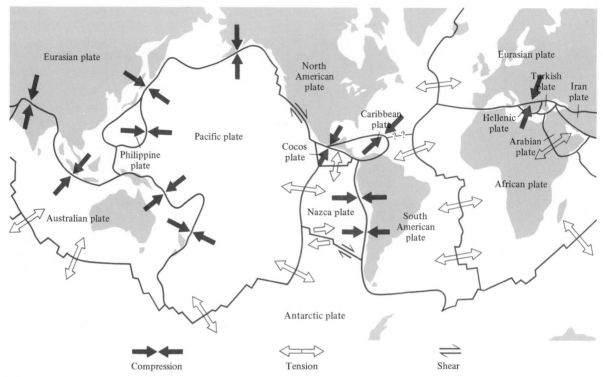

Figure 17.35 Stresses in the modern earth.

angles to the compressional stresses that created them (Figure 17.36). Thus, if geologists can measure the attitudes of the axial planes of regional-scale folds, they can determine the direction of the compressional stresses at the time of deformation. For example, axial planes of the major folds in the Appalachian Mountains strike roughly northeasterly along the entire mountain system. Deforming forces must have acted at right angles to these axial planes—from the northwest and southeast.

However, some folds need not be compressional, and care must be taken in interpreting the stresses responsible for their formation. Folds produced by deck-of-cards movement can be caused by forces parallel to their axial planes

Figure 17.36 Compression as a cause of folding. Note that the axial planes are formed at right angles to the deforming forces.

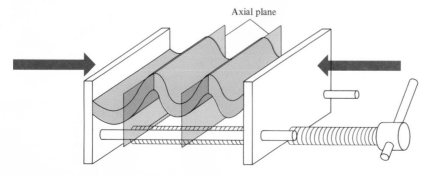

rather than perpendicular to them (Figure 17.37). Therefore, geologists must determine the mechanism by which folding took place before attempting to recreate the stress fields at the time of deformation.

Unraveling the record of ancient deformation

In the early part of the twentieth century, geologists working in Western Europe and in North America found many similarities in the deformational histories of their respective continents. Each continent has been subjected to periods of intense deformation and mountain building that geologists call **orogenies,** and the timing of European and American orogenies has proved to be very similar. For example, in the early part of the Paleozoic Era, two orogenic events occurred in eastern North America (the Taconic and Acadian orogenies) that seemed to correspond with episodes of what is called the Caledonian orogeny in Europe. All three were strongly compressional. Late Mesozoic and early Cenozoic times saw further compression: the Nevadan and Laramide orogenies in the American West, the Alpine orogeny in central Europe, and the deformation that produced the Himalayas in Asia and the Andes in South America. It appeared that relatively simple worldwide stress patterns had been operable in the past.

More recent studies have reversed this conclusion. It has long been known that the east coast of the United States was subjected to strong tensional stresses in Triassic times, resulting in large-scale normal faulting, horst and graben structures, and avenues of access for eruptions of flood basalts. At the same time, however, intense compression affected Japan, causing tight folding in an orogeny accompanied by high-grade regional metamorphism and the intrusion of granites. Tension and normal faulting persisted along the Atlantic Coast of North America during the late Mesozoic Era when compressional orogenies began in the Rockies, Alps, and Andes mountain systems.

Thus, compression and tension have operated simultaneously in the earth in the geologic past just as they do today. The expanding, shrinking, and pulsating earth models are in-

validated in light of this evidence. Of the five hypotheses proposed in Chapter 1, only the plate tectonic model successfully fits all the data we have obtained from studies of the earth. In the next chapter we shall use the plate tectonic hypothesis as a frame of reference within which the major mountain systems of the earth can be discussed.

GEOLOGIC STRUCTURES AND HUMAN BEINGS

The study of deformed rocks is more than just a topic for academic curiosity. Some of the world's most important economic resources either are created by the processes involved in deformation or are found in deformed rocks. Geologic structures also can pose hazards to human beings if they are not recognized during selection of building sites, and as a result most major construction projects involve preliminary geologic examination of the terrain.

Petroleum and Natural Gas

The most avid students of deformed rocks are companies involved in exploration for petroleum and natural gas because much of the world's supply of these commodities comes from rocks in

Figure 17.37 Folds produced without compression. Axial planes are not perpendicular to the deforming forces.

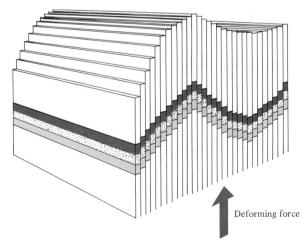

Deforming force

which they are trapped by geologic structures. Two types of structure are particularly well suited for finding these energy resource materials: anticlines and salt domes.

Oil and gas frequently are concentrated at the axial areas of anticlines (Figure 17.38a). Both oil and gas have lower specific gravities than water and will float above water. In a porous and permeable layer of rock, oil and gas also will be displaced upward by water present in the pore spaces. In an anticline, this migration will be toward the axial area from both limbs, as shown in Figure 17.38a. If the folded strata include alternating layers of relatively permeable and impermeable rock, the oil and gas will be confined to the permeable layers, just as water is confined to aquifers and restricted from aquicludes as described in Chapter 12. Oil and gas migrate upward into the axis of an anticline within porous and permeable beds and become trapped there by the interlayered impermeable rocks. Geologists drilling into the axial areas of anticlines thus can find oil at several levels within the structures.

Entrapment against impermeable rock is responsible for oil and gas accumulations near salt domes as well. During deformation of halite beds, viscous flow of the salt takes place as shown in Figure 17.10. As the salt flows upward, it pushes the overlying rock into an arch, or domelike structure, and actually may break through some beds like an intrusive magma (Figure 17.38b). Some oil and gas can collect in the domelike structures formed over the salt if interbedded permeable and impermeable rocks are present. Salt itself is impermeable so that oil and gas migrating upward also can be trapped against the halite.

Geologic Structures as Hazards

Failure to recognize the structures in a region may have catastrophic results if buildings or dams are constructed in dangerous areas. For example, building directly on an active fault seems intuitively foolish, yet many important structures (hospitals and schools) sit directly on the San Andreas Fault in California. This fault has been the site of major earthquakes several times in the past 100 years, yet building near and on it continues. We shall study this problem more in our discussion of earthquakes and earthquake damage in Chapter 19.

Sometimes the problem involves deciding whether a fault is to be considered active. After all, if deforming forces in an area stopped hundreds of millions of years ago, ancient faults may prove relatively safe to build across. However, we must remember that human frames of reference treat time in terms of decades, whereas geologic time involves millions of years. Thus, much debate preceded the construction of the Indian Point nuclear power plant in southeastern New York State because the location was on the

Figure 17.38 Oil and gas traps. Oil and gas rise upward in porous and permeable beds until trapped by overlying impermeable beds *(a)* or near a salt dome *(b)*. Gas (light color) always lies above oil (dark color).

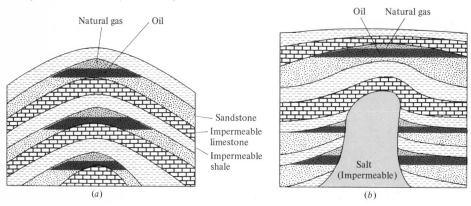

Ramapo Fault. Debates about whether the fault was active or inactive came to an abrupt end in 1978, when the fault proved conclusively that it is still capable of motion. The power plant had been built by that time, but thankfully there was no significant damage.

Even inactive faults may pose a problem in construction because foundations anchored in the fractured and pulverized rock of a fault zone tend to be less stable than those anchored in un- fractured bedrock. Special precautions there- fore must be taken to anchor buildings in faulted rocks, although it would be best to build them in other locations. The porosity and permeability that make gouge zones useful as collectors of water-transported minerals also can cause prob- lems during and after construction. Gouge per- meability facilitates seepage of groundwater into building basements or leakage of surface water from reservoirs.

SUMMARY

Forces acting to deform rock produce compres- sional, tensional, and shear stresses as they are resisted by the cohesiveness and mineral strengths of the rock involved. When the de- forming forces exceed the rock's internal strength, deformation occurs. The amount of deformation, or response to the deforming forces, is called strain. There are three types of behavior in stressed rock, with each taking place over a spe- cific range of stress and strain for every rock type.

Before permanent deformation occurs, most stressed rocks behave elastically. If stress is relaxed, elastically deformed rocks return to their unstressed, and undeformed state. During elastic deformation, the amount of strain is pro- portional to the amount of applied stress. The upward rebound of the continents during and after the melting of continental glaciers is an ex- ample of elastic behavior that is going on today.

Ductile behavior involves a permanent change in size and shape. It occurs by solid- state flow of material and generally occurs in rock subjected to stresses greater than those under which elastic deformation takes place. Ductile behavior is expressed in rocks by folds and by salt domes, in which halite flows and can inject itself into surrounding rock.

Brittle behavior occurs when the cohesive- ness of a rock is overcome by the deforming forces and breakage takes place. Zones of breakage include random fractures, systemati- cally aligned joints, and those breaks along which movement has taken place—faults.

Different types of folds and faults can be re- lated to the different types of stress in the earth. Most folds are the result of compression, but some may form by an upwardly directed force. Faulting is caused by all three types of stress. Tension produces normal faults and results in a local expansion of the crust. Compression causes reverse and thrust faults and the resulting crustal contraction. Shear stress resulting in strike-slip and transform faults causes lateral offsets but involves neither expansion nor con- traction of the earth's crust.

Different stresses currently are active at different parts of the earth. Tension has been detected at the ocean ridge crests, compression at the island arc-trench systems, and shear in the oceanic fractures and in major strike-slip faults on the continents. Some areas appear to be relatively stable, or quiet in terms of deforma- tion.

Modern stresses are not arranged in a simple global pattern of all compressional or all ten- sional stress, as would be expected in an ex- panding or shrinking earth. Neither, it seems, were ancient stresses responsible for folding and faulting of rocks in major mountain ranges. Thus, while intense folding was brought about by compression in the Japanese Islands, tension- induced rifting was taking place along the east coast of North America. The complex stress distribution patterns of modern and ancient earth do not fit any of the earth models except the plate tectonic model, and as a result most geologists today believe that this is the most valid model.

Geologic structures have proved both helpful and hazardous to humans. Migration of petroleum and natural gas through permeable layers leads to entrapment of these fluid resources at the axial regions of anticlines and both above and along the flanks of salt domes. Fault zones tend to be dangerous construction sites because of the difficulties in anchoring foundations in fractured rock, the possibility of subsequent fault displacement, and the possibility of rockbursts.

QUESTIONS FOR REVIEW AND FURTHER THOUGHT

1. The movement of ice in a glacier was discussed in Chapter 13, and both brittle and ductile processes were involved. Explain the movement of ice in terms of what you have learned about the behavior of rock during deformation.

2. Which type of rock would provide the most complete record of deforming forces: competent or incompetent? Explain.

3. What role does water play in deformation?

4. Two rocks subjected to the same conditions of pressure, temperature, and stress are seen to deform differently. How is this possible?

5. An outcrop of slate has a strongly developed cleavage, and the cleavage itself is folded. With your knowledge of cleavage, what can you say about the deformation history of this rock?

6. What changes would you expect in the deformational behavior of a given rock type if it were to be buried deeper and deeper in the earth's crust?

7. How could sedimentary features that help tell tops of beds from bottoms of beds be useful in distinguishing between the two different interpretations shown in Figure 17.16?

8. What criteria may be used to determine the relative movement of rocks on opposite sides of a fault?

9. Why is it difficult to identify folds in an igneous body?

10. Why is oil not found in the axial regions of synclines?

ADDITIONAL READINGS

Few readings in structural geology are intended for students at the introductory level. Some are very theoretical and concentrate on the mechanics of folding and faulting, but those listed below contain excellent descriptive sections and photographs that illustrate structural features.

Beginning Level

Clark, S. P., Jr., *Structure of the Earth,* Prentice-Hall, Englewood Cliffs, N.J., 1971.
(A general view of the earth's interior, including magnetism, gravity, and seismology. Chapters 2 and 4 deal with structural features and their application to plate tectonics)

Sumner, John S., *Geophysics, Geologic Structures, and Tectonics,* W. C. Brown, Dubuque, Iowa, 1969.
(A somewhat more rigorous description of the ways in which geophysics and structures can be used to study the earth's interior. Chapters 9 and 10 deal with structures and the mountain-building processes during which they form)

Intermediate Level

Billings, M. P., *Structural Geology,* 3d ed., Prentice-Hall, Englewood Cliffs, N.J., 1978.
(An updated version of the most widely used textbook in structural geology. Contains excellent descriptive sections on each major type of structure, with examples from throughout the world)

Dennis, J. G., *Structural Geology,* Ronald Press, New York, 1972.
(An excellent textbook in structural geology, with descriptions of deformation in the oceans as well as on the continents)

Advanced Level

Hobbs, B. E., W. D. Means, and P. F. Williams, *An Outline of Structural Geology,* Wiley, New York, 1976.
(An up-to-date, high-level treatment of rock mechanics and rock structures, including detailed discussions of stress, strain, structural features, and tectonic implications of structural features)

CHAPTER 18

Mountains and Mountain Building

To a climber, a mountain is an adventure, a challenge that pits human brain and muscle against rock and gravity. To geologists, mountains are a different kind of challenge. We have seen in preceding chapters that nearly every earth process leaves some kind of record by which it can be detected. Mountains are formed by global-scale internal processes and are the places where the most intense volcanism, deformation, and metamorphism are concentrated. The challenge presented to the geologist by every mountain is to interpret as much as possible about the global-scale forces from the rocks exposed at the surface.

The five hypotheses outlined in Chapter 1 have been the responses of the geologic community to this challenge over the past century. During this period we have learned that there are several different kinds of mountains and that there are mountains in the oceans where none had ever been suspected. In Chapter 17 we evaluated the five hypotheses and saw that only the plate tectonic hypothesis explains the earth's deformational history satisfactorily. In this chapter we will study the different kinds of montains and see how each fits into the plate tectonic model.

WHAT IS A MOUNTAIN?

A **mountain** is any part of the earth's crust that rises more than 600 m (2000 ft) above the surrounding countryside. Similar but smaller topographic highs are called **hills,** and extensive flat areas that stand prominently above their surroundings but have little internal relief are called **plateaus.** Some mountains, such as Mt. Katahdin in Maine and Stone Mountain in Georgia, are solitary peaks that stand alone, projecting majestically above relatively flat surroundings (Figure 18.1). However, most mountains occur in groups called **mountain ranges** (Figure 18.2). Individual peaks in a mountain range are produced by the same processes and at about the same time as the other peaks. The peaks of the Cascade Range in the American Northwest—mountains such as St. Helens, Lassen, Shasta, Baker, Hood, and Rainier—are all stratovolcanoes of Tertiary and Holocene age.

Several ranges may be associated into a larger grouping called a **mountain system.** For example, the Appalachian mountain system consists of several ranges including the Smoky, Blue Ridge, Catoctin, Taconic, Berkshire, and Green Mountains, while the Rocky mountain system contains even more ranges, including the San Juan, Sangre de Cristo, Uinta, Teton, Beartooth, and Tobacco Root ranges. Both the Appalachian and Rocky mountain systems have long histories of sedimentation, volcanism, deformation, plutonism, and metamorphism. An individual range within either system may represent only a small part of that history.

WHERE ARE MOUNTAINS FOUND?

There are mountains on every continent and in every ocean basin (Figure 18.3). Mountains

Figure 18.1 Mt. Katahdin, Maine, a solitary mountain standing above surrounding lower terrain. *(Maine Publicity Bureau.)*

can be found at the edges of some continents (the Andes), near the edges of others (the Appalachians), and near the centers of still others (the Urals, Rockies, and Himalayas).

They are also found throughout the oceans: near the centers (Mid-Atlantic Ridge, the Hawaiian Islands) or near the margins (the Japanese or Aleutian Islands, the Greater Antilles).

Figure 18.2 A mountain range. A portion of the Canadian Rockies north of Banff showing several peaks and some of the characteristic structures. *(Courtesy of John S. Shelton.)*

More important in our study of mountain-building processes is where mountains are located on lithosphere plates. Here too there is great variety. Many mountains are located at plate boundaries. For example, the Mid-Atlantic Ridge *is* the boundary between the North American and Eurasian plates, and the Andes mountain system is at the boundary between the Nazca and South American plates. However, some mountains lie well within lithosphere plates, far from the nea.est plate margins. The Appalachian Mountains are on the North American plate, thousands of kilometers from the Mid-Atlantic Ridge; the Ural Mountains are in the very center of the Eurasian plate; and the Hawaiian Islands are near the center of the Pacific plate.

TYPES OF MOUNTAINS

Mountain systems are very complex, and any attempt to classify them must take into account several factors, only one of which is location relative to plate boundaries. Geographic location—continental versus oceanic—is an important factor because we have seen in several earlier chapters that the composition of the rocks and the processes active in continents and ocean basins are different. Remember, for example, that regional metamorphism and granitic/rhyolitic igneous activity are restricted to the continents. This suggests that there probably are major differences between continental and oceanic mountains. We shall examine these differences later in this chapter.

Another factor is the nature of the rocks that make up the mountains. For example, the Mid-Atlantic Ridge is composed almost entirely of basalt, whereas the Alps contain a wide variety of sedimentary, volcanic, plutonic, and metamorphic rocks. The Cascades are made of volcanic rocks, but unlike the Mid-Atlantic Ridge, they are composed of andesites and rhyolites as well as basalts. The Adirondack Mountains are a mixture of plutonic and metamorphic rocks, whereas the Sierra Nevada Mountains are dominantly plutonic. Such different rock contents indicate different mountain-building processes.

Mountains also differ in the amount and type of deformation exhibited by their rocks. The basalts of the Hawaiian Islands and the andesite-rhyolite-basalt suite of the Cascades essentially are undeformed, while the rocks of the Appalachians, Rockies, and Andes are tightly folded and intensely faulted in places. Rocks of some mountains are folded by compressional stresses whereas others are block-faulted, the results of tension.

Table 18.1 is a simplified classification scheme based on these variables. We will discuss each mountain type shown in Table 18.1 and try to show how the plate tectonic model explains the many differences between them. One of the most attractive aspects of the plate tectonic hypothesis is that it can explain nearly all the complexities of the world's mountains in a simple way.

TABLE 18.1   Classification of mountain systems

Rock types	Oceanic mountains		Continental mountains	
	Deformed	Undeformed	Deformed	Undeformed
Volcanic	Island arcs Ocean ridges	Hawaiian-type islands; seamounts	Andes-type mountains	Cascades-type mountains
Sedimentary, Igneous, Metamorphic	None	Atolls	Fold mountains (Appalachians) Fault-block mountains (Sierra Nevada)	Upwarped (erosional) mountains (Adirondacks, Colorado Plateau)

Figure 18.3 Mountain ranges of the world.

There is one other variable in the classification of mountains that is not included in the table: *time*. Some mountains, such as the Andes, Himalayas, and midocean ridges, are young. The forces that formed them are still active, and they are growing even as we study them.

Other mountains, such as the Appalachians and Urals, are older. The forces that produced them are no longer operating, and they are being worn down slowly by erosion. The Precambrian shields, vast but relatively flat areas of regionally metamorphosed rock found on every continent,

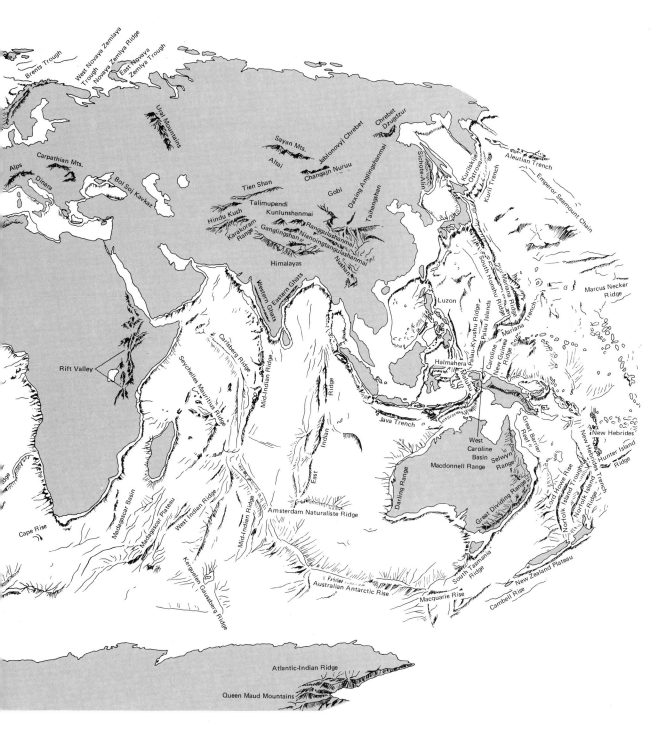

are probably all that remain of mountain ranges that once rivaled the Alps or Andes in height. These shields are the traces of the oldest mountains, the deep roots of mountain systems whose upper parts have been removed by literally billions of years of erosion.

Oceanic Mountains

The geologic study of mountains began on the continents, but our understanding of mountain-building processes is perhaps better for the mountains of the oceans because the oceanic mountains are protected from weathering and erosion. As a result, little material has been removed from the oceanic mountains, and their shapes reflect the processes that formed them. Interpretation of mountain building is thus easier for oceanic mountains than for continental mountains, where erosion has removed large amounts of geologic evidence and where the rugged mountainous shapes characteristic of the Alps and Himalayas are due in part to erosion and in part to the forces that built the mountains.

There are three kinds of oceanic mountains, each corresponding to a major physiographic feature of the sea floor described in Chapter 16: ocean ridges, volcanic island arcs with associated deep-sea trenches, and Hawaiian-type islands and seamounts. All three types are composed mainly of volcanic rocks, but whenever a volcano is built above sea level, erosion of the volcano provides sediment called **volcanoclastic** debris as well. The three differ in the kinds of volcanic rock of which they are made and in the type and intensity of deformation involved in their formation.

Ocean ridges

Calling these mountains ridges is somewhat like referring to the Empire State Building as tall. They form the longest, largest mountain system in the world, over 60,000 km long. As described in Chapter 16, ocean ridges are found in every ocean basin, at the centers of some ocean basins such as the Mid-Atlantic Ridge, and near the margins of others, such as the East Pacific Rise (see Figure 16.9).

The Mid-Atlantic Ridge is typical of this type of mountain system. Over most of its 21,000-km length the crest of the ridge is below sea level, but in some places individual peaks form islands such as Iceland in the North Atlantic, the Azores, and St. Helena and Tristan de Cunha in the South Atlantic. The ridge is a vast outpouring of basaltic lava, and eruptions occasionally form new islands when a volcano manages to build its cone above sea level.

Basaltic rocks of the ocean ridges undergo brittle deformation in two different places. As described in Chapters 16 and 17, the ridge crest is characterized by a deep graben-like valley in

which current earthquake activity and volcanism are concentrated. The central valley, by analogy with continental rift valleys, is thought to be the site of normal faulting caused by intense tensional stresses. In addition, ocean ridges are cut into small segments by the oceanic fractures, and these are interpreted as transform faults produced by shear stresses. Core samples of ocean ridge basalts reveal intense fracturing and dynamic metamorphic recrystallization near fracture zones, but metamorphism does not play a major role in the formation of the ridges.

Origin of ocean ridges — Sea-floor spreading According to the plate tectonic hypothesis, ocean ridges represent one of the three types of plate boundaries — the type at which new ocean crust is formed. They form when strong tensional stresses are applied to the lithosphere from below, causing rifting. As the lithosphere is split into what eventually will become two separate plates, basaltic magma rises into the rift zone and accumulates to form the ridge (Figure 18.4). As rifting continues, more lava rises to the surface and the older lavas are carried away from the rift zone in conveyor-belt fashion. The freshly extruded hot lavas form the ridges; but as the older lavas are carried away from the zone of rifting they cool, contract, and become denser. Eventually they subside and become the basement upon which the abyssal plain sediments accumulate.

This, according to plate tectonicists, is how ocean basins form. As rifting proceeds, the region of the earth's surface underlain by the basaltic rocks grows larger, resulting in an

Figure 18.4 Origin of the ocean ridges by sea-floor spreading.

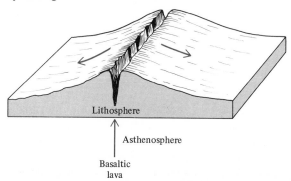

Lithosphere

Asthenosphere

Basaltic lava

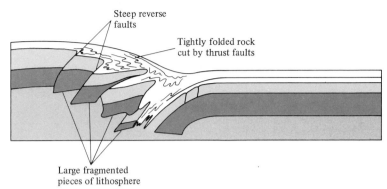

Figure 18.5 Tight isoclinal folding and thrusting in an island arc-trench system as envisaged by a plate tectonic model involving subduction.

ocean. This process of ocean growth is called **sea-floor spreading,** and ocean ridges commonly are referred to as **spreading centers.** If the plate tectonic model is correct, spreading is still going on along the Mid-Atlantic Ridge, moving Europe and North America farther apart. In Chapter 20 we will see how the rate of spreading can be estimated. The process is relatively rapid as geologic processes go; during a typical college semester, New York City and London move about 2 cm farther apart.

Volcanic island arcs

Volcanic island arcs bordered on their convex sides by the oceanic trenches represent a different type of mountain system. Island arcs are not as long as ridges and tend to be concentrated near the ocean margins, as shown in Figure 16.12. For example the northern and western margins of the Pacific Ocean consist of a series of island arcs that includes the Aleutian, Kurile, Japanese, and Philippine Islands. Despite its great size, the Atlantic Ocean has only a single island arc, the Greater Antilles. This arc separates the Atlantic Ocean from the Caribbean Sea and includes the islands of Puerto Rico, Hispaniola, and Cuba.

Island arcs are composed of volcanic and volcanoclastic rocks, but they have a more varied composition than is found in ocean ridges. Andesite is the most abundant rock type, with lesser but significant amounts of rhyolite and basalt. Thick sequences of volcanoclastic debris flank the island arcs and are interlayered with lavas and tephra. The rocks are deformed intensely by *compression* and as a result exhibit

tight folds and both reverse and thrust faults (Figure 18.5). Regional metamorphism accompanies the deformation.

In mature arcs — those in which the mountain-building forces have operated for a long time — plutonic rocks are present as well. In particular, one group of rocks known as the **ophiolite suite** is found in many arcs. This suite consists of several rock types (Figure 18.6). Ultramafic rocks at the base (peridotites) pass upward into mafic intrusives (gabbros) and closely spaced parallel basaltic dikes called **sheeted dikes.** These are capped by pillow basalts and commonly by a thin sequence of deep-sea sedimentary rocks (cherts and black shales).

Origin of island arcs by subduction In the plate tectonic model, island arc-trench systems are plate boundaries at which oceanic crust is returned to the asthenosphere by the process of subduction (Figure 18.7). During subduction,

Figure 18.6 The ophiolite suite.

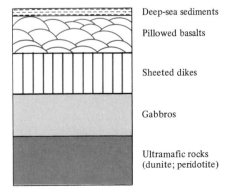

Deep-sea sediments

Pillowed basalts

Sheeted dikes

Gabbros

Ultramafic rocks
(dunite; peridotite)

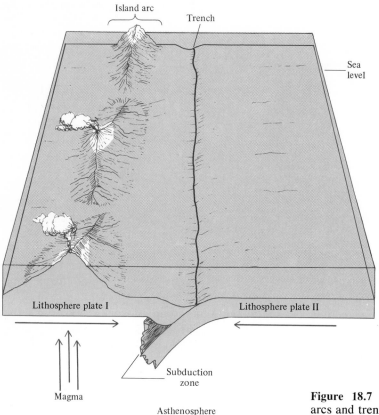

Figure 18.7 Formation of island arcs and trenches by subduction.

one lithosphere plate is thrust beneath another. Where the two plates meet, they buckle, producing the trenches. Heat generated by the radioactive elements of both lithosphere plates and by friction during thrusting rises through the upper plate and causes the volcanism which eventually builds the volcanic island arc. Compression caused by the collision of the two plates produces the folding and thrusting.

Ophiolites also are explained by the collision. The sequence of rocks of the ophiolite suite is quite similar to that found in cores recovered from the sea floors by the Deep Sea Drilling Project, and the ophiolite suite probably represents a cross section through typical ocean crust. It is transported into the island arc during thrusting by a process called **obduction**. During obduction, the overriding plate acts like the blade of a carpenter's plane, literally scraping off a section of the lower plate (Figure 18.8).

The detached (obducted) segment then is plastered onto the upper plate as the ophiolite suite.

Hawaiian-type islands and seamounts

Hawaiian-type islands and seamounts also are volcanic in origin but are associated with neither central graben nor trenches. In Chapter 16 we saw another important difference, the fact that many seamount chains show a progressive change in the age of individual peaks from one end to the other. This type of oceanic mountain is unique in the composition and state of deformation of its volcanic rocks.

Hawaiian volcanoes erupt mostly basalt, and much of the lava is similar to that of the ocean ridges. However, some Hawaiian lavas contain a higher proportion of the alkali elements sodium and potassium and are called **alkali basalts**. These differences may sound insignifi-

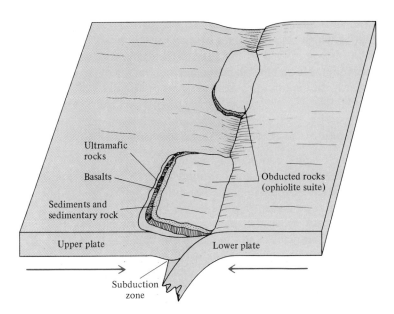

Figure 18.8 Obduction. The upper plate acts like the blade of a carpenter's plane and scrapes up slivers of the crust and mantle of the subducted plate.

Ultramafic rocks

Basalts

Sediments and sedimentary rock

Upper plate

Obducted rocks (ophiolite suite)

Lower plate

Subduction zone

cant, but experimental petrologists have found that alkali basalts come from a different source area than the ocean-ridge basalts, probably from deeper in the mantle.

The volcanic rocks of Hawaiian-type islands essentially are undeformed. Of all the oceanic mountains, only this type has escaped the tension, compression, and shear that play a major role in the formation of ridges and island arcs.

Origin of Hawaiian-type islands—Hot spots in the mantle These islands have escaped deformation because they form within plates rather than at plate boundaries. It is plate collision, rifting, or the shearing in transform faults that causes folding and faulting. Interiors of plates tend to be tectonically quiet. How then are Hawaiian-type mountains formed?

Plate tectonicists believe that there may be areas of high heat flow in the mantle below the asthenosphere. These areas, called **hot spots**, supply heat to the upper mantle, causing formation of magma in the lithosphere. If the lithosphere above the hot spot changes position during plate movement, the site of melting shifts because new material is then brought over the hot spot (Figure 18.9). As the plate moves, active volcanoes become extinct because they are removed from their sources of heat. Long-term

Figure 18.9 Origin of Hawaiian-type islands by movement of a lithosphere plate over a hot spot in the mantle.

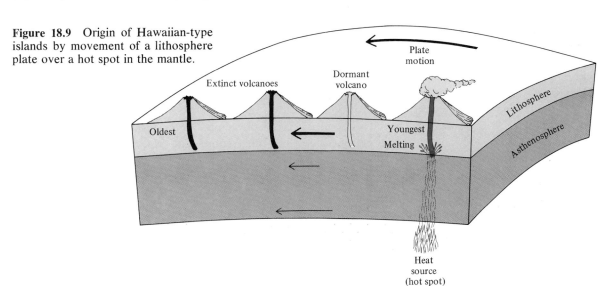

Plate motion

Extinct volcanoes

Dormant volcano

Oldest

Youngest

Melting

Lithosphere

Asthenosphere

Heat source (hot spot)

plate movement results in a long train of volcanoes whose age decreases toward the location of the hot spot, just as the ages of the Emperor Seamount–Hawaiian Island volcanoes decrease toward the south and southeast (see Figure 16.13).

Continental Mountains

Although some continental mountains are volcanic and similar to oceanic mountains, most are more varied than those of the sea floor. Typical continental mountains are made of a broad spectrum of rock types that include volcanic rocks; clastic, chemical, and biogenic sedimentary rocks; plutonic rocks of mafic through felsic composition; and variably metamorphosed equivalents of all of these. Continental mountains composed largely of nonvolcanic rocks are grouped conveniently into classes based on their deformational histories: **fold** mountains (compressional), **fault-block** mountains (tensional), and **erosional** mountains (upward or undeformed).

We must be careful in classifying continental mountains because their histories tend to be longer and more complex than those of oceanic mountains. As an example, consider the Adirondack Mountains, part of the Appalachian mountain system in northern New York State. Its rocks experienced intense compression, very high grade regional metamorphism, and plutonism in Late Precambrian times, and they probably belonged to a fold mountain system 1 billion years ago. Subsequent erosion leveled these mountains by the beginning of the Paleozoic Era, approximately 600 million years ago. Much later uplift followed by stream and glacial dissection has once again created a mountain range in northern New York, but the earlier history is ignored in classifying it. The modern Adirondacks are the result of uplift and erosion. They therefore are classified as upwarped or erosional mountains.

Volcanic mountains

Vast outpourings of lava on the continents generally form plateaus, such as the Columbia and Snake River Plateaus described in Chapter 5, rather than mountains. Indeed, while all oceanic mountains are volcanic, very few continental mountain systems are, and those few are located near ocean basins at the continental margins. We will examine briefly two types of continental volcanic mountains: the Andes and Cascades.

The Andes mountain system The Andes Mountains extend more than 6000 km along the west coast of South America and contain some of the highest peaks in the world. The Andean system is similar to oceanic island arcs in so many ways that it almost certainly must have formed by the same type of process: plate collision and subduction. The similarities include the following:

1. The Andes are situated at a plate boundary.
2. They are bounded on the west by a trench (the Peru-Chile trench).
3. They are composed largely of andesitic, rhyolitic, and basaltic lavas; tephra; and volcanoclastic sedimentary rocks.
4. The rocks are tightly folded, and thrust faults are common in some parts of the system.

Most geologists believe that the Andes are the result of subduction of the Nazca plate beneath the South American plate (Figure 18.10).

However, there are important differences between the Andes and the oceanic island arcs. In addition to the volcanic and volcanoclastic rocks, limestones, dolomites, and old (Precambrian) gneisses and schists are present. An extensive batholith composed to a great extent of granitic rocks also occupies a prominent place in the Chilean and Peruvian portions of the system. Normal and strike-slip faulting are more prominent in Andean evolution than in most island arcs. This difference simply may reflect basic differences between continents and oceans. The causal mechanism — subduction — is probably the same, but continental rocks seem to respond somewhat differently than oceanic rocks.

The Cascade Range The origin of the Cascade Range is less clear. We saw in Chapter 5 that this range consists of several large andesitic

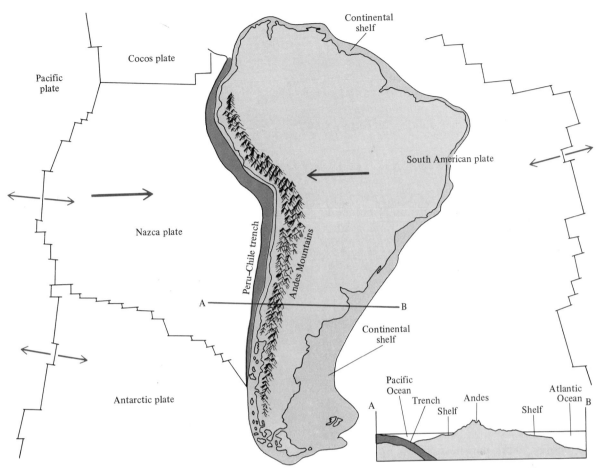

Figure 18.10 Origin of the Andes mountain system by subduction of the Nazca plate beneath the South American plate.

stratovolcanoes in Washington, Oregon, and northern California. Rhyolitic and basaltic rocks also are abundant so that the composition of the range is similar to that of an island arc, although the resemblance stops there. The Cascades are near the western edge of the North American plate, but there is no trench offshore as there is in South America near the Andes. Furthermore, the rocks of the Cascades are undeformed, showing none of the folding and thrusting found in island arcs or in the Andes. Compositionally, then, the Cascades are similar to oceanic island arcs, but structurally they are more like Hawaiian-type mountains.

The compositional similarities are so great that many geologists suggest a history of sub-duction for the Cascades. They postulate that the North American continent may have ridden in some way over the subduction zone and that a trench and a region of intensely deformed rock are buried beneath the continental rocks. This explanation is by no means universally accepted, however. The origin of the Cascade Mountains is one problem that is not answered so satisfactorily by the plate tectonic model as others.

Fold mountains

Most of the world's greatest continental mountain ranges are the result of compressional deformation involving a more varied rock assemblage than those of the volcanic mountains. Thus, the Himalayas, Alps, Urals, Caucasus, Pamirs, Tien Shan, Pyrenees, and Appalachians are characterized by tightly folded rocks and owe their height to buckling produced by enormous compressional forces. Fold mountains are the most complex type of mountain, and each fold mountain system has unique features not found in others. To illustrate the nature of fold mountains, we will examine the Appalachian Mountains, one of the most intensely studied mountain systems in the world. The Appalachian Mountains extend along the east coast of North America from Alabama to Newfoundland, a distance of 3000 km. The visible part of the system is nearly 600 km wide, but sediments of the Atlantic and Gulf Coastal Plains cover the Appalachian system on the south and east, masking its full extent.

Rocks of the Appalachian Mountains   Nearly every type of rock can be found somewhere in the Appalachians. Clastic, chemical, and biogenic sedimentary rocks of all kinds; marine and terrestrial deposits; and ultramafic, mafic, intermediate, and felsic plutonic and volcanic rocks all crop out in the system. In addition, nearly every grade of regionally and contact-metamorphosed rocks can be found. Not all these rocks occur in all places, however. Some regions, such as western Pennsylvania, New York, and Alabama, contain only unmetamorphosed

sedimentary rocks. Others, such as southern New Brunswick, New England, and parts of Virginia, are made up largely of metamorphosed sedimentary and volcanic rocks.

The Appalachian Geosyncline   In addition to the great *variety* of rock types, one of the unique features of fold mountains is the great *thickness* of the stratified rocks. Estimated thicknesses of 10,000 to 20,000 m were recognized early in the study of the Appalachians. In one of the first geologic studies in the United States, James Hall showed in 1857 that the thickness of rock units deposited during a given interval of geologic time increased sharply from west to east across New York and New England. This thickening is shown for rocks of Ordovician age in Figure 18.11.

Changes in sedimentary facies and amount of deformation accompany the change in thickness. The thinner sequence to the west consists of quartzose sandstones, limestones, and dolomites that are deformed only slightly. The thicker rocks to the east are interbedded graywackes, shales, and volcanic rocks that have been folded tightly.

Similar observations have been made all along the western side of the Appalachian system and in other fold mountains. Before deformation, the process by which fold mountains are made apparently includes a stage in which an immense depositional basin fills with extremely

Figure 18.11   Thickening and change in rock type in middle and upper Ordovician rocks of New York State. The change from limestones to shales and "flags" (siltstones and graywackes) is typical of geosynclines. *(Modified from C. O. Dunbar and J. Rodgers, Principles of Stratigraphy, © 1957, with permission of John Wiley & Sons, Inc.)*

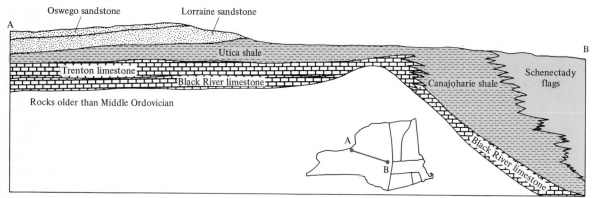

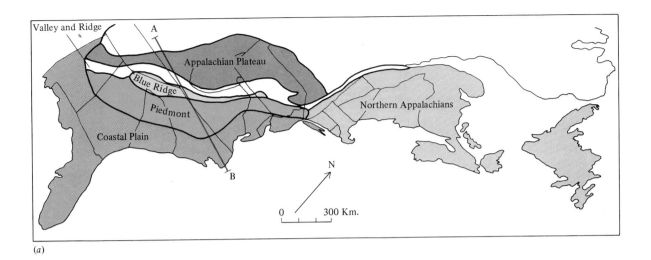

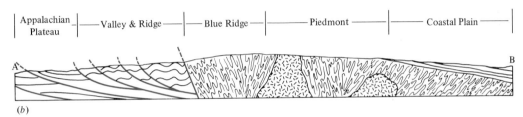

Figure 18.12   The Appalachian mountain system. *(a)* Regions. *(b)* Cross-sectional view; thrust faults are shown in color.

thick accumulations of sedimentary and volcanic rock.   These depositional basins are called **geosynclines,** and the one now represented by the Appalachian Mountains is called the **Appalachian Geosyncline.**   When deformation occurs, rocks of the geosyncline are the most intensely affected part of the mountain system.

Deformation   The type and intensity of deformation are variable in the Appalachians, with some regions being folded tightly while others apparently are undeformed.   A broad subdivision of the Appalachians based on the type, age, and deformation of rocks illustrates the complexities of fold mountains (Figure 18.12).

The **Blue Ridge province** is a region of the Appalachians that is made up of Precambrian rocks that have been deformed multiply (folded and then refolded several times) and recrystallized intensely during regional metamorphism.

Sedimentary and volcanic rocks of Paleozoic age showing more variable metamorphism constitute the **Piedmont province.**   In contrast, the **Valley and Ridge province** contains sedimentary rocks that have been folded tightly but metamorphosed only slightly.   Anticlines and synclines several kilometers across can be traced along strikes for tens of kilometers (Figure 18.13).   At the western margin of the Appalachian mountain system are the flat-lying, unmetamorphosed sedimentary rocks of the **Appalachian Plateau province.**   These are among the youngest rocks of the entire mountain system, ranging in age from the middle to late Paleozoic Era.   Surface mapping suggests that these rocks are undeformed, but data from wells reveal that even these flat-lying beds have undergone large-scale thrust faulting (look back at Figure 17.27).

The deformational history of the Appalachians is both long and complex, consisting of

Figure 18.13 Large-scale folds in the Appalachian Mountains. Strata that are resistant to erosion stand up in relief in this satellite photograph and trace out a pattern indicative of tight folding (compare with Figure 17.19). *(NASA photo, research by Grant Heilman photography.)*

major mountain-building events called orogenies spread out over nearly 800 million years of earth history. This history includes major compressional events in late Precambrian (Grenville orogeny), Ordovician (Taconic orogeny), and Devonian times (Acadian orogeny), and a climactic event near the end of the Paleozoic Era (Alleghenian orogeny). Each orogeny consisted of several episodes of folding, usually accompanied by regional metamorphism, plutonism, and contact metamorphism.

Ophiolites in the Appalachians In the Appalachians and many other fold mountain systems, rocks of the ophiolite suite are thrust into contact with highly deformed metasedi-

mentary and metavolcanic rocks. The intense shearing and metamorphism, and in some instances later episodes of folding, make identification of the complete suite difficult. However, there are enough details to show that the ophiolites follow a systematic distribution pattern throughout the Appalachian system (Figure 18.14). Ophiolites in the oceanic island arcs can be explained by a combination of subduction and obduction, but how can these pieces of the sea floor be found in the midst of mountains hundreds of kilometers from the sea and thousands of kilometers from the nearest trench?

Origin of fold mountains Plate tectonicists believe that the intense compression needed to produce fold mountains can come only from the collision of two lithosphere plates. We saw earlier that collisions between two oceanic plates result in island arc mountains and that collisions between an oceanic plate and continental plate produce Andean-type mountains. When two continental plates collide, the product is a fold mountain belt.

Imagine what would happen if the Mid-Atlantic Ridge were to stop spreading and other forces were to push North America and Europe toward one another (Figure 18.15). Rocks of the Atlantic floor would be subducted as the ocean shrank, and an island arc would develop (Figure 18.15*b*). Sediments derived from the continents would be deposited on the continental shelves of both continents, but the volcanoclastic debris from the island arc would be added to normal-turbidity current deposits on the slopes, rises, and abyssal plains. This produces the thick sequence of sedimentary and volcanic rocks and the facies changes characteristic of geosynclines.

As the Atlantic continued to close, the volcanic and turbiditic sediments would be folded strongly and thrust-faulted (Figure 18.15*c*). This deformation would continue until the shelves of the two continents actually collided. At that point, Europe and North America would be welded, or sutured, into a single giant continent. Fragments of obducted sea floor would be preserved as ophiolites along the zone of suturing. The final collision that formed the current Appalachians probably took place in the late Paleo-

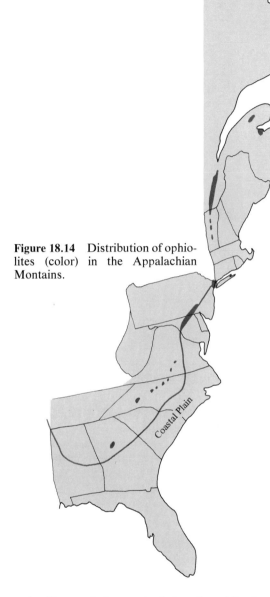

Figure 18.14   Distribution of ophiolites (color) in the Appalachian Montains.

once again separates North America and Europe. Sea-floor spreading apparently began again in Triassic and Jurassic times, rifting the "supercontinent" that had formed by plate collision. The break took place to the east of this suture so that part of ancient Europe was left behind to become part of the modern North American continent (Figure 18.16).

Fault-block mountains

Continental mountains that form by large-scale normal faulting are called **fault-block** mountains. The rocks that make up these mountains may be of any type and may have had complex histories prior to the tension that caused the faulting. We will look at two examples of fault-block mountains from the American Southwest—the Sierra Nevada Range of California and the Basin and Range structural province of Utah and Nevada.

Sierra Nevada Range, California   The Sierra Nevada Range is a fault-block mountain range that formed by the tilting of a single large crustal block (Figure 18.17). This block is nearly 600 km long and 100 km wide and is capped by mountain peaks rising over 4000 m above sea level. The prefaulting events were similar to those found in fold mountains. A geosyncline formed, filled, and was deformed intensely during late Paleozoic and early Mesozoic times. Folding was followed by the repeated emplacement of granitic plutons, which resulted in the formation of the huge Sierra Nevada batholith. The region then became tectonically quiet and was eroded.

zoic Era and is marked by the Alleghenian orogeny.

Similar histories are envisaged for other fold mountains that are far from the sea. Geologists point to the Ural Mountains in the Soviet Union as a result of suturing Asia to Europe and point to the Himalayas as a product of collision between India and Asia. The evolution of the Appalachians is more complicated than that of these other ranges because today an ocean

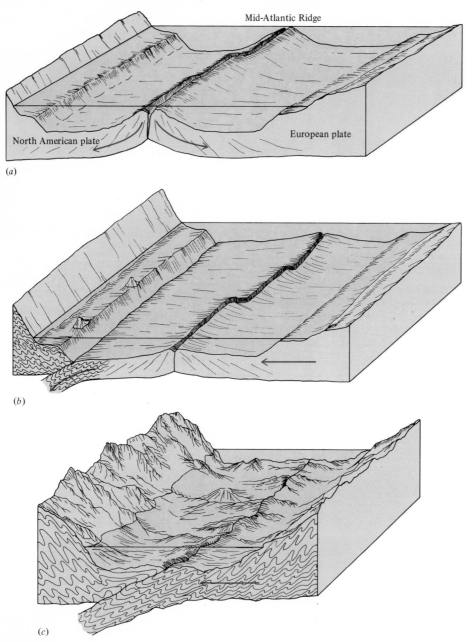

Mid-Atlantic Ridge

North American plate   European plate

(a)

(b)

(c)

Figure 18.15 Hypothetical suturing of Europe and North America to produce a single continent. *(a)* Current spreading state along the Mid-Atlantic Ridge. *(b)* Spreading stops and subduction begins. *(c)* Continental plates collide and ocean crust disappears entirely.

During the Pliocene Epoch, after millions of years of orogenic inactivity, normal faulting began along what is now the eastern margin of the range. A steep scarp formed along the east flank of the range where uplift was greatest, and the land sloped more gently toward the west into what is now the Great Valley of California. Erosion of the tilted block produced the rugged topography of the range, topped by Mt. Whitney.

Basin and Range province, Utah and Nevada The Basin and Range province is a series of horsts and graben typical of areas subjected

Figure 18.16 The Paleozoic suture between North America and Europe in the New England area. *(Based on data from Osberg, 1978.)*

to tension (Figure 18.18). A complex orogenic history preceded the faulting as in the Sierra Nevadas, but it was the normal faults that caused the present topography.

The tensional forces responsible for the structure and topography were first applied in late Tertiary times and continued into the Quaternary Period. Volcanism accompanied the faulting in some parts of the province, particularly in northwestern Nevada. The lavas and tephra used deep-seated fractures formed during faulting to reach the surface, but there are no outpourings comparable to those found at tensional sites in the oceans (the ridges). The volcanic rocks are more varied than those of the ocean ridges, characteristically consisting of abundant rhyolites as well as basalts. Andesites are absent. Not all the deformation is tensional. Some of the major faults appear to have had episodes of strike-slip offset as well as dip-slip movement.

Origin of fault-block mountains Uplift and tension in the oceans are indicated by the ocean ridges, but similar forces seem to produce very different results on the continents. This may be another indication of the differences between continents and oceans. However, it may be the result of a continental lithosphere plate sliding over an active ocean ridge spreading center. The East Pacific Rise is now truncated by the San Andreas Fault system in the Gulf of California, but at one time an active segment of this ridge may have been located *beneath* the continent. Basin and Range and Sierra Nevada–type structures may be the surface expression of rifting after it was transmitted through the entire thickness of the North American plate.

Erosional (upwarped) mountains

Some mountains owe their relief and elevation to erosional dissection of a region that is being uplifted without apparent deformation. This gentle uplift is called **epeirogeny** to distin-

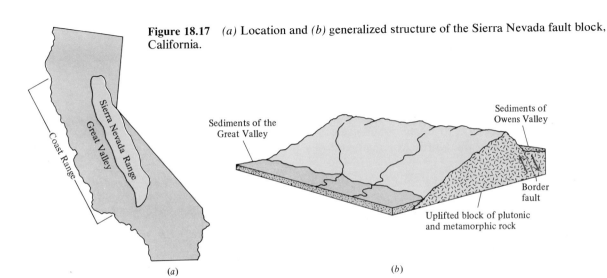

Figure 18.17 *(a)* Location and *(b)* generalized structure of the Sierra Nevada fault block, California.

Sediments of the Great Valley

Sediments of Owens Valley

Border fault

Uplifted block of plutonic and metamorphic rock

(a)

(b)

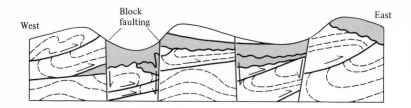

West   Block   East
                        faulting

Figure 18.18 Diagrammatic sketch of the structure of the Basin and Range province, southwestern United States. Early folding (dashed black lines) and thrust-faulting (heavy black lines) episodes were followed by block faulting (color lines).

guish it from orogeny, the intense deformation that produces fold mountains. As stated earlier, the modern Adirondack Mountains are upwarped mountains. Processes similar to those which formed the Adirondacks may be operating now under the Colorado Plateau. Extremely rapid uplift has elevated this region to heights thousands of meters above sea level, but a mountainous topography has not developed because of the climate. The region is so arid that neither stream nor glacial dissection has operated as in the Adirondacks. However, some dissection has taken place, and the major through-flowing stream, the Colorado River, has been able to carve the Grand Canyon through the plateau.

The Rocky Mountains constitute one of the most complex mountain systems in the world. Their evolution has included aspects of nearly every type of continental mountain including the geosynclincal sedimentation, deformation, and metamorphism of fold mountains; and the steep-sided fault-bounded blocks of fault-block mountains. However, the latest activity appears to be regional upwarping. In some instances this upwarping has reactivated earlier faults because these were regions of crustal weakness.

The forces that cause epeirogeny are not understood entirely. They may be part of large-scale compressional or tensional stress fields. Uplift of the Rockies thus may be related genetically to block faulting of the adjacent Basin and Range province, or to some grand-scale compression involving the North American and Pacific plates.

Dome mountains

The Black Hills of South Dakota do not fit into any of the categories of continental mountains listed in Table 18.1. They are one of the

best examples in the United States of **dome mountains** — mountain ranges with generally circular outlines that cannot be explained easily by the plate tectonic hypothesis.

Dome mountains form by a pistonlike uplift of rocks in a localized area, as shown in Figure 18.19. The upper layers are arched into a domal structure, and erosion strips away the cover rocks to expose the older units in the core of the structure. In the Black Hills, rocks of Paleozoic and Mesozoic age are arched over the Precambrian core of the dome. It is in the Precambrian core that Mt. Rushmore (part of a granitic pluton) and the famous Homestake Gold Mine (in metamorphic and plutonic rock) are located.

Figure 18.19 Origin of dome mountains. The dome forms by pistonlike upward movement that arches overlying rocks. The overlying strata originally were horizontal.

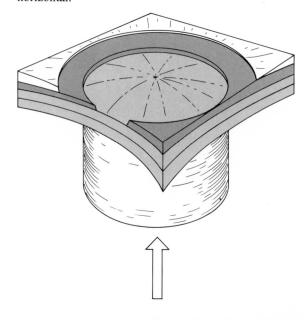

Dome mountains do not fit into the plate tectonic model because they appear to be a result of dominantly vertical forces acting in the center of a lithosphere plate. The other types of mountains discussed earlier form by dominantly lateral forces (compression or tension) at plate boundaries or by volcanism in midplate areas. In Chapter 21 we will examine models of mountain building that involve vertical forces and see whether they can be reconciled with (or replace) the plate tectonic model.

COMPARISON OF OCEANIC AND CONTINENTAL MOUNTAINS

Although plate tectonic processes of sea-floor spreading, subduction, and plate movement seem to be responsible for all mountain building, it has become apparent that basic differences between continents and oceans are responsible for differences in their mountains. Those differences are more fundamental than can be explained simply by the lack of erosion in the oceans. They are inherent in the processes, composition, and ages of the mountains.

Ages of Continental and Oceanic Mountains

The oceanic mountains are invariably young. The oldest rocks of the ocean basins are of Jurassic age so that there can be no older oceanic mountain. The tensional forces that created the ocean ridges probably began during the Jurassic Period (possibly the Triassic), but the rocks at the ridge crests today are of Holocene age. The ridges are still growing (still being rifted). So too are the Hawaiian Islands, as indicated by the recent eruptions of Kilauea, and the island arc mountains of Japan and the Aleutians, as indicated by earthquakes and volcanic activity.

Continental mountains show a much wider age range. The volcanic mountains appear to be relatively recent features, and the Andes still are being uplifted. Fault-block activity continues in some areas but has ceased in others. Fold mountains are of varying ages. The Himalayas, site of the highest peaks on earth, are young. They and neighboring ranges such as the Pamirs and Tien Shan Range still are being

uplifted and deformed. In contrast, the Appalachians are a mature system. The last major compression to have affected them ceased about 250 million years ago, and since that time erosion has been wearing them away steadily.

Thus, the oceanic mountains have been produced over the past 130 million years, but mountains such as the Appalachians are the result of nearly 800 million years of tectonic activity. The forces that built the Appalachians stopped nearly 120 million years before the modern ridges began to form.

Several of the continents appear to have a core of ancient rock surrounded by progressively younger mountain ranges (Figure 18.20). It appears that the younger rocks are added onto the continental nuclei as elongate fold mountain systems and that continents grow larger with time. This process, known as **continental accretion,** requires that the youngest continental mountains be situated at the margins of the continents. We shall examine the consequences of this hypothesis in Chapter 21.

Volcanism and Deformation

All oceanic mountains are volcanic, and the largest mountains on the entire planet were formed by extrusion of lava along the tensional fractures that we call the ocean ridges. In contrast, few continental mountains are volcanic, and those few seem to be associated with compression rather than tension. The volcanism associated with tensional mountains on the continents (the fault-block mountains) is nowhere nearly so extensive as that in the ocean ridges. Is it easier for large volumes of magma to be extruded in the oceans than on the continents? If so, why?

The greatest continental mountains are compressional features, but the mountains produced by compression in the ocean basins are smaller and typically have extensive volcanic activity. The highest fold mountains on the continents—the Himalayas—have no volcanic activity. Is compression more important in continental mountain building than in oceanic mountain building? Why are sedimentary rocks so common in the continental fold mountains and so insignificant in the oceanic mountains?

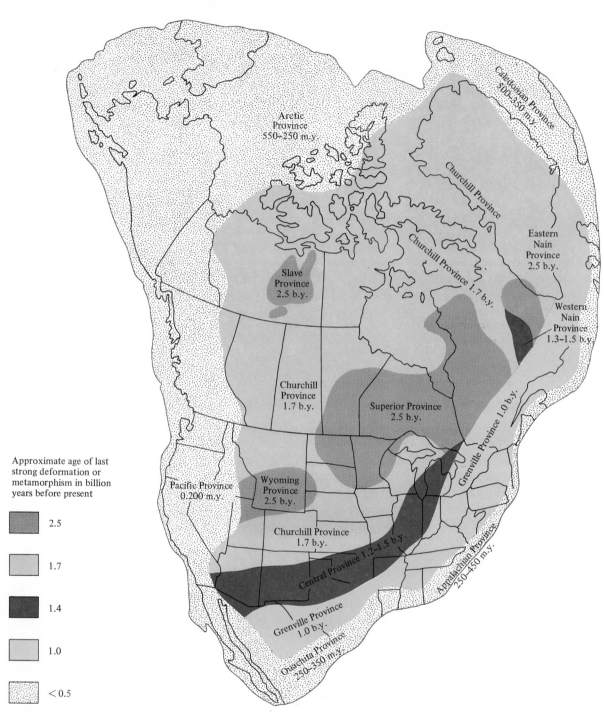

Figure 18.20 Continental accretion in North America. The oldest rocks in North America apparently are concentrated in the center of the continent, surrounded by successively younger rocks. (b.y. = billion years before present; m.y. = million years before present.)

Explaining the Differences

In order to explain the differences, we must know something about the composition of the crust and mantle beneath the oceans and continents. We also must learn whether processes are indeed different beneath these two regions or whether it is the compositional differences alone that are responsible for the variations described above. Evidence from several areas of geology—igneous activity, metamorphism, deformation, and now mountains—points toward the solution to the problem, but only direct examination of the planetary interior can produce conclusive solutions.

These solutions cannot be discussed until we see how geologists interpret the nature of the materials beneath continents and oceans. The methods used will be described in detail in Chapters 19 and 20. With the background from those chapters we can return to the questions raised here, and we will answer them in Chapter 21. The data about the earth's interior also help explain the causes of plate tectonic processes, and these too will be discussed in Chapter 21.

SUMMARY

Mountains are topographic features that rise more than 600 m above their surroundings. They occur singly or in groups called mountain ranges and mountain systems. Mountains are found in the centers and margins of continents and in similar positions in the oceans, and they may be found in any location on lithosphere plates.

Mountains are of several types. Oceanic mountains are all volcanic, composed of lava, tephra, and volcanoclastic sedimentary rocks. Ocean ridges are extremely long volcanic mountain systems composed of basalt. The ridges have a central valley whose shape resembles that of graben-like rift valleys on the continents. This morphology, the concentration of earthquakes, and the extrusion of basalt in this valley indicate that the ridges are sites of active tension. The ridges are sea-floor spreading centers—plate boundaries where new oceanic crust is formed.

Volcanic island arcs are bordered on their convex sides by the oceanic trenches. They are composed of andesitic, basaltic, and rhyolitic lavas; tephra; and volcanoclastic debris; and they have been subjected to intense compressional deformation. Parts of the oceanic crust, called ophiolites, are thrust into the pile of accumulated volcanic rocks during deformation. Island arcs form by subduction of one lithosphere plate beneath another. The subduction causes the compression responsible for the folding, and the heat energy from the two plates causes the volcanism.

Hawaiian-type islands and seamount chains are associated with neither tension nor compression. They are undeformed piles of basaltic lava that form by melting of lithosphere materials over a hot spot in the mantle. As the lithosphere plate moves, a volcano is removed from its heat source, and a new volcano forms over the hot spot. The chain of volcanoes thus marks the path of plate movement.

Mountains on the continents are more varied in composition. Some, such as the Cascade Mountains, are dominantly volcanic and may show no signs of deformation. Others, such as the Andes, also are volcanic but are deformed intensely like the rocks of island arcs. Volcanic mountains on the continents probably are the result of subduction.

Nonvolcanic mountains are classified by the type of deformation present. Fold mountains, such as the Appalachian mountain system, form by intense compression that results in large-scale folds and thrust faults. Fold mountains typically contain extremely thick deposits of sedimentary and volcanic rocks deposited in what are called geosynclines. Ophiolites are found in fold mountains, even though the mountains may be thousands of kilometers from the nearest trench. Fold mountains are a result of the collision of two continental lithosphere plates and in some instances mark the region along which two continents have been sutured together.

Fault-block mountains form by tensional forces acting on continents. Some, such as the

Sierra Nevada Range, are the product of simple tilting, whereas others result from complex horst-and-graben structures. Volcanism accompanies many fault-block mountains because magma can rise easily along the fault-related fractures. Epeirogeny—regional upwarping or uplift without deformation—produces the erosional or up-warped mountains.

Although both oceanic and continental mountains are the results of plate tectonic processes, there are significant differences between the two. These differences probably reflect fundamental differences in the composition of the lithosphere beneath the two kinds of surface region.

ADDITIONAL READINGS

Beginning Level

Dietz, Robert, "Geosynclines, Mountains, and Continent-Building," *Scientific American* Offprint 899, March 1972, pp. 30–38.
(A discussion linking the older concepts of geosynclines with more recent ideas of plate tectonics)

James, David, "The Evolution of the Andes," *Scientific American* Offprint 910, August 1973, pp. 60–69.
(A plate tectonic explanation of the formation of the Andes mountain system by subduction of oceanic plates beneath the South American continent)

Molnar, P., and P. Tapponier, "The Collision between India and Eurasia," *Scientific American* Offprint 923, April 1977, pp. 30–41.

(The formation of several of the world's highest mountain ranges is discussed in terms of the collision between two continental plates)

Sullivan, Walter, *Continents in Motion: The New Earth Debate,* McGraw-Hill, New York, 1974, 393 pp.
(A superbly written book describing the background and theory of plate tectonics in language for the layperson. Accounts of the challenges inherent in scientific research highlight the text. Written by the science editor of *The New York Times*)

Wyllie, Peter, *The Way the Earth Works,* Wiley, New York, 1976, 296 pp.
(The application of plate tectonics to mountain building in a very well written form)

CHAPTER 19

Earthquakes and Seismology

On Sunday, May 31, 1970, at 3:23 P.M., a major earthquake occurred in the Pacific Ocean about 25 km from the Peruvian coast. Shaking ground demolished nearly every adobe building in villages along the coast for a distance of 70 km and in cities 60 km inland. Only modern, well-built structures survived. More than 100 km from the earthquake site, ground motion triggered an avalanche of debris composed of mud, sand, and boulders weighing up to 1000 metric tons. The debris moved down from mountains as a wall 80 m high and swept over two towns with a velocity of over 300 km/h. In all, this earthquake killed or injured 120,000 people, destroyed 200,000 buildings, and left 800,000 homeless.

Mystery and fear surround earthquakes, feelings that are reinforced whenever a great one such as the 1970 Peruvian quake occurs. However, some earthquakes barely can be felt by humans, and many are so weak that they cannot be detected without sensitive instruments. Surprisingly, there *is* a beneficial side to earthquakes. It is mainly from the study of energy released during earthquakes—a branch of geology called **seismology**—that we have been able to infer the composition of the earth's interior. Seismologists have also shown how the record of earthquake energy reveals the mechanics of subduction and plate motion; they were among the early supporters of the plate tectonic model. In this chapter we will look first at the hazards associated with earthquakes and then at the types of information that seismologists have learned about the earth.

EARTHQUAKES

What Is an Earthquake?

An earthquake is a shaking of the ground. This shaking can be barely perceptible, a feeling similar to the vibration felt when a heavily loaded truck rolls by, or it can be so violent that it throws us out of bed or knocks us off our feet. Ground movement during the 1970 Peruvian earthquake leveled houses, and during the 1906 San Francisco earthquake, buildings, bridges, and water mains were broken. The ground moves in response to pulses of energy called **seismic waves** that are generated within the earth. Sudden displacements of rock, as in faults, release energy that passes through the earth until it reaches the surface as seismic waves, just as ripples radiate outward from a rock dropped into a pond (Figure 19.1). When the energy reaches the surface, the ground moves.

Causes of Earthquakes: The Elastic-Strain Hypothesis

After the 1906 San Francisco earthquake, geologist H. F. Reid made a detailed study of rocks along the San Andreas Fault that led him to propose the **elastic-strain hypothesis** as an explanation for earthquakes (Figure 19.2). According to Reid, deformation along the fault was mostly an accumulation of what he called *elastic* strain, which existed until rupture took place. Beds that had bent elastically prior to faulting (Figure 19.2b) returned to their unbent state after faulting (Figure 19.2c). Reid believed that en-

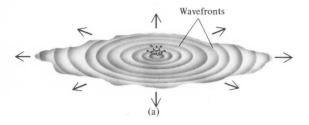

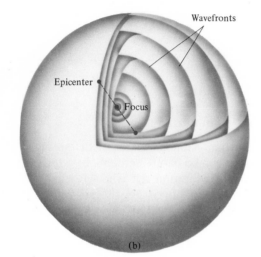

Figure 19.1 Generation of waves in *(a)* a pond and *(b)* the earth.

many kilometers below the surface in the area where the earthquake is felt. For example, the focus of the 1970 Peruvian earthquake was at a depth of approximately 40 km. In reporting the location of an earthquake, seismologists refer to the point on the earth's surface directly above the focus; this point is called the earthquake's **epicenter.**

Most earthquakes have shallow foci—within 60 km of the surface (Table 19.1). The elastic-strain hypothesis explains the origin of many of these earthquakes satisfactorily but runs into difficulty with the deeper-focus events. We saw in Chapter 17 that the deeper rocks are buried, the greater is their tendency to behave ductilely during deformation. It is highly unlikely that rocks deformed at the depths of deep-focus earthquakes would be able to store any large amount of elastic strain. Instead, they would deform ductilely. Recent studies of rocks along the San Andreas Fault show that there is a significant amount of ductile deformation associated with the faulting, indicating that Reid's model may be too simplistic. However, it does appear to explain many earthquakes.

ergy is stored in rocks while elastic strain accumulates before faulting, and is released during rupture. It is this energy that travels to the surface as seismic waves and causes the ground to shake.

The actual site of fault displacement, called the earthquake **focus** (plural, **foci**), thus may be

TABLE 19.1 Frequency of earthquakes during 1963–1966 based on depth of focus

Type of focus	Depth of focus, km	Number
Shallow	0–60	10,855
Intermediate	60–300	3,492
Transient	300–450	257
Deep	>450	739

Figure 19.2 The elastic-strain hypothesis of earthquakes. *(a)* Unfaulted vertical beds before displacement. *(b)* Beds bend elastically as displacement along the fault begins. *(c)* Rocks rupture and are offset along the fault. Elastic strain is released as seismic waves, and the beds return to their original, unbent shape.

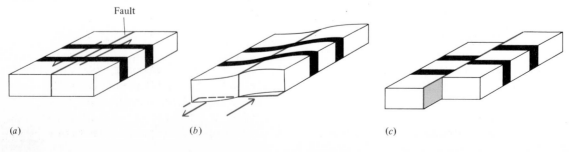

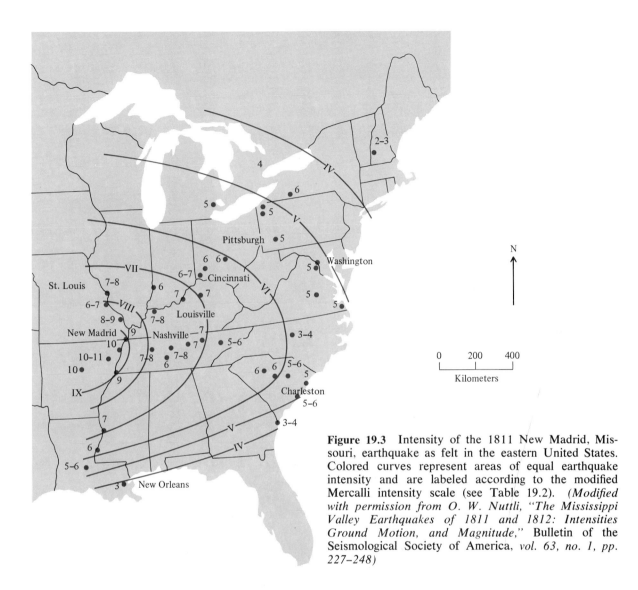

Figure 19.3 Intensity of the 1811 New Madrid, Missouri, earthquake as felt in the eastern United States. Colored curves represent areas of equal earthquake intensity and are labeled according to the modified Mercalli intensity scale (see Table 19.2). *(Modified with permission from O. W. Nuttli, "The Mississippi Valley Earthquakes of 1811 and 1812: Intensities Ground Motion, and Magnitude,"* Bulletin of the Seismological Society of America, *vol. 63, no. 1, pp. 227–248)*

Measuring the Strength of an Earthquake

The size or strength of an earthquake depends on how much energy had been stored in the deformed rocks before faulting. If a fault undergoes displacement by very small increments of motion, relatively small amounts of elastic strain will be involved, and numerous small-scale earthquakes will result. However, if elastic strain is built up over a long period of time before rupture, a single large displacement may occur, accompanied by a single release of all the stored energy in a catastrophic earthquake. Seismologists

measure the strength of an earthquake in two ways, referring to the intensity or magnitude of the event. Intensity is a measure of how much damage the earthquake causes, whereas magnitude is based on the amount of energy released at the epicenter.

Intensity

The **intensity** of an earthquake is a measure of the extent to which human-built structures are damaged. Earthquake damage generally is greatest close to the epicenter and decreases with distance from the epicenter (see Figure 19.3).

TABLE 19.2　Modified Mercalli intensity scale

Intensity value	Description
I	Not felt. Marginal and long-period effects of large earthquakes.
II	Felt by persons at rest, on upper floors, or favorably placed.
III	Felt indoors. Hanging objects swing. Vibration like passing of light trucks. Duration estimated. May not be recognized as an earthquake.
IV	Hanging objects swing. Vibration like passing of heavy trucks, or sensation of a jolt like a heavy shell striking the walls. Standing motor cars rock. Windows, dishes, doors rattle. Glasses clink. Crockery clashes. In the upper range of IV, wooden walls and frames creak
V	Felt outdoors; direction estimated. Sleepers wakened. Liquids disturbed, some spilled. Small unstable objects displaced or upset. Doors swing, close, open. Shutters, pictures move. Pendulum clocks stop, start, change rate.
VI	Felt by all. Many frightened and run outdoors. Persons walk unsteadily. Windows, dishes, glassware broken. Knickknacks, books, etc., off shelves. Pictures off walls. Furniture moved or overturned. Weak plaster and masonry D* cracked. Small bells ring (church, school). Trees, bushes shaken visibly, or heard to rustle.
VII	Difficult to stand. Noticed by drivers of motor cars. Hanging objects quiver. Furniture broken. Damage to masonry D, including cracks. Weak chimneys broken at roof line. Fall of plaster, loose bricks, stones, tiles, cornices, also unbraced parapets and architectural ornaments. Some cracks in masonry C. Waves on ponds; water turbid with mud. Small slides and caving in along sand or gravel banks. Large bells ring. Concrete irrigation ditches damaged.
VIII	Steering of motor cars affected. Damage to masonry C; partial collapse. Some damage to masonry B; none to masonry A. Fall of stucco and some masonry walls. Twisting, fall of chimneys, factory stacks, monuments, towers, elevated tanks. Frame houses moved on foundations if not bolted down; loose panel walls thrown out. Decayed pilings broken off. Branches broken from trees. Changes in flow or temperature of springs and wells. Cracks in wet ground and on steep slopes.
IX	General panic. Masonry D destroyed; masonry C heavily damaged, sometimes with complete collapse; masonry B seriously damaged. General damage to foundations. Frame structures, if not bolted, shifted off foundations. Frames cracked. Serious damage to reservoirs. Underground pipes broken. Conspicuous cracks in ground. In alluviated areas sand and mud ejected, earthquake fountains, sand craters.
X	Most masonry and frame structures destroyed with their foundations. Some well-built wooden structures and bridges destroyed. Serious damage to dams, dikes, embankments. Large landslides.

Water thrown on banks of canals, rivers, lakes, etc. Sand and mud shifted horizontally on beaches and flat land. Rails bent slightly.

XI   Rails bent greatly. Underground pipelines completely out of service.

XII   Damage nearly total. Large rock masses displaced. Lines of sight and level distorted. Objects thrown into the air.

* Key to 1956 revision prepared by Charles F. Richter, *Elementary Seismology,* W. H. Freeman, San Francisco, 1958, pp. 137–138.
Masonry A   Good workmanship, mortar, and design; reinforced, especially laterally, and bound together by using steel, concrete, etc.; designed to resist lateral forces.
Masonry B   Good workmanship and mortar; reinforced, but not designed in detail to resist lateral forces.
Masonry C   Ordinary workmanship and mortar; no extreme weaknesses such as failing to tie in at corners, but neither reinforced nor designed against horizontal forces.
Masonry D   Weak materials, such as adobe; poor mortar; low standards of artisanship; weak horizontally.
Source: Modified with permission from H. O. Wood and F. Neumann, "Modified Mercalli Intensity Scale of 1931," Bulletin of *the Seismological Society of America,* vol. 21, no. 4, pp. 277–288.

The **modified Mercalli intensity scale** used by seismologists and civil defense personnel is outlined in Table 19.2. The 1811 New Madrid, Missouri, earthquake was felt throughout most of North America settled by European colonists at the time. It barely was noticed in New England but caused extensive damage near the epicenter. No single value can be given to the intensity of an earthquake since intensity depends on distance from the epicenter.

Well-constructed buildings can survive more shaking than those which are constructed poorly. This was the case in the 1970 Peruvian earthquake (see Figure 19.4). A modern three-story building survived the earthquake with little damage, while the surrounding adobe and wood buildings were destroyed totally. The modified Mercalli scale takes these differences into account by classifying construction materials into the four categories listed in the footnote to Table 19.2.

Figure 19.4 Damage in the 1970 Peruvian earthquake. This three-story modern building survived the earthquake with only minor damage even though the surrounding adobe-and-wood houses were totally destroyed. *(Courtesy of Lloyd S. Cluff, Director, Geosciences and Vice President, Woodward-Clyde Consultants, San Francisco.)*

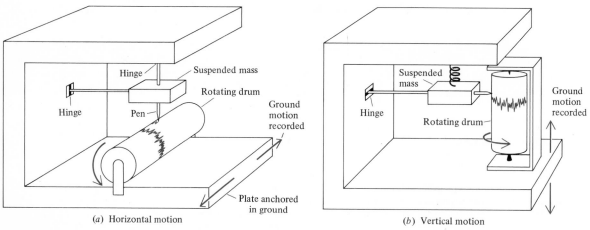

(a) Horizontal motion   (b) Vertical motion

Figure 19.5   Seismographs.   The suspended mass remains motionless, while the rest of the instrument, anchored to the ground, moves as the ground moves.   A record of the ground's horizontal (a) or vertical (b) movement is traced on a drum driven by a clock gear.

Magnitude

The **magnitude** of an earthquake is a measure of the energy released at its epicenter.   It is determined by measuring bedrock motion when the energy arrives at the epicenter in the form of seismic waves.   As the waves pass through rock, they cause it to vibrate, and the vibrations are detected and recorded by sensitive instruments called **seismographs**.   The principles by which seismographs operate are shown in Figure 19.5.

Seismographs are anchored in bedrock so that they can measure earth motion directly; they consist of a recording device (a pen and paper in Figure 19.5) attached to a suspended mass.   When the earth moves, the instrument moves with it; but because of its inertia, the sus-

Figure 19.6   A seismogram.   This is the actual seismogram record of the August 11, 1965 New Hebrides earthquake made by a north-south (horizontal) seismograph at Akureyri, Iceland.   *(NOAA National Geophysical and Solar-Terrestrial Data Center)*

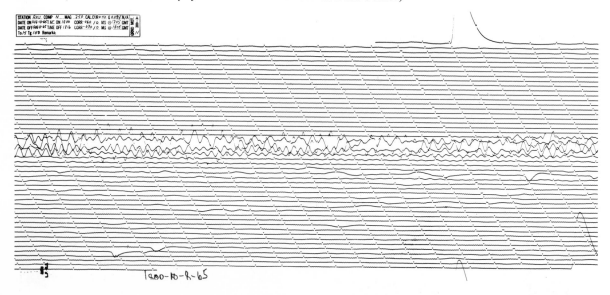

pended mass remains motionless. As the instrument moves, the pen attached to the mass traces out the movement in a record called a **seismogram** (Figure 19.6). Either horizontal (Figure 19.5a) or vertical (Figure 19.5b) motion can be measured, depending on how the mass is suspended.

Magnitude is calculated from the seismogram by measuring the amplitude (height) of the recorded waves. A large energy release produces high amplitudes and hence large magnitudes. Magnitudes are described according to the **Richter magnitude scale**, a system named after the American seismologist Charles F. Richter. In this system, magnitudes are expressed as whole numbers and decimals. The 1970 Peruvian earthquake, for example, measured 7.7 on the Richter scale; and the 1964 "Good Friday" Alaskan earthquake, 8.3.

The scale is designed so that each increase of one integer indicates a 10-fold increase in wave amplitude and an increase in energy release of 31 times. An earthquake of Richter magnitude 5.5 struck southern California in November 1979. The 1964 Alaskan earthquake was 3 magnitude units higher, indicating that its wave amplitudes were 1000 times greater (10 × 10 × 10) and its released energy nearly 30,000 times greater (31 × 31 × 31) than the southern California earthquake.

Earthquakes with Richter magnitudes greater than 7.0 are catastrophic when they affect heavily populated areas; they are called great earthquakes. Earthquakes with magnitudes lower than 5.0 probably would not cause much

damage to modern, well-constructed buildings, although they could destroy poorly or primitively built structures. An earthquake with magnitude lower than 2.0 would not even be noticed without a seismograph. Fortunately, the vast majority of earthquakes are of low magnitude, as shown in Table 19.3. The largest earthquakes ever recorded have magnitudes of about 9.0, and it is unlikely that there can be any much more powerful than these because rocks have finite strengths and can store only a certain amount of strain energy before rupturing.

Earthquake Hazards

Earthquakes cause millions of dollars worth of damage and the loss of thousands of lives each year. Most of the damage is caused by ground vibration at the epicenter, but, as in the case of the Peruvian earthquake, damage may take place hundreds and even thousands of kilometers away. In order to minimize damage, we must know the ways in which ground motion, rock and soil type, and location relative to the epicenter affect the type and amount of destruction.

Ground movement

Buildings are rigid structures. They are designed to move a little in response to the wind —the twin towers of the World Trade Center in New York City actually sway several centimeters in a strong wind—but they can bend only so much before breaking. Strong ground motion can destroy a building if it was not constructed to be earthquake-resistant. The taller the building, the greater the potential damage will be because the building's motion is magnified with height. To demonstrate this, construct a simple framework building with an Erector set or Tinker Toy and shake the base. The upper floors move much more than the lower ones. In a real building the upper floors thus are more prone to damage than the lower ones.

Ground motion can cause building failure in several ways. Witnesses to the December 1978 earthquake in Mexico City told of such great building motion that the upper floors of adjacent buildings, normally separated by a few meters,

TABLE 19.3 Frequency of earthquakes of different magnitudes (January 1963–June 1966)

Richter magnitude	Number
1–2.99	>20,000
3–3.99	1,101
4–4.99	9,937
5–5.99	3,918
6–6.99	299
>7.00	3

Figure 19.7 Earthquake destruction caused by the Santa Barbara earthquake California, 1925. *(NOAA, NSDGC.)*

actually collided. Ground shaking in the disastrous 1755 earthquake in Lisbon, Portugal, nearly leveled the city. Massive masonry in churches gave way and fell on worshipers, and both palaces and hovels collapsed. When the ground shakes, walls may collapse (Figure 19.7), objects may be thrown from shelves, and some buildings may fail completely. Some supposedly well-built concrete-floor structures are destroyed by a phenomenon called **pancaking** (Figure 19.8). The poured-concrete floors remain intact but are separated from the building walls. The upper floors fall onto the floors below, causing them to collapse until the entire building has been razed.

Earthquake damage often is more extensive in areas underlain by unconsolidated sediment than in those underlain by bedrock. Sediment lacks the rigidity of bedrock and as a result undergoes more complex and potentially more damaging motion. A bowl of jello illustrates the phenomenon. A sharp tap on the side of the bowl produces only slight vibration in the plate itself, but the jello quivers erratically. Soil and debris lying on bedrock behave in a similar way.

Liquefaction

Liquefaction—the sudden loss of strength of a′ mass of water-saturated sediment—was dis-

cussed in Chapter 10. The vibration caused by the passage of seismic waves through such deposits sometimes is enough to cause liquefaction, resulting in extensive damage to structures built on the sediment.

During the 1964 earthquake in Niigata, Japan, most of the poorly constructed buildings in the city were destroyed by ground motion. In addition, several modern buildings that had been especially designed to be earthquake-resistant had to be abandoned, even though they had not been destroyed (Figure 19.9). The sediments beneath these buildings liquefied, lost their structural strength, and caused the buildings to rotate and sink into the ground.

Sediments which are deposited rapidly and materials used in artificial landfill often retain large amounts of water in their pores and are particularly susceptible to liquefaction. Thus, the housing developments built on land reclaimed from the San Francisco Bay are particularly prone to damage by this process in a future earthquake.

Mass movements

Vibrational movement of both bedrock and unconsolidated sediments also can trigger mass movements at great distances from earthquake

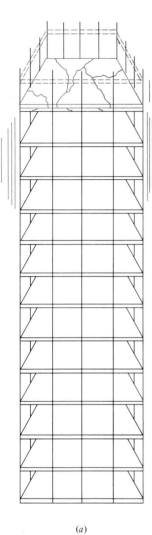

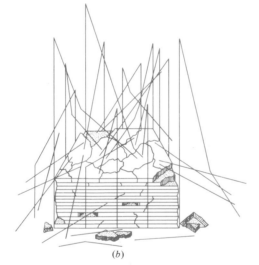

Figure 19.8 Collapse of a building by pancaking. *(a)* Top floor separates from walls and falls on the floor below. *(b)* Entire building collapses from the weight of falling floors.

(a)   (b)

Figure 19.9 The results of liquefaction. This building in Niigata, Japan, survived a 1964 earthquake intact but liquefaction of its substrate caused it to rotate and settle. *(Niigata Nippo Sha, Niigata City, Japan.)*

The Huascarán Debris Avalanche, Peru

One of the most destructive effects of the 1970 Peruvian earthquake was a massive debris avalanche that killed nearly 30,000 people when it buried the towns of Yungáy and Ranrahirca. This is an example of the danger that earthquakes pose even to areas far removed from the epicenter. The two towns were more than 100 km from the epicenter.

The course of the avalanche is shown in Figure 1. Ground shaking loosened debris composed of boulders, till, and glacial ice from the steep slopes of the north peak of Nevado Huascarán. The debris mixed with meltwater and flowed approximately 11 km from the mountain to the two towns in less than 3 min. From eyewitness accounts of the tragedy, the average velocity of the debris avalanche is calculated to have been over 300 km/h.

As the debris rushed downhill, it was split in two by a 250-m ridge. The main part of the flow overwhelmed the town of Ranrahirca, while a much smaller lobe buried Yungáy. However, it was more than sufficient to destroy the village. Only a few people in Yungáy managed to survive. Two hundred children watching a circus were spared when the avalanche roared past them, only a few meters from their grandstand seats. Ninety-three people managed to run up to the top of Cemetery Hill, the highest point in Yungáy and the only part of the village not completely buried by the debris.

The speed, violence, and horror of the event are shown in Figures 2 and 3. Perhaps the clearest picture of the effect on the area is given by one of the few survivors in the following account:

It was a Sunday afternoon, the 31st of May, 1970. I was in Yungáy taking friends on a tour of this area. At about 3:00 P.M. we stopped to take photographs of Nevado Huascarán. We photographed Huascarán

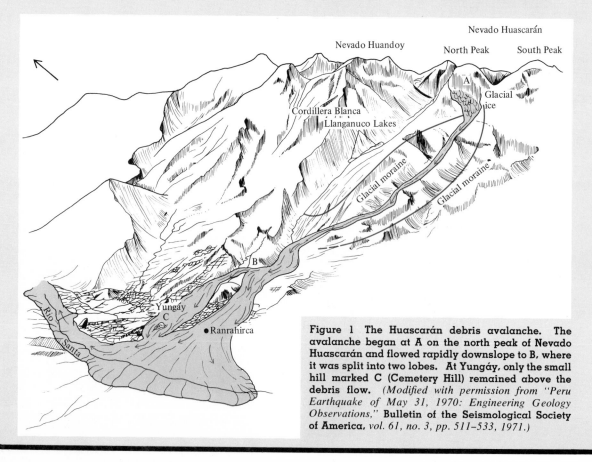

Figure 1 The Huascarán debris avalanche. The avalanche began at A on the north peak of Nevado Huascarán and flowed rapidly downslope to B, where it was split into two lobes. At Yungáy, only the small hill marked C (Cemetery Hill) remained above the debris flow. *(Modified with permission from "Peru Earthquake of May 31, 1970: Engineering Geology Observations,"* **Bulletin of the Seismological Society of America**, *vol. 61, no. 3, pp. 511–533, 1971.)*

Figure 2   Plaza de Armas, Yungáy, Peru before the disastrous 1970 debris avalanche. *(Courtesy of Lloyd S. Cluff, Woodward-Clyde Consultants, San Francisco.)*

Figure 3   The area formerly occupied by Plaza de Armas following the debris avalanche.
(Courtesy of Lloyd S. Cluff, Woodward-Clyde Consultants, San Francisco.)

through the telephoto lens and, at the time, commented about the prominent cracks in the glacier near the summit of the north peak. We talked about the danger of ice avalanches. We then drove the car along the main road between Yungáy and Ranrahirca, in a southerly direction through Yungáy. As we drove past the cemetery, which is situated on a hill, the car began to shake. The car rocked, and my first impression was that something was mechanically wrong with the car. It was not until I had stopped the car that I realized we were experiencing an earthquake. We immediately got out of the car and observed the effects of the earthquake around us. I saw several homes as well as the small bridge crossing the creek near Cemetery Hill collapse. My first thoughts were reflected back to remembering past destructive debris and mudflows that have occurred in this region as the result of bursting morainal dams in the upper valleys. It was, I suppose, after about one half to three quarters of a minute when the earthquake shaking began to subside. At that time, I heard a great roar coming from Huascarán. In looking up, I saw what appeared to be a cloud of dust and it looked as though a large mass of rock and ice was breaking loose from the north peak. My immediate reaction was to run for the high ground of Cemetery Hill situated about 150 to 200 m away. I began running and noticed that there were many others in Yungáy that were also running toward Cemetery Hill. About half to three quarters of the way up the hill, the wife of my friend stumbled and fell and I turned to help her back to her feet. At that time, the debris avalanche had reached

the point above Yungáy where a ridge separates Yungáy from the Llanganuco stream channel. The crest of the wave had a curl like a huge breaker coming in off the ocean. I estimated the wave to be at least 80 m high. At that time, I observed literally hundreds of people in Yungáy running in all directions and many toward the high ground at Cemetery Hill. I continued to run toward the top of the hill and was almost there when I felt a strong turbulent blast of wind. All the while there was a continuous loud roar and rumble. I reached the upper level of the cemetery near the top just as the debris flow struck the base of the hill and I was probably only about 10 seconds ahead of it. At about the same time, I saw a man just a few meters downhill carrying two small children toward the hilltop. The debris flow caught him and he threw the two children toward the hilltop out of the path of the flow to safety, although the debris flow swept him down the valley never to be seen again. I also remember two women who were no more than a few meters behind me and I never did see them again. I saw a man who was caught in the mudflow but was able to struggle out and he looked like a monster covered with thick brown mud. In looking around, I counted 92 persons who had also saved themselves by running to the top of the hill. Most of these were teenagers who were more agile and able to run; apparently, the older, the disabled, and the very young could not run fast enough to escape. It was the most horrible thing I have ever experienced and I will never forget it.

epicenters. The Huascarán debris avalanche described in the Perspective in this chapter is the most extreme type of earthquake-induced mass movement. Large-scale slumping is somewhat less violent but much more common. During the 1964 earthquake in Alaska, massive slumps destroyed much of the Turnagain Heights area of Anchorage, and slump-induced subsidence

caused extensive damage to downtown Anchorage (Figure 19.10).

Tsunamis (seismic sea waves)

Residents fleeing from the destruction caused by the 1755 earthquake in Lisbon, Portugal, felt safe on newly built stone wharves

Figure 19.10 Damage in Anchorage following the 1964 earthquake. Subsidence affected the downtown part of Anchorage as unconsolidated sediment shifted position during the earthquake. *(Alaska Earthquake, U.S. Geological Survey.)*

along the Tagus River. After all, there were no buildings to fall on them, and they could always escape from the stricken city by boat. But their safety was imaginary, and hundreds were killed when massive walls of water called tsunamis swept over the wharves. It could have been far worse. Tens of thousands of unsuspecting people were killed by tsunamis that slammed into the eastern coast of the Japanese island of Honshu in 1896 and along the coast of Java after the 1883 eruption of Krakatoa. Tsunamis are particularly dangerous because they often strike cities thousands of kilometers from the epicenters of the earthquakes that cause them, and no one is prepared.

Tsunamis may be generated when faults with vertical displacement affect parts of an ocean basin. A seismic wave is produced in the ocean that is similar to some seismic waves in rocks. These waves have amplitudes of only a few meters but wavelengths that may exceed 200 km. This kind of wave is so flat that a ship in midocean would never be aware that a tsunami had passed beneath it, even though tsunamis can travel faster than 500 km/h. Tsunamis are slowed by friction when they enter shallow water, but in bays and narrow channels the water

may be funneled into a solid wall up to 20 m high (Figure 19.11). The kinetic energy of such a wave enables it to do extensive damage, as shown in Figures 19.11 and 19.12.

There are warnings of tsunamis. Just before the onslaught of a seismic sea wave, water commonly recedes from the shoreline to far below the low-tide level. This is a warning to get to high ground quickly, but too many curious onlookers see it as a chance to collect stranded fish and shells. All too often they are drowned when the tsunami returns the water to the shore. Another warning: Tsunamis are not just single waves but are part of a wave train that includes many crests and troughs. It often takes hours for the energy to dissipate, and many people have returned to their homes after one wave only to be killed by later ones.

Other hazards

Fires accompanying earthquakes have accounted for more damage and loss of life than ground shaking and liquefaction. In areas where homes are made of wood and open fires are used for cooking, the collapse of wooden walls into fire pits often leads to conflagrations that raze

Figure 19.11 Bore from a tsunami (April 1, 1946) in the mouth of the Wailuku River, Hawaii. One segment of the bridge was destroyed by an earlier wave of the same tsunami. *(Courtesy of Shigeru Ushijama.)*

entire sections of cities. Before modern times, when nearly everything was built principally of wood, fires were the chief fear when the ground shook.

It is not much different today. We now use brick, steel, and concrete in buildings, but there is still much that is flammable and wood is still one of the most important building materials. Advanced technology lets us pipe natural gas to urban centers and distribute it underground to homes, but it also has made us vulnerable to fires that start when gas escapes from ruptured mains. Since water mains also are susceptible to breaking during earthquakes, fire fighting can be limited severely. During the 1906 San Francisco earthquake, broken water mains made it nearly impossible to fight the widespread fires.

Earthquakes may destroy dams, leading to extensive flooding. Highways can be closed when overpasses are knocked down. Earthquakes also may have profound psychological effects. The devastation during the 1755 Lisbon earthquake, coming as it did on a day when so

Figure 19.12 Tsunami damage, Kodiak, Alaska. Tsunamis associated with the Anchorage, Alaska earthquake of 1964 forced many boats onshore, like these shown berthed in the midst of Kodiak. *(Alaska Earthquake, U.S. Geological Survey.)*

Figure 19.13 San Francisco city hall following the 1906 earthquake and fire. *(NOAA, NSDGC.)*

many people were in church, affected religious and philosophical arguments in Europe for decades.

Earthquake Prediction

Much of the terror associated with earthquakes results from the fact that they occur so unexpectedly. A city can be transformed in minutes from a thriving metropolis to a shattered burning wreck, as was San Francisco in 1906 (Figure 19.13). Broken by the earthquake and then scorched by fire, the city took years to be rebuilt. But must earthquakes be so unexpected? If seismologists could learn to predict the time, epicenter location, and intensity of an earthquake, people could be evacuated to safe places, shelves could be cleared of fragile merchandise, windows could be boarded up, and fire-fighting teams could be deployed.

After decades of study, seismologists have had some success at predicting the three necessary factors: location, timing, and intensity. A few hours before it took place, Soviet scientists accurately predicted the time and magnitude of the November 2, 1978, earthquake in the Fergana Valley near the Tadzhikistan-Kirghizia border. In February, 1975, Chinese seismologists gave 24-h warning of a severe earth-

quake in the Manchurian province of Liaoning. Although it is cold in Manchuria in February, people slept in tents, animals and vehicles were taken out of barns and garages, and everyone waited. A severe earthquake indeed affected the region and brought down many houses and other buildings. Thousands of lives were saved.

These predictions were successful because seismologists have learned to recognize a few warning signals that are precursors of earthquakes. We do not yet understand fully how some of them are related to earthquakes, but they have proved helpful.

1. Groundwater flow seems to be disrupted just before severe earthquakes. Artesian wells lose pressure, and deep wells suddenly run dry.

2. Radon, one of the inert gases, is released from the ground along fault zones just before earthquakes. The Soviet and Chinese scientists who predicted the two earthquakes reported that radon emissions nearly doubled before the quakes.

3. There are detectable changes in the electrical conductivity of rocks in fault zones before actual rupture.

4. Specially designed surface antennas detected unusual and unexplained radio static just before the Fergana Valley earthquake.

5. Minor seismic activity increases and then decreases slightly just before a severe earthquake.

6. The ground surface tilts in some cases before earthquakes.

7. Many animals seem to sense imminent earthquakes by means that are as yet unknown. Witnesses to many different earthquakes report that dogs and horses become very restless and even frightened. Birds, including ground-dwellers such as chickens, leave the ground and roost in trees.

In areas where earthquakes are concentrated along active fault zones, seismologists can measure small changes in the relative positions of landmarks on opposite sides of the fault (Figure 19.14) to reveal the buildup of elastic strain. Strain meters anchored across the fault zones record the storage of strain as rocks are deformed before faulting. By knowing both the rate at which strain is increasing and the ultimate rupture point of the rocks involved, seismologists can estimate the time of faulting. For example, studies of the San Andreas Fault have shown that movement is not uniform along the trace of the fault. Small displacements have taken place along some segments of the fault, but others seem to be locked. These segments have not moved, even though adjacent segments have, and they are accumulating strain for what is potentially a major earthquake. Unfortunately, San Francisco is close to one such locked segment.

Earthquake prediction is still very much in its infancy, and seismologists are not able to forecast accurately when the locked segment of

Figure 19.14 Detailed surveying to determine earth motion in California along the San Andreas Fault. Survey lines are shown in color and faults in black. Changes in the lengths of five of the survey lines shown in (a) are illustrated in (b). These can be related to motion along the fault. (*Modified with permission from B. A. Bolt, Earthquakes, A Primer, © 1978 by W. H. Freeman and Co.*)

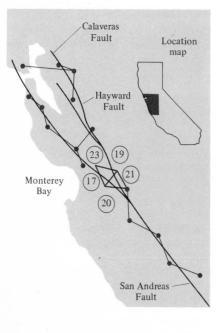

(a)

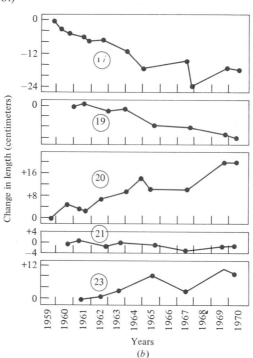

(b)

the San Andreas Fault that menaces San Francisco will be activated. Although there have been some successes, there have been failures as well. The 1979 southern California earthquake and those in northern California in January and May 1980 took place in one of the most intensely studied areas of the world and still were unexpected.

Earthquake Prevention

If earthquake prediction remains a few steps ahead of us, the possibility of actually preventing earthquakes from occurring is miles away and is little more than a dream. However, it may not be an impossible dream. We believe that severe earthquakes require the storage of large amounts of elastic-strain energy. If we can find some way to release that energy in small amounts instead of all at once, we may provide a safety valve that can stop great earthquakes from affecting heavily populated metropolitan areas.

The first step in the direction of earthquake prevention came about in the spring of 1962, when the Denver area was rocked by hundreds of low-magnitude earthquakes. Denver has had its share of seismic activity, but this seismicity was entirely abnormal for the region. Alarmed by this sudden increase in seismicity near a major population center, geologists and seismologists searched for an explanation.

When seismologists plotted the earthquake epicenters on a map, they appeared to be clustered in a circular area surrounding the U.S. Army Rocky Mountain arsenal. Chemical warfare weapons were manufactured at the arsenal, and a deep well had been drilled to dispose of the contaminated water used in the process. The Army started to pump waste water into the ground under high pressure just before the earthquakes began. When pumping was suspended for a year, seismic activity also subsided; when pumping resumed, so did the rash of small earthquakes.

The possibility that humans could make earthquakes happen was studied in a unique experiment by the U.S. Geological Survey under the supervision of Dr. Barry Raleigh. An abandoned oilfield in Rangely, Colorado, was turned into an underground seismologic laboratory. Geologists pumped water down into the wells to see whether they could repeat the Denver experience. As soon as water pressure reached a specific level, earthquake activity near Rangely increased; when the process was reversed and water was pumped out of the ground, the earthquakes ceased. The correlation between pumping and seismicity was too good to be a coincidence. For the first time in history, humans could turn earthquakes on and off at the flick of a switch.

What exactly did this switch do? Water increases fault activity in two ways. As water enters the ground, it passes into small cracks and previously existing faults, reducing friction. This lubrication may permit fault offset under the conditions of strain already present. Furthermore, the injection of water at high pressure increases the pore pressures within the fractured rocks, thus adding to their accumulated strain and promoting rupture. We cannot cause earthquakes to occur in unfaulted rocks, but we can make it easier for them to happen in rocks that already are fractured and in a state of stress.

Consider how this technique could be applied to the locked segments of major faults. After a severe earthquake has relieved most of the elastic strain in the rocks, periodic addition of water could cause small-scale displacements along an active fault zone. This effectively would unlock the fault segment and release the elastic-strain energy in many small increments rather than in a single large event. As a result, it is hoped that there would be no high-magnitude earthquakes but only a series of very small ones.

There are several reasons why this method cannot be used to save San Francisco from its next major quake. Since 1906 the fault segment that includes San Francisco has been accumulating elastic-strain energy steadily. To release that energy now would do exactly what we are trying to avoid—create a severe earthquake. Second, even after most of the energy has been released in the next earthquake, there is no guarantee that the injection of water will cause only small displacement. Only after further experimentation in sparsely inhabited regions can we begin even to think about using this technique near cities.

SEISMOLOGY

Our ideas about the composition and structure of the earth's interior come largely from seismology because despite all our technological advances, we have not yet been able to drill one-half of 1 percent of the way to the center of the planet. As seismic waves carry energy to the surface, they also carry information about the rocks they pass through. We study these buried rocks with seismographs the way a doctor uses a stethoscope to study the inner workings of the human body—from the outside. To understand how seismographs permit interpretation of the earth's interior, we first must examine the nature of seismic waves.

Types of Seismic Waves

In Chapter 16 we saw how waves move across the surface of the ocean because of the interaction of individual water molecules activated by the wind. Similarly, seismic waves move through the earth by the interaction of particles of rock. There are four different types of seismic waves. Two types (P waves and S waves), called **body waves**, transmit energy through the earth; the other two types (Love and Rayleigh waves), called **surface waves**, transmit energy along the surface. In each type, every particle of rock through which the wave passes moves in a path that causes it to collide with other particles, and then each particle returns to its original position.

P waves (longitudinal waves)

The passage of P waves through a rock is similar to the transmission of energy through that familiar toy, the Slinky (Figure 19.15). Rest a Slinky on a table and stretch it slightly to place its coils under tension. Then strike one end sharply. Pulses of energy can be seen passing through the coils as some are squeezed together (compression) and others are moved farther apart (dilatation) (Figure 19.15a). Energy passes from one end to the other as the zones of compression and dilatation move through the spring. The motion of individual

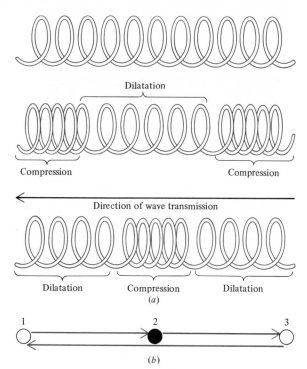

Figure 19.15 P waves (longitudinal waves). *(a)* Transmission of a P wave through a Slinky. *(b)* Motion of an individual particle.

particles in the spring is in a straight line parallel to the direction in which the wave moves—sometimes in the same direction as the wave, sometimes in the opposite direction (Figure 19.15b).

A wave of this type is known as a **longitudinal wave** because the back-and-forth motion of particles is *along* the direction of wave transport. In the earth these waves are called **P waves** because they are the first (the *Primary* wave) to reach a sesimograph after an earthquake.

S waves (transverse waves)

A simple experiment shows the movement of particles during passage of an S wave. Tie one end of a clothesline to a doorknob and hold the other so that there is some slack in the line. A sharp snap of the wrist will send energy through the rope in a wave resembling a writhing snake (Figure 19.16a). The motion of particles in the rope is in a straight line—a back-and-

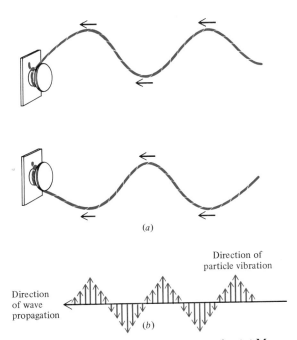

Figure 19.16 S waves (transverse waves). *(a)* Movement of a rope during passage of a transverse wave. *(b)* Particle motion in the rope is back and forth, perpendicular to the direction of wave propagation.

forth motion perpendicular to the direction in which the wave moves, not parallel to that direction as in P waves (Figure 19.16*b*)

Waves of this type are known as **transverse waves** because the particle motion is at an angle to the direction of wave transport. Seismologists call these waves **S waves** because they are usually the *Secondary* waves which are recorded at a seismograph.

Surface waves

Earthquake energy is transmitted along the surface by two different kinds of surface waves, each named after a pioneer seismologist. **Love waves** are similar to S waves (Figure 19.17*a*), but their transverse particle motion always lies *in* the surface along which the wave is traveling. **Rayleigh waves** differ from all other seismic waves in that the particle motion involved is not in a straight-line path but rather in a circular orbit (Figure 19.17*b*). This is similar to the movement of water in an ocean wave, but in the opposite direction.

Properties of Seismic Waves

During an earthquake, all these waves travel simultaneously to and along the earth's surface, where they are recorded by seismographs. Experiments using artificially generated seismic waves show that the four types of waves behave differently in different earth materials. We study seismograms in an attempt to reconstruct the behavior of the waves and, hence, the types of rock through which they have passed. Certain properties of seismic waves are used to interpret the nature of the earth's interior: ease of identification, velocity, and wave reflection and refraction, among others. We will look at these properties and then show how they are used to determine the structure of the earth.

Figure 19.17 Surface waves. *(a)* Love waves are a form of transverse wave in which the particle motion is in the surface along which the energy is transmitted. *(b)* Rayleigh waves are transmitted by a retrograde circular orbit of rock particles.

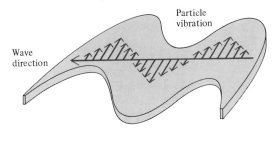

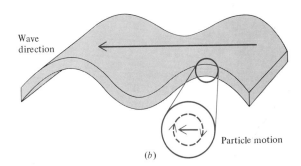

Identification of seismic waves

The first step in interpreting a seismogram is identifying each of the different waves by the trace it produces. Fortunately, each type of wave produces a characteristic record that enables it to be identified (Figure 19.18). The waves differ in their amplitude, wavelength, and period, as indicated by the ground movement recorded on the seismogram. For example, P waves and S waves have shorter periods than either type of surface wave, and the surface waves have higher amplitudes than the two body waves.

Velocity

The velocity of a seismic wave depends on the density and rigidity of the rocks through which it passes, as shown for P waves by the relationship

$$V_P = \sqrt{\frac{\text{rigidity}}{\text{density}}}$$

Although it seems from this relationship that seismic velocity should decrease with increasing rock density, measurements show that the opposite is true (Table 19.4). This puzzling phenomenon is caused by the fact that as the density of a rock increases, its rigidity also tends

TABLE 19.4 Seismic wave velocities in representative igneous rocks

Rock type	Density, g/cm^3	P wave, km/s	S wave, km/s
Serpentinite	2.60	5.67	2.81
Granite	2.64	5.94	3.53
Quartz diorite	2.92	6.46	3.69
Basalt	3.01	6.77	3.77
Dunite	3.26	7.54	4.28

to increase, and the increased rigidity more than counters the effect of density in the equation. Seismic waves do not all travel at the same velocity. P waves travel faster than S waves, as shown in Table 19.4, and thus reach seismographs before S waves.

Wave reflection and refraction

Like light and sound waves, seismic waves are reflected from the surfaces of objects (Figure 19.19). In the case of seismic waves, these surfaces mark the boundaries between materials of different densities, such as bedding planes. Seismic reflection studies also have provided the picture of the ocean floor materials shown in Figure 16.10.

Figure 19.18   Simplified seismogram showing distinction between seismic waves.

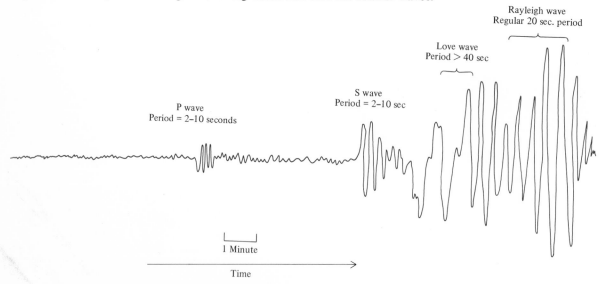

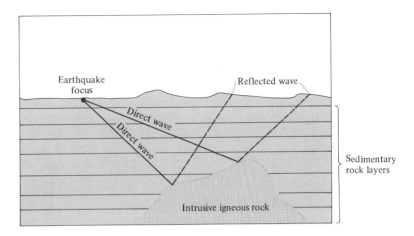

Figure 19.19 Seismic wave reflection.

Like light waves, seismic waves may bend (be refracted) when they cross boundaries between substances of different densities (Figure 19.20). The greater the density contrast between the materials, the greater the amount of refraction. As a result, seismic waves passing through layered rock do not travel in simple straight-line paths.

Seismic waves in liquids

Because of the transverse nature of particle motion, S waves can pass only through relatively rigid substances and cannot be transmitted through a liquid. If particles are free to move, as they are in liquids, they will not return to their position once displaced by the onset of the S-wave energy, and they will not transmit the energy along the direction of wave transport. Longitudinal waves, however, can pass through liquids.

Seismic Waves and the Earth's Interior

When an earthquake occurs, its seismic waves are recorded by hundreds of seismograph stations throughout the world. If the earth were perfectly homogeneous, we would be able to predict exactly how long it would take for P and S waves from an earthquake to reach different seismographs. These predictions are woefully inaccurate, even when we consider the fact that temperature and pressure changes with depth cause changes in the density and rigidity of rock. Thus the earth is not homogeneous.

There seems to be a global-scale layering based on rock density in which the densest rocks are found at the center of the earth and the least-dense rocks are found at the surface. After detailed examination of seismic wave behavior, seismologists divided the earth into three large regions—crust, mantle, and core.

Figure 19.20 Seismic wave refraction.

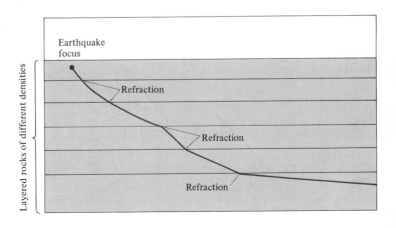

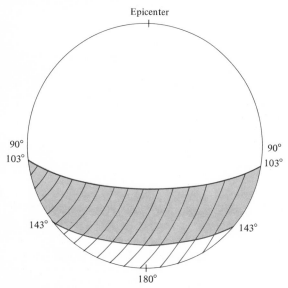

Figure 19.21   No direct S waves arrive in the S-wave shadow zone (color stripes), and no direct P waves arrive in the P-wave shadow zone (gray shading).

The mantle and core

To see how seismologists first deduced the existence of a sharp boundary between the mantle and core, let us examine the arrival of P and S waves from a single earthquake at seismographs around the world (Figure 19.21).

At angular distances of 0 to 103° from the epicenter, P waves predictably arrive at seismographs ahead of S waves because they are faster than S waves. Between 103 and 143°, however, neither P waves nor S waves arrive directly from the focus, as if this part of the surface were shielded from the waves by something within the earth. Beyond 143°, P waves "reappear" but S waves do not. The zones between 103 and 143° and between 103 and 180° are known as the **P-wave shadow zone** and the **S-wave shadow zone,** respectively.

Figure 19.22 shows the internal structure thought to be responsible for the shadow zones. From 0 to 103°, P and S waves follow smooth curved paths, indicating continuous refraction through rocks of gradually changing density. At a depth of about 2900 km, a boundary is postulated between an upper region of relatively low density rocks (the **mantle**) and a lower region of far denser material (the **core**). P waves entering the core are refracted strongly because of the great density contrast at the boundary. The paths followed by the P waves are controlled by the density differences in such a way that no P wave can emerge from the core on a path that will bring it to the surface between 103 and 143° from the earthquake epicenter. Refraction at the mantle-core boundary thus produces the P-wave shadow zone.

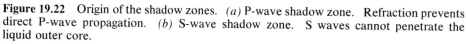

Figure 19.22   Origin of the shadow zones. *(a)* P-wave shadow zone. Refraction prevents direct P-wave propagation. *(b)* S-wave shadow zone. S waves cannot penetrate the liquid outer core.

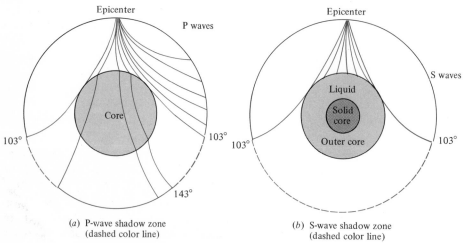

(a) P-wave shadow zone
(dashed color line)

(b) S-wave shadow zone
(dashed color line)

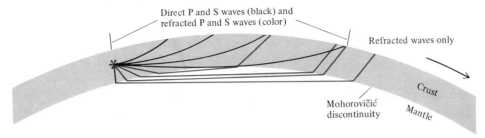

Figure 19.23 The crust-mantle boundary and the explanation of the Mohorovičić discontinuity.

The effect on S waves is more profound and reveals the nature of the outer core. S waves should behave just like P waves, but they do not. This cannot be explained by their differences in velocity, but it does make sense if the core is liquid, as shown in Figure 19.22b. Since S waves cannot penetrate a liquid, they do not pass through the core. Very detailed analysis of P waves that pass through the core indicates that only the outer core is liquid.

Composition of the core and mantle Seismology can tell us the size and density of the core and mantle but not what they are made of. For example, the core has a density of 10 to 11 g/cm³, far higher than that of silicate minerals. We have never seen rocks from the core and can only speculate on what they might be, but our speculations must involve material of appropriate density. Both the liquid and solid parts of the core probably are composed of a mixture of iron and nickel. Evidence from meteorites supports this hypothesis (see Chapter 23). Meteorites are thought to be fragments of an earthlike planet or group of planetoidal bodies, and many are similar in composition to earth rocks. Some, however, are composed of a nickel-iron alloy that has all the appropriate properties of rigidity and density indicated for the core by seismic studies. The presence of the earth's magnetic field (see Chapter 20) also can be explained if the core is a nickel-iron alloy.

The mantle is about 2900 km thick and ranges in density from 5.5 g/cm³ at the core to 3.3 g/cm³ at the contact with the crust. Xenoliths of ultramafic rock in diatremes (see Chapter 4) and in some basalts are thought to be derived from the upper mantle. Experimental mineralogic studies suggest that the upper mantle is made up of ultramafic rocks composed of high-density silicate minerals such as olivine, pyroxenes, and garnet. These minerals may be converted to even denser polymorphs in the lower mantle.

The crust

The crust is the outermost layer of the earth, that which we live on and are most familiar with. It is separated from the mantle by a boundary discovered by the Yugoslav geologist Andrija Mohorovičić in 1910 after a study of P and S waves from shallow-focus earthquakes. Mohorovičić discovered that two separate sets of P waves and S waves are recorded by seismographs located within 800 km of the epicenters of shallow-focus (less than 50 km) earthquakes, but that only a single set of these waves is recorded at greater distances (Figure 19.23).

To explain this phenomenon, Mohorovičić postulated a boundary between rocks of different densities that could cause refraction of the two body waves. According to this explanation, one set of body waves passes directly through the low-density crustal rocks, following a smooth curved path. The other pair of waves arrives later because it follows a longer path. Both P and S waves in the second set penetrate the crust-mantle boundary twice and are refracted each time. This boundary must be a sharp discontinuity, but the density contrast is not as great as at the mantle-core boundary. Furthermore, since S waves pass through the crust-mantle boundary, there is no liquid involved. In honor of its discoverer, the boundary between the crust and mantle is called the **Mohorovičić discontinuity.** It is more commonly referred to by the shortened form, **moho.**

Composition of the crust We have seen in earlier chapters (see Chapters 4, 5, and 7) that there are major differences between the materials beneath the continents and oceans. Seismologists have been able to explain the differences. The crust beneath continents consists of two parts—an upper layer with a density of 2.7 to 2.85 g/cm³ and a lower layer with a density of 2.85 to 3.1 g/cm³ (Figure 19.24). The upper layer is made of igneous, metamorphic, and sedimentary rocks with an average composition similar to that of granite, and it is this material that melts to form granitic magma. The upper crust layer is rich in *si*lica, SiO_2, and *al*umina, Al_2O_3, and is referred to as the **sialic layer** of the crust, or simply as **sial.** The lower layer is similar to basalt in composition and is rich in *si*lica, iron, and *ma*gnesium. It is called **sima** or the **simatic layer** of the crust.

In contrast, oceans are underlain by crust made of a single layer of simatic rock (Figure 19.24). This explains the absence of granite from the oceans (where there is no sialic material to melt and form granitic magma) and confirms the model proposed in Figure 4.26.

The thickness of the crust varies from region to region but in general is greatest beneath mountain ranges on the continents. It is of intermediate thickness beneath plains and continental shields, thinner beneath continental shelves, and thinnest in the ocean basins near ocean ridges. The sialic layer of the continental crust extends beneath the continental shelves but does not continue under the continental rise or slope. The true edge of a continent thus is the edge of its continental shelf rather than the shoreline.

Seismology and Plate Tectonics

Earthquakes are a sign of an active, dynamic planet, and most are caused by modern plate tectonic processes. Seismologists have made major contributions to plate tectonic theory, including the location of plate boundaries, the location and nature of subduction zones, the manner in which plates move across the earth, and the existence of transform faults.

Plate boundaries

When the locations of earthquake epicenters are plotted on a world map, a clear pattern emerges (Figure 19.25). Most of the world's

Figure 19.24 The crust beneath the continents and ocean basins.

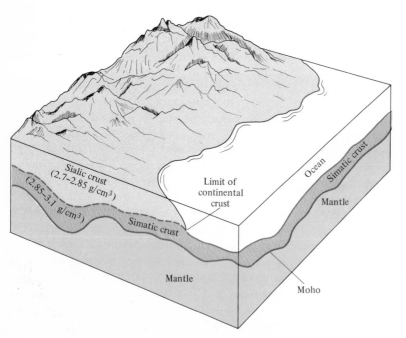

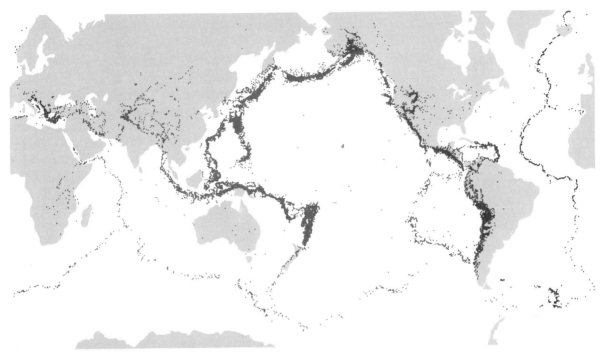

Figure 19.25 Distribution of earthquake epicenters throughout the world. *(Modified with permission from M. Barazangi and J. Dorman, "World Seismicity Maps Compiled from ESSA, Coast and Geoditic Survey, Epicenter Data, 1961–1967,"* Bulletin of the Seismological Society of America, *vol. 59, no. 1, pp. 369–380.)*

earthquakes occur in narrow belts that separate broad regions which are relatively aseismic (devoid of earthquakes). The belts of intense seismicity coincide with regions of volcanic activity (see Chapter 5) and deformation (see Chapter 17). According to plate tectonic theory, most seismic activity should occur at plate boundaries where lithosphere plates collide, grind past one another, or are rifted apart. Accordingly, the world's plate boundaries are delineated by the belts of seismic activity. The interiors of plates, far removed from intense deformation, are the aseismic areas.

Locating earthquakes In order to construct diagrams such as Figure 19.25, seismologists must be able to locate earthquake epicenters accurately. To do so, they take advantage of the different velocities of seismic waves to construct **travel-time curves** that show how the delay in the arrival of S waves and surface waves increases with the distance of the epicenter from the seismograph (color curves in Figure 19.26). They

then can match the observed delays between different types of waves on the seismograph (black lines) against these standard curves to determine distance from the epicenter. The method is quite simple. Imagine two cars setting out on a journey, one traveling at 100 km/h and the other at 80 km/h. The faster car travels the first 100 km in 1 h, but the slower car lags 15 min behind. After 200 km the fast car is 30 min ahead; after 300 km, 45 min. With each increase of distance, the slower car (seismic wave) falls farther behind in the time of its arrival at the destination (the seismograph). Since the two velocities are fixed relative to one another, the distance traveled can be calculated if the interval between arrivals is known.

By using the travel-time curve as shown, the distance of an epicenter from a seismograph can be determined, but this does not locate the earthquake because the epicenter can be in any direction from the instrument. To locate the epicenter, data from a minimum of three stations must be used, as shown in Figure 19.27.

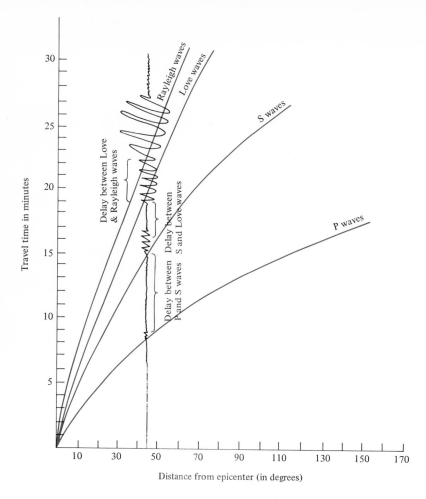

Figure 19.26 Travel-time curves. The distance of a seismograph from an earthquake epicenter can be calculated by matching the intervals between arrivals of the different seismic waves (black) with the travel-time curves (color). For the instrument shown, the epicenter is 45° of arc from the station.

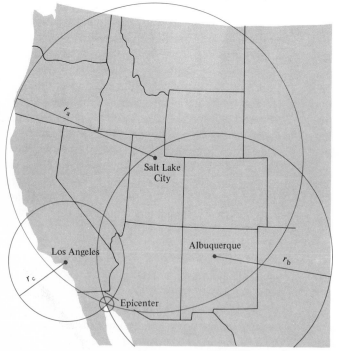

Figure 19.27 Location of an earthquake epicenter. A circle is drawn about each seismograph station, with the radius (r_a, r_b, r_c) equal to the distances from the epicenter determined from the travel-time curve. The intersection of the three circles marks the location of the epicenter.

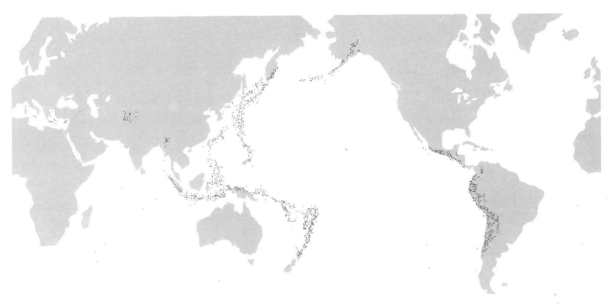

Figure 19.28 Distribution of deep-focus earthquakes (greater than 100 km). *(After Barazangi and Dorman, 1969, Bulletin of the Seismological Society of America 59, p. 369)*

Subduction zones

Deep-focus earthquakes not only are rare, as shown in Table 19.1, they also are restricted to a single tectonic setting (Figure 19.28). They are concentrated near ocean trenches and develop beneath island arcs and coasts where a continent is adjacent to a trench. Another pattern emerges when earthquake foci from a single island arc are plotted on a cross section of the arc-trench system (Figure 19.29). The foci increase in depth from the trench across the island arc, and plot in a thin band called a **Benioff zone.** Most geologists interpret Benioff zones as regions where subduction is taking place. The progressively greater focal depths mark the downward progress of the subducted slab as it is thrust beneath the overlying plate.

Plate motion: The asthenosphere

One of the questions raised by opponents of the plate tectonic theory was how huge plates could move through the solid earth. The answer apparently lies in a seismically unique region in the upper mantle. Seismic velocities increase with depth throughout most of the earth, but

P- and S-wave velocities appear to decrease at depths between 100 and 350 km (Figure 19.30). This region therefore is called the **low-velocity**

Figure 19.29 Benioff zones. The concentration of deep-focus earthquakes in regions called Benioff zones (color) is thought to indicate movement of subducted plates.

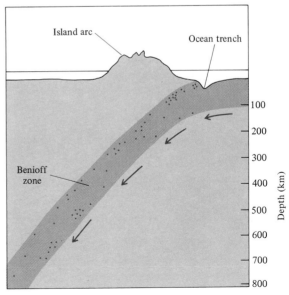

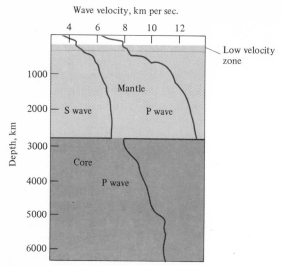

Figure 19.30 below and to left:

Wave velocity, km per sec.

Figure 19.30   Changes in P- and S-wave velocities with depth.

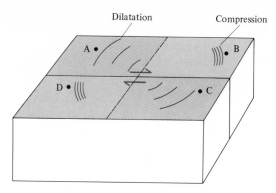

Figure 19.31   P-wave first-motion studies of earthquakes. The nature of displacement can be determined from the geographic distribution of compressional and dilatational first motion. Seismographs B and D (and others in the areas shaded in color) receive compressional first motion; A and C (and others in gray-shaded areas) receive dilatational first motion.

zone.   There is no evidence to suggest a sudden composition or density change at these depths, and the region is probably one of unusually low rigidity.

In fact, the rigidity is so low and the rocks so ductile that this is thought to be the zone on which the plates are gliding as they move. Rocks of the low-velocity zone are referred to as the

Figure 19.32   A seismic reflection record. The salt dome at the far right has penetrated and deformed the surrounding sedimentary rocks. It shows up clearly by its reflective properties. *(Exxon.)*

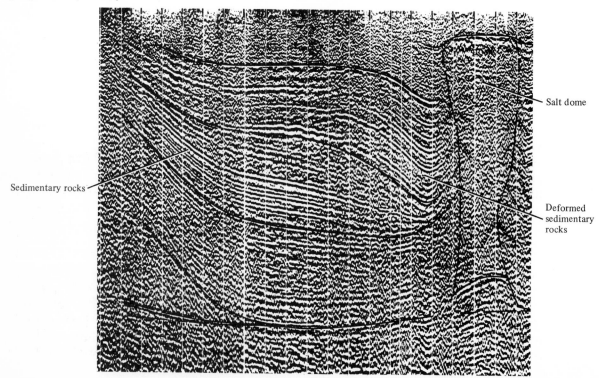

asthenosphere to distinguish them from the more rigid rocks of the moving plates above (the lithosphere). The low rigidity of the asthenosphere probably is due to the widespread presence of small amounts of basaltic magma intermixed with the solid rock.

Transform faults

Seismologists have learned how to determine the direction of fault motion and in so doing have confirmed the existence of transform faults. The method involves analysis of the first motion of the ground when P waves arrive at a seismograph. Consider a vertical fault along which lateral displacement occurs as in Figure 19.31.

When P waves arrive at seismographs B and D, the first motion will be compressional since the fault blocks are moving toward the instruments. At seismographs A and C, the motion is dilatational since the fault blocks are moving away from the instruments. If the attitude of the fault is known (as it would be if it were an oceanic fracture), the nature of the displacement can be deduced from the pattern of compressional and dilatational first motions.

Seismology and Resource Exploration

Today seismic exploration is an important tool in the search for oil and gas and in the exploration of the oceans (see Figure 16.20). The simplest method uses seismic-wave reflection. An energy source is used at or close to the surface to generate seismic waves. This source may be a dynamite blast, the shock of a large weight dropped to the ground from a truck, or the ignition of an explosive mixture of gases. The waves pass through the crust and are reflected back to the surface when they encounter the upper surface of a rock type that acts as a reflecting horizon.

Arrival times of the reflected waves are recorded carefully at the surface, using miniature seismometers called **geophones**. These data are fed into a computer along with information about rock types known to be present in the area. The computer is programmed to estimate the depth to the reflecting horizon, and if the depth changes because the bed is tilted or folded, this will be revealed in the analysis. Structural oil and gas traps such as anticlines, salt domes, and faults can be located by this method (Figure 19.32); and the sites best suited for exploratory drilling then can be selected.

SUMMARY

Earthquakes are episodes of ground motion caused by the transmission of energy from underground faulting to the surface. Ground motion causes buildings to collapse, and vibration may cause liquefaction and failure of the materials on which buildings have been erected. Indirect effects include damage by tsunamis (seismic sea waves), landslides, disruption of traffic, and fire.

There are four basic types of seismic waves: P waves, S waves, Love waves, and Rayleigh waves. Each is characterized by a unique style of particle vibration in the rocks through which it passes. Careful analysis of seismic waves permits geologists to speculate on the structure and composition of the earth's interior.

The earth is subdivided into crust, mantle, and core. The core consists of an inner solid part and an outer liquid part. Both inner and outer core seem to be composed of a mixture of nickel and iron. The mantle probably consists of mafic and ultramafic silicate rocks that increase in density with depth. Within the upper mantle is a zone of decreased seismic-wave velocities called the low-velocity zone. This represents a region of decreased rigidity (the asthenosphere) on which the overlying lithosphere plates are moving.

The crust is made up of familiar igneous, sedimentary, and metamorphic rocks and is separated from the mantle by an abrupt density change called the Mohorovičić (moho) discontinuity. The crust under the continents consists of upper sialic (approximately granitic) and lower simatic layers, whereas the oceanic crust is simatic only.

Earthquake epicenters are concentrated in linear belts that mark regions of greatest crustal instability and thus delineate plate boundaries. Deep-focus earthquakes are associated only with the oceanic trenches in what are called Benioff zones. These zones correspond to subduction zones and they mark the path of the descending plate.

QUESTIONS FOR REVIEW AND FURTHER THOUGHT

1. Describe the different types of particle motion associated with the transmission of the different seismic waves.

2. Discuss the different things that can happen when a P wave reaches a boundary between rock types of markedly different density.

3. What are the differences between the discontinuities that define the mantle-crust and mantle-core boundaries?

4. Why is the S-wave shadow zone larger than the P-wave shadow zone?

5. When an S wave reaches the outer core, what happens to the energy that it represents?

6. When P waves and S waves from a deep-focus earthquake reach a seismograph, one wave causes mostly up-and-down motion of the ground, while the other causes side-to-side motion. Which is which? Explain the differences by relating ground motion to the particle movement associated with the two waves.

7. In a long block of attached apartment houses, earthquake damage caused by ground motion was greatest in the two end buildings. Suggest an explanation.

8. An earthquake felt by people living on the forty-fifth floor of a high-rise apartment house might not be felt by a family living in a split-level ranch house. Why not?

ADDITIONAL READINGS

Beginning Level

Bolt, B. A., *Earthquakes: A Primer*, W. H. Freeman, San Francisco, 1978, 241 pp.
(An excellent paperback combining well-explained theoretical principles with examples of major earthquakes)

Oakeshott, G. B., *Volcanoes and Earthquakes: Geological Violence*, McGraw-Hill, New York, 1976, 143 pp.
(A brief but very clear treatment of the origin and effects of earthquakes)

Thomas, G., and M. M. Witts, *The San Francisco Earthquake*, Stein and Day, 1971.
(A book written for the nonscientist describing the effect of the 1906 San Francisco earthquake on the city and its people)

Intermediate Level

U.S. Geological Survey, *The Alaska Earthquake, March 27, 1964 – Field Investigations and Reconstruction Effort*, USGS Professional Paper 541, 1966, 111 pp.
(An analysis of the Alaskan earthquake with emphasis on the causes of damage and remedial measures to correct it and prevent recurrence of similar failures)

Advanced Level

Gutenberg, B., *Internal Constitution of the Earth*, Dover, New York, 1951, 439 pp.
(An advanced paperback showing how seismology can be used to learn details of internal earth structure, composition, and processes)

CHAPTER 20

Gravity and Magnetism

The earth is surrounded by invisible fields of gravity and magnetism—forces that attract objects to the planet and affect their movement on its surface. In earlier chapters we saw that gravity is the major cause of mass movement, erosion, and circulation of the atmosphere and oceans. Magnetism also plays a vital role in our lives, although its influence is not so obvious. It is the magnetic field that controls movement of a compass needle, giving us a sense of direction through trackless deserts and oceans. Even more important, the earth's magnetic field acts as a shield, protecting every living thing from harmful solar radiation.

Gravity and magnetism operate quietly, without the drama associated with earthquakes, but in their own quiet way the two forces of attraction supply vital information about internal earth processes and composition that cannot be obtained from seismology. Today, magnetic evidence provides some of the best support for the plate tectonic model. In this chapter we will look closely at the gravitational and magnetic fields and show how they can be used to study the earth's interior.

GRAVITY

The use of gravity as a research tool is different from the use of seismic waves. Seismic activity is sporadic, and seismic waves affect rocks on a relatively local basis. Gravity, in contrast, always is operating and affects all rocks regardless of their location. Although many people think of gravity as being the same everywhere on the earth, it is not, and even very small differences in the strength of the gravitational field can tell a lot about the earth's internal composition and active tectonic processes.

Isaac Newton showed that there is a force of attraction between any two objects, regardless of their composition. The force of gravity, G, depends on only the masses, m_1 and m_2, of the two objects and the distance between them, r:

$$G = \gamma \frac{m_1 m_2}{r^2}$$

where γ is a constant determined by the system of units of measurement. The force of gravitational attraction exerted by the earth on an object resting on its surface thus depends on (1) the mass of the object, (2) the mass of the earth, and (3) the distance from the object to the center of the earth.

Measuring Gravity

The strength of the gravitational field is measured with an instrument called a **gravimeter** (Figure 20.1). In a simplified gravimeter, a weight hung from a spring is pulled downward by gravity, stretching the spring. The change in the length of the spring reflects changes in the force of gravitational attraction on the weight and thus measures the strength of the gravitational field.

Gravimeters are calibrated so that they measure the acceleration that would be produced

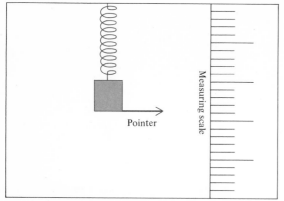

Figure 20.1 A gravimeter is used to measure the earth's gravitational field. In its concept, a gravimeter is a simple instrument. A mass suspended from a spring is attracted by gravity. The stronger the attraction, the longer the spring becomes.

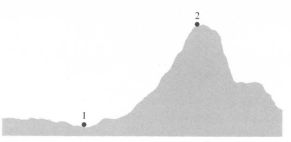

Figure 20.2 Change in gravity with change in elevation. Point 1 is closer to the center of the earth than point 2 so that r in the gravity equation is smaller for point 1 than for point 2. The gravitational attraction expected at point 1 thus should be greater than that at point 2.

by gravity on a falling body. At the equator, that acceleration is 978.049 cm/s/s. An acceleration of 1 cm/s/s is called 1 gal, after *Gal*ileo, who performed simple but important experiments concerning the acceleration of falling objects. The earth's gravitational field at the equator is thus a little more than 978 gal. Modern gravimeters can measure differences in gravitational field strength on the order of *milligals* (thousandths of a gal), and thus measure to within one millionth of the total field strength.

Variations in the Gravitational Field

To an astronaut approaching the earth from the moon, the earth's gravitational field would seem essentially equal everywhere. To a shot-putter or high jumper, the differences are so slight that they would not be noticed at all. However, to a geologist, even very small variations may reflect major differences in rock type or tectonic activity.

Latitudinal variations

The force of gravitational attraction at the poles is greater than that at the equator by about 5 gal. The difference is due to the fact that the earth is not perfectly spherical. Because of the centrifugal force of its rotation, the earth bulges at the equator and is somewhat flattened at the poles so that r in the gravity equation is largest at the equator. Since r is in the denominator in the equation, G will be smallest where r is largest. The strength of the gravitational field decreases systematically from the poles toward the equator so that the strength at any latitude could be predicted from a simple equation if it were not for other variables such as elevation and rock type.

Elevational variations

The higher you climb on a mountain slope, the farther you are from the center of the earth (Figure 20.2). The r^2 term in the gravity equation increases, and G, the force of gravity, de-

Figure 20.3 Differences in gravity due to altitude and elevation. Points 1 and 2 are the same distance from the center of the earth, but G measured at point 2 is greater than G measured at point 1. The gravitational attraction of the rocks in the mountains at point 2 is greater than that of the air beneath point 1. Since the value of G would be the same everywhere along line AB, G at point 2 would be greater than G at point 1 because of the mass of the rocks in the mountains.

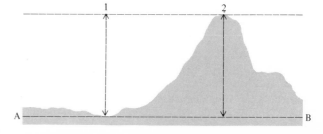

creases. This means that it should be easier to climb the last few hundred meters of a mountain than the first few, but the decrease is so small (less than 0.5 milligal/m) that you could not feel it.

Consider Figure 20.3, however. Points 1 and 2 are at the same distance from the center of the earth, but G measured at the two points would not be the same. Although r is the same, the effective mass of the earth is not. At point 2, the local mass of the mountains contributes its attractive force to that of the whole earth, but point 1 is underlain by air, not mountain, so that the effect of the local mass is not appreciable. As a result, G would be greater at point 2 than at point 1.

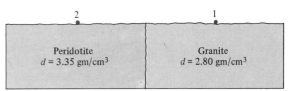

Figure 20.4 The effect on G of different types of bedrock. G measured at point 1 will be less than at point 2 because the mass of the peridotite is greater than the mass of the granite.

Rock type

It is possible for gravimeters located at the same elevation and latitude to record different values for G because even if the topography is identical in the two places, the rocks beneath the surface need not be (Figure 20.4). If one

Weighing the Earth

To determine the mass of the earth, we need only (1) a huge scale, (2) large (several hundred kilograms) weights, and (3) Newton's gravity equation.

Step 1: Place equal weights, m_1, on both pans of the scale as shown in Figure 1. The scale is balanced because the force of gravity acting on pan A is the same as that acting on pan B. Thus,

$$G_a = \gamma \frac{m_1 m_{earth}}{r^2} = G_b$$

where r is the distance from the pan to the center of the earth.

Step 2: Now unbalance the scale by placing a small weight, m_2, on pan B. The scale can be balanced again by placing a very large weight, m_3, *beneath* pan A. Balance is achieved because the gravitational attraction of the earth acting on m_2 is counterbalanced by the attraction of m_3 upon m_1 in pan A.

Step 3: Because the scale is balanced once again, gravity acting on pan A, G_a, must equal the force of gravity acting on pan B, G_b. These forces are shown by the equation:

$$\underbrace{\gamma \frac{m_1 m_{earth}}{r^2} + \gamma \frac{m_1 m_3}{d^2}}_{G_a} = \underbrace{\gamma \frac{m_1 m_{earth}}{r^2} + \gamma \frac{m_2 m_{earth}}{r^2}}_{G_b}$$

By removing terms that appear on both sides of the equation,

$$\frac{m_1 m_3}{d^2} = \frac{m_2 m_{earth}}{r^2}$$

from which

$$m_{earth} = \frac{m_1 m_3 r^2}{d^2 m_2}$$

If you have carried out this experiment successfully, your answer should be approximately 6.0×10^{27} g.

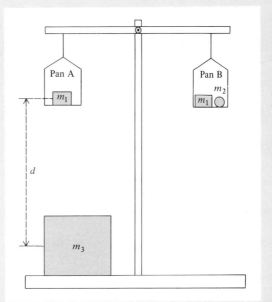

area is underlain by granite with a density of 2.65 g/cm³ and the other by peridotite with a density of 3.35 g/cm³, the denser rock, because it contains more mass per unit volume, would exert a stronger force of gravitational attraction than the rock with lower density would.

Studies of the Earth's Interior Using Gravity

Density stratification

Once Newton postulated his Law of Gravitational Attraction, it was applied to the study of the earth. One of the first uses was to calculate the total mass of the earth (see the Perspective in this chapter), a value of 6.0×10^{27} g. The volume of the earth already had been calculated to be 1.08×10^{27} cm³ by applying the formula for the volume, V, of a sphere:

$$V = 4/3\pi r^3$$

Once total mass and volume were known, the average density of the earth could be determined easily, and it proved to be 5.5 g/cm³. This simple calculation had far-reaching importance. Remember that rocks collected from the crust and even supposed mantle materials such as diamond and peridotite have densities far lower than 5.5 g/cm³.

This means that the inner earth must be composed of materials far denser than those of the crust, and it suggests that the earth has evolved in such a way as to concentrate the denser elements in the interior of the planet and the lighter elements in the outer portions.

Mountains have roots

In the mid-nineteenth century, British engineers making a detailed topographic survey of India encountered unexpected difficulties just south of the foothills of the Himalaya Mountains (Figure 20.5). These difficulties led eventually to significant conclusions about the earth's internal structure and processes.

The distance between two points—one at Kalianpur on the Ganges Plain, the other at Kaliana in the Himalayan foothills—was calculated by two methods: astronomic sighting on a star and standard plane-table surveying. These produced two slightly different results, off by only 150 m (500 ft) for an overall distance of nearly 600 km (375 mi). This small error seems barely worth mentioning, but the state of surveying, even in the mid-1800s, required better accuracy than that, and the discrepancy was not understood.

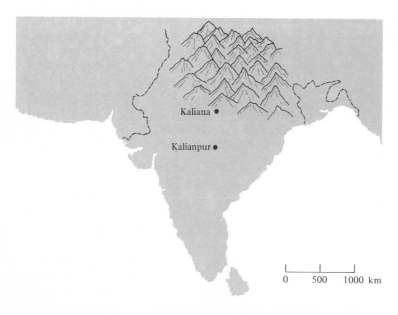

Figure 20.5 Map of India showing locations of the points between which the surveying error was detected.

Kaliana ●

Kalianpur ●

0   500   1000 km

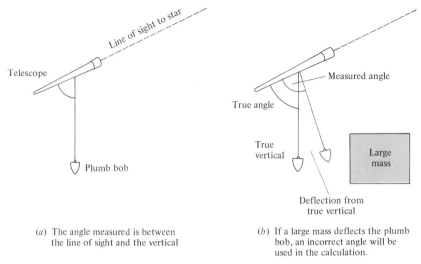

Telescope

Line of sight to star

Plumb bob

Measured angle

True angle

True vertical

Large mass

Deflection from true vertical

(a) The angle measured is between the line of sight and the vertical

(b) If a large mass deflects the plumb bob, an incorrect angle will be used in the calculation.

Figure 20.6   Gravity as a source of error in the Indian star sighting.

An explanation was soon suggested by J. H. Pratt. Star sightings involve measuring the angle between a vertical line and a line of sight to a particular star (Figure 20.6). Pratt suggested that the plumb bob that was used to determine the vertical line in the Himalayan foothills might have been attracted by the local mass of the mountains, thus producing an error in the astronomic readings. This would solve the problem neatly, and Pratt quickly calculated the amount of gravitational attraction expected from the mountains. Surprisingly, he found that the deflection of the plumb bob had been one-third of what it should have been. Now the problem left the realm of surveying and entered the world of geology. Apparently, the mountains exerted far less gravitational pull than expected.

Two competing and contrasting hypotheses were proposed to explain this phenomenon, one by Pratt (Figure 20.7a) and the other by G. B. Airy (Figure 20.7b). Pratt suggested that the mountains and plains of India are underlain by rocks with different densities. In his hypothesis, tall columns of low-density rock are found beneath mountains, shorter columns of high-density rock beneath the plains. Both types of material float on a fluid material of extremely high density. The total volumes of the different materials are in such a proportion that at the base of the col-

umns the effective pressure is the same. Thus, although the *volume* of the material that makes up the mountains is far greater than that of the material beneath the plains, the discrepancy in total effective *mass* is far less; this accounts for the apparently anomalously small amount of deflection of the plumb bob.

Airy agreed that the upper crustal rocks essentially are floating in a medium of higher density but believed that mountains and plains are underlain by the same type of rock. He proposed that beneath every mountain there is a "root" of low-density rock extending downward into the denser medium (Figure 20.7b). As in Pratt's hypothesis, the volume of rocks beneath the mountains is greater than that beneath the plains, but the increased volume involves low-density materials. Thus, the plumb-bob deflection can be explained in this fashion also.

Modern seismic and gravity surveys reveal the presence of roots beneath most of the world's major mountain ranges, indicating that Airy's model is correct. The boundary between the low- and high-density materials in this model appears to be the moho (see Chapter 19), and so Airy's ideas fit the earth's crust-mantle configuration. Pratt was not entirely wrong, however. Compared with the oceans, the continents are made up of large volumes of low-density

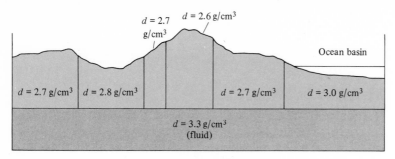

(a) Pratt's hypothesis

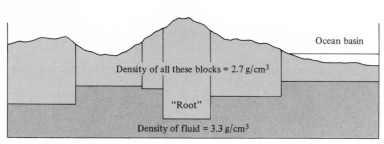

(b) Airy's hypothesis

Figure 20.7 Explanation for the apparent lack of mass in mountainous regions.

material, and if a continental "column" and an oceanic "column" are traced downward, there is a level at which the pressures will be equal. This turns out to be the boundary between the lithosphere and asthenosphere. As is commonly the case in science, the truth turns out to be a combination of two conflicting ideas.

Isostasy

Both hypotheses postulate a buoyant behavior for the earth, with light material floating on denser material. This buoyancy is called **isostasy** when applied to earth behavior. It is similar to the behavior of a cargo ship in water (Figure 20.8). As the ship is loaded, it sinks lower in the water. As it is unloaded, it rises again.

The elastic behavior displayed by the northern continents after melting of the Pleistocene glacial ice is thought to represent this unloading and sometimes is called **isostatic rebound**. The earth apparently acts to achieve isostatic equilibrium as loading and unloading of the crust takes place, and this has significance far beyond adjustments to glaciation. As a mountain is eroded, mass is removed and redistributed on the surface. Debris shed from the Rocky Mountains forms the Great Plains of the American Midwest; debris from the Appalachian and Ouachita Mountains forms the Atlantic and Gulf Coastal Plains. If isostatic equilibrium is to be maintained, the excess buoyancy produced by the mountain roots should cause eroded mountains to rise, while the new mass on the nearby plains should cause a sinking effect. This adjustment keeps stream gradients steep and promotes erosion. The adjustment continues until mass is distributed equally between the mountain and plain, at which point there is no mountain and the subsurface root has disappeared.

Isostasy also has an effect on subduction. During subduction, rocks of relatively low density are forced downward into high-density materials, and strong forces are required to overcome the buoyancy of the subducted slab. Because of its low-density sialic layer, the buoyancy of continental crust is so great that it cannot be overcome by these forces; as a result, *continents cannot be subducted*. Simatic oceanic crust is denser than continental crust, has less buoyancy, and can be subducted.

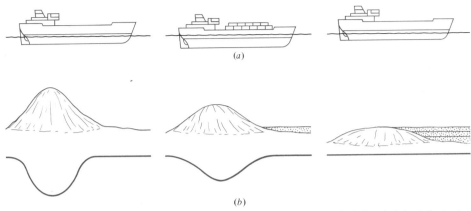

Figure 20.8 Isostasy likens a continent to a ship. *(a)* As the ship is loaded, it sinks in the water; as it is unloaded, it rises. *(b)* Mountain roots act like the keel of a ship. As mountains are eroded, they are buoyed upward isostatically.

Use of Gravity in Exploration for Resources

Any research method that gives information about subsurface structure and composition can be useful in the hunt for mineral and energy resources. A local gravity anomaly suggests the presence of rocks of unusually high or low density beneath the surface, and many economically valuable mineral deposits are associated with such rocks.

We have seen that petroleum often is trapped against or above salt domes (Figure 17.38). Halite's density of 2.16 g/cm³ is much lower than that of the typical sedimentary rocks with which it is associated, and insertion of halite during doming brings low-density rocks into higher-density materials. This results in a negative gravity anomaly (Figure 20.9). But many igneous rocks also have lower densities than the rocks they intrude. How does a petroleum exploration team know that it has found a salt dome? Computers are fed data on the shapes, sizes, and density contrasts of typical salt domes and are programmed to compute the theoretical gravitational field that would be measured over such bodies. The variables are modified in a series of calculations until the theoretical gravity field matches the field that actually is measured.

High-density materials potentially responsible for positive anomalies include many of the ore minerals such as magnetite, galena, and sphalerite and bornite. High-density mafic and ultramafic plutons also can be located, along with their potential chromium ores. Contacts between intrusive igneous rocks and their host rocks commonly are sites of mineral deposits. If the igneous rocks are of a strikingly different density from that of their hosts, the shapes of the subsurface contacts may be mapped out by gravity surveys.

Gravity exploration methods do not guarantee a petroleum field or a mineral deposit. They merely show where subsurface geologic relationships may be appropriate for the deposits to have formed. Only actual drilling can detect the resources themselves.

MAGNETISM

For centuries the peculiar ability of a rock called lodestone (magnetite) to attract objects made of iron was little more than a curiosity. The Chinese were probably the first to realize that small fragments of lodestone themselves were affected by an invisible field of attraction; they had discovered the earth's magnetic field. Fifteenth- and sixteenth-century European explorers used the newly invented magnetic compass to navigate across the oceans, but it was not until 1600 that Sir William Gilbert suggested that the earth behaved as if it were a magnet. In the past few decades, the study of earth magnetism has pro-

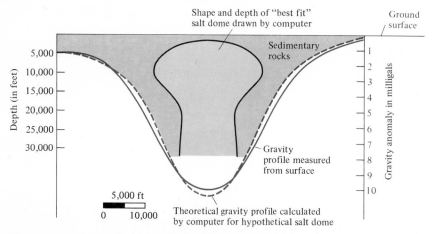

Figure 20.9 A gravity anomaly profile measured with a gravimeter (solid color line) is nearly matched by the gravity profile (dashed color line) that would be generated by a salt dome of the size, shape, and depth shown.

gressed to the point where it is one of the most powerful tools of geologic research.

What Is Magnetism?

Magnetism is a force of attraction that affects only certain substances. Metals such as iron and nickel are affected strongly by magnetism and may themselves become permanently capable of attracting other substances. Minerals such as magnetite and pyrrhotite may act like natural magnets, and others, such as hematite, may be affected strongly by magnetism. Materials that become permanently magnetic are called **ferromagnetic**. **Paramagnetic** materials are those, such as copper and oxygen, which are affected by a magnetic field but do not retain any magnetism once the field is removed. **Nonmagnetic** substances are not affected at all by a magnetic field.

The Earth's Magnetic Field

It is easier to describe the earth's magnetic field than to explain it. The field of attraction surrounding magnetic objects such as a simple bar magnet may be thought of as a series of lines of magnetic force that define the shape and extent of the field. These lines of force are invisible, but their existence can be demonstrated by shaking fine iron filings onto a sheet of paper placed over a bar magnet (Figure 20.10). The filings

become oriented parallel to the lines of force, demonstrating that the lines emanate from the opposite ends (poles) of the magnet. Bar magnets are said to generate a **dipolar** (two-pole) magnetic field.

To a first approximation, the earth behaves as if it too were a dipolar bar magnet, with lines of force issuing from its north and south magnetic poles (Figure 20.11). These poles are not the same as the *geographic poles* (the ends of the axis about which the earth rotates), but are displaced by approximately 11.5° from them.

Measuring the magnetic field

Magnetic surveying is more difficult than gravity surveying because we must measure both the intensity and direction of the field. The gravitational field is directed basically down toward the center of the earth—but the magnetic field has both horizontal and vertical components (Figure 20.12).

Inclination The angle between the magnetic lines of force and the surface of the earth is called the magnetic **inclination**. Inclination ranges from 90° at the magnetic poles to 0° at the **magnetic equator,** where the lines of force are parallel to the surface (Figure 20.13*a*). Inclination is measured with a dip needle—a magnetized needle that is free to swing in a vertical plane (Figure 20.13*b*). The needle swings until it

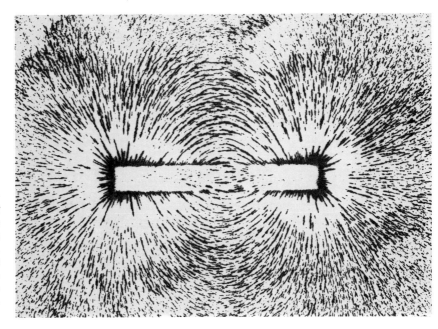

Figure 20.10 Lines of magnetic force surrounding a bar magnet are indicated by the orientation of iron filings sprinkled over the magnet. The lines emanate from the two ends (poles) of the magnet, indicating a dipolar magnetic field (two-pole). *(Education Development Center.)*

comes to rest in a position parallel to the direction of the local lines of force.

Declination   Compass needles swing in a horizontal plane and point toward the magnetic poles, thus indicating the local *horizontal* com-

ponent of the earth's magnetic field.   In most instances this direction is either east or west of true north (Figure 20.14).   The difference in degrees and direction between true north and magnetic north at a given point on the surface is the **magnetic declination** at that point.   Com-

Figure 20.11 Magnetic lines of force around the earth.   The lines of force emanate from the north and south magnetic poles.   The magnetic poles are displaced by about 11.5° from the rotational, or geographic, poles.

Figure 20.12 Directional properties of the gravitational (black arrows) and magnetic (color arrows) fields.

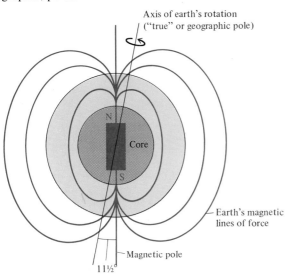

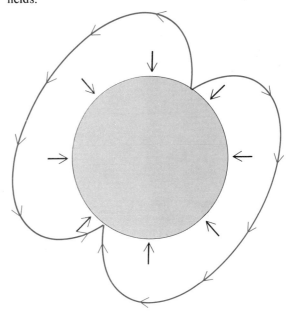

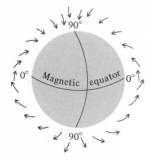

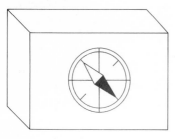

(a) Measured inclination on the earth

(b) A dip needle is used to measure inclination

Figure 20.13   Magnetic inclination.

passes on the east coast of the United States point west of true north, while those on the west coast point east of true north. Locations along a line in the midcontinent have declinations of

0°. There, a compass needle points to true and magnetic north simultaneously.

Magnetic field intensity

The strength of the magnetic field is measured with a **magnetometer** (Figure 20.15). In the primitive device shown here, a small bar magnet is allowed to orient itself with the magnetic field, and a force then is applied to it to swing it perpendicular to the field. The field's strength can be measured because the amount of force needed to swing the needle depends on how strong the attractive force is. The earth's magnetic field is measured in *gauss* (after the man who first measured it accurately) and *gammas*, where 1 gamma = 10^{-5} gauss. A field of 1 gauss operating on a bar magnet 1 cm long would require a force of 1 dyne to be exerted on each end of the magnet to rotate it 90°. The earth's magnetic field is extensive, reaching 60,000 km above the surface, but its strength is only about 0.5 gauss, approximately one-tenth that of a magnet bought in a toy store.

Changes in the magnetic field

Each of the three measured field parameters (intensity, declination, and inclination) varies with time. Intensity fluctuates daily, but these

Figure 20.14   Magnetic declination in the United States. *(After Isogonic Chart of the U.S., U.S. Coast and Geodetic Survey, 1960.)*

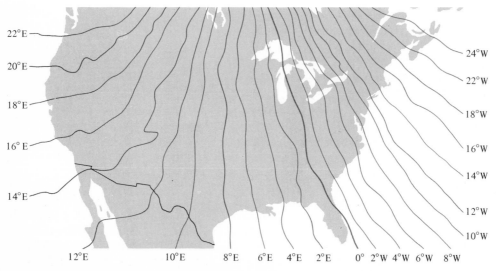

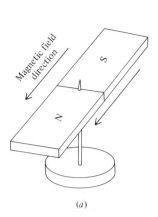
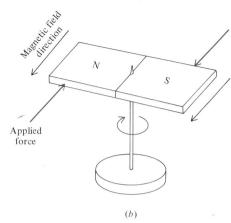

(a)   (b)

Applied
force

Figure 20.15 A simple magnetometer. *(a)* A magnet is permitted to align itself in the magnetic field. *(b)* Force is then applied to rotate the magnet 90°. The amount of force needed is proportional to the strength of the magnetic field.

are temporary changes amounting to perhaps 10 to 20 gammas and probably are related to tidal effects in the atmosphere. Intense solar activity often causes "magnetic storms" that bring about changes in the field by as much as 1000 gammas. These changes also are short-lived, lasting for only a few days.

In addition to these short-term fluctuations, there appear to be long-term, internally produced changes called **secular variations.** Since Gauss first measured it in 1830, the magnetic field has decreased by about 6 percent. Should this rate of decrease remain constant, there would be no magnetic field 2000 years from now. This rate of change in a major worldwide phenomenon is remarkably rapid when compared with most other geologic processes.

Changes in declination also are rapid in some areas. At London, where records have been kept for nearly 400 years, declination has varied from east to west and then back to east again for a total of more than 40° of change.

Perhaps even more astounding is a different type of change. Studies of the ancient magnetic field (see below) indicate that the polarity of the field actually may reverse from time to time. This means that the end of the compass needle that points north would spin 180° and point to what is now the south magnetic pole.

Why Does the Earth Have a Magnetic Field?

There are several theories for the origin of the earth's magnetic field, but none is accepted fully by scientists. This remains one of the major unsolved problems in modern earth science, one that is receiving a large amount of theoretical and experimental attention. Magnetic fields can be generated in different ways, and by studying them we can decide which are most applicable to the earth.

**Hypothesis 1: The earth is
a natural magnet**

Natural permanent magnets, such as lodestones, are composed of atoms grouped by small-scale magnetic fields into **magnetic domains** (Figure 20.16). Within each domain individual atoms exert a force on neighboring atoms, compelling them to align magnetically. If a ferromagnetic substance is placed in a magnetic field, the domains themselves become aligned, and a strongly magnetic object is created. Perhaps the earth contains a large enough amount of ferromagnetic materials so that the sun's magnetic field can turn it into a large permanent magnet.

This hypothesis has two flaws. Pierre Curie showed that magnets lose their magnetism when heated to specific temperatures that vary with the material of which the magnet is made. These temperatures, called the **Curie points** of the materials, reflect the amount of heat energy that is needed to shift the atoms from their aligned positions in the magnetic domains to a magnetically random position. The Curie points for the most abundant ferromagnetic materials are relatively low: iron, 760°C; hematite, 680°C; mag-

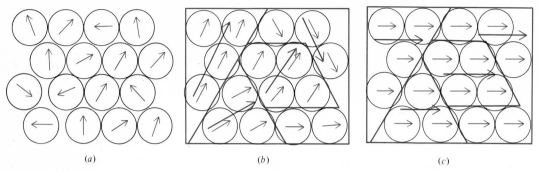

(a)   (b)   (c)

Figure 20.16 Magnetic domains in ferromagnetic minerals. *(a)* At temperatures above the Curie point, the magnetism of each of the atoms is directed randomly. *(b)* At the Curie point, atoms exert forces on their neighbors, and small domains form in which the magnetism of the constituent atoms is aligned. *(c)* If a ferromagnetic mineral cools below its Curie point in the presence of the earth's magnetic field, the domains themselves are aligned.

netite, 580°C. Even with an average geothermal gradient of 25 to 35°C/km, these temperatures would be exceeded within 30 km of the surface—within the earth's crust. The vast majority of the iron atoms in the earth are in the core, where it would be far too hot for them to be magnetic.

Perhaps the field is generated in the upper 30 km. However, the rapidity of changes in the magnetic field conflicts with the rigidity of the lithosphere and with the far slower rates of most geologic processes in the upper part of the earth. The simplest theory thus proves invalid.

Hypothesis 2: The field is caused by electric currents

Faraday showed that every electric current generates a magnetic field. A wire through which a current is passing is surrounded by a cylindrical magnetic field (Figure 20.17). If the earth's field were also generated by an electric current, the current would have to be incredibly strong, on the order of a billion amperes. Such a current cannot be created within the earth by any methods that we know and has not been detected by even the most sensitive instruments.

Hypothesis 3: The earth is an electric dynamo

Electricity is generated commercially by a device known as a dynamo (Figure 20.18) in which an electrical conductor is rotated through

a magnetic field. As it cuts the lines of force, an electric current is generated. Elsasser and Bullard suggested that the earth acts as a special kind of dynamo and that this might cause the magnetic field, as shown in Figure 20.19.

1. *Assume* that there is a weak magnetic field (from the sun?) that envelops the earth.

Figure 20.17 Magnetic lines of force surrounding a wire carrying an electric current. A cylindrical magnetic field forms around the wire, as indicated by the lines of force.

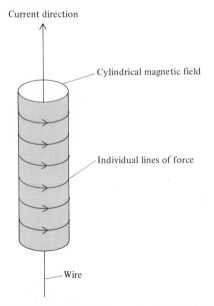

Current direction

Cylindrical magnetic field

Individual lines of force

Wire

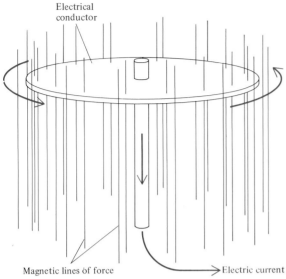

Figure 20.18 An electric dynamo. An electrical conductor that rotates through a magnetic field so that it cuts the lines of force generates an electric current.

2. *Assume* that the outer core is a molten iron-nickel alloy, as discussed in Chapter 19.

3. The liquid outer core certainly must be in a state of complex motion because of convection currents (caused by heat differences between the lower mantle and the solid inner core) and the earth's rotation and tides. The motion of molten iron — an excellent electrical conductor even when it is too hot to be magnetic — within the surrounding weak mag-

netic field generates an electric current within the earth. In so doing, the earth has behaved like a dynamo.

4. This current then generates a magnetic field. If the directional properties are correct, this magnetic field may reinforce the initial surrounding magnetic field. The resultant combined field is what we measure as the earth's magnetic field.

Theoretical and experimental studies suggest that the proposed mechanism is possible. The hypothesis calls upon fluid motion as a causal mechanism, and such motion can be far more rapid than other types of earth processes. This can explain the high rates of change of the field measured at the surface. The entire hypothesis depends on the two initial assumptions. The first is the more difficult to prove; although theory and experiment indicate the feasibility of the mechanism, they cannot explain the details of the earth's field satisfactorily. Most geologists see the good points clearly outweighing the bad. The most likely explanation of the earth's magnetic field is the dynamo model.

Paleomagnetism: A Key to the Earth's Evolution

Many rocks and minerals "remember" the earth's ancient magnetic field. Just as primary

Figure 20.19 The earth as a dynamo. Motion of the fluid outer core caused by convection and by the rotation of the earth acts like the rotating disk in Figure 20.19 and cuts lines of force from the sun's (?) magnetic field. This generates an electric current that in turn produces the earth's measured magnetic field.

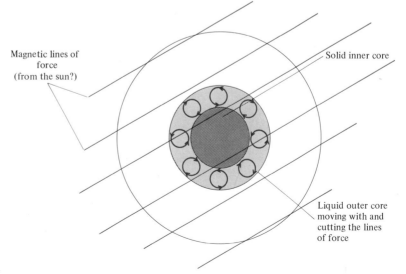

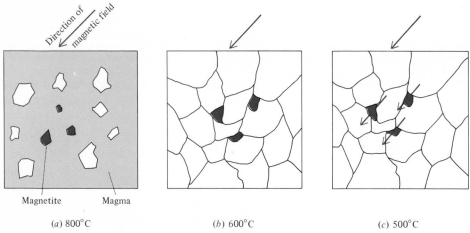

(a) 800°C   (b) 600°C   (c) 500°C

Figure 20.20 Thermoremanent magnetism. *(a)* Magma begins to crystallize, including minor amounts of magnetite. Temperature is far above the Curie point of magnetite. *(b)* Rock has crystallized completely but is still above the Curie point. *(c)* Rock has cooled below the Curie point of magnetite. Magnetic domains in the magnetite are aligned in the magnetic field, recording its directional properties.

features such as mud cracks and ripple marks record the environments and processes of deposition of sedimentary rocks, ferromagnetic minerals record ancient magnetic events. The study of ancient magnetism, called **paleomagnetism,** has been developed into an important field of research in the past 30 years. This "magnetic memory" has revealed the existence of earth processes that previously had not even been imagined.

Remanent magnetism (magnetic memory)

How do rocks remember ancient magnetism? Imagine a magma chamber in which magnetite crystallizes as a minor accessory mineral in an igneous rock (Figure 20.20). As the solid rock cools below 580°C, the Curie point of magnetite, the magnetic domains in all the magnetite crystals are aligned in the earth's magnetic field. Each crystal has become a permanent magnet that records the magnetic field direction that existed when it cooled below the Curie point.

That direction will remain in the mineral's magnetic memory unless the rock is heated again above the Curie point of magnetite. The earth's magnetic field may change intensity or direction, or it may even reverse its polarity, but the mag-

netism recorded in the magnetite is immune to those changes. The traces of the initial magnetism are called **remanent magnetism,** and the type just described is called **thermoremanent magnetism.**

There are other kinds of remanent magnetism. Magnetite grains eroded from the igneous rock described above would retain their magnetism but would be tumbled about and rotated during transport. If the grains were to settle through quiet water during deposition, they could be aligned physically by the earth's magnetic field. The sedimentary rock which included them as clasts would exhibit **depositional remanent magnetism.** Metamorphic rocks or hematite cement in sedimentary rocks may exhibit **chemical remanent magnetism,** a type of magnetic memory that is imprinted when chemical reactions during metamorphism or diagenesis produce ferromagnetic minerals at temperatures below their Curie points.

Reversals of the magnetic field

Early paleomagnetic studies produced a startling result. In sequences of sedimentary rocks, it was found that the direction of the magnetic field in some layers was opposite that in others and opposite that of the earth's present

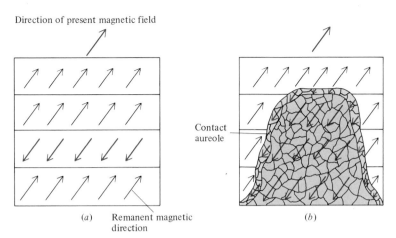

Figure 20.21 Evidence for magnetic reversals. *(a)* In a sequence of sedimentary or volcanic rocks, some layers may show reversed polarity. *(b)* Igneous rocks may show a remanent magnetic field opposite in polarity to that of the country rocks they intrude. If contact-metamorphic temperatures are high enough, minerals in the contact aureole may be heated through their Curie points and be aligned in the new field.

Direction of present magnetic field

Contact aureole

(a)   Remanent magnetic direction

(b)

magnetic field. Studies of thermoremanent magnetism in sequences of lava flows yielded similar results (Figure 20.21*a*), as did magnetic studies of several contact aureoles (Figure 20.21*b*). Geologists realized that the earth's magnetic field must be capable of reversing its polarity—that the north and south magnetic poles can, in effect, change places. We do not know how or why such reversals happen. They constitute yet another piece in the unfinished puzzle of the earth's magnetism.

Radiometric dating of rocks showing normal and reversed polarity has resulted in a picture of a complex magnetic history in which there are long-term *epochs* and short-term *events* of normal and reversed polarity (Figure 20.22). The reversals seem to take place at irregular intervals ranging from 25,000 to a few million years. It is possible that the intensity of the magnetic field decreases gradually until it reaches zero at the moment of reversal. Perhaps this is what the decreasing strength of the modern magnetic field will culminate in. If so, there would be serious inplications for human beings, as we shall see below.

Paleomagnetism and Plate Tectonics

The sea floors are better suited for paleomagnetic studies than are the continents because the volcanic and sedimentary rocks on the sea floors have not been subjected to the deformation and metamorphism found on the continents. As a result, the physical orientation of the rocks has

not changed since their deposition or eruption, and there has been little chance for the minerals to be reheated above their Curie points. The magnetic history of the oceans thus is relatively simple and is unraveled more easily than that of the continents. Indeed, it was studies of ocean-floor magnetism that "proved" conclusively that the continents do change position on the earth's surface. Paleomagnetic data represent perhaps the strongest evidence for the plate tectonic model and explain why this model has been accepted widely by geologists.

Magnetic stripes and sea-floor spreading

In the 1950s, ocean research vessels discovered elongate belts of magnetic anomalies that were tens of kilometers wide and hundreds of kilometers long (Figure 20.23). Like gravity anomalies, they represent areas of unusually high (positive) and low (negative) field intensity. Like gravity anomalies, they must reflect local variations in the field, and their small size compared with that of the entire earth suggests a relatively shallow-seated source. The linear anomalies, called magnetic "stripes," always showed the alternation: positive-negative. Each band might be of a unique width, but they always are arranged symmetrically on opposite sides of the ocean ridges, as shown in Figure 20.23. In every ocean, the anomalies parallel the ocean ridges.

In the mid-1960s, the same ingenious explanation for the stripes was devised indepen-

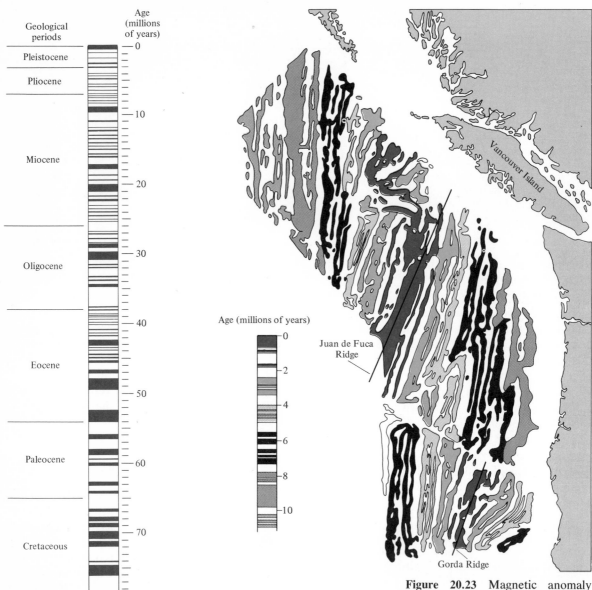

Geological periods

Age (millions of years)

Pleistocene

Pliocene

Miocene

Oligocene

Eocene

Paleocene

Cretaceous

Age (millions of years)

Juan de Fuca Ridge

Vancouver Island

Gorda Ridge

Figure 20.22 The magnetic time chart. Reversals of the magnetic field are shown for the past 76 million years, based on data from the oceans. Color = normal polarity. White = reversed polarity. *(Modified with permission from J. R. Heirtzler et al., "Marine Magnetic Anomalies, Geomagnetic Field Reversals, and Motions of the Ocean Floor and Continents,"* Journal of Geophysical Research, *vol. 73, p. 2119, 1968.)*

Figure 20.23 Magnetic anomaly "stripes" in the Pacific Ocean southwest of Vancouver Island. Colored and gray stripes represent positive anomalies; white stripes are negative anomalies. Note particularly the symmetrical arrangement of anomalies about the Juan de Fuca and Gorda Ocean Ridge segments. *(Redrawn after Fig. 4 from F. J. Vine, "Magnetic Anomalies Associated with Mid-Ocean Ridges," in Robert A. Phinney (ed.),* The History of the Earth's Crust, © 1968 *with permission of Princeton University Press.)*

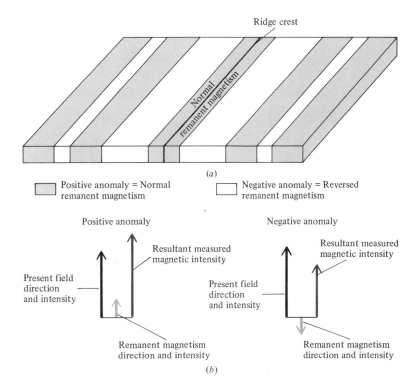

Figure 20.24 The Vine-Matthews-Morley explanation of the magnetic stripes. *(a)* The stripe pattern. *(b)* The total field intensity (dark color arrows) is the vector sum of the present field intensity (black arrows) and the remanent magnetic intensity (light color arrows). If the present and remanent fields have the same direction, the resultant field intensity is greater than the present intensity (the anomaly is positive). If the present and remanent fields are in opposite directions, the resultant measured intensity is less than at present (the anomaly is negative).

dently by L. Morley, a Canadian geophysicist, and the British team of F. Vine and D. Matthews. They suggested that the materials underlying the positive and negative anomalies actually had the same intensity but that the *direction of the remanent magnetism* was different in the rocks of the positive-anomaly areas from that of the rocks of the negative-anomaly areas (Figure 20.24*a*).

The measured magnetic intensity over a given area is the sum of the earth's present field and the rocks' remanent magnetism. If the rocks had attained their remanent magnetism in a period of normal polarity (such as today's field), the remanent magnetism would reinforce the present field and produce a positive anomaly. If the remanent magnetism had been acquired in a period of reverse polarity, the remanent magnetism would oppose the present field, yielding a lower-than-normal intensity and hence a negative anomaly (Figure 20.24*b*).

Sea-floor spreading The mechanism proposed by Vine, Matthews, and Morley for the origin of the striped pattern explains the anoma-

lies and at the same time provides one of the strongest pieces of evidence in support of the plate tectonic model (Figure 20.25).

In their mechanism, the ocean ridges are sites of tension where the oceans are being rifted apart and where basaltic magma derived from the mantle rises to the surface and crystallizes. As the basalt cools below the Curie point of its ferromagnetic minerals, it is magnetized in the earth's magnetic field. When rifting continues, this cooled basalt is split and carried away from the ridge crest in both directions, to be replaced by new upwellings of magma. If the magnetic field reverses during this process, only the magma cooling at the ridge crest will adopt the reversely magnetized field direction. This process, known as sea-floor spreading, explains the parallel alignment of the anomalies and the ridge crests as well as the alternation of positive and negative anomalies.

To check this hypothesis, magnetic profiles have been made across most of the ocean ridges. If the hypothesis is valid, the spreading phenomenon should be seen in all of them. Compari-

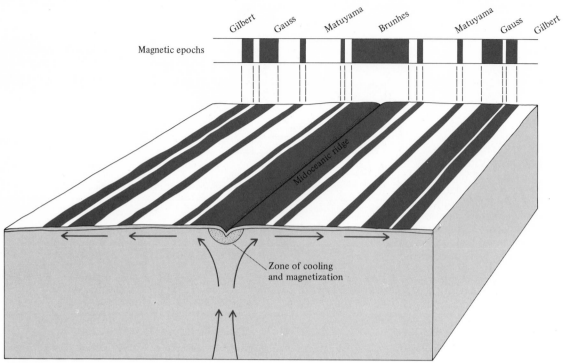

Figure 20.25 Sea-floor spreading and origin of the magnetic stripes. When basaltic magma cools at the midocean ridge-spreading centers, it adopts the earth's magnetic field direction. This continues as rifting proceeds; but if the magnetic field reverses, the direction of remanent magnetization of basalts erupting at the ridge crest reverses as well. The sea floors thus are a record of ancient magnetic pole reversals.

sons of different ridges (Figure 20.26) show that the pattern of reversals found in one ocean can be found in the others and that the sequence of magnetic epochs and events is recorded in the anomalies associated with every ridge.

Although a single magnetic epoch can be recognized in the anomaly patterns of all oceans, the width of the anomaly in any one ocean is rarely the same as that in another. If the rates of rifting and basalt extrusion were the same at all ridges, we would expect anomaly bands of equal widths. The rate of extrusion probably does not vary much from ridge to ridge so that the discrepancy in width most likely is due to different spreading rates at the different ridges. The faster the spreading, the wider the anomaly.

The magnetic anomalies, the Vine-Matthews-Morley hypothesis, and radiometric dating of ocean basalts combine to give us a measure of the spreading rates. Thus, the Reykjanes Ridge, the portion of the Mid-Atlantic

Ridge that passes through Iceland, seems to be spreading at a rapid rate (4.5 cm per year), whereas the East Pacific Rise is spreading at only about 1 cm per year. The sea floors act like magnetic tape recorders, preserving a record of the spreading behavior.

Polar wandering and continental drift

Even before polarity reversals had been discovered, paleomagnetic data had been cited as evidence for continental drift—the shifting of continents across the face of the earth. When we study remanent magnetism, the ancient inclination and declination can be determined. When these data are compiled for a single country or continent, they show that inclination and declination rarely stay constant over long spans of time. The magnetic paleolatitude of a sample is determined from its inclination. Thus, the British Isles seem to have changed their mag-

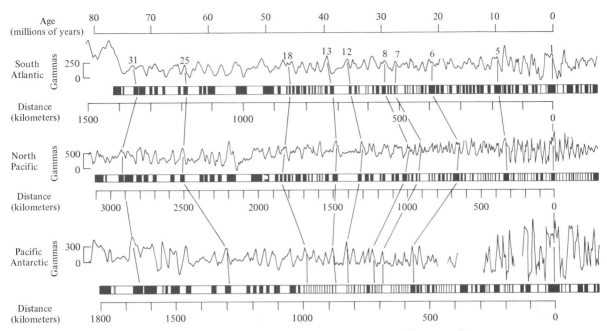

Figure 20.26 Magnetic anomaly profiles across three different ridges. Specific anomalies (numbered) can be found in all three oceans, proving that the magnetic phenomena recorded in the stripes are indeed worldwide events. *(Modified with permission from J. R. Heirtzler et al., "Marine Magnetic Anomalies, Geomagnetic Field Reversals, and Motions of the Ocean Floor and Continents,"* Journal of Geophysical Research, *vol. 73, p. 2119, 1968.)*

netic latitude since the Early Mesozoic because inclination has been increasing continually since that time. This evidence indicates that Britain is now closer to the north magnetic pole than it was during the Triassic Period. Every continent shows some amount of shift relative to the magnetic poles, and some, such as India, seem to have moved across the magnetic equator with time (Figure 20.27).

How does this happen? Three possibilities arise:

1. The magnetic poles have remained fixed, but the continents have moved.

2. The continents have stayed fixed, but the poles have moved.

3. Both continents and poles have moved.

Most dynamo models for the magnetic field require the rotational and magnetic poles to maintain positions relative to one another similar to

Figure 20.27 *(a)* Variation of magnetic inclination with latitude. *(b)* The northward migration of India since the Jurassic based on magnetic paleolatitude data. Map shows the change in magnetic latitude with time.

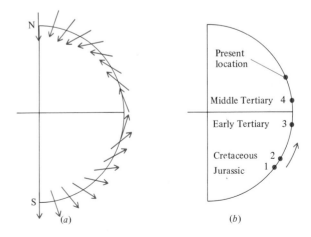

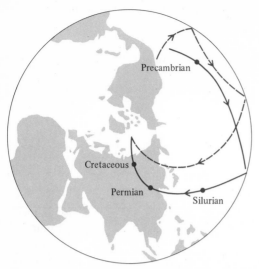

Figure 20.28 The locus of polar wandering based on paleomagnetic evidence from North America is shown by the solid curve. The dashed curve shows the data from Europe. The differences in the curves show that the two continents have moved independently of one another: They have changed their relative positions with time. *(Modified from S. K. Runcorn (ed.),* Continental Drift, *Fig. 20, 1962, with permission of S. K. Runcorn and Academic Press Inc.)*

their present ones. The rotational poles almost certainly remain fixed because the earth behaves like a giant gyroscope, resisting any change in the tilt of its axis. If the rotational axes (and poles) are fixed and the magnetic pole positions must be located nearby, possibilities 2 and 3 appear unlikely. The continents indeed must have moved.

Because we are familiar with modern geography, it is easiest to show the sense of motion by plotting the ancient magnetic pole positions relative to the landmass for which the data are compiled (Figure 20.28). When this is done, the changing position of the pole relative to the landmass is shown as a complex path called a **polar wandering curve**. Remember that it is the landmass, not the pole, that actually has wandered. When data from two continents are compared (Figure 20.28), they show that as the landmasses moved, they did not all follow the same paths. This implies that the relationships of the continents in the past were not the same as their present ones.

MAGNETISM AND HUMAN BEINGS

Even though magnetism does not seem to have as important a role in earth processes as gravity does, it is vitally important to human beings and other animals. The earth's magnetism can be used for resource exploration and protects all living things from harmful solar radiation.

Many of the principles used in gravity surveying also are applicable in magnetic surveying and prospecting. Magnetometer traverses pinpoint magnetic anomalies, just as gravimeter traverses locate gravity anomalies. Indeed, magnetic surveying is easier because a magnetometer can be towed behind an airplane or ship, thus covering large amounts of terrain quickly and inexpensively. Ferromagnetic ore bodies, such as magnetite-rich rocks, can be located, and rock units containing large amounts of ferromagnetic minerals can be traced in the subsurface by means of magnetic surveys.

The earth's magnetic field acts as a shield that protects all living creatures from harmful solar radiation. In addition to the light that reaches the earth from the sun, highly energized particles produced by nuclear reactions in the sun are bombarding the planet constantly. Most are electrons and hydrogen nuclei, and all are traveling fast enough to make them dangerous to creatures on the planet's surface. These particles are trapped by the earth's magnetic field in a region called the **magnetosphere** which extends from approximately 1000 km above the surface to 60,000 km. After orbiting the earth in the magnetosphere, most of these particles escape into space and no longer pose a threat. The magnetosphere was detected by James Van Allen in 1958 during analysis of some of the first data ever returned to the earth by satellites. In his honor, the inner part of the magnetosphere is known as the **Van Allen Belt**.

Although it is comforting to know that the magnetic field protects us from harmful radiation, we are in danger of losing some of that shield. If the magnetic field continues to decrease in intensity, as our measurements indicate it is doing, our protection gradually will decrease to the point at which some of the radiation will strike the planet. This presumably has happened many times before during magnetic reversals, but the long-term effects on human beings are unknown. Radiation-induced mutations would probably increase, possibly accelerating evolutionary processes in primitive organisms, and the radiation could increase the incidence of skin cancers. However, we do not know just what will happen.

SUMMARY

The earth is surrounded by two fields of attraction—gravitational and magnetic. Gravity affects all objects and is the cause of many earth processes. Magnetism affects only certain substances and is not a major causal mechanism for earth processes.

Gravitational attraction is a function of the masses of objects and the distances between them:

$$G = \gamma \frac{m_1 m_2}{r^2}$$

Local variations in the gravity field may be due to latitude, elevation, local underlying rock type, or deforming forces.

Gravity surveying leads to the concept of isostasy: The low-density material of the upper earth floats on a denser material and is buoyed upward much like a ship in water. Mountains, like the superstructure of a ship, must be underlain by a downward projection of low-density rock (a mountain root) that is much like the keel of a ship. As erosion and deposition redistribute mass at the surface and level the mountains, there also must be adjustments in the root zones.

Isostasy exerts some control over the amount of downbuckling or subduction. Gravity anomalies indicate subsurface rocks of higher or lower density than normal, or the operation of deforming forces. Detailed gravity surveys can locate ore and energy resources if they are contained in rocks which have a marked density contrast with the surrounding rocks.

Magnetism is an attractive force that affects only certain substances. It can be envisaged as a group of lines of magnetic force that emanate from the magnetic object exerting the force. The earth is surrounded by lines of force much as a bar magnet is. The magnetic field has both intensity and direction. The directional properties involve the angular relationships between the lines of force and the earth's surface or its rotational poles. Rapid changes take place in the intensity, direction, and polarity of the magnetic field. The origin of the magnetic field is unknown, but some type of natural dynamo probably generates it.

Rocks and minerals possess remanent magnetism—a memory of the magnetic field in existence at the time that they (1) crystallized and cooled from a magma, (2) were deposited as clasts, and (3) crystallized during metamorphism and diagenesis. Remanent magnetism permits reconstruction of the earth's ancient magnetic field, a study known as paleomagnetism. Paleomagnetic studies show that the field has reversed its polarity often in the geologic past, at intervals of 25,000 to a few million years.

Linear magnetic anomalies in the ocean basins ("magnetic stripes") are attributed to crystallization of basalt at ocean ridge crests that are the sites of active sea-floor spreading. Radiometric dating of the anomalies indicates spreading rates of a few centimeters per year. Other paleomagnetic evidence supports the concept of continental drift by demonstrating that the continents have changed their positions relative to one another and to the magnetic poles.

QUESTIONS FOR REVIEW AND FURTHER THOUGHT

1. The sun and moon exert gravitational attractions that cause tides on earth. Explain why the moon exerts the dominant force, even though it is far smaller than the sun.

2. Show how a gravity anomaly may be caused by either an anomalously dense rock type or an active earth process.

3. According to isostasy, what differences would you expect in the depth to the moho under (*a*) a young mountain range such as the Andes and (*b*) a "mature" mountain range such as the Appalachians?

4. Outline the similarities and differences between the fields of gravitational and magnetic attraction.

5. Explain with the aid of a diagram how a compass needle attracted to the north magnetic pole actually may point south.

6. Why is it unlikely for the earth's magnetic field to be caused by a permanent magnet made of iron-nickel alloy in the core?

7. If paleomagnetism is studied more easily in the oceans than on the continents, why must Paleozoic magnetism be studied in continental rocks?

8. If North America were split into two parts and rifted by a spreading center, how could geologists of the future use magnetic data to show that the parts once had been joined?

ADDITIONAL READINGS

Beginning Level

Clark, S. P., Jr., *Structure of the Earth*, Prentice-Hall, Englewood Cliffs, N.J., 1971.
(A general view of the earth's interior and how we interpret its composition. Chapters 3 and 7 deal specifically with magnetism and gravity)

Cox, A., G. B. Dalrymple, and R. R. Doell, "Reversals of the Earth's Magnetic Field," *Scientific American*, vol. 216, 1967, pp. 44–54.
(A discussion of the reversals in magnetic field polarity and their implications for interpreting tectonic evolution of the earth)

Glenn, W., *Continental Drift and Plate Tectonics*, Merrill, Columbus, Ohio, 1975.
(A paperback describing the many lines of evidence that have led to acceptance by most geologists of the plate tectonic theory. Chapters 5 and 6 develop the gravitational and magnetic lines of evidence)

Hurley, P. M., "The Confirmation of Continental Drift," *Scientific American*, vol. 218, 1968, pp. 52–64.
(The use of all lines of evidence to "prove" that the theory is correct, including development of magnetic and paleomagnetic lines of argument)

Sumner, J. S., *Geophysics, Geologic Structures, and Tectonics*, W. C. Brown, Dubuque, Iowa, 1969.
(A detailed study of the principles of gravity and magnetism as applied to tectonic models)

Takeuchi, W., S. Uyeda, and H. Kanamori, *Debate about the Earth*, W. H. Freeman, San Francisco, 1967.
(A series of discussions about plate tectonics, leaning heavily on gravity and particularly magnetism. Contains interesting discussions about the evolution of thought and the development of magnetic lines of evidence)

Advanced Level

Kaula, W. M., "Global Gravity and Tectonics," in Robertson, Hayes, and Knopoff (eds.), *Nature of the Solid Earth*, McGraw-Hill, New York, 1972, pp. 385–405.
(The theory and detailed measurements of the earth's gravitational field and its application to understanding the internal processes and composition)

Runcorn, S., *Continental Drift*, Academic Press, New York, 1962.
(A detailed discussion of paleomagnetism and its application to continental drift. See especially pp. 1–40)

CHAPTER 21

Plate Tectonics:
A Closer Look

A plate tectonic earth is a dynamic, ever-changing planet. The subcontinent of India apparently drifted thousands of kilometers across the surface of this planet to collide with Asia, and in so doing it formed the earth's highest mountains. Plate tectonic processes presumably are reshaping the earth's surface as you read these words. The Atlantic Ocean is growing larger as new crust is generated at the Mid-Atlantic Ridge, while at the same time the floor of the Pacific Ocean is disappearing into the asthenosphere beneath South America, Japan, and the Aleutian Islands. A new ocean probably is being born along the rift valleys of Africa, and as a result East Africa is being torn away from the rest of that continent.

This chapter will examine the plate tectonic earth from several viewpoints. We first will review the basic concepts of the model in light of the geophysical data presented in Chapters 19 and 20. The model then will be tested by seeing how well it explains a variety of phenomena mentioned in earlier chapters. We next will examine the *causes* of plate tectonic processes, an aspect of the model that barely has been mentioned previously. Finally, we will return to the concept of multiple working hypotheses that was described in Chapter 1. We can accept the plate tectonic model only when *all* alternative hypotheses have been discarded. In the preceding 20 chapters we have tested and discarded four models that had been proposed by scientists over the past few hundred years. Others remain, and they must be compared with the plate tectonic hypothesis to see which best explains the manner in which the earth works.

THE PLATE TECTONIC EARTH:
A REVIEW

The plate tectonic model is based on geologic and geophysical studies of the continents, the ocean basins, and the interior of the earth. It draws on all aspects of physical geology and extends them back into the remote geologic past. The model is complex, but a few of its basic concepts explain many geologic processes and phenomena in a simple way. According to the plate tectonic model:

1. The earth's surface is underlain by a small number of relatively rigid rock masses called lithosphere plates (Figure 21.1). Each plate is approximately 150 km thick and is composed of both crust and upper-mantle materials. Boundaries between plates are regions of intense tectonic activity and are the places where most of the world's volcanoes and earthquakes are concentrated. The interiors of plates tend to be relatively quiet tectonically, with the exception of a few midplate volcanoes such as the Hawaiian Islands.

2. The type of crust at the top of each lithosphere plate determines whether the plate will be a continent, an ocean, or a combination of both. Wherever there is a one-layer (simatic) crust, as on the Nazca and Pacific plates, an ocean is found; wherever there is a two-layer crust (sialic plus simatic), as on

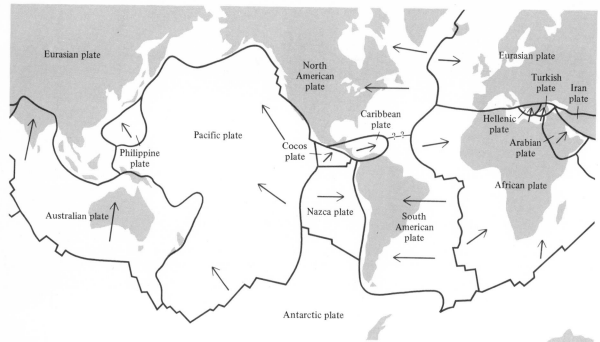

Figure 21.1   The earth's major lithosphere plates.

the Arabian and Iranian plates, there is a continent. Some plates, such as the North American plate, contain both a continent and part of an ocean basin. The two-layer continental crust on such plates extends seaward to the edge of the continental shelf, and a one-layer crust extends from there to the plate boundary.

3. A region of extremely ductile rock material called the asthenosphere separates the lithosphere plates from the lower mantle. It is a zone of detachment that enables the overlying plates to move independently from the deeper parts of the earth. Rocks of the asthenosphere have lost so much of their rigidity that their seismic-wave velocities are anomalously low. As a result, seismologists are able to detect the asthenosphere by locating the low-velocity zone in the mantle.

4. Continents and oceans are merely passengers on the moving plates. When the plates change position relative to one another, the worldwide distribution of continents and oceans also changes. The former positions of continents and oceans can be determined

by paleomagnetic and paleoclimate studies. Paleomagnetic latitudes are preserved in the inclination of a rock's remanent magnetism; by studying the changes in paleomagnetic latitude with time, we can establish the path of continental drift. Fossils of temperate plants and animals in Antarctica, and apparently high-latitude glacial activity in equatorial regions of South America, Africa, and India also indicate that these continents have altered their positions significantly on the earth's surface.

5. There are three different types of plate boundaries (Figure 21.2), and all are sites of tectonic activity. *Divergent* plate boundaries are sites of tension in which plates move apart from one another. These boundaries correspond to the ocean ridges and are the places where new simatic crust is generated from within the mantle in the process of sea-floor spreading. The existence and rate of sea-floor spreading are deduced from the presence and ages of linear magnetic anomaly "stripes" in the oceans.

Convergent boundaries are sites of compression where two plates collide with each other. These boundaries coincide with is-

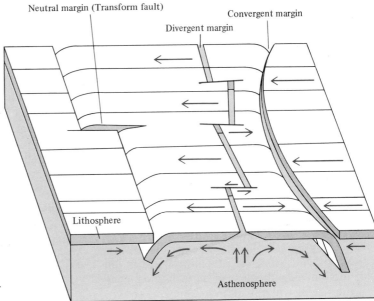

Figure 21.2 Types of plate boundaries.

land arc-trench systems in the oceans, and with Andean-type continental margins, where an oceanic trench is immediately adjacent to the continent. Simatic crust is destroyed at convergent boundaries by subduction of one lithosphere plate beneath another. Subduction zones can be located by the clustering of deep-focus earthquakes in regions called Benioff zones. The dip of the subducted plate is indicated by the dip of the Benioff zone. Fragments of the subducted plate may be scraped off and added onto the upper plate by the process of obduction. Ancient subduction zones may be identified by the presence of the ophiolite suite (obducted ocean crust).

Neutral boundaries involve neither tension nor compression, neither generation of new simatic crust nor consumption of old. These boundaries are the transform faults—sites of shear stress produced as plates slide laterally past each other. Neutral boundaries coincide with the oceanic fracture zones and with some large strike-slip faults on the continents. For example, the San Andreas Fault is probably a neutral boundary between the North American plate and the Pacific plate.

6. Most mountains form at divergent and convergent plate boundaries. The largest mountain system on earth—the ocean ridges—formed by the accumulation of basaltic lavas at divergent plate boundaries. Volcanism and buckling of the lithosphere at convergent plate boundaries create volcanic island arcs and Andean-type mountains.

7. Old fold mountains now situated at the interiors of plates formed by subduction of ocean crust and eventual collision of two continental plates. The two continents then were sutured together to form a single huge continental plate. Thus, the Ural Mountains mark the suture between what originally had been separate European and Asian plates.

8. However, some mountains can form near the centers of plates. These are volcanic mountains, such as the Hawaiian Islands, that form by movement of a lithosphere plate over a hot spot in the mantle.

9. Upwarped (epeirogenic) mountains probably form as a result of large-scale plate motions but without the intense deformation found at plate boundaries.

10. On a dynamic, plate tectonic earth, new subduction zones or ocean ridges may form at any time. When rifting that leads eventually to formation of a new ocean ridge begins beneath a continent, the continent may split

into separate plates. This apparently is happening in East Africa today.

11. As a result of plate tectonic processes, there are no permanent geographic features. New continents may form by the suturing of several smaller ones or by the rifting of a single large one. For example, the Eurasian plate formed by the suturing of three separate plates — Europe, Asia, and India — and the Indian plate formed by rifting from the ancient Antarctic plate.

The sizes and shapes of oceans may change as rates of subduction and spreading change. If the rates of subduction around the margins of the Pacific Ocean become greater than the rates of spreading on all the ridge segments in the Pacific Ocean, the ocean will shrink. If spreading exceeds subduction, as in the case of the Atlantic Ocean, the ocean will expand.

TESTING THE HYPOTHESIS

One of the most attractive features of the plate tectonic model is that it explains many of the aspects of geology that have puzzled scientists for years with a single, unified mechanism. Because plate tectonics is based on geophysical, geologic, and geographic data, it explains many different types of phenomena, including the magnetic anomaly stripes in the oceans (see Chapter 20); polar wandering (see Chapter 20); the worldwide distribution of earthquakes (see Chapter 19), volcanoes (see Chapter 5), and mountain ranges (see Chapter 18); the origin of the major physiographic features of the oceans (see Chapter 16); the location and causes of different types of deformation (see Chapter 17); and the apparent jigsaw-puzzle fit of the coastlines of Africa and South America.

To be accepted as a valid model for the earth, plate tectonics also must answer many of the other questions that were asked in earlier chapters. Several of these questions will now be answered within the framework of the plate tectonic model. Each is a test of the model, and each of the answers that follows is understood to begin with the phrase, "If the plate tectonic model is valid. . . ."

Continents and Oceans

Why are the ocean floors topographically lower than the continents?

The crust beneath the continents is made up of both sialic and simatic rocks, but the crust beneath the oceans is simatic only. The average density of continental crust therefore is lower than that of oceanic crust. The sialic materials of the continents thus are buoyed upward isostatically above the basalts of the oceans.

Why are the ocean floors so young? (Chapters 16 and 18)

The sea floors are composed of basalts that have been erupted at the ocean ridges and have spread laterally by rifting at the ridge crests. However, there are no ocean rocks older than Jurassic age (130 million years).

It is remotely possible, although highly unlikely, that there were no ocean floors prior to the Jurassic. The existence of ancient rocks similar to the sediments found today in the deep oceans argues strongly against this explanation, as does the presence of Paleozoic ophiolite suite rocks in several mountain ranges. It is far more probable that any ocean floor older than Jurassic has been returned to the mantle by subduction. Some of this subduction was accomplished at subduction zones that are active today, but far more took place in ancient subduction zones that no longer operate. Ocean crust presumably once separated Asia from Europe, for example, but a few ophiolite suite rocks in the Ural Mountains are all that remain of that ocean today.

Why are the rocks of the continents so much older than those of the oceans?

If subduction in the oceans has removed all rocks older than Jurassic, why has this not happened on the continents? We saw above that continental crust apparently cannot be subducted because the forces driving one plate beneath another are not great enough to overcome the buoyancy of the low-density continental rocks. Once "old" sialic rocks have formed on a continent, they can be buried beneath younger rocks and thus hidden from direct view, but they cannot be removed from the earth's crust by subduction. The continents are the repository of sialic material, and hence they are the sites of the oldest rocks on the earth.

Why is sial found only on continents, and why is no new sial formed in the oceans?

The answer to this question combines many lines of evidence, including ideas about the origin and early evolution of the earth. What emerges is a hypothesis dealing with the origin and evolution of continents and an explanation for the profound difference between continents and oceans.

In its 4.6 billion years of existence, the earth has evolved into a planet that is stratified (layered) according to density of rock material. Under the influence of gravity, denser elements such as iron and nickel have migrated inward to form the core, while the least dense elements have moved outward to form the atmosphere, hydrosphere, and crust (Figure 21.3). Many geologists believe that early in its history the earth passed through a stage during which it was completely molten. Migration of the elements was most rapid during this liquid stage. Most of the sialic material now found in the continental crust had been concentrated on continents very early in earth history, probably well before the beginning of the Paleozoic Era.

Once it has become part of continental crust, sial cannot be returned to the mantle. Hence, "recycled" sial cannot reemerge at the ocean ridges the way subducted simatic crust can. New continents may form by the rifting or suturing of older ones, but the sialic materials involved appear to be the same as those which were present in the earth's continental crust in the early part of the Precambrian. The formation of continental crust apparently was a one-time process that basically was completed long before the continents attained their present shapes and positions.

This does not mean that segregation of the elements is not still going on. For example, the upwelling of basalt at sea-floor spreading centers is a very efficient mechanism by which low-density magma can be transferred from the mantle to the crust. Modern sea-floor spreading can bring no new sialic materials to the surface

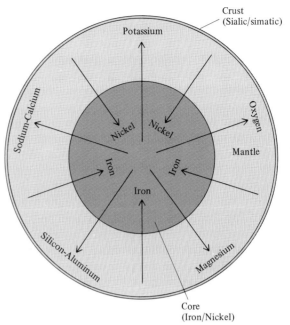

Figure 21.3 Movement of the elements during earth history.

because all the sial was removed from the mantle earlier in the earth's history. The least dense material that now can be separated from the mantle and brought upward is basalt.

Do continents grow by continental accretion?

We saw in Chapter 18 that North America (as well as other continents) appears to have a central core of old rocks surrounded by progressively younger rocks toward the margins. The theory of continental accretion attempts to explain this by showing how new continental crust is added to the margins of a continent with time, but we have just seen that the amount of sialic material available to make continental crust is limited. Is the theory of continental accretion valid?

Erosion of sialic material from a continental interior and deposition of the debris on continental shelves and slopes redistributes materials of the continental crust. Although continent-derived debris may be deposited by turbidity currents on oceanic crust, this does not turn the ocean into a continent instantaneously. If continents and oceans were to remain stable for billions of years, a two-layer crust *could* form at continental margins, and the continent could be extended; but we have seen that few, if any, geographic features are that permanent.

Subduction and plate collisions compress the continent-derived sediments, buckle them, and create new mountains. However, the same compression can weld two plates together to create a supercontinent far larger than would be explained by accretion alone. Furthermore, continental accretion takes little account of the possibility that rifting can make a continent smaller. A continent may grow by the erosion, deposition, and deformation of its rocks, but this growth is modified by plate collision and can be reversed by plate rifting. The theory of continental accretion is far too simplistic in light of modern data.

Mountains and Mountain Building

Why is the west coast of North America more tectonically active than the east coast?

The sites of active tectonism in the world are plate boundaries. Coastlines are the boundaries between ocean water and land, not between plates. Where plate boundaries and coastlines coincide, the land near the oceans will be tectonically active; such boundaries are called **active continental margins.** Most of the west coast of North America is active because it lies close to a neutral boundary (the San Andreas Fault system), a divergent boundary (the East Pacific Rise), or a convergent boundary (the Aleutian Island arc-trench system).

Where coastlines are far from plate boundaries, the land near the shore will be tectonically quiet; such shorelines are called **passive continental margins.** The east coast of North America is a passive continental margin that is located well within the North American plate, thousands of kilometers from the Mid-Atlantic Ridge.

Why are fold mountain systems so complex?

Part of the answer is simply the fact that many fold mountains, such as the Appalachians, have undergone tectonic activity for hundreds of millions of years and thus have had many different things happen to them. A somewhat closer

look at one plate tectonic model suggested for the Northern Appalachians illustrates the complexities involved (Figure 21.4).

During much of Precambrian times, North America and Europe were joined as a single continent, but sea-floor spreading began beneath this continent during the Late Precambrian. The continent split apart, much as Africa is being split now, and an ancient Atlantic Ocean formed. The split perhaps was not clean, but it left small sialic-floored "microcontinents" between ancestral Europe and ancestral North America. The island of Madagascar off the southeast coast of Africa is perhaps a modern microcontinent. As spreading continued throughout the Cambrian Period, the ancient Atlantic Ocean opened wider.

For some reason, spreading ceased and the ancient Atlantic began to close in late Cambrian or early Ordovician times. A subduction zone formed, with its associated island arc preserved as modern volcanic and volcanoclastic rocks in Newfoundland and New Brunswick. As closing persisted throughout the Paleozoic, each of the microcontinents was sutured onto North America; by the end of the Permian Period, Europe and North America had been joined again. The ancient Atlantic Ocean had disappeared, just as the ocean that had separated India from Asia has disappeared in more recent times, but some of the ancient ocean floor is preserved as obducted ophiolite masses. At the end of the Permian, a rugged Appalachian mountain system formed a suture between the European and North American continents, much as the Himalayas now separate Asia from India.

During Triassic times, another episode of sea-floor spreading began. Europe once again separated from North America, but this separation did not occur where the Precambrian separation had. The Atlantic has been opening steadily ever since.

As each microcontinent collided with another or was sutured onto the main North American continent, another orogeny was recorded. The Taconic and Acadian orogenies (see Chapter 18) thus were stages in the formation of the Appalachians that reflect such "minor" collisions, whereas the Alleghenian orogeny marked the final collision of the two major plates.

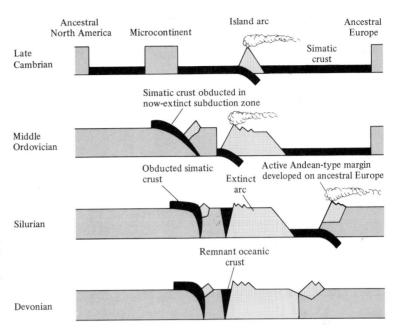

Figure 21.4 A plate tectonic model for the Northern Appalachians. See text for explanation. *(After P. H. Osberg, 1978, Synthesis of the Geology of the Northeastern Appalachians, U.S.A., Geol. Survey of Canada, Paper 78–13.)*

Petrologic Problems

Why do island arcs have different types of basalt than ocean ridges?

Sources of magma beneath a ridge come from a single depth in the mantle below. A relatively homogeneous rock melts or partially melts under a narrow range of pressure and temperature conditions to produce the typical midocean ridge basalt (Figure 21.5).

The basalts of the island arcs are derived by the melting of materials at a variety of depths (and hence pressures) along the subducted plate. The types of rock found at the different depths will not all be the same since each rock's mineral assemblage must represent adjustment to its particular set of physical conditions. When the different types of rock melt, different kinds of basalt result (Figure 21.5). The basalt formed nearest the trench is similar to that found in the ocean ridges and presumably came from approximately the same depth in the mantle. The basalts farthest from the trench are thought to form at the greatest depth.

How do andesites form?

Of the three major igneous rock types discussed in Chapter 4, andesites have the most restricted geographic distribution, being found only in island arcs and Andean-type mountains. Something about the subduction process must cause this particular type of rock to form. Let us review briefly the composition of andesite. Andesite is an intermediate volcanic rock composed largely of a Ca-Na plagioclase feldspar, with amphibole as the dominant ferromagnesian mineral. These minerals crystallize from a magma that must be generated from the rocks present in subduction zones.

Many igneous petrologists believe that andesite magma forms by the partial melting of subducted oceanic lithosphere composed of basalt and water-rich sediments and sedimentary rocks. They feel that the water promotes melting at relatively low temperatures (see Chapter 4) as the plate is subducted, and some of the water is retained in the rock within the structure of the amphiboles. The combination of melted sediment and partially melted basalt produces the appropriate composition for andesite.

Andesites are restricted in their distribution because subduction is a rapid process. It is only in subduction zones that water-saturated rocks of appropriate composition are subjected to the appropriate temperature and pressure con-

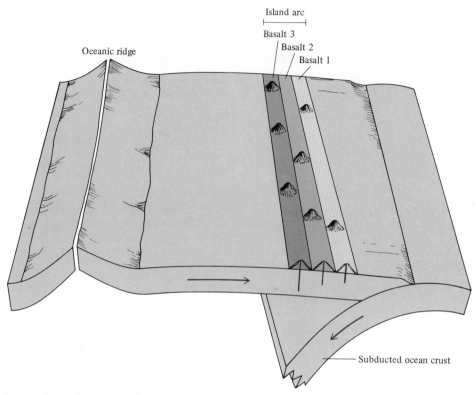

Figure 21.5 Sources of basalt for ocean ridges and island arcs. Ocean ridge basalt and island arc basalt 1 are generated at the same depth in the mantle and are of the same type. Island arc basalts 2 and 3 come from progressively greater depths and are of different types.

Figure 21.6 Metamorphic facies series in southern Japan shown for Late Mesozoic regional metamorphism. The high-pressure-low-temperature facies series forms closer to the subduction zone trench.

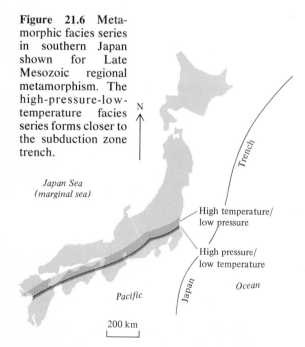

ditions *so rapidly that the water does not escape from the rock.* Instead of escaping, it promotes melting and makes hydrous minerals.

Why are there different facies series of regional metamorphism?

In Chapter 7 different facies series of regional metamorphism were related to different geothermal gradients. Regional metamorphism is associated with fold mountains and hence with convergent plate boundaries. Studies of metamorphism at modern convergent plate boundaries such as the Japanese Islands reveal differences in the facies series of regional metamorphism that clearly are related to distance from the trenches (Figure 21.6).

Rocks of the low-temperature–high-pressure facies series are found close to the trench, those of the high-temperature–low-pressure facies series farther from the trench. Since both the upper and lower lithosphere plates in the sub-

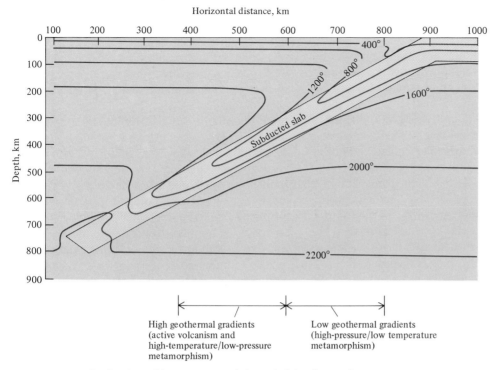

Horizontal distance, km

Figure 21.7   Distribution of heat near a subducted slab of oceanic crust.

duction zone are composed of oceanic crust and upper-mantle materials, both contain the same amounts of radioactive elements and must generate the same amount of heat. Therefore, something about the process of subduction must be responsible for the different geothermal gradients and metamorphic facies series.

An answer lies in the distribution of heat energy around the subducted slab of oceanic lithosphere shown in Figure 21.7. The normal increase in temperature with depth is shown at the left and right sides of the diagram, where the effects of subduction are negligible. However, beneath the island arc, relatively "cold" crust is driven into the hotter mantle, depressing temperatures and geothermal gradients, as shown by the isotherms. Thus, near the trench temperatures are lower than normal, but there are intense directed pressures because of the plate collision. It is there that low-temperature–high-pressure metamorphism occurs. On the arc farther from the trench, directed pressures are less intense, and a more normal geothermal gradient would be

expected. However, the upper plate receives an additional input of heat from the radioactive elements of the oceanic crust in the subducted plate and a small amount of frictional heat from subduction. The result is high-temperature–low-pressure metamorphism.

Paleogeographic and Paleoclimatic Changes

How could there have been continental glaciation near equatorial latitudes and temperate climates at the north and south poles?

In Chapter 13 we saw that a Paleozoic continental glaciation affected parts of Africa, India, Australia, and South America. The affected regions are equatorial or near equatorial, and their climate is the warmest on the earth today. This is a problem because today's continental glaciers are restricted to very high latitudes, such as in Antarctica and Greenland. A related problem concerns ancient plant and animal life. Exploration in the polar regions has revealed the existence of fossils representative of temperate climates.   Coal forests apparently thrived in Antarctica and on Spitzbergen in the Arctic Ocean near the north pole.

Before plate tectonic ideas were proposed, the only reasonable solution to this problem involved worldwide changes in climate linked to changes in the output of heat from the sun. Unfortunately, the required worldwide changes of temperature were proved unlikely because other regions on the earth had climates similar to those they now experience.   Thus, while middle-latitude Pennsylvanian coal forests flourished in what is now Antarctica, Pennsylvanian coal forests were also growing in what are now middle latitudes in North America.   If there had been enough heat to warm the poles to a temperate climate, North America should have been tropical, and it was not.   Similarly, if equatorial Asia, Africa, and South America had been so cold that continental glaciers could survive, the rest of the world would have to have been even colder, and this has been shown to be incorrect.

Plate tectonics explains these phenomena easily: Continents that are now at the equator could have moved there from higher latitudes, and landmasses now at the poles could have moved there from middle or low latitudes.   Indeed, one of the earliest arguments in favor of the mobility of the continents came from evidence of Paleozoic glaciation.   When the distribution of the Paleozoic ice was plotted along with ice directional data from striations, an interesting pattern resulted, as shown in Figure 21.8.   When the directions were reconstructed to produce the radial movement pattern found in Pleistocene ice sheets, the continents could be fitted together in such a way that their coastlines matched fairly well.   If what are now the separate continents of Africa, South America, Antarctica, and Australia, and the subcontinent of India had been joined into a single supercontinent located at or near the pole, all the paleoclimatic problems could be explained.   Geologists have named this hypothetical supercontinent **Gondwanaland**.   Modern paleomagnetic evidence and detailed comparisons of rocks from the southern continents suggest that the reconstruction shown in Figure 21.8 is valid.

THE CAUSES OF PLATE MOVEMENT

Plate tectonic concepts seem to fit geologic and geophysical observations very nicely, both on a large scale and in some small details.   But there is one more step that must be taken before we can accept the model: We must find the *cause* of plate tectonic movements.   Geologists must de-

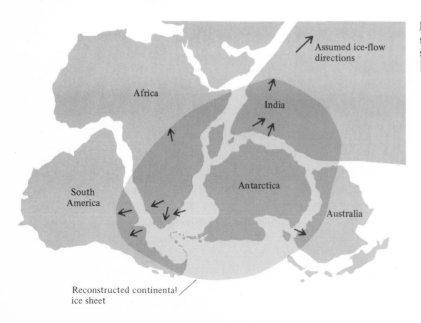

Figure 21.8 Reconstruction of the southern hemisphere continents, showing continental ice sheet and Paleozoic glaciation directions.

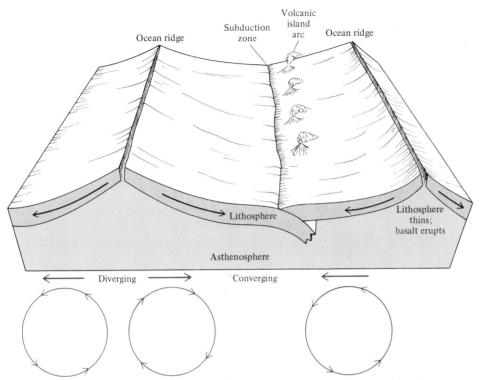

Figure 21.9   Formation of ocean ridge spreading centers and subduction zones by convection cells in the mantle below the asthenosphere.

termine whether there is an energy source great enough to move immense lithosphere plates from one place to another and must find a mechanism by which the energy brings about this motion.

Two different types of hypotheses have been proposed to explain the causes of plate movement. One holds that convection in the mantle produces rifting at the ocean ridges and *pushes* plates against one another in subduction zones. The other states that once lava reaches the sea floor and begins to accumulate, gravity pulls it downward into subduction zones. Subduction thus would be a process caused by gravity *pulling* rather than by convection pushing.

Convection Pushes Plates Downward in Subduction Zones

Many geologists believe that the earth's internal heat energy is the cause of sea-floor spreading and of all plate tectonics. They say that convection (see Chapter 2) is the most practical mecha-

nism by which this heat energy can be transferred to the surface and that large convection cells may exist in the mantle below the asthenosphere (Figure 21.9).

When adjacent convection cells are arranged so that the net movement in their upper regions is divergent, as shown in the left side of Figure 21.9, strong tensional stresses will be produced. In some way these stresses are transmitted through the ductile asthenosphere and applied to the base of the lithosphere to cause rifting and sea-floor spreading. Subduction will occur where plates driven in opposite directions collide, as in the right side of Figure 21.9.

As an example, consider the subduction occurring beneath the west coast of South America. The South American plate is driven westward by spreading along the Mid-Atlantic Ridge, while the South Pacific and Nazca plates are driven eastward by spreading at the East Pacific Rise and the Nazca ridges. When east-moving ocean plates meet the west-moving continental plate, the sima-capped lithosphere of the oceans is subducted beneath the sial-capped continent, as required by isostasy.

Gravity Pulls Plates Downward in Subduction Zones

Some geologists think that the forces involved in convection might be strong enough to cause the initial rifting but not strong enough to cause all sea-floor spreading and certainly not sufficient to cause subduction. These geologists see gravity as the cause of subduction and argue that plates effectively are pulled into the asthenosphere, not pushed.

In this model, basalts rise to the surface at the midocean ridges because the lava is less dense than the surrounding rocks. The basalt cools and is wedged apart by later extrusion of new lava (Figure 21.10). The wedging creates a lateral movement that need not be very rapid, and the force that accomplishes the wedging need not (*could not*) be strong enough to cause subduction thousands of kilometers away. As the basalt cools, it contracts, becomes denser, and hence must sink to achieve isostatic equilibrium. The movement of the basalts would be the vector sum of wedging and sinking—a lateral motion with a significant vertical component. The ridges are by far the highest features of the oceans, and so plate movement essentially would be "downhill" from ridge crest to ocean trench. The plates would be pulled downward by gravity in the same way, although on a much grander scale, that soil creeps downhill.

Push or Pull?

Both "push" and "pull" models for subduction have some validity. There must be some rifting at the ocean ridges to account for the central graben, and at the same time the earth's gravitational field must pull the ocean ridge basalts downward as they cool. Throughout the earlier chapters of this book, gravity and heat energy were shown to be important causes of geologic processes. They probably combine to cause sea-floor spreading and subduction. Plates move because they are pushed and pulled simultaneously.

QUESTIONING THE MODEL

Sometimes it is easy to forget that plate tectonics is not a fact but a hypothesis. It is, to be sure, a very attractive hypothesis because it explains so much about the modern earth and the way the earth has evolved, and today most geologists accept it. Nonetheless, it is just another hypothesis that must constantly be checked, re-evaluated, and revised to fit new observations.

Popularity, simplicity, and comprehensiveness are not valid reasons for accepting any hypothesis, and many popular ideas prove to be incorrect. At various times in human history people believed that the earth was the center of the universe, that it was created in 4004 B.C., that glacial deposits were the result of Noah's flood, and that basalts and granites precipitated from ocean water like chemical sedimentary rocks. These ideas have long since passed into disfavor.

Today plate tectonics is the most popular model for the way the earth works, but it was not always so well received by the scientific community. The "father" of the modern concepts of tectonics was the German naturalist Alfred Wegener. In 1915, he proposed that continents are not fixed but instead can change position and

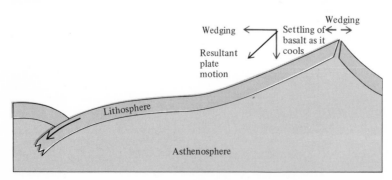

Figure 21.10 Plate motion caused by wedging and gravity settling of cooling basalt.

move across the face of the earth. His ideas were resoundingly rejected for the most part by his "more knowledgeable" contemporaries, and Wegener and his concepts were subjected to ridicule. In less than 65 years the climate of scientific thought has changed to the extent that ridicule is now reserved for those who do *not* accept continental drift.

Today only a small number of scientists are willing to criticize the plate tectonic model. They point out that although many problems are solved by the model, many more remain and new ones are created. Some of these questions deal with the finer details of the model, such as how many microcontinents were involved in the history of the Appalachian Mountains. However, some concern major aspects of plate tectonics.

We will look at some of these questions now, just as we looked at others earlier in this chapter. But this time the questions will be asked with the voice of a skeptic, and each is understood to be prefaced by the phrase, "You say that plate tectonics is correct, *but . . .*"

Convection in the Mantle

Some of the questions that have arisen concern convection. Some raise doubts as to whether convection is possible, while others deal with the manner in which deep mantle convection can affect rocks of the lithosphere.

Can convection cells exist in the mantle?

Convection cells can be seen in beakers of boiling water, and there is strong textural evidence that they also develop in magma chambers, but some scientists do not believe that they actually can form in the solid rock of the mantle. They argue that convection involves the movement of heated material; whereas this movement is relatively easy in a low-viscosity fluid such as water or magma, it is far less plausible in solid rock.

The ductile behavior of solid rock was demonstrated clearly in Chapter 17, but the rapid flow of mantle rock necessary to yield the spreading rates calculated from the ages of the magnetic stripes in the oceans never has been demonstrated. The asthenosphere is composed of extremely ductile materials and can

move rapidly, but the postulated convection cells must operate *below* the asthenosphere in rocks whose seismic-wave velocities indicate significant rigidity. Until the rate and very existence of convection in the mantle can be proved conclusively, the causal mechanism of plate tectonics remains problematical.

How large are the convection cells?

Even if we assume that mantle convection is possible, there still are questions about its application to lithosphere plates. Some geologists believe that convection cells are as large as the plates above them. If this is correct, the diameters of the cells beneath the South American, Nazca, and Pacific plates would have to be very different (Figure 21.11). The three cells then would originate at different depths and presumably at very different temperatures. Physicists studying the distribution of heat in planetary objects find this extremely difficult to explain.

Perhaps there are many small convection cells instead of a few large ones (Figure 21.12a). If this is the case, it is difficult to see how a plate as large as the South American plate could be moved as a single mass without breaking up or why subduction zones have not formed beneath it. Perhaps there are a few small convection cells near the subduction zone itself and none beneath the continent (Figure 21.12b). If so, the small cells might not have enough energy to move the plates.

The problems of mantle convection and its application to the lithosphere obviously are complex. Convection has become such an integral part of the plate tectonic hypothesis that these problems must be solved in one way or another.

Sea-Floor Spreading

Can spreading centers move?

Some variations of the plate tectonic model state that active ocean ridges can move just like the lithosphere plates they separate and might even be subducted or overridden by a continental plate. If ocean ridges do form by the coupling of mantle convection cells with the asthenosphere

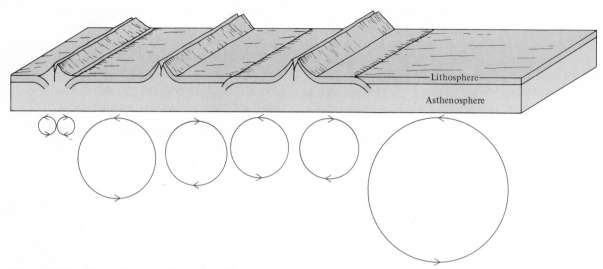

Figure 21.11 Sizes of convection cells. If convection cells are as wide as the plates they drive, they must originate at different depths within the mantle.

and lithosphere, it is difficult to understand how a ridge could move unless the convection cell beneath it also moved. However, the asthenosphere supposedly is a zone of uncoupling between the lithosphere and the lower mantle in which the convection is going on. How can lithosphere plates drift freely over the lower mantle and at the same time be coupled to a specific convection cell that operates in the lower mantle?

Figure 21.12 Alternative mechanisms for convection as a cause of subduction.

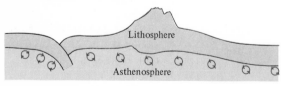

(a) Many small convection cells beneath plates

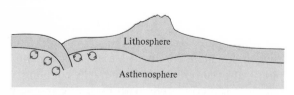

(b) A few small convection cells near the subduction zone

How does sea-floor spreading start? Why does it stop?

In the history of the Northern Appalachians described earlier in this chapter, the spreading center that produced the ancient Atlantic Ocean sprang into existence beneath a continent near the end of the Precambrian Era and mysteriously stopped operating during the late part of the Cambrian Period some 100 million years later. Spreading to produce the modern Atlantic Ocean supposely began during Triassic times, about 135 million years ago, after a 400-million-year hiatus. If convection is the cause of sea-floor spreading, what kind of stop-and-go heat changes are involved in stop-and-go spreading? We do not know.

The manner in which spreading stops perhaps is somewhat easier to understand. If plates driven apart by a spreading center should meet plates driven in the opposite direction by more powerful spreading centers, compressional forces eventually might bring about the extinction of spreading at the weaker center, as shown in Figure 21.13.

Why are some continents completely surrounded by spreading centers?

Africa and Antarctica are completely or nearly completely surrounded by spreading centers (see Figure 21.1) and therefore should

Alternative Hypotheses   487

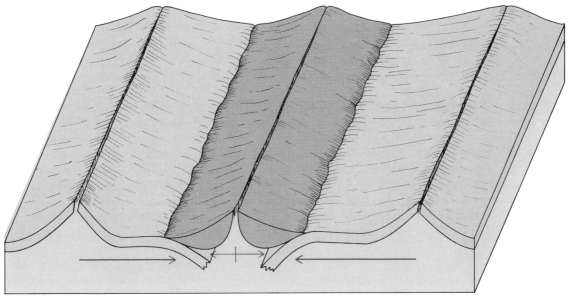

Figure 21.13 One method for stopping sea-floor spreading. Rapid spreading from the two outer ridges (light gray) overwhelms the spreading of the ridge in the center (dark gray), compressing the ridge and eventually bringing spreading to a halt.

be in states of intense compression. Why has subduction not been initiated? For example, parts of the floors of the Atlantic and Indian Oceans ought to have been subducted beneath Africa, but instead there is a new spreading center developing in the East African rift system. This behavior is strange and not readily explained by the simple plate tectonic model we have been studying. Neither is it explained readily by any other model.

An Evaluation

There are three ways to interpret these unanswered questions. One is to say that because they are not answered, they prove that the entire plate tectonic model is incorrect. Another is that some of these questions point out small details of the model that eventually will be ironed out, while others deal with deep earth processes that we still know very little about. This view even could suggest that plate tectonics provides a method for figuring out what some of the deep mantle processes might be. Both views seem somewhat prejudiced.

A third view is more objective. The unanswered questions show that there still is some

doubt about the model and that it should not be accepted dogmatically. How then should plate tectonics be viewed?

At the very least, the plate tectonic model gives geologists and geophysicists an organized framework within which the earth can be studied. It points to the types of questions about earth composition and processes that must be answered if scientists are to understand the origin of mountains and oceans. At best, it may be correct and thus may explain a large part of the internal and external activity of the planet. Correct or incorrect, the development, refinement, and constant testing of the plate tectonic hypothesis has brought challenge and excitement to geology over the past 30 years. It is a peaceful revolution in the earth sciences comparable to the development of ideas about evolution in the biologic sciences during the nineteenth century.

ALTERNATIVE HYPOTHESES

When the concept of multiple working hypotheses was introduced in Chapter 1, five possible mechanisms for mountain building were described and then were tested in subsequent chapters. Now that four have been discarded in the light of geologic and geophysical data, are there any new ideas to compete with plate tectonics? Several alternative explanations for the earth's evolution have been proposed, some dealing only with alternative energy sources for sea-floor spreading, but some refuting the entire model. The ideas vary considerably in their plausibility. Some follow the traditional lines of uniformitarian reasoning, while others call on nonrepetitive "catastrophic" events to explain the course of earth history. Three of these models will be discussed to show the range of ideas.

Vertical Tectonics

Extensive lateral movement of plates in sea-floor spreading is one of the basic concepts in the plate tectonic model. It is also a problem; even though it is supported strongly by paleomagnetic and paleoclimate data, the mechanics and energy for plate motion are not well established. A small number of geologists believe that mountain building need not necessarily involve lateral movements. Instead, they propose that all the major tectonic features on the earth can be explained by large-scale *vertical* movement of rock material, with very little horizontal motion. Their ideas are called hypotheses of **vertical tectonics.**

According to these hypotheses, vertical movements are brought about by changes in density and volume in the mantle associated with the ongoing density stratification of the planet. This model states that elements are being exchanged continually between mantle and core, effectively lowering the density of the lower mantle (Figure 21.14). It is unlikely that this process operates at the same rate everywhere in the mantle so that in some areas rocks may become less dense than those around them. They then would rise isostatically and in so doing bring about mountain-building events in the overlying crust. In a series of classic experiments, Hans Ramberg showed that such "gravity tectonics" can produce many of the features of mountain ranges found on the continents, including folds and thrust faults normally attributed to lateral plate movements near subduction zones (Figure 21.15).

Variations in heat energy in the mantle also can bring about surface tectonic events. Local heat fluctuations in the mantle can cause widespread metamorphism comparable to, and perhaps even larger in scale than, regional metamorphism in the crust. Most of the metamorphic reactions involve either an increase or a decrease in volume. Local increases in volume could cause expansion, uplift, and rifting in the crust and thus could produce ocean ridges, rift valleys, fault-block mountains, and upwarped regions such as the Colorado Plateau (Figure 21.16). Local decreases in volume could cause subsidence in the upper mantle and the formation of a depression (a geosyncline?) in the overlying crust. With continued subsidence, the flow of crustal material into the depression could result in strong compressional forces. Fold mountains, island arcs, and large regional thrust faults could be produced without any of the lateral motion required for plate convergence.

Vertical tectonic models satisfactorily explain simultaneous tension and compression in the earth without calling on problematic convection. They also explain nearly all the major

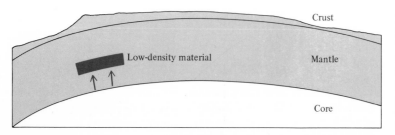

Figure 21.14 Density changes in the lower mantle and vertical tectonics. Transfer of ions from the lower mantle to the core may result locally in formation of anomalously low-density material in the lower mantle. The low-density material rises to an equilibrium level, warping the mantle and crust above it.

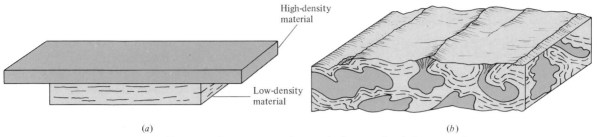

(a)   (b)

Figure 21.15 Formation of fold mountain structures by vertical tectonics (after experimental work by H. Ramberg). *(a)* A block of high-density material is underlain by a low-density material. *(b)* Low-density materials rise, dense materials sink, and the flow of rocks yields fold structures reminiscent of fold mountains.

tectonic features and types of mountains found on the continents, but they are far less satisfactory in explaining features of the oceans. The magnetic anomaly stripes in the ocean basins and the origin of Hawaiian-type islands, for example, are not explained in a vertical tectonic earth. Vertical tectonic models also cannot explain the apparent paleoclimatic anomalies since the models deny large lateral movements of landmasses.

Both the plate tectonic and vertical tectonic hypotheses leave some questions unanswered. Most geologists and geophysicists today favor plate tectonics because they feel that the problems with vertical tectonics are far more serious than those of sea-floor spreading and subduction. However, a small minority still supports vertical tectonic ideas.

An Alternative Energy Source

Some geologists believe that external sources of energy may combine with the earth's internal heat to bring about major mountain-building processes. External energy sources in the form of meteorite impacts certainly seem to have been important in the evolution of other bodies in our solar system. Apollo landings on the Moon and robot missions to Mars, Mercury, and the moons of Jupiter have shown that the surfaces of those bodies owe their appearance to impacts rather than to internally produced mountains. We shall return to a more detailed study of processes on other planets in Chapter 23.

Most meteors burn up in the earth's protective atmosphere, but a large one occasionally strikes the surface, producing a large crater. These few large meteorites may have played an important role in tectonism. For example, we know that a large meteorite crashed into the earth about 750,000 years ago, at the time of a magnetic pole reversal. In some way that we do not yet understand, the energy of the impact may have caused the reversal. An exceptionally large meteorite may even initiate rifting. The conversion of its considerable kinetic energy to heat would create thermal instability and lead to magmatism, while the impact site might be weakened enough to be rifted easily during the return to a normal geothermal regime.

Figure 21.16 Vertical tectonism caused by mantle metamorphism. Dashed black lines show original rock volumes.

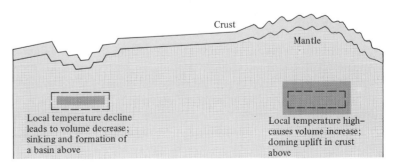

Crust

Mantle

Local temperature decline leads to volume decrease; sinking and formation of a basin above

Local temperature high- causes volume increase; doming uplift in crust above

Meteorite impacts thus are not necessarily the *cause* of sea-floor spreading but might be catalysts that help initiate some of the plate tectonic processes. For example, one of the puzzling facts about magnetic reversals is the irregularity of the time intervals between reversals. This merely may reflect the sporadic nature of meteorite impacts on the earth.

Colliding Planets: A Catastrophic Model

Plate tectonics, vertical tectonics, and even the meteorite-assisted tectonic hypothesis all are based on uniformitarianism. Their supporters feel that processes acting today or in the very recent geologic past can explain all those ancient events for which evidence is preserved in rocks. In these models, earth processes slowly, gradually, but inevitably move forward toward completion, perhaps aided on occasion by collisions with large meteorites.

Other types of hypotheses call upon extraordinarily large, rapid, worldwide events to bring about major changes in the earth. Such abnormal events are called **catastrophes.** For several decades before his death in 1979, Immanuel Velikovsky was the leading spokesperson for the catastrophists. Velikovsky suggested that Earth's normal (uniformitarian) operations were interrupted occasionally by collisions or near collisions with other objects in the solar system, possibly including comets, very large asteroids, or even the planet Venus. More imaginative ideas now postulate collisions or near misses with black holes.

One effect of such a catastrophic event could be to force the earth to tumble in its orbit so that the surface shifts position and new rotational poles are developed. This could explain how coal in Antarctica and glacial deposits near the equator have formed. The vast rearrangement of rotational energy would lead to structural changes, worldwide orogenies, and probably extensive volcanic and plutonic activity.

Velikovsky's ideas are exciting, but few scientists accept them. Now most physicists and astronomers believe that planetary-scale collisions and near misses might have happened early in the history of the solar system but are extremely unlikely during the later stages, as required by Velikovsky to explain relatively recent mountain-building events. Furthermore, most geologists feel that evidence preserved in rocks refutes Velikovsky's ideas. Of the three alternative models discussed in this chapter, this one has received the least support from the scientific community. But *all* hypotheses must be investigated before they can be discarded, and this one, no matter how unpopular or farfetched, must be subjected to the same careful scrutiny as the others.

SUMMARY

The plate tectonic hypothesis envisages the formation of mountains, continents, and oceans by the movement of lithosphere plates. These plates are composed of crust and upper-mantle materials, and they drift across the earth on a layer composed of extremely ductile, low-rigidity rock called the asthenosphere. Most mountains form where plates diverge at ocean ridge spreading centers or converge at subduction zones, but a few volcanic mountains may form in midplate regions. The plate tectonic model satisfactorily explains many diverse features of the earth, including earthquakes, volcanic activity, metamorphism, deformation, and paleomagnetic and paleoclimatic anomalies.

The causes of plate tectonic processes are not understood fully but are thought to involve the earth's internal heat and gravitational field. Mantle convection is thought to be the driving force behind sea-floor spreading, and a combination of convective "push" and gravitational "pull" probably causes subduction.

Even though the plate tectonic hypothesis explains many earth features, it does not explain all of them. In particular, the source of the energy needed for sea-floor spreading and the mechanics of plate movement are problematical. Alternative hypotheses have been proposed that avoid these problems but leave others unresolved. Vertical tectonic hypotheses deny extensive lateral movements in the crust and mantle—one of the important concepts on which plate tectonics is built. Catastrophic models that invoke planetary collisions refute even the basic uniformitarian background of all other tectonic theories.

Most geologists and geophysicists favor the plate tectonic model, at least in principle, because it explains so much about the earth. They attribute its problems to minor imperfections that eventually can be ironed out. Only further testing, data collection, and new methods of study will determine whether plate tectonics, the most popular tectonic model today, is correct.

ADDITIONAL READINGS

Beginning Level

Glenn, William, *Continental Drift and Plate Tectonics,* Merrill, 1975, 188 pp.
(A brief but comprehensive treatment of plate tectonics for the beginner)
Heezen, Bruce, and I. D. MacGregor, "The Evolution of the Pacific," *Scientific American,* November 1973.
(A treatment of the Pacific Ocean in terms of sea-floor spreading and subduction)
Hurley, Patrick, "The Confirmation of Continental Drift," *Scientific American,* vol. 218, 1968, pp. 52–64.
(A marshaling of radiometric, paleomagnetic, paleontologic, seismologic, and structural evidence as "proof" of the plate tectonic model)
Sullivan, Walter, *Continents in Motion,* McGraw-Hill, New York, 1974.
(A treatment of the history of continental drift and plate tectonic ideas, with a description of the evidence that gradually has accumulated in its favor)
Takeuchi, W. S., S. Uyeda, and H. Kanamori, *Debate about the Earth,* Freeman, Cooper, San Francisco, 1967.
(A well-written explanation of the geophysical and geologic evidence that led to the formulation of the plate tectonic model)
Uyeda, S., *A New View of the Earth,* W. H. Freeman, San Francisco, 1971.
(An updated version of the preceding reading that explains major earth features in terms of plate tectonics and presents the evidence for the model)
Velikovsky, I., *Earth in Upheaval,* Dell, New York, 1955, 301 pp.
(The case for the catastrophist model of the earth as made by its chief proponent. A vision of sequences of major catastrophes as the cause of massive extinctions, ice ages, and mountain buildings)
Wegener, Alfred, *The Origin of Continents and Oceans,* Dover, New York, 1966.
(A recent reprinting of Wegener's 1929 book in which he set forth his theory of continental drift)
Wyllie, P. J., *The Way the Earth Works,* Wiley, New York, 1976.
(A well thought out presentation of tectonic processes, including their causes and mechanisms)

Advanced Level

Beloussov, V. V., "Why Do I Not Accept Plate Tectonics," 1979, and Sengor, A. M. C., and K. Burke, 1979, "Some Comments On: Why Do I Not Accept Plate Tectonics," in EOS, *American Geophysical Union,* vol. 60, 1979, pp. 207–210.
(A fascinating exchange of letters between one of the leading proponents of vertical tectonics and two of the strongest supporters of the plate tectonic model. Although many of the points raised are at an advanced level, a beginning student can appreciate the nature of the argument and follow much of it)

CHAPTER 22

Earth Resources

by
David H. Speidel*

Earth resources are those materials formed by geologic processes which have value to humans. There is hardly a surface or subsurface process that does not form or concentrate earth resources. The two most important earth resources are land itself and water, because we cannot live without them. The study of the processes of formation, means of discovery, and geographic distributions of earth resources is called **economic geology. Mineral economics,** on the other hand, stresses the technological, economic, social, and political controls on making such resources available. This chapter will show the interrelation of the two in the utility and availability of land, water, mineral, and energy resources.

RESOURCES

In order for an earth material to be considered a resource, it must be *available* in sufficient quantities and have a present or perceived future *use*. But the use and value of earth materials change with time. Uranium, for example, has been known for about 100 years, but we have been able to use it for less than 50 years. Other earth materials have become obsolete as resources. For example, flint and obsidian were used for tools in prehistoric times but are valuable only as curios today. Technological advances in the last century have enabled us to

find, extract, and utilize a greater variety of materials than ever before.

The increased demand for earth resources results in a decreased supply in many instances. **Depletion** of a resource indicates that its availability for use is decreasing, but there are two ways to view depletion. The *physical* view of depletion sees the earth as having a fixed initial amount of a particular resource that will be consumed eventually. Petroleum from an oilfield and copper in a mine would be expected to have such limits, but we must consider more than just physical depletion in examining resources. The *economic* view of depletion holds that as physical availability of resources shrinks, increasing costs of production will force prices up and choke off demand. Some of the resources remain physically, but they are too expensive to be used.

Only a small part of a particular resource can be extracted profitably with existing technology. In the United States, this portion is called **reserves.** The difference between reserves and resources is shown in Figure 22.1. Note that reserves may expand as geologic exploration yields new deposits or as new technology increases the extractability of the resource. In addition, as profitability increases and decreases, the amount of reserves also changes. Because reserves may increase in the future through additions, depletion times based on the continued usage of the reserves available in a given year are misleadingly short. For example, we have mined twice as much copper in the past 30 years as was estimated to be re-

* Department of Earth and Environmental Science, Queens College, New York.

serves in 1950. New discoveries and technological advances coupled with changing economic conditions have added tremendously to the reserves since that estimate was made.

It is appropriate to use reserves to predict short-term availability, but they cannot be used for long-term calculations. Instead, estimates of total resources (Figure 22.1) are used for such calculations. Estimates of long-term availability assume that we will continue to identify new deposits of the resource and will have the technology and money to make it available when it is demanded. However, recent experience with an energy resource—oil shale—has shown that we cannot count on instant availability. Even though the oil in the shale has an immediate use, technological, environmental, and economic problems limit its availability.

Renewable and Nonrenewable Resources

Renewable resources are those resources for which the process of formation continues to make a supply available. For example, forests can be cut for timber, and after a period of time new trees will grow to renew the resources. Water also is renewable; moisture removed from the atmosphere during precipitation is replaced every 10 days by evaporation from the surface of the earth (Figure 11.2). However, other earth resources take much longer to form. Hundreds of years are needed to form soil, thousands of years to form sand and gravel deposits in a floodplain, and millions of years to form some types of ore deposits.

Nonrenewable resources are resources which are being depleted more rapidly than they form. From the human viewpoint, resources that take several hundred years to form must be considered nonrenewable, even though they are renewable in the framework of geologic time. Water and biologic materials thus are renewable resources, and soil is at the boundary between renewable and nonrenewable. Most of our mineral and energy resources are nonrenewable.

Distribution of Resources

Previous chapters have shown that mountains, volcanoes, oceans, swamps, and rivers are not

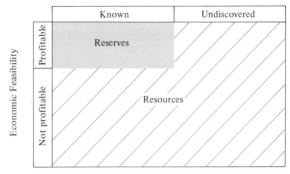

Figure 22.1 The economic and geologic distinction between reserves (color shading) and resources (color stripes).

distributed evenly over the globe. Ore deposits, evaporites, coal, and the sand and gravel that form in these environments therefore will be distributed unevenly as well. Large countries with varied geology thus are likely to have large amounts and a wide variety of earth resources. Countries that fit this description— Australia, Brazil, Canada, South Africa, the United States, and the Soviet Union—are the largest producers of mineral and energy resources in the world (see Appendix D).

Two major resource problems face us today. First, we have already used the richest, most easily extractable resources in our growth and development. Both physical and economic constraints are decreasing the reserves, even though the resources remain large. For most resources, the United States is *economically* rather than *physically* dependent on imports. We have all the copper we need but it is cheaper to import it from Africa or South America. Thus, earth resources and international politics mix, and suppliers may exert a control over our policies that is out of proportion to the quantity of resources they supply. The second problem is one of population.

Population

As population increases, demand for a resource also increases unless the use per person decreases. If the use per person also increases as population grows, the demand for the resource may become so great that we must worry

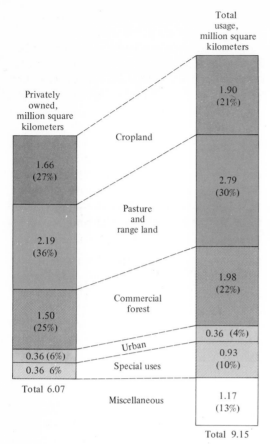

Total
usage,
million square
kilometers

Privately
owned,
million square
kilometers

1.90
(21%)

Cropland

1.66
(27%)

2.79
(30%)

Pasture
and
range land

2.19
(36%)

1.98
(22%)

Commercial
forest

1.50
(25%)

0.36 (4%)

Urban

0.93
(10%)

0.36 (6%)

Special uses

0.36 6%

Total 6.07

Miscellaneous

1.17
(13%)

Total 9.15

Figure 22.2 Land use in the United States. The left column shows privately owned, non-Alaskan land use, and the right column shows the total United States use. Values are in millions of square kilometers. Special uses include land covered by ponds, lakes, and reservoirs; farmsteads and roads; parks and wildlife refuges; military bases; and mines. Miscellaneous land includes marshes, swamps, bare rock areas, deserts, and tundra. The federal government owns more than 50 percent of Idaho, Nevada, Oregon, and Utah and more than 90 percent of Alaska. (*Information from Soil Conservation Service, U.S. Forest Service and the National Commission on Materials Policy.*)

about its depletion. This is the case with energy in the United States today. It is important to recognize how population has changed to understand why the problem of earth resources may be reaching a critical level.

Human population has increased from an estimated 10 million in 4000 B.C. to almost 5 billion as you read these words and is esti-

mated to exceed 8 billion by the year 2100. It is difficult to predict population growth accurately, but it is easy to see that projected increases in total population will require even greater resources just to maintain a constant standard of living. In the following pages we will examine some resources in four major categories—land, water, rocks and minerals, and energy—to see how they are used, how they are being depleted, and what alternatives may be utilized.

LAND

Land provides the living space and the means of growing food for humans, but not all land is suitable for these purposes. Only about 30 percent of the land surface can be used for permanent human settlement; the rest is too cold, dry, or mountainous. The major uses of land are as cropland, pasture and rangeland, and forests; these account for nearly 75 percent of land use in the United States, as shown in Figure 22.2. As with all earth resources, the distribution of *arable* land (land suitable for crops) in this country is uneven. Crops demand moderate temperatures, sufficient water, and relatively flat land—the same conditions that humans find attractive—and human dwellings often compete for available space with the crops that feed people.

Population changes affect the use of land. Increased population requires more land for urban or other uses, and this expansion (often called "urban sprawl") is generally at the expense of arable land. Reduction in arable land is especially serious because new cropland and increased productivity must be developed to provide food for the increased population. To increase cropland, we cut forest (decreasing the timber supply), upgrade pastures and ranges by means of irrigation so that they can support crops, and irrigate dry areas. For example, irrigation in Colorado, Texas, Arizona, and California has increased the food yield from those states greatly. New land can be created by draining swamps and other wetlands, as in the Netherlands, Louisiana, Florida, Wisconsin, and California.

Soil Resources

In cropland and forests, it is not the land itself that is important but rather the *soil cover*. In some places we are in danger of depleting this important resource. Land stripped of soil by erosion is no longer arable. In good cropland, the yearly rate at which new soil forms (4 to 5 tons per acre, or 900 to 1150 metric tons per square kilometer) is well in excess of rates of erosion, but in other areas the rate of erosion exceeds the rate of soil formation. One-third of the cropland in the United States lost more than 6 tons per acre in 1977, and the loss in Hawaii, Mississippi, Missouri, and Tennessee was particularly severe — over 10 tons per acre.

The rate of soil erosion varies with the manner in which the land is used (Table 22.1). As humans alter the land surface for their particular needs, we usually increase the rate at which soil is removed. Soil erosion from cleared cropland is 200 times that from forest, and erosion in strip mines is 2000 times that in forests. Human intervention in the form of land clearing for crops, road and home building, and surface mining is estimated to have increased the rate of soil erosion to about three times that produced by geologic processes alone.

However, we must distinguish between the *rate* and total *amount* of soil erosion. For example, the rate of erosion from strip mines is enormous relative to that from forests, but the total amount of soil eroded from mines is very small relative to total loss. This is because there are huge areas covered with forests from which soil is eroded every day, whereas there are only a few surface mines.

Conversions from one land use to another place a great strain on soil resources. As we convert flatland from agricultural to urban uses, we must use hilly, less-desirable land for crops. Such land is more susceptible to erosion and more expensive to farm than the flatlands. As a result, more hilly land must be used to produce the same amount of crops as flatland, and a vicious cycle thus is created. Energy needs also compete with agriculture for soil resources. Major reserves of coal exist beneath large areas of prime farmland in Indiana and the Dakotas. Surface mining of this coal would occupy many square kilometers of arable land and would increase total soil erosion greatly. There are techniques to minimize soil loss during mining, but they are expensive and use large volumes of water, putting a strain on available water supplies. The land can be reclaimed eventually, but decades are required before soil can develop once again to the point where crops can be grown.

WATER

The earth's surface is dominated by water; in our discussion of this hydrologic cycle in Chapters 11 and 12, we saw that water is in constant movement from oceans to atmosphere, to rivers, and eventually back to the oceans. An inventory of the earth's water reservoirs and the rates of transfer from one reservoir to another is given in Table 22.2.

Over 99 percent of the water is in the world's oceans and ice caps. Less than 0.5 percent is liquid water on the continents, but this

TABLE 22.1   Representative rates of erosion from various land uses

Land use	Metric tons per km^2 per year	Relative to forest = 1
Forest	8.5	1
Grassland	85	10
Abandoned surface mines	850	100
Cropland	1,700	200
Harvested forest	4,250	500
Active surface mines	17,000	2,000
Construction	17,000	2,000

Information from *Erosion and Sediment Control: Surface Mining in Eastern United States,* U. S. Environmental Protection Agency, EPA-625/3-76-006, 1976.

TABLE 22.2   Water: Reservoirs and fluxes

Reservoirs, km^3		Fluxes, km^3 per year	
Ocean	1,350,000,000	Ocean evaporation	425,000
Ice caps and glaciers	27,500,000	Ocean precipitation	385,000
Groundwater and soil moisture	8,270,000	Land evaporation	71,400
Lakes and inland seas	205,000	Land precipitation	111,000
Atmospheric moisture	13,000	Runoff to oceans	39,600
Rivers and streams	1,700		

Values taken from a compilation in D. H. Speidel and A. F. Agnew, *The Natural Geochemistry of our Environment,* Westview Press, Boulder, Col., 1981.

tiny fraction is what we consider as our water resource. The greatest value of ocean water is as the source of the precipitation that produces our freshwater resource, as shown in Table 22.2. More water evaporates from the oceans than returns to it by precipitation—the balance falls on the land, where we can use it. In addition, the ocean provides us with food, a convenient means of transportation, and can be converted into freshwater by desalinization.

Water on the continents is distributed as unevenly as our other resources are. For example, South America is very moist overall, whereas Africa and Antarctica are comparatively dry (see Chapter 14). Location of water resources places a severe constraint on the uses to which water can be put.

Uses of Water

Water always has been an essential resource. Prehistoric settlements clustered around sources of surface water or groundwater, and irrigation of cropland began in Egypt, Mesopotamia, and Peru soon after the start of agriculture. Today, we utilize water in many ways, both **instream** (while it is in the stream channel) and **offstream** (after it has been withdrawn from the channel).

A major instream use of water is for transportation—an inexpensive way to ship large volumes and masses of material. Of all the freight shipped on our inland waterways, 80 percent consists of mineral, rock, and energy resource materials. Another instream use is generation of electricity by hydroelectric power plants. Nearly 15 percent of the total United States electricity output is produced this way, and almost all the

electricity used in the Pacific Northwest is hydroelectricity. Instream uses also include recreation, fish and wildlife management, and removal of wastes. Water is not actually consumed during instream use but continues flowing after we have utilized it.

During offstream use, water is removed from its reservoirs. Some is returned, but much is **consumed**—which is to say it is changed from a usable liquid form by evaporation, plant transpiration, or boiling into steam—and temporarily lost for further use. The four major offstream uses of water are irrigation, domestic use, industrial use, and energy production. The average daily use of offstream water in the United States is shown in Figure 22.3 and broken down by water region.

Agricultural uses

The largest United States use of offstream water is for irrigation. Almost 161 billion gallons of water are withdrawn each day for irrigation,* and a large part of that is consumed (shaded part of A columns in Figure 22.3). Consumption is high because (1) plants absorb water in their roots and release it from their leaves into the

---

* The units most often used internationally for water flow rate are km^3/yr (cubic kilometers per year) as in Table 22.2. In the United States irrigation water is often measured in acre-feet—the amount of water needed to cover 1 acre to a depth of 1 ft. The United States Water Resources Council uses million (or billion) United States gallons of water per day (mgd or bgd). Emphasizing that earth resources are resources because they are used, subsequent water-use values in this chapter are given in gallons per day. The conversion factors are as follows:

$$1 \text{ km}^3/\text{yr} = 811,000 \text{ acre-ft/yr}$$
$$= 723 \text{ mgd}$$
$$= 31,700 \text{ liters/s}$$

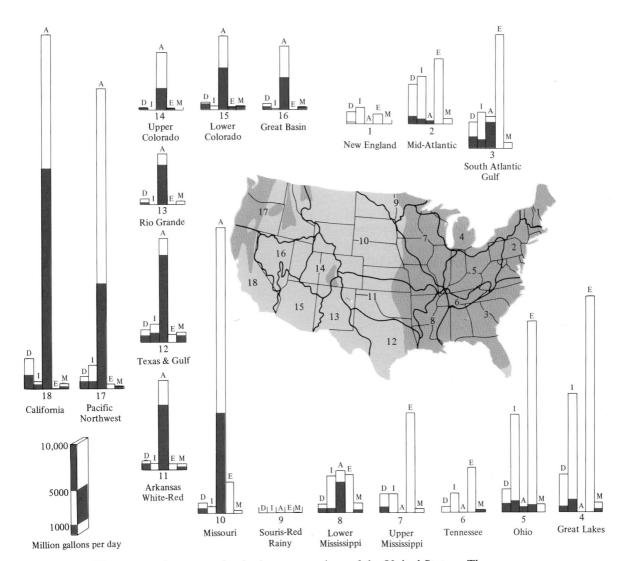

Figure 22.3 Water use and consumption in the water regions of the United States. The map shows the outline of the United States water regions (heavy black lines). The color areas indicate regions in which annual precipitation exceeds annual evapotranspiration. The water use in each district is given by the length of the bars (scale in million gallons per day is given) for domestic (D), industrial (I), agricultural (A), offstream energy production (E), and mineral industry (M) use. The color region within each bar indicates the amount of water consumed. *(Information is from the Water Resources Council, 1978. Redrawn with permission of David H. Speidel.)*

atmosphere (*transpiration*) during photosynthesis and (2) irrigation usually is used in arid areas where water evaporates rapidly from the irrigated fields. The shaded part of the map in Figure 22.3 shows where irrigation is most necessary — the water-deficient Western states in which evaporation exceeds precipitation.

Offstream energy production

In nuclear or fossil fuel power plants, uranium, coal, petroleum, or gas is used to convert offstream water to steam. The steam spins a turbine, which turns a generator to produce electricity. The steam returns to the liquid state, and only about 1 percent of the water used in this manner is consumed. In large, centralized steam-generating plants, the steam is piped through buildings after it has driven the turbines, and it can be used for heating. Power plants in Moscow and New York City make use of steam in this manner. The length of the E columns in Figure 22.3 shows that the major areas which use water for energy production are the industrialized parts of the East and upper Midwest.

Water also is used in the mining and processing of coal and oil shale and in the extraction and refining of natural gas and petroleum. For example, it takes about 350 gallons of water to refine 1 ton of gasoline from petroleum, and about 1000 gallons to produce 1 ton of shale oil by surface mining methods. When we increase coal and oil shale mining in the Missouri and Upper Colorado River regions, water use also must increase. Conflicts about water can be expected because these areas already use large amounts of water for irrigation (Figure 22.3).

Industrial use

The principal uses of water by industry are for cooling (the I columns in Figure 22.3) and removal of industrial wastes. About 2 percent is used by the mining industry (M columns in Figure 22.3), including water used in mining coal and oil shale. Large amounts are used in mining and processing phosphate rock, sand, and gravel. Severe problems of thermal and chemical pollution may arise when such water is returned to its reservoir or to the ground.

When water heated by passage through a factory or power plant returns to a stream or lake, it raises the local water temperature. Hot water holds less oxygen than cool water, but at the same time fish in the water need more oxygen because metabolism increases as temperatures go up. The result is increased mortality of the organisms living in the water. These environmental problems are called **thermal pollution** and are countered most easily if the users recycle the water or cool it before returning it to the reservoir. Chemical pollution occurs when industrial wastes return to the natural reservoirs with the water. Regulations developed in the last 20 years to combat this problem require industry either to recycle its waste water continuously or to find some other method for disposing of it.

Domestic use

We use water at home for drinking, washing, and waste removal, and the single largest use (45 percent) is toilet flushing. We also use water to keep lawns green; in some parts of the United States lawn sprinkling may account for almost two-thirds of domestic water use for single-family dwellings.

Many people cannot afford the luxury of watering lawns. We tend to take access to safe drinking water for granted, assuming that everyone can get water that is free from pollutants and disease-causing organisms. However, the World Bank has estimated that only 60 percent of the people of Latin America, Europe, North Africa, and the Middle East have access to safe water. Only 30 percent of the population of Asia and 20 percent of that of sub-Saharan Africa can rely on a safe drinking supply.

Availability of Water

Even in the United States, there is not always enough surface water for all the competing uses. In the Southwestern states and the High Plains states from Nebraska to Texas, groundwater must be used to supplement the supply. In our efforts to make more water available, problems sometimes arise.

When we withdraw groundwater faster than it is recharged, reduced stream flow, salt-water encroachment in coastal areas, and land subsidence may occur, as discussed in Chapter 12. Dams built to store water in times of high precipitation and to provide flood control act as sediment catchments (see Chapter 11) and also increase the amount of water lost through evaporation. Evaporation from irrigated fields may cause deposition of dissolved salts that are harmful to the plants we are trying to grow. Figure 22.3 shows how we have tried to change patterns of land use by manipulating the available water. We have come a long way from ancient Egypt or Peru, but the availability of water still controls the uses that we can make of land and the types of crops we can grow.

ORIGIN OF MINERAL DEPOSITS

A **mineral deposit** is a concentration of minerals, rocks, or organic matter that can be mined and processed profitably. All reserves are by definition mineral deposits. There are three major categories of mineral deposits: metallic mineral deposits, industrial rocks and minerals, and energy deposits. Minerals that are concentrated to form metallic deposits are called **ore minerals.** From these ore minerals we extract the particular metal we are seeking (for example, iron from magnetite or copper from bornite). Industrial rocks and minerals are valued for their physical and chemical properties rather than as sources of a metal. Energy deposits represent a special type of industrial deposit, but they deserve attention because of their overwhelming importance in our society. Fossil fuels and uranium minerals may be thought of as energy ores.

The key word in the discussion of mineral deposits is *concentration*. No element exists in sufficiently high concentrations in average crustal rock to be extracted profitably. Common minerals and rocks therefore are not generally considered ore minerals. To form ore minerals, the elements must be concentrated to levels well above their average crustal abundances by geologic processes. For example, aluminum must be concentrated to about 4 times its aver-

age abundance, but copper (80 times), lead (2000 times), and mercury (100,000 times) require far more enrichment. Fortunately, enrichment can occur in many ways, and we will examine some of the more important enrichment processes in the next sections.

Mineral Deposits Formed by Surficial Processes

Some minerals are formed, concentrated, or both by processes such as weathering, erosion, and deposition that act at the earth's surface. Weathering in a tropical climate can form lateritic soils (see Chapter 9), and these soils may be enriched in aluminum, iron, nickel, or manganese, depending on the composition of the parent rock. Weathering of primary minerals such as feldspar can yield secondary minerals such as the clay mineral kaolinite. Kaolinite may accumulate in place as a residual regolith or may be eroded and carried by streams to places where it is deposited as beds of kaolinitic clay. Particles carried in a stream may be concentrated by variations in stream dynamics, as discussed in Chapter 6. Minerals of high specific gravity such as gold and cassiterite (the major ore of tin) are concentrated where current velocity decreases or where the bottom is irregular. Deposits of this type are called **placer** deposits. Waves along a coast also sort minerals by specific gravity, removing the lighter ones and concentrating the remainder. Deposits of this type supply most of the world's titanium and zirconium; on the coast of Namibia (Southwest Africa), diamonds brought to the coast by streams are concentrated by wave action along the shore (see Chapter 15).

Evaporation of water causes an increase in the concentration of ions present in solution, and minerals may precipitate to form evaporite deposits (see Chapter 6). Seawater evaporites supply calcite, gypsum, halite, and a mixture of potassium and magnesium salts, whereas evaporites from lakes in arid continental regions (see Chapter 14) are rich in borates, such as the borax of Death Valley. Major Precambrian iron deposits also seem to have been formed by precipitates, but the atmosphere has changed so much since the Precambrian that iron deposits are not forming by this process today.

Mineral Deposits Found in Mafic and Ultramafic Igneous Rocks

Several elements are concentrated by magmatic processes such as crystal settling that are common in mafic magmas (see Chapter 4). The specific gravity of ore minerals generally is greater than that of magma, and as a result early-formed crystals sink through the magma and accumulate near the bottom of the magma chamber. Practically all the deposits of platinum and chromite (the ore of chromium) have been concentrated in this manner. One such deposit — that in the Bushveld igneous complex in South Africa — contains over 75 percent of the world's known resources of platinum.

Other mafic complexes, such as the Sudbury body in Canada, illustrate a second mechanism of concentration — separation of immiscible liquids. Mafic magmas with high sulfur contents may separate into two magmas: one that is sulfur-rich and one that is a typical sulfur-poor silicate magma. Elements such as iron, nickel, and copper are associated preferentially with the sulfur-rich magma and are concentrated in sulfide ore deposits.

Some deposits are found in ultramafic rocks. Serpentine rocks associated with ophiolites often contain small amounts of chromite, talc, and asbestos. These minerals apparently form by the metamorphism to which the ultramafic rocks are subjected during subduction and obduction. The plate tectonic model thus may help us locate more deposits of this kind. Kimberlites, the principal source of diamonds, are ultramafic rocks derived from the mantle and emplaced rapidly as diatremes in the crust (see Chapter 4).

Mineral Deposits Found in Felsic Rocks

Hydrothermal solutions are fluids rich in dissolved metals that form during crystallization of felsic magmas (see Chapter 4). These fluids pass through fractures in the cooler rocks surrounding a felsic pluton and crystallize in veins and cavities. In some instances the concentrations of ore minerals and associated **gangue** (minerals that formed in the same way but are not used) can be so massive that they appear to have completely replaced the original rock, but in other cases the ore minerals occur as small

TABLE 22.3 Average annual value of mineral products (states producing more than $500 million)

State	Average annual value (million $) 1965–1968	Portion of value by type of mineral		
		Metals	Industrial rocks and minerals	Fuels
Texas	5040	1	7	92
Louisiana	3497	0	6	94
California	1669	3	32	65
Oklahoma	967	1	5	94
Pennsylvania	904	15	25	60
West Virginia	886	0	8	92
New Meixco	830	18	15	67
Illinois	617	1	27	72
Michigan	592	34	56	10
Arizona	564	91	8	1
Kansas	561	1	13	86
Minnesota	529	91	9	0
Wyoming	529	10	12	78

Source: *The National Atlas of the USA*, U.S. Geological Survey, 1970.

grains disseminated throughout the host rock. The porphyry copper deposits of the western United States are an example of disseminated deposits. The ore that is mined contains only about 0.1 percent copper because it is so widely disseminated, but the total amount of copper in a single deposit has been estimated at 1 billion tons! Molybdenum, lead, and zinc also occur in disseminated deposits.

Pegmatites—coarse-grained rocks formed from late-crystallizing fluid-rich portions of felsic magmas—commonly contain concentrations of those elements which did not enter mineral structures earlier in the magma's history. Pegmatites supply us with gemstones (tourmaline, beryl, topaz, and others) and with rare metals such as gold and silver.

MAJOR MINERAL RESOURCES

We can evaluate the importance of earth resources either by the actual amount (weight) of the materials used or by the cost of the materials. Figures 22.4 and 22.5 show the weight and dollar values of the major materials used on a per person basis in the United States. Oil cost $12 a barrel at the time of the calculation; it now costs over $40 a barrel so that the economic importance of the energy resources has increased greatly. The diagrams also indicate that renewable resources play a small role in United States consumption compared with nonrenewable resources.

Metals

Each year we consume more than 900 kg of metals (worth about $280) *per person* in the United States (see Figures 22.4 and 22.5). Iron and steel account for 95 percent by weight of the metals used, followed by aluminum, copper, lead, and zinc. All other metals together constitute less than 1%. Table 22.3 indicates that Arizona, Minnesota, Michigan, New Mexico, Pennsylvania, and Wyoming are the chief metal-producing states in terms of dollar value. Production of the most important metals is shown for the United States in Figure 22.6 and for the world in Figure 22.7.

Iron and steel

Iron and steel (an alloy of iron and carbon) are the key metals in industrial nations such as the United States. Minor amounts of other metals such as nickel, cobalt, chromium, and tungsten are added to steel to control its physical and chemical properties. For example, cobalt is needed to produce magnets, tungsten is used to form heat-resistant steel, and chromium is used to form stainless steel. The availability of the minor metals thus determines the types of steel which can be produced.

Although iron ore deposits exist in many geologic environments, by far the largest amounts of iron are found in Precambrian sedimentary iron formations such as those of the Lake Superior region in the United States or those in Brazil, Africa, and India. Other deposits may be a result of volcanic exhalations on the sea floor near spreading centers or may be associated with igneous and contact metasomatic rocks, but these do not compare in volume with the Precambrian sedimentary deposits.

Aluminum

The major source of aluminum is bauxite—a mixture of several aluminum oxides and hydroxides, with small amounts of iron oxides and silica (see Chapter 9). Bauxite forms by intense chemical weathering of aluminum-bearing rock in warm, humid climates in which continuous leaching removes the silica and other soluble components. Most bauxite is found in tropical or subtropical lateritic soils or in ancient deposits that originally formed under those conditions (see Figure 22.7). Some is produced in areas of karst topography as residual clay left behind when limestone was dissolved.

Aluminum is used in place of steel in many instances because it is lightweight and generally noncorrosive. Figures 22.4 and 22.5 illustrate two reasons why this substitution is limited. The first is the sheer quantity of the substitution; we use almost 50 times as much iron and steel as aluminum and would have to build new aluminum production facilities. Second, aluminum costs six times as much as steel, making its products more expensive.

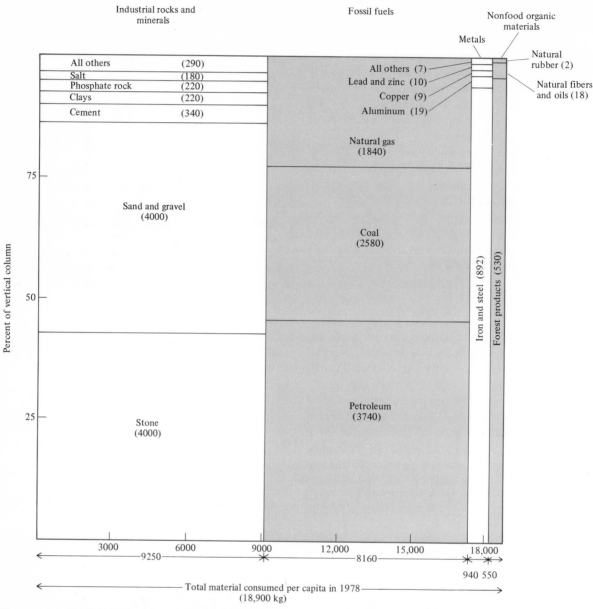

Figure 22.4   Weight of basic materials used per capita in the United States in 1978 (in kg). Values for industrial rocks and minerals and metals and metallic ore are for 1978 consumption, whereas those for fossil fuels and nonfood organic materials are given for 1977 consumption. Weights for industrial rocks and minerals and fossil fuels are to nearest 23 kg (50 pounds), and other weights are to nearest 2.3 kg (5 pounds). Percentage of vertical column is given on the left. *(Information from U.S. Bureau of Mines, U.S. Forest Service, USDA, United Nations and National Materials Commission. Redrawn with permission of David H. Speidel.)*

Copper

About 60 percent of the world's copper resources are in igneous porphyry deposits and associated hydrothermal deposits, such as those in the southwestern United States (see Figure 22.6). The copper ore is removed from huge "open pit" mines such as that at Bingham Canyon, Utah. Another 30 percent is found in

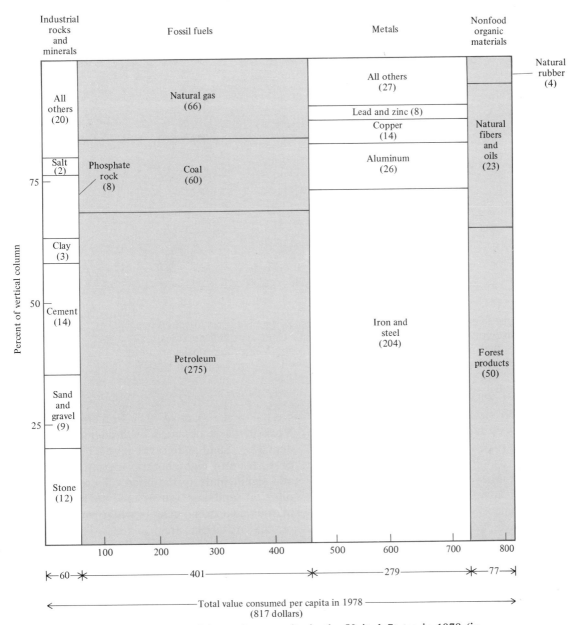

Figure 22.5 Value of basic materials used per capita in the United States in 1978 (in dollars). Values are to nearest whole dollar. Values for 1977 or 1978 consumption and sources of information are the same as in Figure 22.7. *(Redrawn with permission of David H. Speidel.)*

sedimentary rocks such as those of the copper belt of south-central Africa (see Figure 22.7), where the ore minerals are found in dolomitic shales. The copper is thought to have precipitated from sedimentary pore waters during diagenesis. Some copper deposits are massive ores associated with mafic and ultramafic rocks.

These ores generally are rich in nickel as well, and that metal is recovered as a by-product of copper mining.

Lead and zinc

Zinc and lead ores usually are associated with each other and often with ores of copper

Figure 22.6 United States sources of basic materials. Black type indicates locations that provide more than 25 percent of total domestic production. Color type indicates locations that provide more than 10 percent but less than 25 percent. Values for Alaska are not indicated. *(Information is from the Bureau of Mines and the Congressional Research Service.)*

and other nonferrous metals as well. There are two major types of deposit: massive sulfide ores in metamorphic rocks, and deposits in limestone and dolomite beds. The latter type is common throughout the Mississippi Valley. Volcanic rocks in island arcs also have some lead-zinc deposits, presumably formed by hydrothermal activity near subduction zones.

Industrial Rocks and Minerals

We use more industrial rocks and minerals by weight than any other earth material, but the dollar value of these materials is low (see Figures 22.4 and 22.5). Crushed stone, sand, and gravel account for most of the weight (87 percent), followed by cement, clays, phosphate rock, and salt (a total of 10 percent). All other industrial rocks and minerals account for only 3 percent of the weight used.

Sand and gravel and crushed stone

Most of the sand and gravel used in the United States comes from floodplains, river channels, and stratified glacial deposits, but in coastal areas such as San Francisco or New York some is obtained by dredging nearshore deposits. Sand and gravel have a relatively low cost per unit weight, but transportation costs can add significantly to the price. To reduce costs, pits for sand and gravel tend to be most numerous near urban centers where large volumes of the materials are needed. However, as cities and suburbs expand they encroach on the sand and gravel pits, forcing the producers to move, which results in increased costs. This is another example of two different land uses competing for the same space.

Crushed stone is produced in practically all the states. Most is limestone (75 percent), with the remainder composed of granite and "trap-rock," a name used for any dark, fine-grained igneous rock. Several environmental problems arise during surface quarrying of crushed stone, including destruction of the surface, increased erosion and sediment pollution, and pollution of surface water and groundwater. To combat these problems, industry increasingly is turning to underground mining.

Cement and clays

Cement production is concentrated in areas where there are abundant deposits of both low-magnesium limestone and shale. The clay minerals in the shale react with the calcium from the limestone during firing to form "clinker," which, when ground and mixed with small amounts of gypsum, is called Portland cement.

Clay minerals are used in the manufacture of ceramics such as brick, tile, and pottery. Georgia is the principal source of the clay mineral kaolinite, which also is used as a filler in papermaking. The kaolinite flakes align themselves on the paper fibers and make the paper smooth and white.

Phosphate rock and halite

Phosphate rock is an essential source of the phosphorus needed for fertilizers and animal-feed supplements. It is so vital that many sci-entists think that its availability will limit our ability to continue to increase food production. Phosphate rock is obtained from three types of deposits—marine sedimentary rocks, bird droppings on tropical islands, and apatite-rich carbonatite plutons—with most coming from the sedimentary rocks. Portions of Florida are underlain by rich phosphate deposits, but the expanding population now is competing for the same space as the mining operations. Furthermore, surface phosphate mining requires large amounts of water, as does the ever-increasing population.

The United States is the world's largest producer of halite, more commonly known as common table salt. The principal use of halite is by the chemical industry to make chlorine gas, and it is also used as a highway deicer. It is, of course, an essential ingredient in the diets of human beings and animals. Most halite is mined from evaporite deposits, but a small amount is produced by a process unchanged since pre-

Figure 22.7 World producers of important metals (iron, aluminum, copper, and lead and zinc). The location of symbols within regions does not correspond to specific mine locations. Size of symbols is proportional to production. *(Redrawn with permission of David H. Speidel.)*

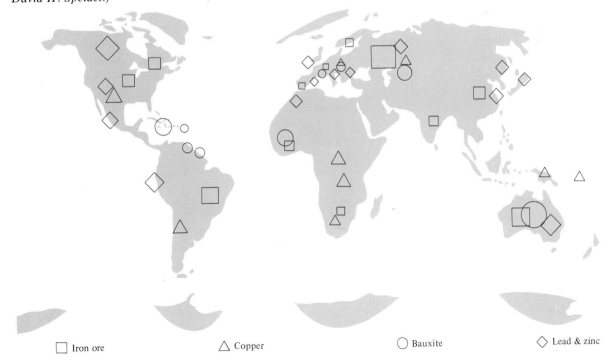

☐ Iron ore   △ Copper   ○ Bauxite   ◇ Lead & zinc

historic times—evaporation of seawater in artificial ponds.

ENERGY RESOURCES

Our energy deposits are the fossil fuels coal, petroleum, and natural gas, and the nuclear fuels uranium and plutonium. The nuclear fuels are not used in sufficient quantity or dollar value to show in Figures 22.4 and 22.5 but are included in this discussion because of the tremendous concern about their use.

Mineral Fuel Deposits

One of today's most pressing problems is the availability of affordable energy. We spend more per person on petroleum than on any other resource (see Figure 22.5), and almost half our total resource expenditure is for fossil fuels. These are so vital to our society that it is hard to believe that we have been using them for only

the last 150 years. Figure 22.8 shows the location of the major producers of mineral fuels.

Coal

Coal forms by the accumulation of nonmarine plant debris (see Chapter 6). The different types of coal—lignite, bituminous, and anthracite—differ in texture and composition. The compositional variables are the amounts of carbon, volatile materials, moisture, and ash (noncombustible inorganic matter). Moisture is an important component because the less moisture coal has, the more energy we can get by burning it. Lignite has the highest moisture content; 3 tons of it must be burned to yield the same energy as 1 ton of bituminous coal. Anthracite has the highest carbon content and the lowest amount of moisture. The type of coal that forms is thought to be due to the amount of tectonic force applied rather than to the kind of original plant material. Anthracite, for ex-

Figure 22.8 World producers of mineral fuels (coal, natural gas, crude oil, and uranium). The location of symbols within regions does not correspond to specific mine locations. Size of symbols is proportional to production. (*Redrawn with permission of David H. Speidel.*)

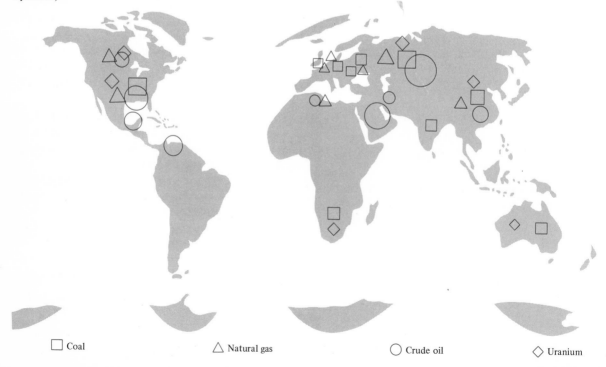

☐ Coal   △ Natural gas   ◯ Crude oil   ◇ Uranium

ample, is concentrated in parts of mountain belts that have experienced tight folding.

Petroleum and natural gas

Petroleum is a mixture of compounds of carbon and hydrogen (hydrocarbons), with minor amounts of sulfur and other elements. It is thought to form when marine or brackish-water organic material is trapped and preserved in a reducing mud on the sea floor or in deltas. Subsequent heat and pressure cause chemical changes that form the hydrocarbons. Petroleum under pressure from the overlying rocks migrates from its source rock to a porous reservoir rock, where it accumulates in structurally or strati-graphically controlled deposits (see Chapter 17). Each petroleum deposit has hydrocarbons with unique chemical and physical properties, the result of different source materials and geologic histories.

Natural gas, composed mostly of methane, CH_4, also forms by the decay of organic material. It commonly is found with petroleum but also may occur in large quantities by itself in what are called gas fields.

We cannot recover all the petroleum in a reservoir rock because of the surface tension between the pores or fractures in the host rock and the petroleum, as discussed in Chapter 10. Indeed, we now recover an average of only 30 percent of the petroleum in a reservoir rock. Natural gas is much more mobile than crude oil because of its lower viscosity, and as a result recovery of natural gas is much greater than that of crude oil.

Uranium

Uranium minerals are found in a wide variety of geologic environments, including pegmatites and igneous vein deposits, sandstones, lignite, and Precambrian placer deposits. Most of the ore now mined comes from the sedimentary deposits, and it apparently formed when the minerals, soluble under oxidizing conditions, became reduced and precipitated. Widespread uranium deposits in the Colorado Plateau probably formed when groundwater rich in dissolved uranium became reduced while passing through sandstones and precipitated the uranium minerals in pore spaces. The reducing conditions in coal-forming swamps also led to the occurrence of uranium with lignite, but we do not know whether the ore was added by groundwater or deposited with the original plant debris.

Uses of Energy

The choice of units to measure use of energy is even more difficult than that for water. Tons of coal, barrels of oil, cubic feet of gas, kilowatt hours of electricity, kilocalories, and Btu's all are used. The British thermal unit (Btu) is the amount of energy needed to raise 16 ounces of water 1°F, and the Quad, equal to 1 quad-rillion (10^{15}) BTU's, is the unit most used in discussing energy resources in the United States.*

The amount of energy used in the United States is about 30 percent of total world consumption. The sources and uses of energy in the United States during 1976 are illustrated in Figure 22.9, a type of representation often called a "spaghetti diagram." The sources are shown on the left, the uses are shown in the boxes, and the pathways show how different energy sources are used. Two relationships are evident. First, fossil fuels so clearly dominate the energy supply that there are serious problems involved in finding substitute energy sources. Second, different energy sources are needed by different users. For example, transportation depends most heavily on petroleum for gasoline and jet fuel. However, petroleum is also the basic source of the hydrocarbons used to make plastics and synthetic fibers (nonenergy use shown in Figure 22.9).

Most of the industrial use of energy (67 percent) is by companies that process other earth resources: petroleum refining, production of iron and steel, and manufacture of chemicals, paper, aluminum, and cement. The type of energy used varies with the product. For example, the production of aluminum uses large

---

* 1 QUADrillion Btu approximately equals the energy content of 500,000 barrels of petroleum per day for a year, 40 million tons of bituminous coal, 1 trillion ft³ of natural gas, 100 billion kilowatthours, 2500 tons of U_3O_8, or about 1 billion slaves working 24 hours a day.

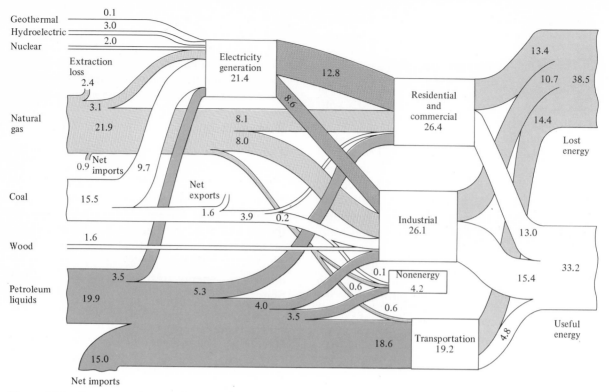

Figure 22.9 Sources and uses of energy in the United States in 1976. Values are in Quads (10¹⁵ Btu). Note that 38.5 Quads of energy is lost (light-gray paths). Of these, 14.4 Quads are lost through conversion losses and transmittal line losses when electricity is used for industrial, residential, and commercial purposes (mixed gray and color pathways). *(Modified from Schurr et al.* [see Additional Readings] *and from D. A. Tillman, 1978,* Wood as an Energy Resource, *Academic Press, Inc.   Redrawn with permission of David H. Speidel.)*

amounts of electricity; to make 1 ton of aluminum, electricity equivalent to the energy from 4 tons of coal is needed. Aluminum therefore is made where there is an abundant and inexpensive source of electrical power.

The major domestic use of energy is for heating, and natural gas and electricity are the most commonly used sources. Use of electricity appears to be the key to problems of adequacy of energy and alternative energy sources, but the problems are complex. While electricity is the most versatile energy form, the transmission and production processes are inefficient, resulting in significant energy loss. Of the 21.4 Quads of electricity produced each year in the United States, 14.4 are lost and only 7 actually are available for use. Another problem is that the use of electricity is growing rapidly so that

several estimates indicate that it will provide 50 percent of all the energy used in the United States in A.D. 2000. A third difficulty is that not all sources of energy can be used to produce electricity, whereas others can be used only for that purpose. Sources such as crude oil are too valuable in producing plastics, fertilizers, and pharmaceuticals to be used *only* for energy. These complexities must be considered when evaluating energy resources and alternatives.

ADEQUACY OF MINERAL AND ENERGY RESOURCES

It is difficult to evaluate the adequacy of our resources, as indicated earlier in this chapter. Should we use physical or economic depletion values in our calculations? What is to be the base of our calculations — reserves or resources? How do we predict future rates of resource consumption? A constant per person rate generally is the lowest rate used in projections of resource use, while a steadily increasing per person rate (based on United States use in the last 40 years) is the highest. Furthermore, we must consider world resources in addition to United States supplies because many of our needs are met today by imports. In 1973, we imported more than two-thirds of the cadmium, nickel, tin, fluorine, asbestos, and platinum-group minerals that we used, and over 95 percent of the cobalt, tantalum, manganese, strontium, niobium, and mica.

Despite the difficulties, projections must be made. Table 22.4 shows estimates of the worldwide adequacy of mineral resources for the year 2000, using the lower rate of growth. The right-hand column lists those materials which will not have problems of worldwide supply by the year A.D. 2050. Columns 2 and 3 list resources that will be in short supply — those for which we will have to find substitutes or decrease our consumption. The uneven distribution of resources is reemphasized in column 4. Worldwide supplies of antimony may be adequate and worldwide coal may be in short sup-

ply, but the reverse is true in the United States.

We worry about energy resources, but there are several mineral resources whose decreased availability also may have a major impact. There is no suitable substitute for phosphate as a fertilizer, and only phosphate rock supplies it in the volumes we need. Any decrease in the availability of fertilizer will limit food production. Fluorite (known commercially as fluorspar) is an essential material in the production of steel and aluminum. Its availability can limit the growth of developing countries and keep production of steel and aluminum well below the mills' capacity.

The adequacy of energy resources is debated often, be it a gasoline crisis or the safety of nuclear power. The focus of the problem in the United States is mainly on dependence on foreign oil and the conversion of oil users to coal. The dependence on foreign oil appears to be necessary, based on statistics that indicate trends in discovery and production. The data show that (1) most petroleum lies in a limited number of giant oilfields, (2) the giant fields are discovered early in the exploration of an area, and (3) increased drilling of wells is most likely to discover only small fields. The discovery and production of petroleum follow a pattern that is

TABLE 22.4   Adequacy of worldwide mineral resources anticipated for A.D. 2000

Adequate or almost so	Inadequate by 10–30%	Inadequate by More than 30%	Adequate discovered U.S. resources	Only worldwide resources not over 50% inadequate in 2050
Antimony	Chromium	Aluminum	Aluminum*	Coal
Barite	Coal	Asbestos	Barite	Sulfur
Cadmium	Cobalt	Copper	Coal	Tin
Iron ore	Lead	Fluorite	Manganese*	Titanium
Lead	Manganese	Gold	Molybdenum	Tungsten
Tin	Mercury	Nickel	Nickel*	Uranium (breeders)
Titanium	Molybdenum	Niobium	Phosphate rock	
Tungsten	Natural gas	Tantalum	Titanium	
Vanadium	Phosphate rock	Petroleum	Vanadium*	
Zinc		Platinum metals	Zircon*	
		Silver		
		Uranium		
		Zircon		

* Indicates not now profitable.
Source: From C. B. Reed, *Fuels, Minerals and Human Survival,* Ann Arbor Science, Mich., 1974, 175 pp.; S. H. Schurr et al., *Energy in America's Future,* Johns Hopkins Press, 1979, 555 pp. Additional information from U. S. Geological Survey and U. S. Bureau of Mines.

expected for any nonrenewable resource. The discovery rate is zero at the beginning of exploration, reaches a maximum, and becomes zero again when there is no material left to be discovered. Production follows a similar trend, delayed about 10 years by the time needed to finance and drill the wells and to build refineries. The rate of discovery in the United States reached a maximum in 1955, and production peaked in 1971. Using these patterns, a 1979 U.S. Department of Energy prediction indicated that the United States had total recoverable resources of 159 billion barrels, of which 117 billion already had been discovered by the end of 1978. The remainder has an energy equivalent of about 250 Quads, but we use about 35 Quads annually and would run out quickly.

The United States has about 40,000 Quads of recoverable coal (Figure 22.10), well in excess of the estimated 2400 Quads needed between now and the year 2000. As petroleum production decreases on a worldwide basis, we will have to turn to coal, and United States resources will have an increased international importance. If electricity is to replace petroleum and natural gas in industrial and domestic uses, the materials that produce electricity (mainly coal) will have to expand. Conversion of coal to liquid fuels (see below) may ease the transportation problem somewhat, but we cannot use coal twice: Either it is made into liquid fuel or it is burned to make electricity.

The anticipated rise in demand for electricity has led to a prediction of increased use of nuclear energy to generate electricity. Nuclear reactors today produce 10 percent of the

Figure 22.10 Recoverable fossil fuel resources. Tons of minable coal, barrels of recoverable crude oil (40 percent of estimated world resources), and billions of cubic feet of recoverable natural gas (50 percent of estimated world resources) are converted into their energy equivalents (10^{15} Btu) for comparison. Coal resources are shown in dark color, oil in light color, and natural gas in white. (*Information from the Congressional Research Service, U.S. Geological Survey, Department of Energy, and the C.I.A. Redrawn with permission of David H. Speidel.*)

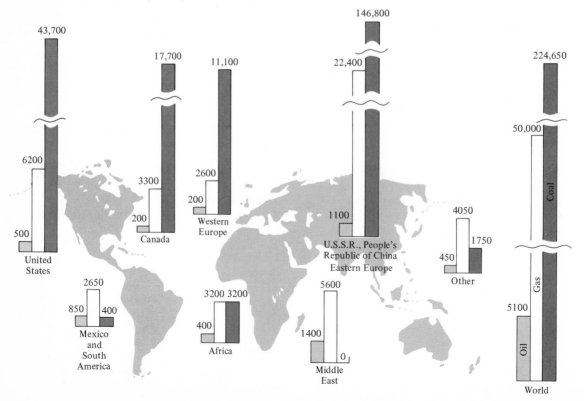

electricity used in the United States, but even at the present low rate of use we only have approximately 50 years worth of uranium resources (1000 Quads). The use of breeder reactors to convert nonfissionable uranium to fissionable plutonium may increase the amount of energy recoverable from uranium to 68,000 Quads and help solve the problem. Others are created, however. Problems of safe disposal of nuclear waste products, and the potential use of plutonium in nuclear weapons, appear to outweigh the environmental problems associated with increased burning of coal so that expansion of coal use is probably the only realistic energy solution for the rest of the twentieth century. France, however, has made a commitment to develop a breeder reactor.

ALTERNATIVE ENERGY AND MINERAL RESOURCES

Very real problems of adequate energy and mineral resources face us today. Two types of approaches have been suggested to cope with the predicted shortages: conservation, and development of alternative sources. Conservation at most can buy us more time in which to develop these new resources because it only delays the inevitable depletion of nonrenewable resources. It cannot *prevent* depletion. Furthermore, even if we can cut the per person use of a resource, the increase in population still will raise the total demand for it. We must develop the alternative sources since they alone provide hope of increased supplies of energy and mineral resources. Unfortunately, this takes time.

Unconventional Hydrocarbon Sources

There are two methods for increasing the available amount of petroleum from existing deposits of energy ores: gasification and liquefaction to convert solid coal into liquid or gaseous fuel, and increased recovery from existing oilfields.

Chemical processes to convert coal to hydrocarbon fluids were developed and used in Germany before World War II, and today there are about 30 commercial coal gasification plants in the world. Gasification and liquefaction processes manufacture **synfuels**—synthetic equivalents of natural gas and petroleum

liquids. South Africa has two major facilities producing gasoline from coal in order to decrease its dependence on petroleum imports, and the United States and the Soviet Union are expected to develop similar facilities soon.

We can increase petroleum supplies by improving on the 30 percent recovery rate from existing oilfields. **Secondary recovery** procedures involve pumping water down into the reservoir rocks to flush out crude oil. In **tertiary recovery** chemicals are pumped into the reservoir rock to reduce surface tension and free yet more oil from the particles to which it adheres. The assumption that these techniques will improve future recovery was used in calculating the future petroleum resources shown in Figure 21.10.

Alternative Oil Resources

Two other sources of petroleum are known but not presently tapped in great quantity: oil or tar sands, and oil shales. **Oil** or **tar sands** are sands or sandstones impregnated with a highly viscous asphalt with little or no gas or liquid petroleum present. They usually are mined by open pit methods and treated with hot water to decrease the viscosity of the hydrocarbons and allow their removal. The Athabasca deposit in Alberta, Canada, is the largest known oil sand deposit in the world and could produce the equivalent of an estimated 145 Quads of recoverable crude oil. The United States has much smaller deposits of tar sands in Utah, amounting to only about 3 percent of the Canadian deposits.

Shales with exceptionally high amounts of organic material are called **oil shales**. They form in reducing conditions on lake bottoms where organic matter is preserved. When such shales are heated, the hydrocarbons can be removed and refined for typical petroleum uses. The richest oil shale resource in the world is in the Green River Formation of Utah, Wyoming, and Colorado. Deposits there that yield a minimum of 25 gallons of oil per ton of shale are estimated to contain the equivalent of about 3000 Quads of energy. The United States dominates this type of resource: Only about 4500 Quads are known in such deposits throughout the world.

Energy Produced by Moving Water

The movement of water also may be used to generate electrical energy, and we already do this in hydroelectric plants. Seventy-five percent of Canada's electricity is produced in this way, compared with only 15 percent of that in the United States. Only 3 percent of the dams in the United States have electricity-producing equipment. The General Accounting Office of Congress estimates that our hydroelectric power can be tripled to about 8 Quads if half the existing dams are reconstructed to make electricity. This is not easy to do, because dams are expensive, construction is a long-range project, and sedimentary filling of reservoirs behind dams limits the useful life of the dam (see Figure 11.4).

Unconventional water-related energy sources currently are being discussed. Tidal motion may be used to drive turbines along coasts where the tidal range is high and currents are rapid. There is only one plant operating now, at the Rance River estuary in France, but Canada is planning a similar one along the Bay of Fundy. It is conceivable, although not yet practical, to get energy from the rise and fall of water in waves or the continual flow of water in large marine currents such as the Gulf Stream.

Solar and Wind Energy Alternatives

Both the direct energy of the sun and the indirect energy found in wind are important renewable energy resources. Solar energy is attractive because it is renewable, does not affect the environment seriously, and allows conservation of other energy sources. Competitive technology already exists for solar water heating and space heating, but we do not know yet how feasible solar electrical power is. The suitability of solar-generated electrical power depends on the amount of sunlight in an area, which in turn determines the size of the solar cell array needed, the amount of storage required, and the economic feasibility. The more sunlight there is and the greater its intensity, the smaller the number of cells and the storage equipment. In 1980, the Department of Energy set a goal of 20 percent of total United States energy by solar power by the year 2000.

Firewood and dried animal dung are the major sources of energy in the nonindustrial world. Indeed, in many parts of the world, the "energy crisis" consists of a lack of firewood to heat the home and cook meals, not how to buy gasoline for the second car. These energy sources, called **biofuels,** are basically solar energy stored within plant and animal tissues and released during combustion. In the United States, industry and homeowners have turned increasingly to wood as an energy source in the past few years. Wood scraps formerly discarded as unwanted by-products of the pulp and paper industry now are saved and burned to generate heat and electricity in pulp mills. If properly managed, tree farms can be a renewable resource and might yield as much as 4.5 Quads of energy per year by A.D. 2000, *an amount comparable to that expected from nuclear power.* Processing of municipal and agricultural wastes also may release hydrocarbons that can be used to generate energy, but the amount of energy that might be produced is intensely debated.

The movement of wind is a result of uneven solar heating of the earth and the earth's rotation. Wind is estimated to contain a worldwide renewable energy supply more than 40 times that provided by hydropower and geothermal heat combined. To be useful, wind must have a sufficient velocity (about 7.2 km/h) and occur regularly. Only about 10 percent of the land surface of the United States meets this requirement, but fortunately the otherwise energy-poor coastal areas of the Northeast are included. About half a million windmills of 200-kilowatt peak power would be needed just to replace the 2 Quads of nuclear electrical energy now used.

Geothermal Energy

The earth's internal heat also can be tapped for energy. The distribution of volcanoes (see Chapter 5) illustrates where steam and hot water from geysers could be used. Geothermal electrical power already is produced in the Larderello region of Italy, the Wairakei region of New Zealand, and The Geysers of California. No other regions seem suitable for geothermal electrical production, although local use for heating is possible, as in Iceland.

A different approach is to use the geothermal gradient itself, a method called **hot dry rock geothermal energy.** Water is pumped down into a hole 10 km deep and converted to steam by the heat of the rocks (about 250°C). The steam can be pumped back up the hole and used to drive a turbine. The technological and economic feasibility of this method remains unproved.

Nuclear Fusion

Energy in the sun is produced by nuclear *fusion,* a process in which hydrogen atoms combine to form helium and release large amounts of energy. We have managed to initiate this fusion reaction in hydrogen bombs but have not yet learned how to control the reaction rate to make it safe. A strong magnetic field may be able to contain the fusion reaction, but it is possible that more energy may be needed to create the magnetic field than would be generated by the fusion. For the time being, nuclear fission appears to be the only practical source of nuclear energy.

What Is the Answer?

The preceding discussion of energy alternatives clearly indicates the immense problems in replacing fossil fuels. All the alternatives require development of new technologies, long start-up times, and large capital costs; and the feasibility of some of the alternatives has yet to be demonstrated. No single alternative other than increased use of coal is capable of meeting the demand for energy over the next 20 years, but this too causes problems. Disease and fatalities among miners, water pollution by acid waste waters from mines, acid rain caused by burning high-sulfur coal, and the unknown effect of CO_2 added to the atmosphere are among the problems we will face. Sometime soon, however, decisions must be made to ensure a continuing supply of energy.

SUMMARY

A resource is something that is of value to humans. To have value, it must have a perceived use and must be available. In addition to the inherent resources of people and their ingenuity, we have the resources of the earth: land, water, minerals, and energy.

Resources are classified as renewable and nonrenewable. If the rate of use is greater than the rate of formation, the resource is nonrenewable. As population increases, the amount of resources used will increase unless there is a decrease in standard of living. In the United States, per capita use of water, minerals, and energy materials has increased continuously, adding to the demand for resources.

Land is a major earth resource. Only 30 percent of the land surface is fit for human habitation, but this resource is not distributed evenly or located conveniently. As population expands, large expenditures of energy, water, and fertilizers are used to enlarge cropland and make it more productive. Loss of cropland soil through erosion is severe in many states.

The most common worldwide use of water is for irrigation in dry areas. Irrigation is also the largest consumer of water in terms of removing it from further surface use. Other major uses of water are industrial, domestic, and for electricity generation.

The weight of mineral and energy resources used per person per year in the United States exceeds 40,000 pounds. Iron and steel constitute 95 percent of the weight of metals used, while sand, gravel, and crushed stone constitute about 90 percent of the industrial rocks and minerals processed. The per person costs of mineral and energy resources are dominated by the costs of the energy resources, especially that of petroleum. Depletion patterns calculated from discovery and production rates emphasize the decreasing availability of petroleum in the United States, our dependence on imports, and the problems of an adequate energy supply. The use of coal — the most abundant energy resource — can be expanded, but not without major strains on land, water, and air resources.

Alternative energy sources such as wind power, geothermal power, and even nuclear power do not appear capable of providing the quantity of energy needed. Sources such as oil shales, solar power, and fusion appear to have significant time lags before they can be implemented as major substitutes. For the United States, coal appears to be the only realistic energy source for the rest of the twentieth century.

Worldwide mineral resources appear to be adequate for that time period. For those cases in which the United States is dependent on other nations to supply raw materials for our economy, we must expect problems. The control of earth materials, on land or sea, has major international and political consequences. The resources of the earth are essential to our lives. The geologist can strive to increase our knowledge and understanding of how we are able to find new resources and better utilize old resources to maintain this supply.

QUESTIONS FOR REVIEW AND FURTHER THOUGHT

1. Comparison in resource use between countries often is done on a per capita basis. Explain why this might give a misleading picture of demand for resources.

2. Why are earth resources not distributed evenly? Use an example from each of the four areas discussed in this chapter.

3. Why is the rate of erosion of soil from cropland greater than that from forests? How can the rate be decreased?

4. Why would you not expect petroleum deposits under the deep ocean basins?

5. Explain why the Andes mountains have major mineral deposits apparently associated with hydrothermal solutions as well as some placer deposits.

6. In the book *Limits to Growth,* depletion of many natural resources was predicted by the year 2000. Reserve estimates were used as a base. Should we worry about depletion by that time?

7. Discuss the major problems for land and water generated by mining lignite in Montana.

8. How much longer will all the alternative energy sources for oil and gas extend our energy resources? Assume a 5 percent increase per capita per year in energy use. Justify your recovery estimates.

ADDITIONAL READINGS

Beginning Level

Brown, Lester, R., *The Twenty-Ninth Day,* W. W. Norton, New York, 1978, 363 pp.
(Written to stress "accommodating human needs and numbers to the Earth resources," it stresses the economic, social, and political aspects of resources)

Burk, Creighton A., and Charles L. Drake, *The Impact of Geosciences on Critical Energy Resources,* Am. Assoc. Advanc. Sci., Sym. 21, Westview Press, Boulder, Colo., 1978, 114 pp.
(Assessment of energy resources in both amount and location is indicated accompanied by overviews of research, government action, and future directions for different energy sources)

Clawson, Marion, *America's Land and Its Uses,* R.F.F., Johns Hopkins Press, Baltimore, 1972, 166 pp.
(A survey of facts and issues about land and land policy focusing on land use in the United States)

Skinner, Brian J., *Earth Resources,* 2d ed., Prentice-Hall, Englewood Cliffs, N.J., 1976, 152 pp.
(An elementary treatment of mineral and energy resources)

Intermediate Level

Brobst, Donald, and W. P. Pratt, Eds., *United States Mineral Resources,* U.S. Geological Survey Prof. Paper 820, 1973, 722 pp.
(An excellent discussion of the meaning of re-

sources followed by an exhaustive description of use, exploitation, geologic environment or origin, and description of resources for industrial rocks and minerals)

Mineral Commodity Summaries, 1979, Bureau of Mines, U.S. Department of Interior, 1979.
(An annual report on the U.S. domestic mineral industry including prices, import-export figures, world production, and major trends for industrial rocks and minerals, metals, and metallic ores)

Congressional Research Service, *Project Interdependence: U.S. and World Energy Outlook through 1990,* Committee Print 95–33, Interstate and Foreign Commerce Committee, U.S. House of Representatives, 1977, 939 pp.
(A massive comprehensive look at U.S. and world energy consumption and supply. A primary source for energy discussions)

Schurr, S. H., J. Darmstadter, H. Perry, W. Ramsay, and M. Russell, *Energy in America's Future, The Choices Before Us, Resources for the Future,* Johns Hopkins Press, Baltimore, 1979, 555 pp.
(A comprehensive study of the technical, economic, institutional, environmental, and health and safety aspects of alternative energy futures)

U.S. Water Resources Council, *The Nation's Water Resources, 1975–2000, Volume 1: Summary,* 1978, 86 pp.
(A national assessment of current and projected water use and supply information by region for the United States. Essential for understanding of the various water problems of the U.S.)

Advanced Level

Dixon, Colin J., *Atlas of Economic Mineral Deposits,* Cornell Univ. Press, Ithaca, N.Y., 1979, 143 pp.
(A discussion of the different types of mineral resources by use of detailed discussion of specific locations for each type of deposit)

CHAPTER 23

Planetary Geology

by
Jeffrey L. Warner*

A new era for geology began on July 20, 1969, when Apollo 11 Astronaut Neil Armstrong took his first step onto the surface of the Moon. Quite literally, new worlds were opened for us to explore. Up until then, our knowledge of the Moon was limited to what could be observed from the Earth by telescope. Now, the rocks, structure, and surface of another planetary body could be examined directly for the first time. Since then, other parts of the Moon have been visited by teams of astronaut-geologists, unmanned probes have landed on the surfaces of Venus and Mars, and spacecraft have flown past Mercury, Jupiter, Saturn, and the satellites of Jupiter and Saturn. We have learned that there are basalt and gabbro on the Moon, wind erosion and deposition on Mars, active volcanoes on one of Jupiter's satellites, and plate tectonics on another of Jupiter's satellites.

Perhaps more important, we have found that we can apply our techniques for studying the earth to learning about our neighbors in the solar system. We are using the principles of superposition and cross-cutting relationships (Chapter 8) to study the sequence of events in the histories of the Moon and Mars, gravity to investigate the possibility of plate tectonics on Mars and Venus, and seismology to study moonquakes and the internal structure of the Moon. With the wealth of new data about the other planets, we can view the earth in its proper perspective as a unit of the solar system that is

*NASA Johnson Space Center, Houston, Texas.

unique in many ways but nevertheless shares a common history with its companions. In this chapter we will examine some of the things that have been learned about the other planets and see how these data have led to ideas about the formation of the solar system.

THE SOLAR SYSTEM

A **solar system** is a family of objects held in orbit around a star by gravitational attraction. Our solar system, of course, is controlled by the gravitational pull of the sun. There are countless other stars, and astronomers agree that there certainly must be many other "solar systems," even though we have never detected planets associated with other stars. Our solar system is quite complex, with an assortment of different types of objects: a star, nine planets, 40 satellites, and thousands of asteroids and comets.

The Sun

Our sun is a star named Sol, from which comes the word *solar*. It is an incandescent spheroid composed mostly of the gases hydrogen and helium, but traces of all the other naturally occurring elements have been identified in it. The Sun's heat and light are by-products of the nuclear fusion reaction by which four hydrogen atoms are combined to make a helium atom (see Chapter 2). The Sun is in many ways the center of the solar system. Compared with even the largest of the planets, it is enormous (see Table

TABLE 23.1 Comparison of the planets of the solar system

Planet		Mean distance from the Sun, AU*	Diameter, km	Bulk density, g/cm^3	Number of satellites
Mercury	●	0.387	4,880	5.4	0
Venus	●	0.723	12,102	5.3	0
Earth	●	1.0	12,756	5.5	1
Mars	●	1.52	6,794	3.9	2
Asteroids		—	—	—	—
Jupiter	●	5.20	143,200	1.3	15
Saturn	●	9.52	120,000	0.7	14
Uranus	●	19.16	51,800	1.6	5
Neptune	●	29.99	49,500	2.2	2
Pluto	●	39.37	2,600	5 (?)	1

* AU is an astronomical unit which equals the mean distance between Earth and the Sun. 1 AU = 149.6 million km.

23.1) and contains more than 99 percent of the total mass in the solar system. Its mass gives it the strong gravitational attraction that keeps the nine planets in their orbits—even Pluto, some 6 billion km away. Its light and heat are vital to life forms and surface processes 150 million km away on the third planet, Earth. The Sun's gravitational attraction not only keeps Earth in its orbit but also produces the tidal effects described in Chapter 2.

Planets

Spheroidal bodies that revolve around the sun in fixed, nearly circular orbits are called **planets.** There are nine such bodies in our solar system; their locations and some of their vital statistics are given in Table 23.1. The nine planets fall into two broad categories based on their sizes and compositions: inner, or terrestrial, and outer, or Jovian, planets.

The inner four planets—Mercury, Venus, Earth, and Mars—are composed mostly of rock and metal and have average densities between 3.9 and 5.5 g/cm^3. They are relatively small, with diameters ranging from 5000 km (Mercury) to 12,750 km (Earth). They are called **terrestrial planets** because of their general similarity to Earth. Pluto, located among the outer planets, appears to have a size and density similar to those of the inner planets and generally is considered a terrestrial planet.

The other outer planets—Jupiter, Saturn, Uranus, and Neptune—are very different from Earth. They are far larger, with diameters between 50,000 and 150,000 km, but are not nearly so dense (0.7 to 2.2 g/cm^3) as the terrestrial planets. These planets are called **Jovian planets** after Jove, another name for Jupiter, and appear to be made of liquids and gases rather than rock and metal.

Satellites

Most planets are orbited by smaller bodies called **satellites.** Satellites come in a wide range of sizes and shapes, from irregularly shaped objects 10 km across to spheroidal bodies more than 5000 km in diameter (Titan, a satellite of Saturn). Our satellite, the Moon, has a diameter of about 3500 km. As shown in Table 23.1, the terrestrial planets have very few satellites (if any), whereas Jupiter alone has 15.

The different shapes of satellites are related to their sizes. The larger satellites are spherical because they contain so much mass that high internal pressures are generated by gravity, and these pressures deform the rocky material into a spherical equilibrium shape. This process, called self-gravitation, also explains why planets are spherical. Many of the largest satellites, such as our moon, really are small-scale planets with the same properties as terrestrial planets. We will discuss them later, along with the planets. In contrast, the smaller satellites do not have enough mass to result in rock deformation and therefore do not develop spherical shapes. Neither do they appear to have experienced internal geologic activity such as that seen in the planets and larger satellites.

Asteroids

Relatively small rock bodies, most of which revolve around the Sun in a position between the orbits of Mars and Jupiter, are called **asteroids.** A cluster of these objects, called the **asteroid belt,** occurs at the boundary between the inner and outer planets. Asteroids are similar to planets in that they revolve only around the Sun, but in size, shape, and history they are more like satellites. The largest asteroids—Ceres (1020 km diameter), Vesta (550 km), and Pallas (538 km)—are spherical because of self-gravitation and probably experienced some internal geologic activity during their evolution. The small asteroids have diameters of only a few kilometers and probably have always been geologically inactive.

Comets

Comets are a totally different class of objects. Unlike the planets and asteroids, which have nearly circular orbits, comets have elliptical orbits. Comets are ice objects containing a minor amount of rock fragments, and they range in diameter from 1 to 10 km. During the parts of their orbits that bring them closest to the Sun, some of the icy material evaporates, forming a temporary atmosphere around the comet and a "tail" of gases that streams out in a direction away from the Sun.

Order in the Solar System

Despite the variety of bodies present, the solar system is a carefully ordered collection of objects. The order involves the systematic location and movement of the objects and strongly suggests that the Sun, planets, satellites, and asteroids are part of a single family that formed at the same time by a single process.

The orbits of all nine planets (and those of the asteroids) lie within a few degrees of the same plane so that the solar system is essentially disk-shaped. This plane, known as the **ecliptic,** is defined simply by the plane traced out by Earth's orbit. Furthermore, within this plane most planets and asteroids revolve in the same direction—counterclockwise when viewed from above Earth's north pole. There is also regularity in the distances of the planets from the Sun. If the asteroids are assumed to represent a "missing" planet, there is a spectacularly regular spacing between the orbits of the planets. Each planet's orbit is 75 percent larger than that of the next inner planet, a relationship known as the **Titus-Bode law.**

Not only are the dynamics of the solar system strictly regulated, but there also appears to be a chemical ordering. The inner four planets are of one type (terrestrial), whereas the next four are quite different in composition (Jovian). Pluto is an apparent exception to this chemical ordering, but we do not know very much about the farthest planet and thus cannot understand its significance.

INVESTIGATING THE SOLAR SYSTEM

In the first 21 chapters of this book, we saw how complicated the study of a single planet can be, with all the different types of data that must be collected and the interrelationships that must be learned. Imagine how much more difficult it must be to study the other planets when we cannot even visit most of them. Despite the difficulties, we have been able to work out a surprisingly complete picture of the terrestrial planets and some large satellites by a combination of remote sensing and direct sampling. Detailed measurements of a planet's size, density, moment of inertia, and magnetic field provide important information about its internal composition and tectonic activity. Photographs taken through telescopes on Earth or by spacecraft give indications of topography and thus of surface processes. Actual samples of extraterrestrial rock, both meteorites found on Earth and lunar samples returned to Earth by astronauts, give us an idea of the minerals, rocks, and rock-forming processes on other planets. We will concentrate on the terrestrial planets because, being earthlike, they are studied the most easily with techniques that were developed to study the Earth. They also hold the greatest promise for telling us about the early evolution of our home

planet. We first will examine the remote sensing methods and in a later section see how actual rock samples can be used to interpret the history of a planet.

Planetary Size

The size of a planet is an important factor in understanding its evolution because size has a large effect on the internal heat available for magmatic and tectonic activity. Size also can be determined easily by astronomic observations from Earth.

Large terrestrial planets tend to be hotter than small ones. Planetary heat comes from the decay of radioactive isotopes and from gravitational effects during the initial formation of a planet. Large planets simply have more matter (to produce gravity-derived heat) and more radioactive atoms (with which to produce heat by nuclear reactions) than small planets. Large planets also cool more slowly than small planets, and so they stay hot longer. All the heat generated within a planet is lost eventually into space by radiation from the planet's surface. Therefore, the greater the ratio of surface area to volume, the faster cooling will be. On a proportional basis, small planets have higher surface area to volume ratios than large planets and thus dissipate their heat more rapidly.

Because large terrestrial planets get hotter and stay hot longer than small ones, they can be more tectonically active and maintain activity longer.   The two largest terrestrial planets (Earth and Venus) are geologically active today and presumably have been active throughout their histories. Two of the three next largest are geologically inactive today but show signs of past activity (see below). However, the smallest bodies apparently were never active internally.

Io, the second satellite of Jupiter, is an exception in that it is very small but also very active. However, its activity is not caused by internal heat but rather by strong tidal interactions with Jupiter. External forces may cause geologic activity on any object, but only the larger terrestrial planets can *maintain* self-induced tectonism.

Planetary Density

The average density of a planet gives information about its chemical composition. Studies of meteorites, moon rocks, and Earth suggest that terrestrial planets are composed of three different types of material: silicate rocks with an average density of about 3.0 g/cm^3, iron-nickel metal alloy with a density of about 8 g/cm^3, and ice (or water) with a density of about 1 g/cm^3. A planet's density is determined from careful observations of its size and orbit and the orbits of its satellites, and it can be used to estimate the relative proportions of the three different types of material.

Each terrestrial object is thought to be composed principally of two of the three materials described above. Planets with densities between 3 and 8 g/cm^3 are considered to consist of a mixture of rock and metal. Earth, with a density of 5.5 g/cm^3, consists of silicate rock in its crust and mantle (two-thirds of the planet) and iron-nickel alloy in the core (one-third). Water and ice are present on Earth but account for less than 1 percent of the planet and thus do not affect the calculations significantly. Mercury, Venus, and Mars have similar densities and are thought to represent somewhat different mixtures of metal and rock.

If a body's density is between 1 and 3 g/cm^3, it probably consists of a combination of rock and either ice or water. The satellites Callisto, Ganymede, and Titan probably are made up of a rock core, a mantle of water, and an icy crust. On the other hand, the satellites Moon and Europa have densities very close to 3.0 g/cm^3 and are considered to be composed almost entirely of silicate rock.

Moment of Inertia

The tendency of a spinning object to keep on spinning is called its **moment of inertia** and basically is controlled by the object's mass and size. However, the distribution of mass within an object controls a function called the moment of inertia factor. For a perfectly homogeneous sphere, the moment of inertia factor is 0.400; for a sphere in which most of the mass is concentrated near the center, the factor is less than 0.400; and for a sphere in which the mass is concentrated near the periphery, the factor is greater than 0.400. The effect of mass distribution can be seen by watching a rapidly spinning skater. By pulling the arms in close to the body, the skater spins more rapidly because mass then is concentrated near the center of the skater's body. Outstretched arms distribute mass farther out from the body's center and slow the spinning.

The moment of inertia factor of a planet tells the size of the planet's core and indicates how much differentiation of elements of the type discussed in Chapter 21 has occurred. The greater the mass concentrated in a planetary core, the smaller the moment of inertia factor will be. As of 1980, the moment of inertia factor was known for only Earth, Mars, and the Moon. The value for Earth is 0.333, indicating the presence of a substantial core — a conclusion supported by the seismic, gravity, and magnetic data discussed in Chapters 19 and 20. The value for Mars, 0.365, suggests that Mars has undergone some differentiation and has a modest-sized core smaller in proportion to its size than Earth's. The Moon apparently has an extremely small core, as shown by its moment of inertia factor of 0.392, which is very close to that of a homogeneous sphere.

Planetary Magnetism

Not all of the terrestrial planets have dipolar magnetic fields like Earth's. As discussed in Chapter 20, a planetary magnetic field is thought to require complex motions in the electrically conductive portion of that planet. For terrestrial planets, this generally means movement of molten iron-nickel metal or sulfide minerals. The presence or absence of a magnetic field thus may tell something about the presence or absence, motion or lack of motion, of specific materials in planetary cores.

Earth and Mercury have strong dipolar magnetic fields, suggesting that Mercury contains a considerable volume of electrically conductive molten material, perhaps in a core like Earth's. On the other hand, the Moon does not have a magnetic field, probably because, as shown by the moment of inertia factor, it lacks a substantial core. Venus apparently has a core similar to Earth's but does not have a magnetic field because its rotation is so slow (once every 243 days) that complex motions in the core have not been produced. The absence of a planetary magnetic field about Mars is not understood fully because Mars has a modest-sized core and rotates fast enough to produce the necessary motions in molten material. If the Martian core is totally solid, however, complex motions would not be produced. Several scientists believe that a solid Martian core is the best explanation for the absence of a magnetic field.

Topography

In studying other planets and satellites, we look for major differences in topographic level, chains of mountains, and volcanoes. We then try to explain these features by referring to similar landforms on Earth for which we know

the origins. In the sections that follow, we will see what the faces of our neighbors in the solar system look like and what their surface features tell about their internal processes. For example, are there transform faults or subduction zones on Mars?

Cratering

Interplanetary rock fragments that collide with planets or satellites are called **meteorites.** Meteorites collide with great force, excavating bowllike pits called **meteorite impact craters** in the surface. The depth of penetration is several times the diameter of the meteorite. The excavated material is thrown upward and outward from the crater and settles on the surface as an **ejecta** deposit. Ejecta deposits extend far from the crater in a series of light-colored **rays** that make the crater look something like a sunburst (Figure 23.1). On airless and waterless

Figure 23.1 The full Moon (photograph taken through a telescope). Notice the differences between the light-colored, cratered highlands and the dark-colored, less cratered maria. *(Lick Observatory, University of California, Santa Cruz.)*

planetary bodies such as the Moon, ejecta deposition is the only type of sedimentation (see the Perspective in Chapter 9).

Meteorites range in size from microspheric particles to massive bodies 100 km or larger in diameter. The diameter of meteorite impact craters is from 100 to 1000 times the diameter of the meteorite that formed them. Craters larger than 20 km tend to have flat rather than bowl-shaped floors; groups of hills called central peaks are usually situated on the crater floors (Figure 23.2).

On planetary bodies with atmospheres, such as Earth, smaller meteorites (less than 10 meters in diameter) burn up or are slowed down by friction with gas molecules to such a degree that they have too little energy to form a crater. Planets and satellites without atmo-

spheres, on the other hand, are bombarded continually with meteorites of all sizes.

Crater densities and the ages of planetary surfaces

The principles of superposition and crosscutting relationships discussed in Chapter 8 also can be used to determine the relative ages of planetary surfaces. Young craters tend to have sharp features, such as knife-edged rims. As the craters age, they become more subdued in form as a result of mass movements triggered by nearby meteorite impacts and by partial burial by younger ejecta deposits. Overlapping crater walls (crosscutting relationships) and superposition of one ejecta deposit upon another can be used to determine the sequence of events in an area.

Figure 23.2 The crater Euler on the moon. Euler is 27 km in diameter and displays features typical of relatively young craters larger than 20 km in diameter. It has a flat floor, central peak complex, terraces along its walls, continuous ejecta blanket, and subtle rays. The floor and ejecta blanket of Euler are cut by younger craters with diameters up to 5 km. The smaller craters are bowl-shaped. *(NASA.)*

Crater density—the number of craters per unit area—also can be used to determine the relative ages of planetary surfaces. Older surfaces have been exposed to more meteorite impacts than younger ones, and so they have a higher crater density.

Radiometric dating of lunar rock samples has pinpointed the ages of cratering events on the Moon, as shown in Figure 23.3. If the rate at which meteorites strike the surface of the Moon remained constant throughout time, the curve in Figure 23.3 would be a simple line because a 3-billion-year-old surface would have three times the crater density of a 1-billion-year-old surface. However, the actual curve shows a sharp increase in crater density for surfaces 4.0 billion years old, indicating that there were far more meteorites impacting the Moon's surface early in its history than in more recent times.

Why were there more meteorites 4 billion years ago than now? Some scientists feel that the 4-billion-year-old cratering marked the last major stage in the formation of the planets and large satellites—a gathering up of loose debris onto the larger bodies by gravitational attraction. If this is correct, the number of meteorites being swept up at any time should have been about the same everywhere in the solar system, and the crater density-time curve can be used to correlate events from one planet to another.

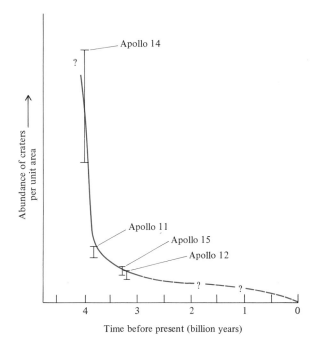

Figure 23.3 Crater density on surfaces of the Moon as a function of age of the surfaces. There are no data to indicate the crater density on surfaces older than 4 billion years. Because this curve is equivalent to the number of impacting meteorites as a function of time, it can be used to correlate the geologic history among planets and large satellites.

THE TERRESTRIAL PLANETS AND LARGE SATELLITES

We know more about the Moon than about any other body in the solar system except Earth itself, and so we will begin this discussion with our own satellite. However, in recent years we have gathered a lot of data about the other inner planets and are beginning to learn more about the Jovian planets. There are several interesting comparisons between Earth and the other terrestrial planets that help us understand the evolution of the solar system.

The Moon

The Moon is unique among the satellites because it is so large in comparison to the planet it or-

bits. Satellites of Jupiter and Saturn are 1000 km larger than the Moon (in radius) but are only a small fraction of the size of their planets, whereas the Moon is more than 25 percent of the size of Earth. Its low density of 3.3 g/cm^3 indicates that it is composed almost entirely of silicate rock, and its moment of inertia factor of 0.392 shows that it has only a very small core. At most, its core accounts for perhaps 5 percent of its total mass, compared with 33 percent for Earth's core.

Lunar topography

Figure 23.1 shows that there are two distinctly different types of terrain on the lunar surface, much as is the case with Earth's continents and oceans. The dark areas are topographically low, as are Earth's ocean basins, and are called **maria** (singular, *mare*, the Latin word for *sea*) because early astronomers thought

TABLE 23.2   Characteristics of the terrestrial planets and large satellites

Planet	Large satellite	Radius, km	Mass, 10^{22} kg	Bulk density, g/cm³	Moment of inertia factor	Global magnetic field	Geologic activity	Atmospheric pressure at surface, bars	Atmosphere constituents Chief	Atmosphere constituents Minor
Earth		6378	598	5.5	.333	Dipole	Active	1.0	N_2, O_2	CO_2, H_2O, Ar
Venus		6051	487	5.3	?	Nil	Active?	93	CO_2	N_2, Ar, H_2O, SO_2, O_2
Mars		3397	64	3.9	.365	Nil	Past activity	0.005 to 0.01	CO_2	N_2, Ar, O_2, H_2O
	Ganymede	2638	15	1.9	?	?	Past activity	Vacuum	—	—
	Titan	2560	14	1.9	?	?	?	1.5	N_2	CH_4, C_2, H_2, HCN, C_2H_4
	Callisto	2410	11	1.8	?	?	Never active	Vacuum	—	—
Mercury		2440	33	5.4	?	Dipole	Never active	Vacuum	—	—
	Io	1816	9	3.5	?	?	Active	Vacuum	—	—
	Moon	1738	7	3.3	.392	Nil	Ancient activity	Vacuum	—	—
	Europa	1563	5	3.0	?	?	Surface activity	Vacuum	—	—

The column headed Geologic activity is a general assessment of internal geologic activity interpreted from surface features.

they were oceans. We know now, of course, that there is no water on the Moon. The maria are seas of a different sort—vast outpourings of basaltic lava, also quite similar to Earth's ocean basins. Topographically high, mountainous regions are called highlands and are light-colored in Figure 23.1. As with the ocean-continent relationships on Earth, maria are significantly younger than the highlands, although the actual ages are far older than on Earth. Highlands rock represents an age of 4.6 to 3.85 billion years (older than any rock dated on Earth), while the maria are "only" 3.9 to 2.5 billion years old, as much as 20 times older than our ocean basins.

Rilles Linear, steep-walled depressions called **rilles** cut both the maria and the highlands (Figure 23.4). Straight rilles cut through both types of terrain and are thought to represent fault-valleys or graben. Meandering valleys called sinuous rilles are restricted to the mare. Early theories that the sinuous rilles were carved by lunar streams were disproved by the absence of water on the Moon; they probably are volcanic features similar to collapsed lava tunnels on Earth (see Chapter 5).

Craters The lunar surface is pockmarked by craters formed by repeated meteorite im-

pacts. Indeed, the lunar mountains are circular in outline and represent the walls of very large craters. Although photographs and the naked eye may suggest that the maria are flat, smooth features, they contain craters that are 1000 times more abundant than impact craters on Earth. Craters in the highlands are more abundant than they are on the maria and are so numerous that they commonly overlap, with the rim of one crater cutting that of another. The greater abundance of craters in the highlands is due to the greater age of the highlands. The surface of the highlands is older than that of the maria so that it has been exposed to meteorite impacts for a longer period of time.

A comparison of the surfaces of Earth and the Moon

The Moon's surface differs markedly from Earth's in several respects. Physiographic features on Earth result from the interaction of two opposite kinds of processes: aggradation and degradation. Aggradation is the building up of positive-relief features and is accomplished by volcanic eruption and orogeny on Earth, whereas topographic highs are worn away (degradation) by weathering and erosion. The Moon has no atmosphere or surface water, and so there can be neither weathering nor erosion

Figure 23.4 The Eastern portion of Mare Imbrium on the moon. Notice the straight and sinuous rilles. The sinuous Hadley Rille in the center meanders across the mare surface passing close to the Apollo 15 landing site. Straight rilles appear in the upper and left portions of this photograph where they strike north-east and cut both mare surface and the Apennine Mountains of the moon's highlands. *(Jet Propulsion Laboratory.)*

on its surface comparable to those on Earth. The Moon also is so small that it has had little if any orogenic activity. Its surface results from meteorite impacts, and its mountains are the walls of large craters and huge impact-created basins. Earth is protected from all but a few meteorites by its atmosphere; most meteors vaporize through frictional heat and never reach the surface of the planet.

Another major contrast is in the ages of the lunar and terrestrial surfaces. Earth's surficial features are relatively new and are forming or being modified as you read these words. Even the present configuration of oceans and continents dates back less than a few million years. The lunar surface is very old, with the youngest rocks being the mare basalts that formed billions of years ago.

A third difference, learned from studies of moon rocks and lunar seismicity, concerns the nature of the planetary crust beneath the topographic lows on the two bodies. The maria are basins a few kilometers deep that have been filled with basalt. The basins themselves were gouged out of the Moon by meteorite impacts, and beneath the mare basalts

lies a crust identical to that found beneath the highlands. Basalts also make up the ocean basin filling on Earth, but, as we saw in Chapters 4, 19, and 20, the crust beneath the oceans is very different from the crust beneath the continents.

Mercury

Mercury is a Moon-sized terrestrial planet that, like the Moon, has no atmospheric envelope. Its average density of 5.4 g/cm³ is comparable to Earth's, indicating that its interior is similar to Earth's—a silicate outer layer surrounding an iron-nickel core. Mercury's dipolar magnetic field is tilted, like Earth's, about 10° from its rotational axis. The presence of the magnetic field indicates that at least part of Mercury's core probably is molten.

While Mercury's interior is similar to Earth's, its surface is much more Moonlike (Figure 23.5). Most of its surface is dominated by overlapping, shoulder-to-shoulder craters reminiscent of the lunar highlands, and there are small dark regions with few craters that may be comparable to the lunar maria. The largest Mercurian craters are basins surrounded by con-

Figure 23.5   Cratered surface of Mercury. Image obtained by the Mariner 10 spacecraft as it flew by Mercury in 1972. N and S indicate the north and south poles. *(Jet Propulsion Laboratory.)*

centric rings of ridges, and these also are comparable with what are called multiringed basins on the Moon. Rilles have not been identified on Mercury, but there are puzzling sinuous cliffs that might be traces of overthrusts.

Venus

In some ways Venus is much like a twin of Earth, similar in size and density but very different in the nature of the processes acting on and inside it. Although its Earthlike size and density suggest that Venus has an iron-nickel core of appreciable size, it has no magnetic field. As stated earlier, this probably is due to the very slow rotation of Venus (243 days versus 24 hours for Earth) which is not fast enough to maintain complex motions in its metal core.

Venus has the densest atmosphere of all the terrestrial planets—some 90 times the density of Earth's atmosphere. It is composed almost entirely (96 percent) of carbon dioxide, with only minor amounts of water vapor, nitrogen, and argon. The lower 31 km of this atmosphere is remarkably clear, but between 48 and 68 km there are three layers of dense clouds that completely surround the planet. This dense atmosphere creates a "greenhouse effect" (see Chapter 13) that traps incoming solar radiation and keeps the Venusian surface at a red-hot 470°C. Thus, even though water vapor is in the Venusian atmosphere, the surface temperature is too high for liquid water to exist, and there can be no erosion or deposition by streams, ocean currents, groundwater, or glaciers.

The clouds that prevent heat from escaping from Venus also prevent our telescopes from seeing what the planetary surface looks like. As a result, we know less about the Venusian surface than about the surface of any other terrestrial planet. Radar, of course, is used for navigation on Earth because it can penetrate clouds. In the late 1970s, radar measurements of Venus were carried out from Earth and by the Pioneer Venus spacecraft that orbited Venus in 1979 and 1980. These studies penetrated the Venusian cloud cover enough to give some idea of the large-scale surface features (Figure 23.6).

Most of the surface of Venus consists of rolling hills with relief of about 1 km. These hills are pockmarked by large impact craters that are more abundant than craters on Earth and about as abundant as those on the lunar maria. Venus also has two continent-sized plateaus that stand about 4 km above the rolling terrain. Both "continents" are broken on their eastern ends by what seem to be rift valleys and fault-block mountains. One of the two continents has mountains that rise an additional 4 km above the plateau surface. The origins of the plateaus and mountain ranges are not known.

Venus has at least one giant volcano. It is a large conical feature 700 km across and 10 km high, with a summit depression similar to that of Earth's shield volcanoes. Measurements of the radioactivity on the flanks of this volcano by a robot Soviet spacecraft suggest that the volcano may be of basaltic composition. The immense size of the volcano suggests that the rocks that form it must have been erupted over a very long period of time. To sustain such a long-term process, a large subsurface heat source is needed, perhaps comparable to a mantle hotspot on Earth. However, plates on Earth move over hot spots so that single huge volcanoes cannot form, and Hawaiian-type island chains result instead (see Chapters 16 and 18). Apparently, the Venusian surface does not move laterally the way Earth's lithosphere does, and so the lava can construct one immense volcano. This suggests that there may be no plate tectonic activity on Venus.

The abundance of craters on the rolling hills of Venus indicates that these terrains must have been stable for hundreds of millions, or perhaps billions, of years. On Earth, only the Precambrian shield areas have been so stable, and they have the thickest lithosphere on Earth. By analogy, the gently rolling hilly terrain on Venus

Figure 23.6 Map of the surface of Venus showing surface elevation. This map was produced by computer from altimetry data obtained from radar on the Pioneer Venus spacecraft that orbited Venus during 1979 and 1980. The light tones are high elevations and the dark tones are low elevations. Note the two continent-sized plateaus and the large conical volcanic mountain. *(Courtesy of Professor Gordon Pettengill.)*

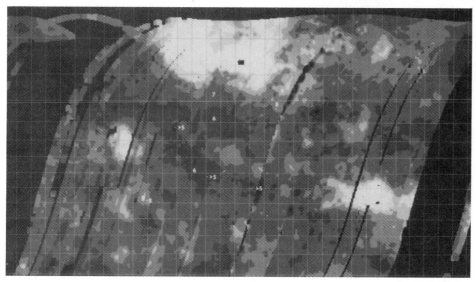

also must have a thick lithosphere. With the high surface temperature on Venus and a normal geothermal gradient, a thick lithosphere can be obtained only if the interior of Venus is very dry.

Mars

Mars is intermediate between Earth and the Moon in many aspects of interior and surficial processes. The Martian density and moment of inertia factor are intermediate between those of Earth and the Moon, indicating that Mars probably has a rocky mantle and an iron-nickel core and that the core is proportionally smaller than Earth's but larger than the Moon's. Some regions of the Martian surface are heavily cratered, suggesting a relatively inactive planet, but there is an atmosphere and some tectonic activity in some portions of Mars.

Figure 23.7 Photomosaic of channels on the Martian surface obtained by the Viking orbiter spacecraft. This photomosaic of the Maja Vallis region shows one of the best developed channels on Mars. This type of surface feature is evidence that water once flowed across the Martian surface. The impact craters that cut the channels indicate that the channels are very old. *(Jet Propulsion Laboratory.)*

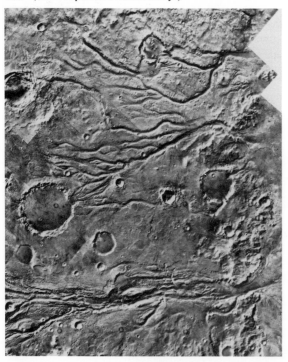

The atmosphere of Mars is only about 1 percent as dense as Earth's and principally is made of carbon dioxide (95 percent), with minor amounts of oxygen, nitrogen, argon, and water vapor. The low atmospheric density cannot cause a greenhouse effect such as that on Venus, and so Mars is a cold planet, with temperatures ranging from 0°C to −130°C. Today, Mars is essentially a cold, arid desert, but this was not always the case. Surface features such as dendritic valley patterns, outflow channels, and streamlined islands in channels indicate that liquid water probably once flowed across the Martian surface (Figure 23.7). The channels are cut by impact craters, suggesting that it has been hundreds of millions of years since water last flowed through them.

The Martian surface consists largely of highlands and plains similar to the highlands and maria of the Moon. Based on the abundance of impact craters, the plains are younger than the highlands. Like the maria they probably are composed of basalt lava flows. Cratering is more abundant even on the plains than it is on Earth, indicating that the Martian landscape is much older than Earth's.

A series of volcanic cones is built on both highlands and plains in several regions of Mars. In the Tharsis region in particular (centered around 30°N Martian latitude, 90°W Martian longitude), the volcanoes have merged into an uplifted volcanic plain. Craters are notably scarce in the Tharsis region, indicating that these volcanoes are among the most recent geologic features on Mars. Unlike the giant Venusian volcano, the Tharsis region could well have been produced in Earthlike fashion by plate movement over a mantle hot spot.

However, most of the evidence suggests that Mars is dominated by vertical tectonism. Volcanoes and broad regions of what appear to be horst-and-graben structure are common. One graben, called Valles Marinarus, is 4000 km long, as wide as 240 km in places, and up to 6 km deep. Unlike rift valleys on Earth, Martian rifts show no evidence of associated volcanism.

Martian volcanoes are not active today, and surface activity is limited to processes of wind and glacial activity somewhat different from those on Earth. Some of the effects of Martian wind activity were discussed in Chapter 14. Vast regions of Mars are covered with dune fields, and every spring a dust storm forms in Mars's northern hemisphere that lasts for several months and spreads over almost the entire planet. Both polar regions of Mars (above 82° latitude) are occupied by residual caps of H_2O ice and regions of layered (glacial? wind?) deposits. Each winter, a seasonal polar "ice" cap composed of CO_2 frost forms and extends from the poles to latitudes as low as 50°. There is no evidence of movement of the ice in Martian ice caps. This apparent absence of plastic flow makes the ice caps on Mars very different from those of Antarctica and Greenland.

The Galilean Satellites of Jupiter

The four largest satellites of Jupiter—Io, Europa, Ganymede, and Callisto—were discovered by Galileo in 1610 and have been named Galilean satellites in his honor. Although they can be observed easily with a telescope, they are so distant from Earth that they were little more than specks of light until the Voyager spacecraft passed through the Jovian system in 1979. Voyager images of these satellites (Plate 23) demonstrated that each is a very different world, with its own geologic features and history.

Ganymede and Callisto

Ganymede and Callisto, the fourth and fifth satellites of Jupiter, are about the size of Mercury but have densities less than 2 g/cm³. This suggests that they are made of ice and rock, probably with a rock core and a lower-density ice-water mantle. The surface relief on both satellites is less than 1 km, probably because an ice-water structure is not strong enough to support larger features. All of Callisto and much of Ganymede exhibit a dark surface, suggesting that a thin layer of rock dust or debris covers the ice.

The surface of Callisto is covered relatively uniformly by overlapping and shoulder-to-shoulder craters. Its craters are more abundant than even those on the lunar highlands, suggesting great age. Callisto also has

several huge circular features that consist of about 20 low concentric ridges that are arranged in bullseye patterns and have very few craters in their centers. These apparently are the icy-planet version of the multiringed basins on Mercury and the Moon.

In contrast, Ganymede has two types of surface terrain: cratered and furrowed (Figure 23.8). The cratered regions are similar to those on Callisto, but the furrowed terrains are lighter-colored and consist of sets of furrows that are 5 to 10 km across and about 100 m deep. The furrowed terrains appear to separate regions that are heavily cratered. In places, the belt of furrowed terrain widens abruptly in a pattern reminiscent of midocean ridges and transform faults on Earth. Some geologists feel that plate tectonic processes may have acted in Ganymede's geologic past. They believe that the furrows are comparable to ocean ridge spreading centers on Earth. In this model, water rises from the mantle of Ganymede into the rift zones and freezes to form the furrowed terrain. Wherever the cratered "plates" spread with uneven rates, a transform fault such as that shown in Figure 23.8 forms.

Are plates still moving on Ganymede (if they ever did at all)? The furrowed terrain itself is cratered with about the same density as the lunar maria, indicating an age of billions of years. This suggests that if plate tectonic activity did exist on Ganymede, it stopped billions of years ago. Although "spreading centers" have been inferred on Ganymede, there is no evidence as yet of subduction zones.

Europa

Europa, the third satellite of Jupiter, is about the size of the Moon. Its density indicates that its interior is made mostly of rock, but it also has a smooth, highly reflective surface that probably is composed of ice. The size and bulk density suggest that the ice layer may be as much as 100 km thick. Europa has a very low abundance of craters, even lower than that of Earth. This indicates a very active surface environment in which the smooth surface ice is renewed regularly to eradicate impact features.

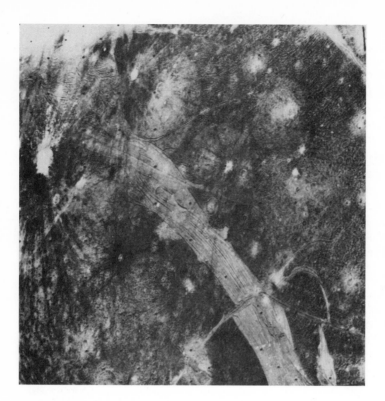

Figure 23.8 Image of Ganymede obtained by the Voyager spacecraft in 1979. The image shows a region 1200 km across that has two terrain types: dark-cratered terrain and light-furrowed terrain. Note the transverse fault cutting the furrowed terrain, suggesting that the furrowed terrain grows somewhat like a midocean ridge on Earth. However, the furrowed terrain is probably ice whereas midocean ridges are basalt. *(Jet Propulsion Laboratory.)*

Io

Io, the second satellite of Jupiter, is one of the most interesting bodies in the solar system. It is Moonlike in size and density but is very unlike our satellite in other ways. The surface of Io is bright-colored, in tones of red, orange, white, and brown; these colors are thought to represent extensive deposits rich in sulfur. There are no impact craters on Io, but its surface is marked with hundreds of volcanic calderas and associated lava flows. We have seen earlier that there are volcanoes on other planets and satellites, but those on Io are unique in one important respect: Some are still active today. Cameras on the Voyager spacecraft detected eight active volcanoes on Io, each ejecting gas and dust as much as 250 km above the surface (Figure 23.9).

The Jovian Planets

Jupiter, Saturn, Uranus, and Neptune are far larger than the terrestrial planets but much lower in density (Table 23.1). The Jovian planets have atmospheres, but they are unlike those of Venus, Earth, or Mars. They consist mostly of hydrogen, with 10 percent helium and minor amounts of methane, ammonia, and other reduced gases. The interiors of the Jovian planets also are very different from those of the terrestrial planets. They are composed largely of shells of liquid hydrogen and helium. The visible surface of these planets are formed by clouds in their atmospheres. Their fast rotation (of about 10 hrs) causes complex patterns in these clouds (Plate 25).

Three of the Jovian planets—Jupiter, Saturn, and Uranus—have a set of rings defined by a plane containing tiny particles in orbit over their equators. Saturn's rings (Plate 24) are the best known because they are light-colored and reflect light brilliantly and thus are easy to detect. Voyager imagery of the Jupiter system showed that there is a different type of ring around Jupiter—one that is made of dark and poorly reflecting particles. It had not been detected in studies from Earth. Saturn's rings probably are made of ice particles, whereas those of Jupiter are composed of fine-grained rock dust and debris. The nature of Uranus's ring still is a complete mystery.

Meteorite Impacts, Climatic Change, and the Extinction of the Dinosaurs

Four billion years ago Earth's surface must have been as cratered as the Moon's surface is now (Figure 23.1). They are very different today because Earth, unlike the Moon, subsequently developed an atmosphere, biosphere, and hydrosphere. Billions of years of weathering and erosion have obliterated most traces of these early craters. In addition, once Earth developed an atmosphere, it was protected from the smaller meteorites.

Despite the processes that act to destroy them, almost 100 meteorite impact craters have been identified on Earth. Meteor Crater in Arizona (Figure 1) and the much-debated Sudbury structure in Ontario are two well-known meteorite craters in North America. The shape of Lake Manicouagan in Quebec, Canada (Figure 2), is the result of a meteorite impact.

Although meteorite impacts are rare today, they may have been of profound importance in the past. The energy released by a meteorite the size of that which formed the Sudbury or Manicouagan craters is about 1000 times the energy released by earthquakes of Richter magnitude 8 (see Chapter 19) and a million times the energy of a volcanic explosion such as Krakatoa (see Chapter 5). In Chapter 21 we saw that some geologists have treated meteorite impacts as the causes of plate tectonic processes.

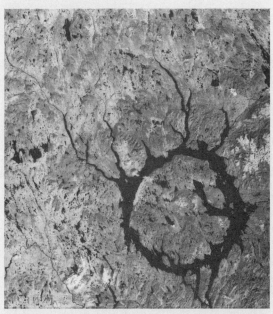

Figure 2 Satellite photograph of the meteorite impact crater at Lake Manicouagan, Quebec. The crater is 75 km in diameter and is outlined by Lake Manicouagan, a reservoir formed by a dam (not shown in photo). Manicouagan Crater was formed 210 million years ago and the ejecta deposits are all eroded away. Structures in the rocks indicative of the impact crater remain. For example, the light region in the middle of the crater is a central peak complex.

Figure 1 Meteor Crater (near Flagstaff, Arizona) is approximately 1220 m in diameter and 170 m deep measured form the rim crest which rises 50 m above the surrounding plain. The outline of the crater is square with rounded corners, not circular, because of the preexisting joint system in the flat sedimentary country rocks. *(Yerkes Observatory, University of Chicago, Williams Bay, Wisconsin.)*

Large and even modest meteorite impact events have the potential to cause serious environmental and ecologic disturbances. For example, a large impact into the oceans could eject enormous amounts of rock dust and water high into Earth's atmosphere. The suspended dust would settle out of the atmosphere after a few weeks or months, but during that time the dust would block out sunlight, directly altering Earth's climate. The suspended water would have an even greater effect on Earth's climate because the water would be trapped high in the atmosphere for years or even thousands of years. Water high in Earth's atmosphere would act as a greenhouse agent (see Chapter 13) indirectly altering Earth's climate by raising the temperature of the atmosphere and oceans. Some scientists have suggested that the extinction of dinosaurs and half the marine creatures at the boundary between the Cretaceous and Tertiary Periods was due to an atmospheric/ecologic disturbance triggered by a large meteorite impact event.

ROCK SAMPLES FROM BEYOND THE EARTH

The analysis of rock samples is as important to planetary geology as it is to the study of Earth. The mineralogy and geochemistry of rock samples provide direct evidence of processes that were important in the internal development of the body from which the samples came. Analysis of samples can indicate the physical and chemical processes involved in large-scale planetary differentiation and in specific smaller-scale events of igneous or tectonic nature. Radiometric dating of the samples can determine the chronologic relationship of events for the parent planet body. To date we have studied two types of extraterrestrial rock samples: meteorites and moon rocks.

Meteorites

Most meteorites originated from collisions between asteroids between the orbits of Mars and Jupiter. The collisions knocked asteroids or parts of asteroids out of orbit and set them on paths that crossed Earth's gravitational field, much like a solar system game of marbles. However, other meteorites probably originated from comets, and some others possibly were derived from the surfaces of planets by major impact phenomena. Meteorites are the richest source of information about the mineralogy, petrology, and geochemistry of bodies other than the Moon and Earth, and they contain data vital to an understanding of the origin of planets. About 3000 different meteorites have been found, and one-third of them were observed as they fell.

Types of meteorites

Chondritic meteorites, the most common type, are stony objects that typically contain small grains called chondrules. Chondrules are small rounded bodies composed of the minerals olivine or pyroxene, with minor amounts of plagioclase feldspar and iron-nickel metal. Some chondritic meteorites contain substantial amounts of carbon and graphite and a complex mixture of organic molecules. These carbonaceous chondrites also contain water which has reacted with the original minerals to form a mixture of clay minerals similar to the clay minerals formed during weathering processes on Earth.

The radiometric age of the chondrites is 4.6 billion years, indicating that chondritic meteorites are preserved samples of material that existed at about the time of the formation of the solar system. Furthermore, their chemical composition is similar to that of the Sun, except for the gaseous elements. This suggests that chondritic meteorites have not experienced any heating, melting, or differentiation and may be so primitive that they are samples of the materials out of which the planets, asteroids, and satellites were formed.

Achondritic meteorites are stony objects that do not contain chondrules. They are similar to mafic and ultramafic igneous rocks such as gabbro and peridotite, and they have bulk chemical compositions significantly different from average solar abundances of the elements. They obviously have undergone processes of igneous differentiation that the more primitive chondrites were not subjected to. **Iron meteorites** consist of coarse-grained iron-nickel metal alloy with minor amounts of the mineral troilite, FeS. **Stony-iron meteorites** consist of an intergrowth of iron-nickel metal with silicate minerals or bits of achondrite meteorite.

Taken together, the three nonchondritic meteorite types provide a sampling of the products that would be expected from the differentiation of a body (such as a large asteroid) composed initially of elements in chondritic abundance. Achondrites, iron meteorites, and stony-irons have ages in billions of years but some are younger than the chondrites and the presumed age of the solar system (4.6 billion years). Earth presumably has undergone differentiation similar to that of the parent body (or bodies) of these meteorites. However, asteroids are much smaller than Earth, and so the processes that differentiated asteroids must have been different from the differentiation processes that operated on the planet Earth.

Moon Rocks

The Apollo missions to the Moon returned 382 kg of rock and regolith samples that contain a treasure of information about the differentiation and subsequent history of the Moon. Many of the Moon's secrets have yielded to the analysis of moon rocks by thousands of scientists.

Rocks from the maria

Rocks returned to Earth from the maria are basalts. Although generally similar to Earth basalts, basalts from the maria have distinctive chemical compositions that set them apart. Moon basalts, like all Moon samples, contain lower abundances than Earth rocks of the volatile elements—those elements with low melting temperatures. In fact, the more volatile an element is, the lower its abundance on the Moon compared with Earth. Water, an extremely volatile compound, is totally absent from lunar rocks. There are other differences. The Moon's mare basalts contain higher abundances of iron, titanium, and nonvolatile elements than Earth basalts. Moon basalts from different regions of the maria crystallized between 3.9 and 3.2 billion years ago. The mineralogy and geochemistry of the lavas suggest

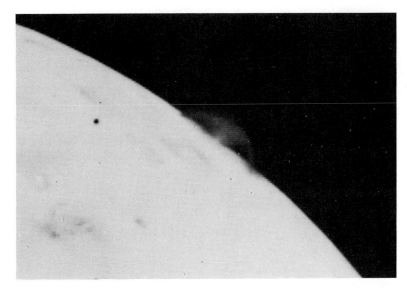

Figure 23.9 Volcanic plumes on Io. View of the plume illustrates its 280-km elevation. This photograph was obtained by the Voyager spacecraft as it passed through the Jovian system. Other Voyager data demonstrate that Io had eight active volcanic plumes during 1979 ranging in elevation from 70 to 280 km. Io is the most volcanically active satellite or planet in the solar system. (*Jet Propulsion Laboratory.*)

partial melting in the Moon's mantle at depths between 150 and 400 km.

Rocks from the highlands

Most samples returned from the highlands of the Moon are impact breccias (Figure 23.10) that formed by lithification of material deposited during meteorite impacts. Impact breccias have fine-grained matrixes that contain preexisting rock and mineral fragments.

The impact debris is a mixture of cold fragmental rock and mineral chips with superheated melt. These two different types of material are mixed thoroughly in a few seconds so that they are disseminated evenly in each other on a scale of less than 1 cm.

The impact breccias generally are basaltic in composition but are distinct from mare basalts by being poorer in iron and titanium and richer in calcium and aluminum. In terms of mineralogy, they are more feldspathic and con-tain fewer mafic minerals. All impact breccias dated thus far give a narrow range in age, indicating that most observable impacts and brecciation on the surface of the highlands took place between 3.9 and 4.05 billion years ago.

Source rock of the breccia On Earth, studies of the composition of grains in a clastic rock permit reconstruction of the source area from which those clasts were derived (see Chapter 6). The fragments in the impact breccias should give an indication of the composition of the surface of the Moon.

An important group of plutonic igneous rocks is found as fragments in impact breccias collected from all over the Moon. These fragments are composed of plutonic rocks such as gabbro, plagioclase-rich gabbro, and anorthosite, a gabbrolike rock consisting of over 90 percent plagioclase feldspar. Although these rock fragments have been crushed during the impact events, their bulk chemical composition

Figure 23.10 An impact breccia from the highlands of the Moon. Note the mixture of fine-grained matrix with rock and mineral fragments. The fragments can be either lighter or darker than the matrix. Impact breccias such as this one were formed by the lithification of impact deposits. *(Johnson Space Center.)*

and the composition of their minerals preserve evidence of their igneous origin. These plutonic rocks have ages between 4.6 and 4.3 billion years, indicating that they crystallized during the first few hundred million years of Moon history. The wide distribution of these rocks over the lunar highlands suggests that the Moon, like the Earth, has a dominantly feldspathic crust. However, Earth's crust contains potassium and sodium feldspars in large amounts, whereas the Moon's crust is composed predominantly of calcium feldspars.

Evolution of the Moon

A history of the development of the Moon can be constructed from the information gained from moon rocks and topographic data discussed earlier in this chapter. The plutonic rock fragments in the impact breccia tell us that the Moon has a feldspathic crust that formed by intrusive igneous activity during the first few hundred million years of Moon history. This plutonic stage of the Moon's evolution was driven by large amounts of heat presumably generated during gravitational accumulation of the materials to form the Moon (see below). If the solar system itself is only about 4.6 billion years old, the initial accumulation and heating of materials and the formation of a crust could have taken only about 100 million years. The end of the Moon's formation is marked by the 4.0-billion-year cratering in the highlands that produced the impact breccias and resulted in a rapid drop in the crater density-time curve.

The intense brecciation of the highlands 4 billion years ago marked the end of geologic activity in that portion of the Moon. Indeed, after the intense brecciation, surface activity was reduced considerably, but there is evidence that geologic activity was continuing at depth. The mare basalts are a series of lava flows that erupted from depths of a few hundred kilometers between 3.9 and 3.2 billion years ago. These were quiet eruptions of the flood basalt type (see Chapter 5), originating from large fissures, and they were not associated with volcanic cones and calderas. Since the last eruption of mare basalt, the surface of the Moon has been inactive except for the formation of a thin surface regolith from sporadic meteorite impacts.

Activity still must have continued at depth, however. Seismic instrument packages left on the Moon by the Apollo teams record "moonquakes" at a depth of about 1000 km. The quakes mark the location of present-day geologic activity, but the overlying lithosphere is far too thick to allow any magma to reach the surface. Since the end of the eruption of mare basalts, the Moon has been cooling. There apparently has been no geologic activity on the surface of the Moon for the past 3 billion years, and there will be none in the future.

ORIGIN AND EVOLUTION OF THE SOLAR SYSTEM

We now know quite a bit about the origin and evolution of two planetary-sized objects (Earth and the Moon) and about the nature of the asteroids. From these data and from remote sensing information about the other planets and large satellites, a comprehensive model has been evolved that explains the stages in the formation of the solar system and the stages in the evolution of a terrestrial planet. This model will be described here, not for each individual planet but rather as a measure against which each planet can be compared.

Formation of the Solar System

Most astronomers and geologists agree that the Sun and planets formed from a rotating disk of gas and dust called a **nebula** (Figure 23.11), although details of this process are not well understood. A nearby supernova (the explosion of a star) caused the nebula to contract rapidly, increasing its internal density, pressure, and temperature. The rapid increase in density provided a center of gravity which accelerated contraction. Within the central part of the nebula, where densities were greatest and temperatures reached millions of degrees, there was spontaneous ignition of the nuclear fusion reaction by which hydrogen atoms join to form helium atoms. This stage marked the birth of the Sun.

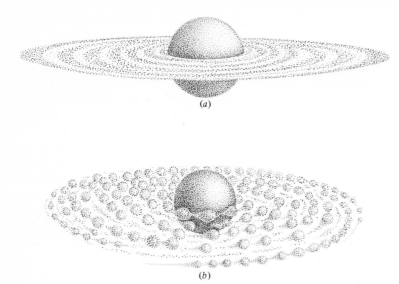

(a)

(b)

Figure 23.11 Formation of the solar system from a rotating gas and dust nebula. *(a)* An early stage in the process when the protosun existed as a spherical region of high density in the nebula and was surrounded by a disk in which the planets and satellites would develop. *(b)* In a later stage the Sun has started to burn hydrogen, and hundreds of protoplanets existed in the disk. The protoplanets continued to accumulate and eventually formed the 9 planets, 40 satellites, and thousands of asteroids and comets.

Meanwhile, in more peripheral regions of the nebula, dust grains collected to form fist-sized clumps, which in turn coalesced to form millions of meter-sized bodies in a disk around the still-growing Sun. This process of gravitational accretion continued with smaller bodies coalescing into larger ones. After about 1 million years, planet-sized objects formed. The planetary objects become heated by a combination of self-gravitational compression, the decay of naturally radioactive elements, the impact of additional materials, and in some cases, internal frictional forces resulting from tidal effects. It is this heat that drives all geologic activity within the planets.

Evolution of Planets

The development of a planet or large satellite can be traced with geologic information from the last stages of its accretion—the gravitational collection of large asteroids. This event, recorded in the brecciation of the lunar highlands, is one of the earliest for which we have actual evidence.

Planetary accretion generates heat that accumulates faster than the planet can radiate it back into space so that the planet heats up. When it has attained its full size, the planet is so hot that there is a shell of magma encircling it to a depth of at least several hundred

kilometers. This shell of molten rock is called a "magma ocean." As the planet cools, a solid lithosphere begins to form at the top of the magma ocean, where cooling is most rapid, and thickens gradually with time.

From this point on, the evolution of the planet is controlled by two factors: the amount of internal heat and the thickness of the lithosphere that develops. All internal geologic activity eventually will be driven by the excess heat and by the tendency for the excess heat to escape from the planet. After all, a volcanic eruption brings hot materials to the surface, where the heat can be lost more effectively to space than would be possible within the planet's mantle. A planet with an abundance of excess heat will display evidence of a large number of internal processes, whereas a planet with little excess heat will have only a few internal processes, and these will cease after only 1 or 2 billion years. As discussed earlier in this chapter, the large planets generally have the greatest amount of heat, the smaller ones the least.

The amount of available heat controls how active a planet can be, but the thickness of the lithosphere determines the nature of the activity. When the lithosphere is less than a few hundred kilometers thick, it still can be broken, buckled, and moved fairly easily. Such a planet will display abundant features of lateral compression and vertical upwarping, buckling, and so forth.

Earth, with its drifting continents, subduction zones, volcanic chains, and uplifted plateaus, is an example of a planet with a thin lithosphere.

When the lithosphere has thickened to the point where it is greater than several hundred kilometers but less than 1000 km, it becomes too strong and massive to be broken or buckled easily. As a result, folds, subduction, and features of lateral deformation do not form, and vertical tectonic regimes become dominant. Magma still can work its way to the surface, and there can be large-scale uplifts and normal faulting. Mars, with its enormous shield volcanoes, volcanic plains, and giant rift valleys, is an example of a planet with a lithosphere of intermediate thickness.

When the lithosphere thickens to 1000 km, the planet displays few features of internal geologic activity. There no longer are active forces of lateral compression or extension, and the magma generated within the planet no longer can work its way to the surface. The lithosphere is so thick, strong, and stable that there is no surface expression of activity. There may well be geologic activity going on below the lithosphere, but this cannot be detected from views of the surface. The Moon, with its most recent volcanic activity occurring about 3 billion years ago, is an example of this stage of planetary development. Moonquakes indicate that there is indeed activity within the body, but it is confined below the stable lithosphere.

SUMMARY

Planets and their satellites, asteroids, and comets are members of the solar system that are related by systematics in their motions about the Sun. These planets and large satellites have similar heat sources—gravitational energy collected during the object's formation and heat from radioactive decay. Each object utilizes similar schemes to dissipate excess heat—differentiation into core, mantle, and crust along with tectonic geologic activity. Larger objects have more heat sources and cool more slowly so that they are more geologically active than small objects. Lithosphere thickness determines the nature of the geologic activity in that a thicker lithosphere promotes crustal stability. The bulk composition of a planet (essentially a mixture of ice-water, rock, and iron-nickel metal alloy) is estimated from its bulk density. From the moment of inertia factor, global magnetic field, and surface morphology, models can be made of the internal structure of a planet or large satellite.

Surface topography is used to decipher the geologic history of planets and large satellites. The Moon and Mercury have surfaces dominated by highlands, which are regions saturated with large meteorite impact craters. These objects have had short geologic histories characterized by low tectonic activity. Mars and Venus have giant volcanoes that attest to their more active geologic development. Io, a satellite of Jupiter, is the most volcanically active object in the solar system. Io is a special case in which the volcanic activity is driven by tidal action, not by internal heat generated from radioactive decay. Although there is no water on the Martian surface today, stream channels indicate the past presence of water and constitute evidence for different past climates.

Meteorite impact cratering is an important process of erosion and deposition on most planets and large satellites in the solar system. The density of craters on a planet's surface is a measure of the relative age (see Chapter 8) of that surface in relation to others. Older surfaces have a higher crater density than younger ones. There was a sharp rise in the crater density 4 billion years ago. Although impact is not a common process on Earth, it may have had big effects on Earth's development. The debris from impact of a large meteorite could have changed Earth's climate drastically for a few years. This may have caused an ecologic disturbance that resulted in the extinction of dinosaurs and half the marine creatures between the Cretaceous and Tertiary Periods.

Rock samples of meteorites and of the Moon provide direct evidence of geologic events in the solar system. Chondritic meteorites are stony objects that date from the very start of the solar system and provide a sample of the original stuff that was used to form planets and large satellites. Achondritic stony meteorites and iron meteorites, taken together, provided the first clues about the products of planetary differentiation and served as a model for early ideas on the internal structure of Earth.

Moon rocks have been essential in understanding the processes of meteorite impact and in unraveling the history of the Moon. Plutonic rock fragments from the lunar highlands demonstrate the existence of a feldspathic crust formed by igneous activity during the first 100 million years of the Moon's history. Four billion years ago the Moon was bombarded by a final burst of meteorites, marking the end of the Moon's accumulation. Quiet eruption of basalts filled the low-lying maria between 4 and 3 billion years ago. Since that time the only activity on the Moon's surface has been sporadic meteorite impact forming a surface regolith.

QUESTIONS FOR REVIEW AND FURTHER THOUGHT

1. Discuss the evidence suggesting that all planets and their satellites are part of a family with a common origin.

2. What are the chief differences between the inner and outer planets?

3. What factors determine the geologic activity of a planet?

4. What can you tell about a planet's bulk composition from measurement of its bulk density?

5. What factors determine the nature of erosion and deposition on a planet's surface?

6. How is the geologic development of Mars similar to that of the Moon and that of Earth?

7. Contrast and compare the geologic development of Venus, Mercury, and Earth.

8. What happens when a meteorite impacts Earth?

9. What role do meteorites play in our understanding of the history and structure of Earth?

10. How have studies of moon rocks changed our ideas about the development of Earth?

ADDITIONAL READINGS

Beginning Level

"The Solar System," *Scientific American*, September, 1975, 145 pp.
(An introduction to each planet in the solar system and to the sun and asteroids)

Pettengill, G. H., D. B. Campbell, and H. Masursky, "The Surface of Venus," *Scientific American*, vol. 243, no. 2, August 1980, pp. 54–65.
(An introduction to the geology of the planet Venus based on the results of the Pioneer Venus spacecraft mission. Contains a topographic map of Venus)

Soderblom, L. A., "The Galilean Moons of Jupiter," *Scientific American*, vol. 238, no. 1, January 1980, pp. 88–100.
(An introduction to the geology of the four large satellites of Jupiter based on the results of the Voyager 1 and 2 spacecraft missions to the Jovian system)

Intermediate Level

Carr, M. H., "The Volcanoes of Mars," *Scientific American*, vol. 235, no. 1, January 1976, pp. 32–43.
(Discussion of the geologic history of Mars with an emphasis on the volcanic history. Discussion of timing of Martian volcanism)

Hartman, W. K., *Moons and Planets: An Introduction to Planetary Science*, Wadsworth, Belmont, Calif., 1972, 404 pp.
(A text covering the formation of the solar system and its planets and asteroids. Emphasis on orbital dynamics and reflectance spectral properties of planetary surfaces)

King, Jr., E. A., *Space Geology*, Wiley, New York, 1976, 349 pp.
(An introductory text to the planets and meteorites of the solar system. Emphasis is on the mineralogy of the objects and their geologic history)

Wood, J. A., *The Solar System*, Prentice-Hall, Englewood Cliffs, N.J., 1979, 196 pp.
(A general introductory treatment of both astronomical and geological aspects of the origin and development of the planets, asteroids, and meteorites of the solar system)

APPENDIX A
Conversion from Metric to English Units

Metric–English			English–Metric		
To convert	To	Multiply by	To convert	To	Multiply by
UNITS OF LENGTH					
Centimeters (cm)	Inches (in.)	0.3937	in.	cm	2.54
Meters (m)	Feet (ft)	3.2808	ft	m	0.3048
m	Yards (yd)	1.0936	yd	m	0.9144
Kilometers (km)	Miles (mi)	0.6214	mi	km	1.6093
UNITS OF AREA					
cm^2	$in.^2$	0.1550	$in.^2$	cm^2	6.452
m^2	ft^2	10.764	ft^2	m^2	0.0929
m^2	yd^2	1.196	yd^2	m^2	0.8361
km^2	mi^2	0.3861	mi^2	km^2	2.590
m^2	Acres	2.471×10^{-4}	Acres	yd^2	4.840
			Acres	m^2	4047.
UNITS OF VOLUME					
cm^3	$in.^3$	0.0610	$in.^3$	cm^3	16.3872
m^3	ft^3	35.314	ft^3	m^3	0.02832
m^3	yd^3	1.3079	yd^3	m^3	0.7646
Liters (l)	U.S. liquid quarts	1.0567	U.S. liquid quarts	l	0.9463
l	U.S. liquid gallons	0.2642	U.S. liquid gallons	l	3.7853
UNITS OF MASS					
Grams (g)	Ounces Avoirdupois	0.03527	Ounces Avoirdupois	g	28.3495
g	Ounces Troy	0.03215	Ounces Troy	g	31.1042
Kilograms (kg)	Pounds Avoirdupois	2.20462	Pounds Avoirdupois	kg	0.45359
UNITS OF DENSITY					
g/cm^3	lb/ft^3	62.4280	lb/ft^3	g/cm^3	0.01602
UNITS OF VELOCITY					
km/h	mi/h	0.6214	mi/h	km/h	1.6093
km/h	cm/s	27.78	mi/h	in./s	17.60
UNITS OF PRESSURE					
kg/cm^2	$lb/in.^2$	14.2233	$lb/in.^2$	kg/cm^2	0.0703
Bars	Atmospheres	0.98692	Atmospheres	Bars	1.01325
kg/cm^2	Atmospheres	0.96784			
kg/cm^2	Bars	0.98067			
UNITS OF TEMPERATURE					
°C	°F	(5/9) (°C − 32)			
°F	°C	(9/5) (°F) + 32			

Degrees	C.	100	90	80	70	60	50	40	30	20	10	0	−10
	F.	210	200	190	180	170	160	150	140	130	120	110	100	90	80	70	60	50	40	30	20	10	0

APPENDIX B
Identification of Minerals

Once the physical properties of an unknown mineral have been determined, identification is made by systematically excluding as possibilities all minerals whose physical properties do not fit those of the unknown mineral. This is easier if the minerals are grouped in categories based on some of the properties.

This appendix is divided into two parts. Table B.1 subdivides minerals first by luster and then by hardness, streak, and cleavage so that these four properties may be used for identification. Table B.2 is an alphabetical list of the more common rock-forming minerals that can be used for further data if the unknown has been narrowed down to a few possibilities.

TABLE B.1   Determinative table

A.   Minerals with a metallic or submetallic luster*

1.   Hardness less than $2\frac{1}{2}$.   (Can be scratched with a fingernail.)

Streak:	Black	Gray	Red	Red-brown	Yellow-brown
	Graphite(1)	Galena(3)	Cinnabar	Hematite	Limonite
	Molybdenite(1)				
	Stibnite(1)				

2.   Hardness between $2\frac{1}{2}$ and 5.   (Cannot be scratched by a fingernail; can be scratched by a knife.)

Streak:	Black	Coppery red	Yellow
	Bornite	Copper	Sphalerite(6)
	Pyrrhotite		Gold
	Chalcopyrite		
	Chromite		

3.   Hardness greater than $5\frac{1}{2}$.   (Cannot be scratched by a knife.)

Streak black: Arsenopyrite   Marcasite   Ilmenite
Pyrite   Magnetite

B.   Minerals with nonmetallic luster*

1.   Hardness less than $2\frac{1}{2}$.

Streak:	Orange	Yellow	White-colorless	
	Realgar	Sulfur	Muscovite(1)	Bauxite
		Orpiment	Chlorite(1)	Sylvite(3)
			Talc(1)	Serpentine
			Gypsum(1)	
			Kaolinite	

2.   Hardness between $2\frac{1}{2}$ and $5\frac{1}{2}$.

Streak:	Red, red-brown	Yellow-brown	Green	Blue
	Cinnabar	Limonite	Malachite	Azurite
	Hematite	Siderite(3)		
		Sphalerite(6)		

Streak colorless:	Halite(3)	Calcite(3)	Dolomite(3)
	Barite(1)	Anhydrite	Serpentine
	Rhodochrosite(3)	Aragonite	Kyanite(2)
	Sphene	Fluorite(4)	Apatite(1)

3.   Hardness between $5\frac{1}{2}$ and $7\frac{1}{2}$.

Streak:	Brown-black		Colorless	
	Rutile	Sillimanite	Kyanite	Andalusite
		Quartz	Cordierite	Garnet
		Turquoise	Potassic feldspar	Plagioclase
		Amphiboles(2)	Pyroxenes(2)	Olivine
		Tourmaline	Cassiterite	Rutile

4.   Hardness greater than $7\frac{1}{2}$. (Harder than streak plate.)

Topaz	Diamond(4)	Corundum	Spinel
Beryl	Zircon		

* Numbers in parentheses indicate number of cleavage directions.

TABLE B.2   Alphabetical listing of rock-forming and ore minerals

Mineral	Composition	H	Specific gravity	Color	Cleavage	Comments	Occurrence*	Uses
Actinolite	$Ca_2(Mg,Fe)_5(Si_8O_{22})(OH)_2$	5–6	3.0–3.2	Green	2/56,124°	An amphibole	M	Fibrous variety as asbestos; variety of jade
Aegirine	$NaFeSi_2O_6$	6	3.4–3.55	Green	2/87,93°	A pyroxene	I	
Alabaster	$CaSO_4 \cdot 2H_2O$	2	2.32	Varied	—	Massive gypsum	Evaporite	Sculpture
Albite	$NaAlSi_3O_8(An_0)$	6	2.62	Gray-white	2/93,87°	A plagioclase; striations	I, M, S	Abrasive, jewelry
Almandine	$Fe_3Al_2(SiO_4)_3$	7	4.25	Red	Conchoidal fracture	A garnet	M (i)	Abrasive, jewelry
Amethyst	SiO_2	7	2.65	Purple	Conchoidal fracture	Variety of quartz	I	Jewelry
Analcime	$Na(AlSi_2O_6) \cdot H_2O$	5	2.27	White		Crystals like garnet; a zeolite	I, M	
Andalusite	Al_2SiO_5	7.5	3.16–3.20	Gray, Pink		Elongate prisms	M	Spark plugs, ceramics
Andesine	$An_{30}\text{-}An_{50}$	6	2.69	Gray	2/87,93°	A plagioclase; striations	I, M	
Andradite	$Ca_3Fe_2(SiO_4)_3$	7	3.75	Brown, Yellow	Conchoidal fracture	A garnet	M	Abrasives, jewelry
Anorthite	$CaAl_2Si_2O_8(An_{100})$	6	2.76	Gray	2/93,87°	A plagioclase; striations	I	
Apatite	$Ca_5(F,Cl,OH)(PO_4)_3$	5	3.15–3.2	Green, Brown		Hexagonal prisms	I, M, S	Source of phosphate fertilizer
Aquamarine	$Be_3Al_2Si_6O_{18}$	8	2.77	Green	3 (poor)	Variety of beryl	I	Jewelry
Aragonite	$CaCO_3$	3.5–4	2.95	White, Yellow		Polymorph of $CaCO_3$	M, S	
Arsenopyrite	$FeAsS$	5.5–6	6.05	Silver		Black streak	Hydrothermal	Ore of arsenic
Augite	$Ca(Mg,Fe,Al)(Al,Si)_2O_6$	5–6	3.2–3.4	Green	2/93,87°	A pyroxene	I (m)	Minor ore of copper
Azurite	$Cu_3(CO_3)_2(OH)_2$	3.5–4.0	3.77	Blue		Fine-grained crusts; reacts with HCl	Alteration of copper ores	
Barite	$BaSO_4$	3.0–3.5	4.5	Clear, White	1 good 2 poor	Unusually high specific gravity for nonmetallic mineral	S; hydrothermal	Source of barium; drilling mud additive
Bauxite	Mixture of clays	—	2.0–2.5	Gray, Brown		In small spheres	Weathering	Ore of aluminum
Beryl	$Be_3Al_2Si_6O_{18}$	8	2.77	Varied		Hexagonal crystals	Pegmatites	Source of beryllium; gemstone
Biotite	$K(Mg,Fe)_3(AlSi_3O_{10})(OH)_2$	2.5–3.0	2.8–3.2	Green, Brown	1	A mica	M,I	
Borax	$Na_2B_4O_7 \cdot 10H_2O$	2.0–2.5	1.7	White	1	Bitter taste	Evaporite	Antiseptic, cleanser
Bornite	Cu_5FeS_4	3	5.07	Bronze		Metallic luster; colorful tarnish	Hydrothermal	Copper ore
Bytownite	$An_{70}\text{-}An_{90}$	6	2.72–2.75	Gray	2/93,87°	A plagioclase; striations	I	
Calcite	$CaCO_3$	3	2.72	Varied	3/*not* 90°	Chief mineral in limestone, marble	S,M (i)	Cement, fertilizer
Carnotite	$K_2(UO_2)_2(VO_4)_2 \cdot nH_2O$	—	—	Yellow		Yellow powder	Alteration of uranium ores	Major source of uranium
Cassiterite	SnO_2	6–7	6.8–7.0	Dark brown		White streak	M,I	Ore of tin
Chalcopyrite	$CuFeS_2$	3.5–4.0	4.2	Brassy yellow		Bronze tarnish; black streak	Pegmatite Hydrothermal	Ore of copper
Chlorite	$(Fe,Mg,Al)_6(Al,Si)_4O_{10}(OH)_8$	2–2.5	2.6–2.9	Green	1 (perfect)	Similar to micas;	M	
Chromite	$FeCr_2O_4$	5.5	4.6	Black		Dark brown streak; submetallic luster	I (ultramafic)	Only ore of chromium
Chrysocolla	$CuSiO_2 \cdot 2H_2O$	2–4	2.2	Blue-green	Conchoidal fracture		Hydrothermal	Minor ore of copper
Cinnabar	HgS	2.5	8.10	Crimson		Scarlet streak	Hydrothermal	Ore of mercury
Copper	Cu	2.5–3.0	8.9	Copper-red		Dendritic masses; malleable, ductile	I	Wire, jewelry
Cordierite	$Mg_2Al_3(AlSi_5O_{18})$	7–7.5	2.65	Blue-gray	1 (poor) Parting common	Vitreous luster	M	Minor gemstone
Corundum	Al_2O_3	9	4.02	Varied		Hexagonal crystals	M, I	Abrasive, gemstone

542

Cuprite	Cu_2O	3.5–4	6.0	Red		Red-brown streak; adamantine luster	Altered copper minerals	Minor ore of copper
Diamond	C	10	3.5	Varied	4		I, M	Abrasives, jewelry
Diopside	$CaMg(Si_2O_6)$	5–6	3.2–3.3	Green	2/87,93°	May show parting	M	Minor gemstone
Dolomite	$CaMg(CO_3)_2$	3.5–4	2.85	White, Pink	3/not 90°	Powder reacts in cold HCl	S	Special cements, ornamental stone
Emerald	$Be_3Al_2Si_6O_{18}$	8	2.7–2.8	Green		Variety of beryl	I, M	Gemstone
Enstatite	$MgSiO_3$	5.5	3.2–3.5	Brown, Green	2/87,93°	A pyroxene	M	
Epidote	$Ca_2Al_3O(SiO_4)(Si_2O_7)(OH)$	6–7	3.4	Apple green		Often in fine granular coatings	I	
Fayalite	Fe_2SiO_4	6.5	4.14		Conchoidal fracture	An olivine		
Fluorite	CaF_2	4	3.18	Varied	4		S; hydrothermal	Flux in steelmaking
Forsterite	Mg_2SiO_4	6.5	3.2	Green	Conchoidal fracture	An olivine	I, M	
Galena	PbS	2.5	7.5	Gray	3/90°	Metallic luster; cubic crystals	Hydrothermal	Major ore of lead
Garnet	$X_3^{2+}Y_2^{3+}(SiO_4)_3$	7	3.5–4.3	Varied	Conchoidal fracture	Mineral family	M (i)	Abrasives, jewelry
Glaucophane	$Na_2(Mg,Fe)_3Al_2(Si_8O_{22})(OH)_2$	6.5	3.1	Blue-gray	2/56,124°	An amphibole	M	Ore of iron
Goethite	$HFeO_2$	5–5.5	4.37	Earthy brown		Earthy luster; yellow-brown streak	Weathering	
Gold	Au	2.5–3	15–19.3	Yellow		Malleable	I	Jewelry, dentistry
Graphite	C	1–2	2.3	Black	1 (perfect)	Greasy feel	M	Lubricant, pencils, electrodes
Grossularite	$Ca_3Al_2(SiO_4)_3$	7	3.53	Green, Yellow, White, Red	Conchoidal fracture	A garnet	M	Abrasives, jewelry
Gypsum	$CaSO_4 \cdot 2H_2O$	2	2.32	Colorless, White	1		Evaporite	Construction material (plaster, plasterboard)
Halite	$NaCl$	2.5	2.16	Colorless, White	3/90°	Cubic crystals	Evaporite	Nutrient, chemical industry
Hematite	Fe_2O_3	5.5–6.5	5.26	Red-brown, Black		Red-brown streak	I, M, S.	Major ore of iron
Hornblende	$Ca_2Na(Mg,Fe)_4(Al,Fe)_3(Si_8O_{22})(OH)_2$	5–6	3.25	Green	2/56,124°	A common amphibole	I, M	
Hypersthene	$(Mg,Fe)SiO_3$	5.6	3.4–3.5	Bronze	2/87,93°	Bronze pyroxene	I, M	
Ice	H_2O	1.5	0.92	Colorless, White		Hexagonal crystals	Snowfall	Snowballs, cooling
Idocrase	$Ca_{10}(Mg,Fe)_2Al_4(Si_2O_7)_2(SiO_4)_5(OH)_4$	6.5	3.40	Brown		Striated four-sided crystals	M	Minor gemstone
Ilmenite	$FeTiO_3$	5.5–6	4.7	Black		Black streak; may be magnetic	M, I	Ore of titanium
Iron	Fe	4.5	7.3–7.9			Magnetic; metallic luster; malleable	Meteorites	Manufacture of steel
Jadeite	$NaAlSi_2O_6$	6.5–7	3.4	Green	2/87,93°	A pyroxene	M	Jade
Kaolinite	$Al_2Si_2O_5(OH)_4$	2–2.5	2.62	White, Gray	1 (perfect)	A claylike sheet silicate	Alteration of feldspars	Brick, tile, ceramics
Kyanite	Al_2SiO_5	5 and 7	3.61	Blue-gray	2/90°	Directional hardness is diagnostic	M	Refractory porcelains (spark plugs)
Labradorite	An_{50}-An_{70}	6	2.71	Dark gray	2/87,93°	Play of colors	I, M	A plagioclase feldspar
Leucite	$KAlSi_2O_6$	5.5–6	2.45	Grayish tan		Garnetlike crystals; a feldspathoid	I	
Limonite	$FeO(OH) \cdot nH_2O$	–	–	Yellow-brown		A mineraloid; amorphous	Alteration of ferromagnesians	Pigment (yellow ochre)
Magnetite	Fe_3O_4	6	5.18	Black		Black streak; octahedral crystals	I, M	Major iron ore
Malachite	$Cu_2(CO_3)(OH)_2$	3.5–4	3.9–4	Green		Reacts with HCl	Altered ores of copper	Ore of copper

543

TABLE B.2 continued

Mineral	Composition	H	Specific gravity	Color	Cleavage	Comments	Occurrence*	Uses
Marcasite	FeS_2	6–6.5	4.89	Silver, Gray		Black streak; metallic luster	S; hydrothermal	Source of sulfur for sulfuric acid
Microcline	$KAlSi_3O_8$	6	2.55	Gray, Flesh, Salmon, Green	2/90°	A potassic feldspar	I, M	Porcelain, ceramics
Molybdenite	MoS_2	1–1.5	4.67	Black	1 (excellent)	Greasy feel; black streak	I, M	Ore of molybdenum
Muscovite	$KAl_2(AlSi_3O_{10})(OH)_2$	2–2.5	2.76–3.10	Colorless	1 (perfect)	A colorless mica	I, M	Electrical insulator
Nepheline	$(Na,K)AlSiO_4$	5.5–6	2.55	Colorless, White	3/not 90°	Often greasy luster	I	Glass, ceramics
Oligoclase	$An_{10}–An_{30}$	6.1	2.65	Gray	2/87.93°	A plagioclase	I, M	Ceramics
Olivine	$(Mg,Fe)_2SiO_4 [Fo_0–Fo_{100}]$	6.5–7	3.27–4.37	Green	Conchoidal	A mineral family	I, M	Minor gemstone
Opal	$SiO_2 \cdot nH_2O$	5–6	1.9–2.2	Varied	Conchoidal fracture	Milky or fiery luster	Cavity filling in I, S	Gemstone
Orpiment	As_2S_3	1.5–2	3.49	Yellow	1	Resinous luster; yellow streak	Hydrothermal	Source of arsenic
Orthoclase	$KAlSi_3O_8$	6	2.57	Varied	2/90°	A potassic feldspar	I, M	Ceramics, porcelain
Perthite	Plagioclase plus potassic feldspar						I, M	
Platinum	Pt	4–4.5	14–19	Gray			I (ultramafic)	Chemical apparatus, jewelry
Prehnite	$Ca_2Al_2(Si_3O_{10})(OH)_2$	6–6.5	2.8–2.9	Green			I, M	
Pyrite	FeS_2	5	4.3	Brassy yellow		Black streak; in cubes	I, M, S	Source of sulfur for sulfuric acid
Pyrope	$Mg_3Al_2(SiO_4)_3$	7	3.51	Deep red	Conchoidal fracture	Twelve-sided crystals; a garnet	M (i)	Abrasive, jewelry
Pyrrhotite	$Fe_{1-x}S(x=0–0.2)$	4	4.62	Bronze		Black streak; magnetic	I (m)	Source of iron and associated nickel
Quartz	SiO_2	7	2.65	Varied	Conchoidal fracture	Hexagonal crystals	I, M, S	Glass, gemstones, optical lenses
Realgar	AsS	1.5–2	3.48	Red-orange		Orange streak	Hydrothermal	Ore of arsenic
Rhodochrosite	$MnCO_3$	3.5	3.45–3.6	Pink	3/not 90°	Soluble in hot HCl; a carbonate	Hydrothermal	Minor ore of manganese
Ruby	Al_2O_3	9	4.02	Red	Parting common	Six-sided crystals (variety of corundum)	M	Gemstone
Rutile	TiO_2	6–6.5	4.18–4.25	Red		Submetallic luster; four-sided prisms	I, M	Coatings for welding rods
Sapphire	Al_2O_3	9	4.02	Blue	Parting common	Variety of corundum	M	Gemstone
Serpentine	$(Mg,Fe)_3Si_2O_5(OH)_4$	2–5	2.2	Green, Yellow		Platy or fibrous	M	Asbestos
Siderite	$FeCO_3$	3.5–4	3.83–3.85	Brown, Yellow	3/not 90°	Soluble in hot HCl	S, I	Ore of iron
Sillimanite	Al_2SiO_5	6–7	3.23	White	1	Prismatic or fibrous	M	
Silver	Ag	2.5–3	10.5	White		Malleable, ductile	M	Jewelry, coinage
Sphalerite	ZnS	3.5	3.9–4.1	Varied	6 (rare)	Light-yellow streak	Hydrothermal	Ore of zinc
Spessartine	$Mn_3Al_2(SiO_4)_3$	7	4.18	Red-brown	Conchoidal fracture	A garnet	M	Abrasive, jewelry
Sphene	$CaTiSiO_5$	5–5.5	3.4–3.55	Brown, Green, Black, Yellow	2/not 90°		I, M	Ore of titanium
Staurolite	$Fe_2Al_9O_7(SiO_4)_4(OH)$	7–7.5	3.65–3.75	Red-brown		Commonly in crosslike crystals	M	Gemstone
Stibnite	Sb_2S_3	2	4.52–4.62	Gray-black	1	In slender prisms	Hot springs	Ore of antimony
Sulfur	S	1.5–2.5	2.07	Yellow	Conchoidal fracture		S; volcanoes	For producing sulfuric acid

Mineral	Formula	Hardness	Specific gravity	Color	Cleavage/Fracture	Other properties	Occurrence	Uses
Sylvite	KCl	2	1.99	Varied		Bitter salty taste	Evaporite M	Source of potassium
Talc	$(Mg,Fe)_3Si_4O_{10}(OH)_2$	1	2.7–2.8	Green			I	Lubricant
Topaz	$Al_2(SiO_4)(F,OH)_2$	8	3.5	Varied	1	Elongate prisms	I, M	Gemstone
Tourmaline	$XY_3Al_6(BO_3)_3Si_6O_{18}(OH)_4$	7–7.5	3.0–3.25	Blue, Red, Green, Black, Yellow		Striated trigonal crystals; varicolored		Gemstone
Tremolite	$Ca_2Mg_5Si_8O_{22}(OH)_2$	5–6	3.0–3.33	White, Pale green	2/56,124°	Sometimes fibrous; an amphibole	M	Fibrous variety for asbestos
Turquoise	$CuAl_6(PO_4)_4(OH)_8 \cdot 4H_2O$	6	2.6–2.8	Blue-green		Often as crusts	Alteration of volcanic rock	Gemstone
Uraninite	UO_2	5.5	9.0–9.7	Black		Black streak	I	Uranium ore
Uvarovite	$Ca_3Cr_2(SiO_4)_3$	7	3.77	Bright green	Conchoidal fracture	A garnet	M	Gemstone
Vanadinite	$Pb_5Cl(VO_4)_3$	3	6.7–7.1	Red, Orange, Yellow, Brown		Rounded crystals; globular masses	Hydrothermal	Ore of lead; vanadium
Wolframite	$(Fe,Mn)WO_4$	5–5.5	7.0–7.5	Black	1		Pegmatites	Ore of tungsten
Wollastonite	$CaSiO_3$	5–5.5	2.8–2.9	White		White streak	M	Ceramics, porcelain, tile
Wulfenite	$PbMoO_4$	3	6.8	Red-orange	2/90°		Hydrothermal	Minor ore of molybdenum
Zircon	$ZrSiO_4$	7.5	4.68	Varied		Four-sided prisms	I, M	Gemstone; ore of zirconium

* I, M, S: Common in igneous (I), metamorphic (M), or sedimentary (S) rocks.
i, m, s: Rare in igneous (i), metamorphic (m), or sedimentary (s) rocks.

APPENDIX C
Soil Classification

Two major soil classifications are in use in the United States. The older classification, based on soil orders and groups, has been in use for many years. The newer classification, **The Seventh Approximation,** has been adopted by the U.S. Department of Agriculture and many state soil surveys. Proponents of the new classification claim that it is superior to the old one because it is more precise, with 10 major orders and 47 suborders permitting a detailed classification of soils. Opponents of the new classification claim that it is not needed and is much too complex for use by anyone except soil specialists. Needless to say, the topic of soil classification is controversial at the present time. In the interests of objectivity, we present both major classifications. We have cross-referenced the two classifications to each other and have indicated which soil types in each classification fall under the major soil types, the pedalfers and pedocals, described in Chapter 10.

TABLE C.1 Soil classification by orders and groups

ALL SOILS

ARID, SEMIARID, AND SUBHUMID CLIMATES (PEDOCALS)

Azonal order[a]

Lithosols
Stony, thin mountain soils that either have had little time to develop or are on slopes steep enough so that thicker sections of soil cannot accumulate.

Regosols
Very poorly developed soils with no horizon development that have formed on recently deposited sediments.

Intrazonal order[b]

Saline soils
Soils without horizons that contain an excess of soluble minerals. Formed during the desiccation of fine-grained sediments and saline waters in basins with internal drainage.

Zonal order[c]

Chernozem soils
Dark surficial layer (A) of highly organic soil derived from decay of grass parts. Underlain by a B layer which is brown to yellowish brown. Reduced rainfall results in little leaching and formation of calcium carbonate nodules in the B horizon. Common in the eastern parts of the Dakotas. Nebraska, central Kansas, and western Oklahoma and Texas.

Chestnut soils
Similar to chernozem soils but form in a drier climate and are less organic, lighter in color, and have more abundant calcium carbonate nodules in the B horizon. Common in eastern Montana, the western parts of the Dakotas and Nebraska, northeastern Colorado, and western Kansas.

Brown soils
Drier versions of the chestnut soils. Have a light-brown color and a B zone with a distinctive columnar structure. Common on the plains abutting the Rocky Mountains.

Gray desert soils
Sandy, pale grayish- to reddish-gray soils with little organic material. Form in cooler mid-latitude desert areas of Nevada, Arizona, Utah, and New Mexico. Grayish-red soils characterize subtropical deserts such as those in southern Arizona and New Mexico.

Red desert soils
Red-colored soils in tropical deserts.

(← Increasing aridity →)

Prairie soils *(transitional between pedocals and pedalfers)*
Soils transitional between the pedocals (chernozems) and pedalfers (podzols). Among the most fertile soils because they contain sufficient organic material, such as in the chernozems, and form in sufficient rainfall, such as in the podzols. Common in northern Illinois, Iowa, northwestern Missouri, eastern Kansas, Oklahoma, and north-central Texas.

HUMID CLIMATES (PEDALFERS)

Zonal order[c]

Podzols
Soils with organic-rich accumulations in the top of the A zone underlain by a strongly leached, ash-white horizon in the lower part of the A zone. The B zone is clay-rich. The type of soil that forms in humid subarctic climates. Common in northern Wisconsin, Minnesota, Michigan, and northern New England.

Gray-brown podzols
Similar to the podzol but less intensely leached. The base of the A zone is grayish brown rather than ash-white as in podzols. Thick, dark-brown B zone rich in clay minerals. Widespread in the humid temperate northeastern United States.

Red-yellow podzols
Similar to gray-brown podzols but has a B zone enriched in hydroxides of iron and aluminum. Common in the humid subtropical climates of the southeastern and Gulf Coastal states.

Latosols
Deep brownish-red surface deposits with only a thin cover of organic debris. Soil lacks horizons and becomes lighter in color with depth. Forms by intense chemical weathering in hot humid climates. Silica is leached out, and the soil becomes enriched in hydroxides of iron, aluminum, or manganese depending on the composition of the parent material.

(← Increasing temperatures →)

Intrazonal order[b]

Bog soils
Dark-brown water-saturated and partially decomposed peaty material. Can form in arctic, temperate, or tropic climates wherever plant material accumulates under standing water.

Meadow soils
Dark, organic-rich upper layers beneath which is a bluish-gray clay horizon.

Planosols
Thick and very dark organic-rich soil formed on flat surfaces between stream valleys where soil erosion is limited.

Tundra soils
Soils which form in areas underlain by permafrost. Mixture of organic material and physically weathered rock debris. Chaotic structure resulting from perennial freezing and thawing. Common in arctic and subarctic climates.

---

[a] Soil horizons are poorly developed or absent.
[b] Soil characteristics are determined by local conditions such as poor drainage.
[c] Well-developed soil horizons corresponding to the climatic and vegetative zones in which they are found.
Source: Modified with permission from, A. N. Strahler, *Physical Geography*, New York, Wiley, 1960.

TABLE C.2 Soil classification (Seventh Approximation)
U.S. Dept. of Agriculture Soil Conservation Service (1960)

Soil order	Characteristics	Some areas where it is common	Approx. equiv. in great soil groups
Entisols	Soils without horizons; soil-forming processes have not had sufficient time to produce horizons.	Wide geographic range from desert sand dunes to frozen ground of sub-arctic zones.	Azonal soils
Inceptisols	Weakly developed soil horizons; soil horizon A is developed.	Wide geographic range wherever soils have just begun to develop on newly deposited or exposed parent materials such as volcanic or glacial deposits.	Lithosols (mountain soils) Regosols (recently deposited sediments)
Spodosols	Humid forest soils with a gray leached A horizon and a B horizon enriched in iron or organic material leached from above. Commonly under coniferous forests.	New England, northern Minnesota, and Wisconsin.	Podzolic and brown pod-zolic soils (pedalfer)
Alfisols	Soils with clay enrichment in the B horizon. Lower organic content than mollisols. Medium to high base supply. Commonly under deciduous forests.	Western Ohio, Indiana, lower Wisconsin, northwestern New York, central Colorado, western Montana.	Gray-brown podzolic soils (pedalfer)
Mollisols	Grassland soils with a thick, dark organic-rich surface layer. High base supply (calcium, sodium, and potassium).	Widespread in central and northern Texas, Oklahoma, Kansas, Nebraska, North and South Dakota, and Iowa.	Chestnut, chernozem, and prairie soils (pedocal)
Aridisols	Desert and semiarid soils; low organic content along with concentration of soluble salts within soil profile.	Widespread in desert and semiarid areas of Nevada, California, Arizona, New Mexico.	Desert soils (pedocal)
Ultisols	Deeply weathered red and orange clay-enriched soils on surfaces that have been exposed for a long time.	Humid temperate to tropical soils. Widespread in southeastern United States east of Mississippi Valley.	Red and yellow podzolic soils, certain lateritic soils (pedalfer)
Oxisols	Intensely weathered soils consisting largely of kaolin, hydrated iron and aluminum oxide. Bauxite forms in these soils.	Warm tropical areas with high rainfall.	Most lateritic soils (pedalfer)
Histosols	Organic soils and peat.	Mississippi delta, Louisiana; Everglades, Florida; local bogs in many areas.	Bog soils (pedalfer)
Vertisols	Swelling soils with high clay content which swell when wet and crack deeply when dry.	Southeast Texas; local areas	Swelling clays

Glossary

Aa   Highly viscous, blocky lava.

Abrasion   The grinding of mineral and rock particles against each other or against bedrock.

Abyssal hills   Extinct submarine volcanoes partially buried by sediments on the abyssal plains.

Abyssal plains   Flat parts of the ocean basins covered by deep water.

Achondritic meteorites   Stony meteorites that do not contain chondrites.   They are similar in composition to gabbro and peridotite.

Active continental margins   Coasts where shoreline and plate boundaries coincide so that the area is tectonically active.

Active volcano   A volcano that is erupting currently.

Aeration zone   That part of the regolith and underlying rock in which the pore spaces are not completely filled with water.

Agate   A banded variety of quartz.

Agglutinates   Particles of the lunar soil fused together by the heat of micrometeorite impacts on the moon's surface.

Aggradational stream   A stream that is filling its valley actively with alluvium because of a rise in base level.

Albedo   The reflectivity of a body—a measure of the amount of solar radiation that is reflected back into space rather than being absorbed.

Alkali basalts   Basalts rich in the alkali elements sodium and potassium.

Alluvial basin   A topographically low area underlain by sediments deposited by streams.   (See **alluvium**.)

Alluvial fan   A triangular-shaped stream deposit extending out from the bases of mountain ranges, most commonly in arid and semiarid regions.

Alluvium   The general name given to all sediment deposited by streams.

Alpha decay   A nuclear reaction in which a particle composed of two neutrons and two protons (alpha particle) is expelled from the nucleus of an atom.

Alpha particle   A particle composed of two protons and two neutrons that is ejected from a nucleus during a nuclear reaction.

Alpine glacier   See **mountain glacier.**

Amber   Hardened tree sap in which insects often are preserved.

Amethyst   A purple variety of quartz.

Amino acid dating   Dates determined by changes in the ratio of two different forms of amino acids with time.

Amphiboles   A group of minerals containing double chains of silicon-oxygen tetrahedra.

Andesite   A fine-grained intermediate igneous rock with the same composition as diorite.

Anhydrite   A chemical sedimentary rock composed of anhydrite crystals, $CaSO_4$.

Angle of repose   The maximum slope angle to which particles of a given size can be built up without slumping.

Anion   A negatively charged ion formed when an atom gains one or more electrons.

Annual ring   The portion of a tree which has grown in one year. Counting successive annual rings can give an age for the tree.

Anthracite coal   A hard, black coal with a semi-metallic luster and a carbon content between 92 and 98 percent.   Anthracite is the highest-ranking coal and burns with a short blue flame and without smoke.

Anticline   A fold in which the limbs dip away from one another so that old rocks are arched upward in the axial region of the fold.

Aphanitic   An igneous texture so fine-grained that individual mineral grains cannot be seen with the naked eye.

Aquiclude   A sediment or rock through which groundwater cannot pass.

Aquifer   A sediment or rock through which groundwater can move.

Arable land   Land suitable for raising crops.

Arenite   A sandstone with less than 15 percent matrix, in which the particles are held together by a crystalline cement.

Arête   A narrow, knife-sharp ridge formed as an erosional remnant between two cirques.

Arkose sandstone   A sandstone that has greater than 25% feldspar.

Artesian spring   A spring under artesian pressure which flows on the land surface or from the bottom of a body of water.

Artesian well   A well in which the water rises to the surface under its own pressure, without being pumped.

Asbestos   The name given to fibrous varieties of serpentine or an amphibole.

Assemblage   The combination of minerals that makes up a metamorphic rock.

Assimilation   Incorporation in a magma of ions from surrounding host rock or xenoliths.

Asteroids   Bodies of rock a few kilometers to 1000 km in diameter that orbit the Sun between the orbits of Mars and Jupiter.

Asteroid belt   Clusters of asteroids between the orbits of Mars and Jupiter.

Asthenosphere   A region in the mantle 100 to 250 km beneath the surface; it is composed of materials of abnormally low rigidity and corresponds to the seismic low-velocity zone.   In the plate tectonic model, this is the region of decoupling of lithosphere plates and the lower mantle.

Atmosphere   The gaseous envelope composed mostly of nitrogen and oxygen that surrounds the solid earth.

Atoll   A circular coral reef and islands with a lagoon in the center.

Atom   The smallest particle which possesses the chemical properties of an element; it is composed of a central nucleus and orbiting electrons.

Atomic mass   The number of protons and neutrons in the nucleus of an atom.

Atomic number   The number of protons in the nucleus of an atom; the number of electrons orbiting the nucleus of an atom.

Authigenic minerals (and sediments)   Minerals that form in place in the sedimentary basin of deposition.

Axial plane   An imaginary plane that connects fold axes from all the different layers of rock in a fold.

Btu   The British thermal unit—a unit of energy equivalent to that necessary to raise the temperature of 1 pound of water 1°F.

Backshore zone   The area between the foreshore and the cliff (in rocky beaches) or dunes (in barrier island beaches).

Bajada   A sloping plain at the base of a mountain front which is formed by the merging of adjacent alluvial fans.

Bank-full discharges   Stream discharges sufficient to fill a stream channel completely.

Bar   A ridgelike accumulation of coarse sediment deposited by flowing water.

Barchan   A crescent-shaped sand dune, the tips of which point downwind.

Barrier reef   A reef separated from the land by a body of water.

Barrier island   An elongate body of coastal sand separated from the mainland by a body of water.

Basalt   A fine-grained mafic igneous rock with the same composition as gabbro.

Base flow   The portion of stream discharge that is supplied by groundwater.

Base level   The lowest level to which a stream can erode its channel.

Basin and range   See **horst-and-graben.**

Batholith   A large, irregularly shaped body of intrusive igneous rock with an outcrop area of at least 75 km^2.

Bauxite   A mixture of aluminum oxides and hydroxides formed during extensive chemical weathering of aluminum-rich rocks in hot, humid climates.

Beach   A gently sloping accumulation of coarse sediment along a lake or ocean shoreline. The beach extends from the low-water line landward to where there is a change in morphology, such as a cliff, dunes, or permanent vegetation.

Bed   (*a*) A layer of sedimentary rock more than 1 cm thick.   (*b*) The bottom of a stream channel.

Bedding plane   A discrete physical break in a sedimentary rock which separates adjacent beds and allows them to be distinguished from each other.

Bed load   All sedimentary particles which are carried at the bottom of a moving fluid such as water or air.

Benioff zone   The region of concentration of deep-focus earthquakes associated with ocean trenches. Focal depths increase systematically from the trench toward and beneath the adjacent island arc or continent.

Bentonites   Clays that have the ability to absorb great quantities of water and to swell to as much as eight times their original volume.

Bergschrund   A large crevasse that separates a cirque glacier from its headwall.

Beta decay   A form of nuclear reaction in which an electron (beta particle) is expelled from the nucleus of an atom.

Beta particle   A particle ejected from the nucleus of an atom during a nuclear reaction; essentially the same as an **electron.**

Biofuels Organic materials, such as animal dung and firewood, which release energy on combustion.

Biogenic rock Lithified biogenic sediment.

Biogenic sediment Sediment produced by organic processes or composed of organic remains.

Bituminous coal A dark-brown to black coal with about 15 to 20 percent volatile matter. Bituminous coal ranks between lignite and anthracite in the coal-forming process and is the most abundant form of coal.

Block fields Areas underlain by frost-wedged bedrock masses up to 10 m on a side, produced by periglacial processes.

Body waves Seismic waves that transmit energy through the earth (P waves and S waves).

Bonding The electrostatic attachment of one atom to another or to several others.

Boss A small irregularly shaped body of intrusive igneous rock.

Bottomset beds The fine-grained, organic-rich, and horizontally layered beds which extend out from the base of a delta.

Boudins Features of deformed rocks that result when incompetent and competent rock types are interlayered. Named for their appearance, which is similar to a string of sausages (*boudin* in French).

Boulder A sedimentary particle with a diameter greater than 256 mm.

Bowen's Reaction Series A diagrammatic representation of the sequence in which igneous minerals crystallize form a magma.

Brackish water Water with less than 17 ppt of dissolved solids, generally formed in estuaries and lagoons by the mixing of stream and ocean water.

Braided stream A stream whose channel splits into numerous intersecting channels separated by islands or sand bars.

Breaker zone The portion of the nearshore zone where wave velocity and wavelength decrease and the waves get higher and become progressively asymmetrical.

Breccia (*a*) Sedimentary rock composed of angular particles of pebble, cobble, or boulder size. (*b*) Angular chunks of rock in a fine-grained matrix produced by faulting (fault breccia).

Brittle behavior Breaking of rock during deformation to produce faults, fractures, and joints.

Brittle zone The upper zone of a glacier in which motion of ice produces breakage.

Brown clays Pelagic deposits which have a reddish-brown color produced by the oxidation of iron in the sediment.

Burial metamorphism Metamorphism caused predominantly by lithostatic pressure in a thick pile of sedimentary or volcanic rocks or both.

Butte A flat-topped erosional remnant in dry regions which is similar in origin to a mesa but has a less extensive summit area.

Calcareous ooze Biogenic sediment composed of the skeletons of microscopic organisms made of calcium carbonate.

Calcareous rocks Limestones, dolomites, and their metamorphosed equivalent, marble.

Calc-silicate (*a*) A mineral containing calcium and silica, e.g., wollastonite, $CaSiO_3$. (*b*) A rock containing those minerals formed by metamorphism of quartzose limestones and dolomites.

Caldera Huge, irregular volcanic craters formed by explosion, collapse, or a combination of the two.

Caliche A type of pedocal soil in which calcium carbonate accumulates in the upper part of the soil as a hard crust.

Capacity The ability of a transporting agent to carry material as measured by the amount of load carried at a given point per unit of time.

Capillary fringe A thin zone at the very top of the saturation zone where water is held by surface tension in some of the pores against the force of gravity.

Carbonaceous chondrites Chondritic meteorites that contain substantial amounts of carbon and graphite and a complex mixture of organic molecules.

Carbonate compensation depth The depth in the oceans at which calcareous sediment dissolves as it falls because of the greater solubility of $CaCO_3$ in the colder bottom waters.

Carbonation A type of chemical weathering in which the parent material reacts with carbon dioxide dissolved in water (i.e., a weak carbonic acid solution).

Carbonization The process of fossilization in which organic material is transformed into carbon imprints, preserving the form of the original material.

Cast A cavity within a sediment or rock that preserves the external features of a plant or animal fossil which formerly filled the cavity.

Cataclastic metamorphism See **dynamic metamorphism.**

Catastrophism A view of earth history that states that major earth changes, such as orogenies, are produced by sudden, abnormal processes rather than by slow evolution, as suggested by uniformitarianism.

Catchment areas Large paved areas on slopes used to funnel rain water into storage basins.

Cation A positively charged ion formed when an atom loses one or more electrons.

Cave A natural underground open space. The most

common type of cave is formed by groundwater solution of soluble rocks, principally limestone.

Cavern   An underground space similar to a cave but larger in size.

Cementation   Precipitation of crystalline material in pore spaces during lithification.

Central rift valley   A steep-sided valley found at the crest of every ocean ridge and presumably formed by normal faulting (tension).

Chalk   A soft, white limestone composed mostly of the skeletons of microscopic marine plants and animals.

Chemical load   Sediment which is carried as dissolved ions in water.

Chemical reactions   Interactions between atoms involving their outer electrons.

Chemical remanent magnetism   The record of the earth's ancient magnetism imprinted on minerals that form below their Curie points during diagenesis or metamorphism.

Chemical rocks   Sedimentary rocks which are formed from the precipitation of crystals from solutions.

Chemical weathering   The decomposition of rock by chemical reactions in which some of the minerals are destroyed, new ones may form, and soluble ions are removed by water flowing through the weathered material.

Chert   A chemical sedimentary rock composed of silica, SiO_2.

Chondritic meteorites   Stony meteorites that contain small grains called chondrules.

Chondrules   Small rounded bodies composed of the minerals olivine or pyroxene, with minor amounts of plagioclase feldspar.

Cinder cone   A volcano made entirely of tephra.

Cirque   A bowl-shaped rock basin which was eroded by a mountain glacier above the snow line.

Cirque glacier   A small alpine glacier which occupies a cirque basin above the snow line.

Cisterns   Underground storage tanks into which rain water falling on roofs is channeled.

Clastic   (*a*) A type of particle formed by weathering of preexisting rocks. (*b*) A type of sedimentary rock composed of mineral or rock particles, and matrix or cement or both.

Clay   (*a*) Sedimentary particles with diameters less than $\frac{1}{256}$ mm. (*b*) Minerals with a layered structure that may be hydrous silicates or nonsilicates.

Cleavage   (*a*) The breakage of a mineral along smooth flat surfaces parallel to zones of weak bonding in the mineral structure. (*b*) Close-spaced fractures developed in folded rocks, generally parallel to the axial planes of the folds.

Climatic desert   An area that is arid because it is located at a place where atmospheric circulation patterns result in descending dry air sweeping across the surface.

Coastal aquifers   Aquifers that lie beneath coastal regions and are infiltrated partially by salt water.

Coastal deserts   Areas that are arid because they are located at coastlines bordered by cold oceanic currents that generate cool, moisture-deficient winds.

Cobble   A sedimentary particle with a diameter between 64 and 256 mm.

Cohesion   The ability of rock and mineral particles to attract and hold one another.

Col   A mountain pass formed when cirque glaciers erode through opposite sides of an arête.

Colluvium   Sediments deposited by mass movements.

Columnar joints   Systematic fractures produced in fine-grained igneous rocks during cooling that result in polygonal smooth-sided shapes.

Comets   Bodies of rocky dust and ice approximately 1 km in diameter that orbit the sun at greater distances and with more eccentricity than planets.

Compaction   Rearrangement of particles and decrease in volume caused by the weight of overlying material during lithification.

Competence   A measure of the largest size particle that can be carried by a transporting agent.

Competent rock   Rock that resists changes in shape or volume during deformation until applied stresses become intense.

Composite cone   A steep-sided volcano composed of alternating layers of lava and tephra.

Compound   A substance formed by the chemical joining of atoms.

Compression   A type of stress produced by deforming forces that squeeze rocks from opposite sides.

Concentration   The amount of dissolved ions per volume of water.

Conduction   Transfer of heat from one object to another object with which it is in contact.

Cone of depression   A conically shaped depression in the water table which occurs when groundwater is pumped out of a well at a faster rate than it is replenished.

Confined aquifers   Aquifers in which the groundwater is under pressure and therefore rises toward the surface without pumping.

Conglomerate   A sedimentary rock composed of rounded particles of pebble, cobble, or boulder size.

Congruent melting   A melting process in which a solid passes directly into the liquid state at a single temperature.

Consumption Uses of water in which the water is evaporated, transpired from plants, incorporated into plants or animals, boiled into steam, or otherwise changed from a usable liquid form.

Contact metamorphism Metamorphism around a body of intrusive igneous rock caused by the heat given off during magmatic cooling.

Continent A large body of land underlain by a two-layer (sial and sima) crust.

Continental accretion A hypothesis that states that continents appear to grow outward from an old, stable region.

Continental glacier A mass of glacial ice that covers most of the topographic features of a region of continental or subcontinental proportions.

Continental rise A gently sloping part of the sea floor at the base of the continental slope formed by deposition of continent-derived sediment.

Continental shelf A very gently sloping surface that extends outward from the shoreline and is underlain by a two-layer crust (sial and sima).

Continental slope A relatively steeply sloping part of the sea floor between the continental shelf and the continental rise or abyssal plain.

Continuous melting A melting process common in solid solution series in which a mineral melts gradually over an interval of temperature.

Convection Transfer of heat from one place to another by the movement of the heated material.

Convection cell (a) The circular path followed by materials transferring heat. They rise when hotter than their surroundings, sink when cooler. (b) The possible source of movement of lithosphere plates.

Convergent plate boundaries Subduction zones where two plates collide with one another and lithosphere material is subducted back into the asthenosphere.

Coordination number The number of ions that can surround ions of the opposite charge.

Coquina A sedimentary rock composed largely of abraded and transported fossil debris.

Cordilleran Ice Sheet A Pleistocene continental glacier which was centered in the Rocky Mountains and spread westward toward the Pacific and eastward beyond the Rocky Mountains.

Core The innermost chemical region of the earth; a combination of very high density materials—probably a mixture of iron and nickel.

Coriolis effect The deflection of moving water or wind toward the left in the southern hemisphere and toward the right in the northern hemisphere because of the earth's rotation.

Correlation The demonstration that rock units from different areas were formed at the same time.

Coupled ionic substitution Simultaneous substitution of two ions brought about to maintain the electrical neutrality of a mineral structure.

Covalent bonding The joining of atoms by combining their outermost electrons in a special molecular orbit.

Crater (a) A bowl-shaped depression at the summit of a volcano. (b) A landscape feature formed by impact of a meteorite.

Crater density The number of meteorite craters per unit area.

Creep Very slow downslope movement of surface soil and rock material under the influence of gravity.

Crevasse A fracture formed in the brittle zone of a glacier during glacial movement.

Cross-bedding (also cross stratification) A type of stratification in which the layers within the bed are not horizontal but are inclined to the horizontal upper and lower surfaces of the bed.

Crosscutting relationships A principle of relative age dating that states that any geologic feature that cuts through another must be younger than the feature that it truncates.

Crust The outermost layer of the solid earth. It is composed of low-density silicate and nonsilicate minerals.

Cryptocrystalline texture Crystals too small to be seen without magnification.

Crystal A mineral specimen occurring in a regular geometric shape that reflects its internal structure.

Crystal settling Separation of early-formed minerals from magma by sinking to the bottom of the magma chamber.

Crystalline structure The ordered internal atomic structure characteristic of minerals.

Cubic packing The type of particle packing which is the loosest and has the greatest pore space.

Curie point The temperature at which a substance loses its magnetism when heated.

Current ripple marks Asymmetrical ripple marks formed by currents of air or water moving along the sediment surface.

Darcy's Law A law that expresses the rate of water migration through sediments or rocks as $V = Kh/l$, where V is the velocity of water migration, K is a coefficient of permeability which varies for different materials, and h/l is the vertical drop in the water table h over a given horizontal distance l between any two points.

Daughter element The element formed as the product of a nuclear reaction.

Debris avalanches Very rapid to extremely rapid downslope movements of relatively dry material.

Debris flow   Rapid downslope movement of relatively dry material.

Decay rate   The rate at which atoms of radioactive isotopes break down.

Declination   The angular difference at a given location between the geographic north and the magnetic north.

Deep currents   Subsurface ocean water circulation caused by seawater density differences. Deep currents involve vertical movement as well as horizontal displacement.

Deflation   The erosion of sediment particles from the ground by wind.

Dehydration   The removal of water from a mineral or a sediment.

Delta   A deposit formed when a stream flows into a large body of standing water.

Dendritic stream pattern   A stream network similar to the branching pattern of the veins in a leaf. Dendritic stream patterns are common in areas underlain by rocks that have the same general resistance to erosion.

Depletion   The decrease in availability of a resource.

Depositional remanent magnetism   A record of ancient earth magnetism preserved by alignment of grains during deposition of sediment.

Desert   A land area that receives less than 25 cm of precipitation per year.

Desert pavement   A layer of coarse particles left at the surface in arid regions after wind, water, or both have removed all the finer grains.

Desert varnish   Dark-colored iron and manganese oxide coatings that form on rocks exposed to the air in dry climates.

Diagenesis   Changes that take place in sediment after its deposition and both during and after lithification.

Diatremes   Cylindrical plutons composed of ultramafic rock that form by explosive upward boring from great depths.

Differential weathering   Weathering that occurs at different rates as a result of variations in composition of the parent materials.

Dike   A flat, tabular mass of intrusive igneous rock that cuts across previously existing structures in its host rock.

Diorite   A coarse-grained intermediate igneous rock composed of a CaNa plagioclase feldspar and ferromagnesian minerals such as amphiboles and pyroxenes.

Dip   A measure, in degrees, of how much a planar geologic structure is displaced from the horizontal, and the general direction of the tilt. It is used with strike to describe the three-dimensional attitude of planar structural features.

Dipolar field   A magnetic field defined by lines of force that emanate from two ends (magnetic poles) of an object.

Dip-slip fault   A fault in which the displacement has been largely up or down the dip of the fault plane; i.e., there is little or no horizontal motion.

Directed pressures   Pressures that are greater in some directions than others when applied to rocks; they generally are due to tectonic activity.

Discharge   The volume of water passing a point in a stream or aquifer in a given amount of time.

Discontinuous melting   See **incongruent melting.**

Distributary drainage pattern   A network pattern in which the streams branch out in the downstream direction.

Divergent plate boundaries   Sea-floor spreading centers (ocean ridges) in which two lithosphere plates move apart from each other and new lithosphere is formed.

Dolomite   A sedimentary rock composed of the mineral dolomite, $CaMg(CO_3)_2$.

Dolomitization   A diagenetic change during which original calcite grains are converted to dolomite by exchange of ions between minerals and groundwater.

Dormant volcano   A volcano that has not erupted in about 100 years but has a history of eruption in human record.

Drainage basin   The area drained by a stream.

Drainage divide   A topographically high area that separates the drainage basins of two streams.

Drift   A general term for glacial deposits.

Driving forces   Those factors which tend to aid in mass movements.

Drumlin   A streamlined hill composed largely of till deposited by continental glaciers.

Dry valley   A valley not currently occupied by a stream because of climatic change, stream capture, or the diversion of surface water underground in areas of soluble rocks.

Ductile behavior   A type of deformation in which the strain produced is *not* proportional to the amount of stress applied. Ductile deformation is permanent; when stresses are relaxed, the rock retains its deformed appearance.

Dunes   Streamlined hills composed of sand-sized particles deposited by wind.

Dynamic equilibrium   A system in which changes in one part are balanced by changes in another.

Dynamic metamorphism   Metamorphism caused by directed pressures in fault zones.

Ebb currents Tidal currents that move out from estuaries and wetlands into the ocean as the tidal level is falling.

Ecliptic The plane traced out by the earth's orbit.

Economic geology The study of the process of formation, means of discovery, and geographic distribution of earth resources.

Eddies Swirling, whirlpoollike masses of water which extend down through the streamflow and move with the current.

Effective porosity The percentage of the volume of the sediment or rock that is composed of interconnected pore space.

Effluent streams Those streams which receive a portion of their discharge from groundwater flow.

Ejecta Debris produced by meteorite impact that surrounds impact craters.

Elastic behavior A type of deformation in which stress and strain are proportional and in which the effects of deformation disappear once the deforming forces are relaxed.

Elastic rebound See **isostatic rebound**

Electric dynamo A device used to generate electricity; a rotating electrical conductor cuts magnetic lines of force and has a current induced in itself as a result.

Electron A negatively charged subatomic particle which orbits the nucleus.

Electron capture A form of nuclear reaction in which an electron is taken into the nucleus of an atom from the inner electron shells.

Electron shell A group of atoms that are approximately at the same distance from the nucleus and contain the same energy.

Electrostatic forces Forces of attraction and repulsion between positively and negatively charged ions or subatomic particles.

Element An element consists of all atoms that have the same atomic number.

Emerald The deep-green gem-quality variety of beryl.

End moraine A moraine deposited near the terminus of a glacier.

Energy The capacity to do work or cause activity.

Enrichment ratio The degree of concentration needed to form ore mineral deposits.

Epeirogeny An episode of gentle regional uplift without extensive folding or faulting.

Ephemeral stream A stream which flows for only a short time after a rainfall before the water evaporates or infiltrates into the ground.

Epicenter The point on the earth's surface directly above the focus of an earthquake.

Epoch A span of geologic time. (See **era**.)

Era The longest subdivision of geologic time. Geologic eras are distinguished from one another on the basis of marked evolutionary changes in the fossil record. Eras are subdivided into **periods,** which in turn are subdivided into **epochs.**

Erosion The removal of sediment, rock, or both by gravity, ice, water, or wind.

Erosional mountains Mountains formed by broad regional upwarping and subsequent deep erosion.

Erratics Rock fragments carried by glaciers far from their source and deposited on bedrock of a completely different type.

Esker Stratified drift deposited by meltwater streams in channels carved into the surface of a glacier or in tunnels beneath a glacier and left as a narrow sinuous ridge after the ice has melted away.

Estuary A branch of the sea with tidal flow and saline to brackish water.

Eustatic changes in sea level Variations in the ocean surface elevation resulting from increasing or decreasing volumes of ocean.

Evaporation Conversion of water to water vapor.

Evaporites Chemical sedimentary rocks deposited from concentrated solutions.

Evapotranspiration Return of water vapor to the atmosphere after it has passed through the life processes of plants.

Exfoliation The separation of curved sheets of rock from rock exposures at the earth's surface brought about by the release of pressure.

Exfoliation domes Rounded landforms that result from exfoliation.

External drainage A drainage system in which streams draining high-precipitation areas traverse a desert.

Extinct volcano A volcano which has not erupted in historic time.

Extrusive rock See **volcanic rocks.**

Facies (*a*) In sedimentary rocks, a group of rocks with distinctive characteristics that were deposited in a specific sedimentary environment. (*b*) In metamorphism, all rocks subjected to a particular set of pressure-temperature conditions.

Facies series The sequence with which metamorphic facies show increased metamorphic grade in a given region.

Fall A mass wasting process which involves sediment and rock moving through the air and accumulating at the base of the slope.

Fault A fracture formed when rocks break and materials on opposite sides of the fracture are displaced relative to one another.

Fault-block mountains Mountains formed by tension-induced normal faulting.

Faunal succession The change in fossils from the bottom of a sequence of rocks to the top that reflects the evolution of life during the time span represented by the rocks.

Feeder conduit A cylindrical channel through which lava rises to the surface to form a volcano.

Feldspar A large family of framework silicate minerals, including the plagioclase and potassic feldspar groups.

Feldspathic sandstone A sandstone that has a feldspar content between 10 and 25 percent.

Felsic rocks Igneous rocks composed largely of potassic and plagioclase feldspar, often with quartz.

Ferromagnetic materials Substances that become permanently magnetic once they are exposed to a magnetic field.

Fetch The distance over which a wind blows to generate waves. The greater the fetch, the higher the waves which may be generated.

Filter pressing Separation of crystals from magma when a magma chamber is deformed and the liquid is forced through narrow fissures, leaving the solid mineral grains behind.

Firn A granular type of snow formed by partial melting of the edges of snowflakes and refreezing of the meltwater.

Fission track dating Dates obtained by determining the track density on unit areas of crystals.

Fjords Steep-sided U-shaped valleys that have been drowned by the postglacial rise in sea level.

Flank eruption Eruption of lava from the side of a volcanic slope rather than at the summit.

Flood currents Tidal currents moving from the ocean into estuaries and wetlands as tidal level is rising.

Floodplain The area adjacent to a stream channel that is covered periodically with water when the stream overflows its banks.

Flow A mass wasting process in which sediment or rock exhibits a continuity of motion in a plastic or semifluid state.

Flowlines The paths individual particles make in moving through a fluid such as air or water.

Flowstone A banded form of calcite deposited as ribbons, curtains, and wall coatings in caves and caverns.

Fluorescence The color of a mineral under ultraviolet light which differs from its appearance under normal light.

Flux The annual rate of transfer of water from one reservoir to another.

Focus The actual site of fault displacement and source of seismic-wave energy.

Fold A flexure in rock formed when rocks bend in response to external forces.

Fold axis A line that traces the maximum curvature of a folded layer and separates the two limbs.

Fold mountains Mountains that form as a result of intense compression and hence are characterized by tightly folded rock.

Fold nose The convergence of bands of rock that results when plunging folds are eroded.

Foliation Parallel alignment of planar minerals in a rock.

Footwall block The fault block that lies beneath an inclined fault plane.

Foreset An inclined layer of sandy material deposited upon or along an advancing and relatively steep frontal slope, such as the outer margin of a delta or the lee (downwind) side of a dune.

Foreshore zone The sloping front face of the beach which is exposed at low tide and covered at high tide.

Fossils The preserved remains or traces of plants and animals that lived in the geologic past.

Fracture (*a*) Breakage in minerals that occurs in random directions relative to the mineral structure. (*b*) Irregular, randomly oriented breaks in rocks.

Fringing reef A reef built up against the coast.

Frost creep The downslope movement of particles resulting from repeated freezing and thawing of slope materials.

Frost heaving Upward movement of soil masses by freezing and subsequent expansion of pore water.

Frost wedging A process of physical weathering in which water in rock openings freezes and expands, fracturing the rock.

Fumarole A volcanic vent that erupts gases.

G The strength of the earth's gravitational field at any point.

Gabbro A coarse-grained mafic igneous rock composed of plagioclase feldspar and ferromagnesian minerals such as olivine and pyroxenes.

Gal An acceleration of 1 cm/s/s. It is used to measure the gravitational field of the earth (1 milligal = 0.001 gal).

Galilean satellites Four large, planet-sized satellites of Jupiter (Callisto, Ganymede, Europa, and Io).

Gangue Minerals formed and associated with ore minerals that have no economic value and are discarded during separation of the ore mineral.

Garnet A family of minerals containing independent silicon-oxygen tetrahedra: $(Ca, Fe^{2+}, Mn^{2+}, Mg^{2+})_3$ $(Al, Fe^{3+}, Cr^{3+})_2(SiO_4)_3$

Geochronology The science of obtaining radiometric ages by the analysis of the radioactive elements in a mineral, rock, or organic material.

Geode A spherical and often partially hollow and/or mineral-filled sedimentary structure which forms after the deposition of the sediment.

Geologic structures Features that reveal the deformational history of a rock, such as folds or faults.

Geometric variables The set of variables which describe the shape of a stream channel.

Geomorphology The study of surface landforms and the processes that form them.

Geophysics The application of physics to the study of the earth, including seismology, gravity, magnetic, and heat flow studies.

Geosyncline Large elongate depositional basins that accumulate great thicknesses of sediment and volcanic rock prior to the deformation that produces a fold mountain system.

Geothermal energy Energy obtained by tapping the heat contained within the earth.

Geothermal gradient The rate at which temperature in the earth increases with depth.

Geyser Groundwater that emerges from the ground as a jet of steam and hot water.

Ghyben-Herzberg ratio A ratio describing the static relation of fresh groundwater and saline groundwater in coastal areas. For each meter of freshwater above sea level, the saltwater surface is displaced 40 m below sea level, i.e., in a ratio of 1:40.

Glacial economy The balance between accumulation of snow and ice in the snowfield, and loss of snow and ice along the length of the glacier. The glacial economy determines whether glaciers will advance, retreat, or stagnate.

Glacial groove A deep, wide, and usually straight furrow cut in bedrock by the abrasive action of a rock fragment embedded in a glacier.

Glacial striations Elongate scratches carved in rock surfaces by rock particles carried at the base of a glacier.

Glacier A large mass of ice formed by the compaction and recrystallization of snow that flows under the pressure caused by its own weight.

Glass An igneous rock that cooled so quickly that no minerals had a chance to form.

Gneiss Metamorphic rock that is both layered and foliated.

Gneissosity A feature of layered metamorphic rocks in which alternating layers are foliated and nonfoliated and commonly color-banded as well.

Goldich Stability Series A scale which expresses the relative weatherability of rock-forming silicate minerals.

Gondwanaland A supercontinent composed of what are now the southern hemisphere continents plus India.

Gouge Crushed and broken rock held together by a finer-grained matrix. Gouge results from the grinding of two fault blocks against one another during deformation.

Graded bedding A variety of stratification in which the particle size decreases systematically upward in the bed.

Gradient The difference in elevation of a streambed or slope over a given horizontal distance.

Granite A coarse-grained felsic igneous rock composed mainly of quartz, potassic feldspar, and plagioclase feldspar, with minor amounts of ferromagnesian minerals (biotite, hornblende) and muscovite.

Granoblastic texture Metamorphic texture in which grains are oriented randomly; i.e., there is no foliation or lineation.

Granofels Metamorphic rock with a granoblastic (nonfoliated and nonlineated) texture.

Granule A particle of sediment with a diameter between 2 and 4 mm.

Gravimeter An instrument used to measure the strength of the earth's gravitational field.

Gravity A force of attraction that exists between all substances because of their mass.

Gravity anomaly An area of the earth's surface where the measured force of gravitational attraction is significantly higher (positive anomaly) or lower (negative anomaly) than the average.

Greenschist (*a*) Foliated, low-grade metamorphosed mafic rock. (*b*) Low-grade facies of regional metamorphism.

Greenstone Metamorphosed mafic rock with no foliation.

Groins Rock, concrete, or wood structures which are built into the surf to trap sand and build out beaches.

Ground moraine Till deposited as a sheet beneath a moving glacier.

Groundwater Water that is accumulated in sedimentary pores and rock fractures beneath the ground surface.

Groundwater potential The relative ability of a rock or sediment to store groundwater.

Guyot   A flat-topped seamount formed by subsidence and wave erosion of a volcano.

Gypsum   A chemical sedimentary rock composed of gypsum, $CaSO_4 \cdot 2H_2O$.

Half-life   The amount of time necessary for 50 percent of the total number of parent atoms to decay into daughter atoms in nuclear reactions.

Halite   A chemical sedimentary rock composed of crystals of halite, NaCl; commonly called rock salt.

Hanging valley   A preglacial tributary valley not eroded as deeply by valley glaciers as the main stream valley, and as a result left isolated well above the main stream valley when the ice melts.

Hanging wall block   The fault block that lies above an inclined fault plane.

Hardness   The resistance of a mineral to being scratched.

Hawaiian-type islands   Basaltic island chains not associated with ocean ridges or trenches.

Headwall   The steep, near-vertical slope formed at the upslope side of a cirque.

Headward erosion   Erosion of previously undissected areas at the upstream end of a stream or the upvalley end of a glacier.

Heat energy   A form of kinetic energy in which the movement is on the atomic scale.

Helical flow   The corkscrewlike flow of fluids such as water and air.

Hill   A topographic high standing less than 600 m above its surroundings.

Horn   A steep-sided faceted mountain peak formed by headward erosion of several cirque glaciers.

Hornfels   Granoblastic rock produced by contact metamorphism.

Horst-and-graben   A type of structure developed in regions of tension in which uplifted blocks (horsts) are separated from down-dropped blocks (graben) by normal faults.

Hot dry rock geothermal energy   Energy obtained by pumping water down into holes as deep as 10 km, where it is heated by the geothermal gradient. The resulting steam is used to drive a turbine.

Hot spots   Fixed sources of heat in the mantle below the asthenosphere that may be the cause of volcanism in Hawaiian-type islands.

Hydration   The addition of water to a mineral or rock, producing new minerals.

Hydraulic variables   Variables such as velocity and discharge that describe streamflow.

Hydrologic cycle   The continued transformation of water into its different forms by evaporation, melting of ice, and so forth.

Hydrolysis   The reaction between minerals and the H^+ and $(OH)^-$ ions formed during dissociation of water.

Hydrosphere   The liquid outer covering of the earth, encompassing the oceans, rivers, and lakes.

Hydrothermal solutions   Hot water formed either by heating of groundwater by plutons or by expulsion from a cooling magma. Such solutions contain ions that may be deposited in rocks to form economical mineral ores.

Hypersaline water   Water with more than 40 ppt of dissolved solids.

Iceberg   A floating mass of ice broken off from the terminus of a glacier along a coast.

Ice cap   A glacier so thick that it covers most of the topographic features of an entire region.

Ice centers   Huge snowfields that spawned continental glaciers.

Ice-contact stratified drift   A deposit of stratified drift that originally was deposited in contact with glacial ice.

Ice-rafting   Transport of sediment into the deep ocean basin by debris-laden icebergs that then melt and drop the sediment to the sea floor.

Ice sheet   A very large and thick glacier covering an area of subcontinental or continental size.

Ice shelf   A mass of floating ice consisting of those parts of a glacier which have advanced beyond the shoreline and float because of their low specific gravity.

Ice-wedge cast   A structure that is formed when an ice wedge melts during postglacial thawing and leaves an open space (cavity) which subsequently is filled with washed-in fine sediment.

Igneous rock   Rocks formed by solidification of molten rock material.

Impact craters   Bowl-like pits excavated in the surface as a result of meteorite impacts.

Impact metamorphism   Metamorphism caused by meteorite impact.

Inclination   The angle between magnetic lines of force and the earth's surface, ranging from 90° at the magnetic poles to 0° at the magnetic equator.

Inclined fold   A fold with an axial plane that is neither horizontal nor vertical.

Inclusions, principle of   A principle of relative age dating which states that a rock containing fragments of another rock must be younger than that rock.

Incompetent rock   Rock which changes shape easily during deformation, with only little stress added.

Incongruent melting   A melting process in which a substance first melts partially to form liquid and a new mineral, followed by gradual melting of the residual mineral.

Index fossil The remains of an organism that lived over a broad geographic area but for only a relatively short period of time so that its presence in a rock is indicative of a specific time span.

Index mineral A metamorphic mineral whose appearance is indicative of a particular grade of metamorphism.

Indicators Rock fragments with a unique source area that can be used to determine the path of glacial advance.

Infiltration capacity The rate at which water may pass through surface materials.

Influent stream A stream that is above the local or regional water table and thus contributes water to the groundwater reservoir. Compare with **effluent stream.**

Initial porosity Porosity that develops before or during rock formation.

Injection wells Wells in which water is pumped *into* the ground to recharge aquifers.

Inselberg A mountain left as an erosional remnant in arid regions; it is surrounded by gently sloping pediments.

Instream use Human use of water within a stream channel.

Intensity A measure of the damage to human-built structures caused by an earthquake; it is measured by the modified Mercalli intensity scale.

Interglacial A relatively warm period of time separating episodes of continental glaciation.

Intergranular slip A mechanism of deformation in which grains slide past one another to new positions, changing the shape of the deformed substance.

Intermittent streams Streams which are dry for part of the year but receive part of their flow from the water table when it is high enough.

Internal drainage A type of drainage system in which streams flow into topographic lows bounded by mountains and as a result never connect with the ultimate base level—the oceans.

Intragranular gliding See **intracrystalline slip** (below).

Intracrystalline slip A mechanism of rock and ice deformation in which there are tiny amounts of displacement between planes of atoms in a crystal structure; similar to intragranular gliding.

Intrusive rock See **plutonic rock.**

Ion A charged particle formed when an atom gains or loses an electron in a chemical reaction.

Ionic bonding The joining of atoms that have become electrically charged by gaining or losing electrons.

Ionic substitution The ability of an ion to take the

place of another in a mineral structure because the two have similar sizes and charges.

Iron meteorite A meteorite composed predominantly of an alloy of nickel and iron.

Island arc An arc-shaped volcanic island chain found on the concave (landward) side of oceanic trenches.

Isoclinal fold A fold whose limbs dip in the same direction at the same angle.

Isograd The boundary between two metamorphic zones along which a new index mineral appears. Ideally, metamorphic conditions along an isograd are equal.

Isostasy The buoyancy of low-density rock material in higher-density material.

Isostatic rebound Uplift of the earth's crust due to unloading by (*a*) removal of an ice sheet, or (*b*) erosion of a mountain range.

Isotope Isotopes of an element are atoms that have the same atomic number but different atomic masses.

Jade A semiprecious stone that is either the amphibole tremolite or the pyroxene jadeite.

Jetties Rock or concrete structures built into the surf zone on either side of a tidal inlet to prevent the inlet from closing through deposition of sand.

Joint Systematically aligned fractures along which there has been no displacement. Joints are produced by cooling in igneous rocks, unloading due to erosion, and deformation.

Jovian planets The fifth through eighth planets—large, low-density bodies composed of hydrogen, helium, ammonia, and methane (Jupiter, Saturn, Uranus, and Neptune).

Kame A conical hill composed of stratified drift formed by meltwater deposition in the crevasses of a stagnating glacier.

Kame moraine A discontinuous ridge composed of a number of kames that were deposited at the terminus of a stagnating glacier.

Kame terrace A flat-topped meltwater stream deposit formed between a stagnating valley glacier and the valley walls.

Karst topography Topography characterized by sinkholes, underground drainage, caverns, and a few hills that exist because they are resistant to groundwater solution.

Karst towers Residual masses of limestone that stand above the surrounding lowlands.

Kettle A bowl-shaped depression in outwash plains formed by the melting of a block of stagnant ice that had been buried by outwash. It is called a kettle lake when filled with water.

Kinetic energy   The ability of a moving object to induce activity in other objects.

Laccolith   A pluton that is flat-bottomed and has a dome-shaped top that arches up the overlying rocks.

Lag time   The time difference between the center of mass of rainfall and center of mass of the stream discharge.

Laminae   Rock and sediment layers less than 1 cm thick.

Laminar flow   A type of fluid flow characterized by parallel, horizontal movement with little or no vertical movement of the flowlines.

Lateral continuity   A principle of stratigraphy which states that as a sedimentary rock layer is deposited, it extends outward horizontally until either it thins out and disappears or it terminates against the boundaries of the basin in which it accumulates.

Lateral erosion   Erosion of materials on the side of a stream or valley glacier.

Lateral moraine   Rock fragments carried at the sides of a valley glacier and deposited as a low ridge when the ice melts.

Laterite   A soil enriched in oxides and hydroxides of iron or aluminum.

Laurentide Ice Sheet   A Pleistocene continental glacier which covered most of Canada and extended deeply into the northern U.S.

Lava   Magma extruded onto the earth's surface.

Lava domes   Bulbous bloblike shapes characteristic of the eruption of extremely viscous lava.

Lava plateau   A broad upland region underlain by lava.

Leaching   The removal of soluble material (generally from soil) by water percolating through the soil.

Left-lateral fault   A strike-slip fault in which displacement of one block has been to the left as viewed from the other block.

Lignite   A brownish-black coal that is an intermediate stage between peat and subbituminous coal in the coal-forming process.

Limbs   The opposite sides of a fold.

Limestone   A sedimentary rock composed of calcite, $CaCO_3$.

Lineation   Parallel alignment of rod-shaped minerals in a rock.

Liquefaction   The loss of cohesion and structural strength in sands or clays brought about by water coating grain surfaces and reducing friction drastically.

Lithic wacke   A sandstone composed of rock fragments and matrix (more than 15 percent).

Lithification   The process of transforming a sediment into a sedimentary rock.

Lithosphere   (*a*) The uppermost 100 km of the earth, composed of relatively rigid rock material incorporating both crust and part of the upper mantle. (*b*) The rock material of which moving plates are made.

Lithostatic pressure   Pressure that is equal in all directions on a rock buried beneath a cover of other rocks and sediments.

Load   The sedimentary materials carried by a transporting medium. The physical load is carried as particles; the chemical load is transported as ions in solution.

Local section   A series of rocks that record the history of deposition in a local area.

Loess   Silt derived from deflation of glacial outwash or desert sediments and deposited downwind of the source.

Longitudinal dune   A massive linear sand dune oriented parallel to the dominant wind direction.

Longitudinal profile   A cross section showing the gradient of a stream from its headwaters to a point downstream.

Longitudinal wave   A seismic wave (P wave) propagated by the straight-line back-and-forth motion of particles parallel to the direction of wave propagation.

Longshore currents   Currents moving parallel to the coast in the nearshore zone.

Longshore drift   Particles of sediment that are moved along a shoreline by wave action and longshore currents.

Lopoliths   Funnel-shaped intrusive igneous rock bodies generally composed of mafic rock.

Love wave   A seismic surface wave propagated by straight-line transverse particle vibration that lies in the surface along which the wave is moving.

Low-velocity zone   A region in the upper mantle in which seismic-wave velocities decrease, probably because of decreased rigidity of the mantle material.

Mafic rocks   Igneous rocks rich in ferromagnesian silicate minerals such as olivines and pyroxenes but low in potassic feldspar.

Magma   Molten rock material.

Magma chamber   The region within the earth where magma crystallizes.

Magmatic differentiation   Crystallization of magma during which early-crystallized minerals are separated from the remaining melt.

Magnetic anomalies   Areas of the earth's surface where the magnetic field strength is abnormally high (positive anomaly) or low (negative anomaly).

Magnetic dating   Dates determined from changes over

time in the intensity and polarity of the earth's magnetic field.

Magnetic domains Regions within a substance characterized by parallel alignment of the magnetic fields of atoms.

Magnetic equator An imaginary line connecting all those points on the earth's surface at which the magnetic lines of force parallel the surface.

Magnetic reversals Periods during which the earth's magnetic field reversed its polarity.

Magnetometer An instrument used to measure the intensity of the earth's magnetic field.

Magnetosphere A region thousands of kilometers above the earth's surface in which solar particles are trapped by the earth's magnetic field and prevented from striking the planet's surface.

Magnitude A measure of the amount of energy released at the epicenter of an earthquake, based on the amplitude of seismic waves. It is calculated according to the Richter magnitude scale.

Manganese nodules Authigenic concretions formed largely from manganese precipitated from seawater.

Mantle The intermediate region of the earth's interior, lying between the crust and the core. It is composed largely of high-density silicate minerals.

Marble Metamorphic rock formed by recrystallization of limestone or dolomite.

Mare A large, dark-colored, relatively flat basin on the moon filled with basalt (plural, *maria*).

Marginal sea A shallow sea that lies between a continent and an island arc.

Massive bedding Uniform beds deposited under constant environmental conditions.

Mass wasting (or **mass movement**) Downslope movement of regolith and rock at the earth's surface caused by gravity.

Matrix (*a*) Fine-grained particles that fill spaces between coarse-grained particles in a sedimentary rock. (*b*) Fine-grained crystals in a porphyritic igneous rock.

Mature soil A soil which has developed sufficiently so that it exhibits several soil horizons.

Meander A curve in the course of a stream.

Meandering stream A stream with a channel composed of a series of sinuous curves.

Medial moraine A low ridge of rock debris formed when the lateral moraines of two merging valley glaciers join.

Melting The change from the solid state to the liquid state.

Melting point The temperature at which a substance passes from the solid to the liquid state.

Mesa A flat-topped erosional remnant bordered on the sides by steep slopes or cliffs and capped with resistant and nearly horizontal rocks (commonly lavas).

Metallic bonding Joining of atoms by the sharing of all outer shell electrons among all the nuclei present.

Metamorphic grade A measure (e.g., low-grade) of the intensity of metamorphism.

Metamorphic rocks Rocks formed within the earth when previously existing rocks are subjected to temperatures and pressures different from those under which they first formed.

Metamorphic zone A mapped area in which metamorphic intensity was approximately the same, as shown by the presence of a metamorphic index mineral.

Metamorphism Changes in the mineralogy, texture, and composition of a rock caused by physical or chemical conditions different from those under which the rock first formed.

Metasomatism Metamorphism in which the composition of the affected rock changes significantly by addition or removal of ions.

Meteorite A rock fragment from interplanetary space that has collided with the earth or other planetary body.

Milankovitch theory A theory to explain cyclical fluctuations in Pleistocene climate by changes in the ellipticity of the earth's orbit and the inclination of the earth's rotational axes.

Milligal See **gal.**

Mineral A naturally occurring inorganic solid with an ordered internal arrangement of atoms or ions and a relatively fixed chemical composition.

Mineral deposit A concentration of minerals, rocks, or organic materials such that mining and processing them is physically and economically possible.

Mineral economics The study of the technological, economic, social and political controls on making earth resources available.

Mineral fuels Fuels produced by geologic processes and stored in rocks and minerals (e.g., coal, oil, natural gas, and uranium).

Mineraloid A substance similar to a mineral but lacking either an ordered internal structure or a relatively fixed chemical composition (e.g., opal or limonite).

Moho See **Mohorovicic discontinuity.**

Mohorovicic discontinuity The boundary between the earth's crust and mantle, identified by a sudden increase in seismic-wave velocity in the mantle.

Mold A three-dimensional impression of a fossil organism. Casts are formed when sedimentary particles or crystalline material fills a mold.

Moment of inertia   The tendency of a spinning object to keep on spinning.

Moonstone   A translucent, semiprecious gemstone variety of either albite or orthoclase.

Moraine   A landform, usually composed largely of till, which is deposited at the margin of a glacier.

Mosaic texture   Metamorphic texture in which equal-sized grains interlock with no preferred orientation; a variety of granoblastic texture.

Mountain   A topographic high that stands more than 600 m (2000 ft) above its surroundings.

Mountain glacier   Any glacier in a mountain range except an ice cap or ice sheet.   It usually originates in a cirque and may flow down into a valley previously carved by a stream.   Also referred to as an **alpine glacier.**

Mountain range   A group of mountain peaks that have been formed at approximately the same time by the same process.

Mountain root   The downward projection of relatively low-density rocks beneath major mountain ranges.

Mountain system   A group of mountain ranges.

Mud cracks   Polygonal patterns of downward-closing fractures formed by the shrinking which occurs in the drying out of a clayey sediment.

Mudflow   Downslope movement of fine-grained material that contains large (up to 30 percent) amounts of water.

Mylonitic texture   A texture produced in dynamic metamorphism in which original rock grains have been stretched, flattened, crushed, and smeared out into a well-developed foliation.

Natural bridge   An archlike rock formation. One type of natural bridge forms when a stream abandons a meander and cuts through the narrow meander neck; in a limestone area, a natural bridge represents the remnant of the roof of an underground cave or tunnel that has collapsed.

Natural levees   Low ridges of sediment that parallel stream channels and are deposited when streams overflow their banks.

Nearshore zone   The body of water between the beginning of the breaker zone and the beach; includes the breaker and surf zones.

Nebula   A rotating disk of gas and dust.

Neptunists   People who believed that all rocks formed by precipitation from ocean water.

Neutral plate boundaries   Transform faults along which two plates slide past one another with neither creation of new lithosphere nor destruction of old.

Neutron   An unchanged subatomic particle found in the nucleus of an atom.   It is composed of a proton and an electron.

Nonmagnetic materials   Substances that are not affected by a magnetic field.

Nonplunging fold   A fold with a horizontal axis.

Nonrenewable resource   A resource which is being depleted more rapidly than it is being formed.

Normal fault   A dip-slip fault in which the footwall block has moved upward relative to the hanging wall block as a result of tensional stresses.

Nuclear energy   Energy released from a nucleus during a nuclear reaction.

Nuclear reactions   Alterations of atoms involving changes in their nuclei.

Nucleus   The central part of an atom in which the protons and neutrons are located.

Nuée ardente   A fiery cloud composed of tephra and gases that rises from a volcano and moves rapidly downslope like a superheated density current.

Oasis   An area of vegetation and standing water in a desert formed by artesian springs rising from any break in a confined aquifer below.

Obduction   The scraping off of part of a subducted plate and its incorporation in the upper plate by thrusting during subduction.

Obsidian   A glassy igneous rock with a composition similar to that of granite.

Obsidian hydration layer dating   Dates determined by finding the depth of the hydrated layer in obsidian.

Ocean   A large body of water whose basin is underlain by a one-layer, simatic crust.

Ocean fractures   Linear topographic lows that segment and offset the ocean ridges.

Ocean ridges   Large elongate mountain ranges found in every ocean that are sites of earthquake activity and basaltic volcanism.

Oceanic trenches   Deep arc-shaped depressions in the sea floor in which the greatest water depths of the oceans are recorded.   They are the sites of extensive earthquake activity.

Oceanography   The study of the ocean, including its physical, chemical, biologic, and geologic aspects.

Offstream use   Human use of water that has been taken from its natural reservoir and transported to its place of use (e.g., for cooling an atomic power plant).

Oil   A liquid composed of compounds of carbon and hydrogen which is believed to form from diagenetic changes in marine plant and animal matter that accumulated in fine-grained sediments.

Oil sands Sands or sandstones with pore spaces filled with highly viscous asphalt.

Oil shales Shales with an exceptionally high content of organic matter that yield oil when heated.

Ooliths Spherical, concentrically banded grains believed to form from inorganic precipitation of calcium carbonate in an agitated marine environment.

Oolitic limestone A rock containing ooliths of calcium carbonate cemented together by calcite.

Open fold A fold with a large angle between its two limbs.

Ophiolite suite A sequence of ultramafic through basaltic plutonic and volcanic rocks capped by deep-ocean sediments that is found in mountains produced by intense compression. It is thought to result from obduction of oceanic crust.

Ore minerals Minerals that are concentrated to form metallic deposits.

Original horizontality A principle of relative age dating which states that sediments are deposited in layers that are horizontal or nearly horizontal at the time of their formation.

Orogeny An episode of mountain building characterized by extensive folding and faulting.

Oscillation ripple marks Symmetrical ripple marks formed by wave action.

Outlet glacier A fast-moving valley glacier fed by an ice cap or ice sheet which achieves high velocities when the ice is forced to flow through narrow openings.

Outwash Sediments deposited by meltwater streams well beyond the terminus of a glacier.

Outwash plains Flat and gently-sloping plains underlain by meltwater stream deposits built beyond the margins of glaciers.

Overgrowths Dissolved material deposited around the rims of sedimentary particles in a rock.

Oxbow lake An arc-shaped lake formed by water filling cutoff meander loops.

Oxidation A type of chemical weathering in which oxygen from the atmosphere interacts with minerals and rocks.

P wave A longitudinal seismic wave that because of its high velocity of propagation is generally the first wave to reach a seismograph after an earthquake occurs.

Packing The density of aggregation of particles in sediment.

Pahoehoe Highly fluid, ropy lava.

Paleo A prefix meaning *ancient,* as in **paleocurrent** (ancient current), **paleoclimate** (ancient climate), **paleomagnetism** (ancient magnetism), or **paleowind** (ancient wind direction).

Pancaking A form of earthquake-induced building collapse in which the upper floors of a building fall onto the lower ones.

Pangaea A single "supercontinent," composed of all the separate continents as we now know them, which may have existed prior to the Jurassic opening of the modern Atlantic Ocean.

Parabolic dune A crescent-shaped sand dune whose tips point upwind.

Paramagnetic materials Substances that are affected by application of a magnetic field but do not act magnetically once the field is removed.

Parent element A radioactive element before it decays to form a new element.

Parent material A term used to describe the original material (rock, sediment, or soil) which is being weathered.

Passive continental margins Coastlines where the shoreline is located far from the nearest plate boundary so that earthquake and volcanic activity are absent.

Patterned ground Surface features such as bands, stripes, polygonal networks of rock debris, and polygonally fractured soil formed by the freezing and thawing of the upper part of the permafrost.

Peat An unconsolidated deposit of semicarbonized plant remains with a high moisture content; an early stage in the development of coal. Carbon content is about 60 percent and oxygen content about 30 percent.

Pebble A particle of sediment with a diameter between 4 and 64 mm.

Pedalfer A type of soil produced in humid regions characterized by the absence of calcium carbonates and the accumulation of clay minerals and iron oxides in the B horizon.

Pediment A gently sloping surface carved into the bedrock at the base of a mountain range in arid or semiarid environments.

Pedocal A type of soil common in arid regions where calcium carbonate is accumulated rather than leached out of the soil as it is in humid climates.

Pegmatite An extremely coarse grained igneous rock.

Pelagic ooze Very fine grained continent-derived oceanic sediment.

Pelagic sediments Deposits which form from particles which have settled through the water.

Pelitic rock Rock containing large amounts of alumina, Al_2O_3, e.g., shales, slate, and schist.

Peneplain A large, featureless, flat land surface caused by extensive stream erosion.

Perched water table A local zone of saturation which exists at a level higher than the regional water table because of the presence of an underlying layer of impermeable rock or sediment.

Perennial stream   A stream which flows year-round and receives most of its discharge by groundwater flow.

Peridot   The gem variety of olivine.

Periglacial climate   The cold climate that occurs at the periphery of continental glaciers.

Period   A span of geologic time. (See **era.**)

Permafrost   Perennially frozen ground.

Permeability   A measure of the ease with which a fluid can move through a rock or sediment.

Petrifaction   The replacement of the soft parts of fossil animals by minerals precipitated from solutions circulating through the enclosing sediment.

Petrology   The study of rocks.

Phaneritic   An igneous texture coarse enough for individual mineral grains to be seen with the naked eye.

Phenocryst   See **porphyritic.**

Phosphorite   A chemical sedimentary rock composed of crystals of complex phosphates of calcium.

Phosphorite nodules   Phosphate-rich nodules which form on the sea floor.

Photic zone   The upper few tens of meters of the oceans to which sunlight can penetrate.

Phyllite   Medium-grained foliated metamorphic rock wnose aligned micas produce a dull sheen.

Physical load   The sediment carried as discrete particles within a fluid.

Physical properties   Those attributes of minerals which are caused by mineral composition and structure (e.g., hardness, color, density, and so forth).

Physical weathering   The disaggregation of a rock into progressively smaller pieces without altering the chemical composition of the minerals present.

Piedmont glacier   A large glacier formed when two or more trunk glaciers merge at the foot of a mountain range and flow across the adjacent plains.

Piezometric surface   The height to which artesian water would rise if a confined aquifer were tapped.

Pillar   A column formed by the merging of a stalactite and a stalagmite.

Pillows   Globular masses of volcanic rock formed when lava flows into a body of water.

Placer   A surficial mineral deposit concentrated by water flow in streams or by wave action in the near-shore zone.

Plagioclase feldspar   A complete solid-solution series between the framework silicates albite, $NaAlSi_3O_8$, and anorthite, $CaAl_2Si_2O_8$.

Planet   A spheroidal body that revolves around the sun in a fixed, nearly circular orbit.

Planetology   The study of planets, their satellites, and other bodies in the solar system.

Plastic zone   The lower zone of movement in a glacier in which the pressure from the overlying snow and ice enables the glacier to move by plastic flow.

Plateau   An extensive flat area with little internal relief that stands prominently above its surroundings.

Plate tectonics   A model for earth behavior in which the outer 100 km of the planet (the lithosphere) consists of a small number of individual masses called plates which can move independently of one another. Separation and collision of plates results in the formation of ocean basins and mountains.

Playa lake   A temporary lake formed in the low areas of desert basins after a rainfall.

Plucking   A process of glacial erosion in which meltwater at the base of a glacier flows into rock crevasses, refreezes, and loosens the rock so that it is incorporated in the advancing ice.

Plug   A small, irregularly shaped body of intrusive igneous rock.

Plunging fold   A fold whose axis is not horizontal.

Pluton   A body of intrusive igneous rock.

Plutonic rock   An igneous rock formed by the solidification of magma within the earth.

Pluvial climate   Cool and wetter climates that affected presently arid areas during the Pleistocene.

Pluvial lakes   Lakes that formed in what are now arid areas during periods of pluvial climate.

Point bar   Arc-shaped sand bars deposited on the inner parts of meander loops.

Polar desert   A cold desert which forms in polar areas where there is low precipitation.

Polar molecule   A compound, such as water, H_2O, that has positive and negative ends, even though the compound is electrostatically neutral.

Polar wandering curve   A diagram showing the apparent migration of the magnetic poles with time, relative to a landmass.

Polymorphs   Minerals that have the same chemical composition but different internal structures (e.g., diamond and graphite).

Pools   Relatively deep water portions of streams with straight channels.

Pore   A space between grains in a rock or sediment.

Porosity   That portion of the volume of a sediment or rock which is made up of open spaces.

Porphyritic texture   An igneous texture in which there are a few coarse grains (phenocrysts) found in a generally finer-grained groundmass.

Porphyroblast   A large metamorphic mineral surrounded by finer grains.

Potassic feldspar   Any one of the framework silicate

minerals (orthoclase, microcline, and sanidine) in which potassium is the dominant large cation, $KAlSi_3O_8$.

Potential energy Energy in storage prior to being used.

Potholes Circular cavities in a streambed carved by rock particles that have been swirled around in eddies.

Precision Depth Recorder (PDR) An instrument that measures water depth by timing the passage of sound energy from a ship to the sea floor and back.

Preferred orientation Alignment of particles in a specific direction.

Pressure solution A mechanism of deformation in which minerals dissolve along grain contacts and are reprecipitated in pore spaces or regions of relatively low pressure.

Primary minerals Those minerals in a parent material which survive weathering virtually unaltered.

Primary sedimentary structure See **sedimentary structure.**

Proglacial lakes Lakes fed by glacial meltwater which accumulated behind dams of ice or glacial deposits.

Prograde metamorphism Metamorphism that occurs as heat and pressure increase toward their maximum values for a given metamorphic event.

Progressive metamorphism Gradual readjustment of a rock to continuously changing temperature, pressure, and chemical conditions.

Proton A subatomic particle found in the nucleus with 1832 times the mass of an electron, and a positive charge.

Pumice A vesicular (porous), glassy, light-colored igneous rock with a composition similar to that of granite.

Pyroclastic texture Igneous texture of rock made of broken particles ejected into the air from a volcano. (See also **tephra.**)

Pyroxene A group of minerals containing single chains of silicon-oxygen tetrahedra.

Quad One quadrillion (10^{15}) Btu's.

Quartz arenite A sandstone composed of quartz grains and crystalline cement with less than 15 percent matrix. (See also **arenite.**)

Quartz wacke A sandstone composed of quartz grains and more than 15% matrix. (See also **wacke.**)

Quartzite Metamorphosed quartz sandstone.

Quickclay Clay deposits that undergo rapid liquefaction.

Radial pattern A stream network in which streams either drain away in all directions from a central high point or drain into a low point from all directions.

Radiant energy Energy forms transmitted as electromagnetic waves, e.g., x-rays or light.

Radiation Transfer of heat energy by conversion to a form of radiant energy (commonly infrared energy).

Radioactivity The spontaneous nuclear breakdown of an atom.

Radiogenic A term used to describe the product of a nuclear reaction.

Radiometric age The age determined for a mineral rock or organic material based on an analysis of the radioactive elements present.

Rays Ejecta deposits extending out radially from meteorite impact craters.

Rayleigh wave A seismic surface wave propagated by a retrograde circular orbital motion of rock particles.

Reaction rim A coating of one mineral surrounding another formed by discontinuous reactions during magmatic crystallization.

Recessional moraine An end moraine which is deposited by a glacier that melts back from a terminal moraine.

Recharge The natural or artificial addition of water to the groundwater system.

Rectangular pattern A stream network in which tributaries flow into main streams at right angles. Streamflow generally is controlled by fractures in the bedrock.

Recumbent fold A fold whose axial plane is horizontal.

Reef A wave-resistant organic structure.

Regelation Melting of glacial ice under excess pressure to produce a basal meltwater layer that refreezes when pressure is reduced.

Regional metamorphism Widespread metamorphism caused by directed and lithostatic pressures, heat, and chemically active fluids.

Regional section A series of rocks that record the history of deposition over a regional area.

Regolith The unconsolidated layer of soil, sediment, and rock debris that partially or completely covers bedrock.

Rejuvenated stream A stream that begins downcutting actively after prolonged lateral cutting as a result of a lowering of its base level.

Relative age The age of a rock or a geologic event in comparison (i.e., older or younger) to that of others.

Relict texture A feature in metamorphic rocks inherited from the original rocks.

Relief The relative difference in elevation between the highest and lowest points in an area.

Remanent magnetism Traces of ancient earth magnetism preserved in a rock or mineral.

Renewable resource   A resource for which the processes of formation regenerate the material and thus make it continually available.

Replacement   A process of fossilization in which the hard parts of a plant or animal are replaced by minerals precipitating out from solutions circulating through the enclosing sediment.

Reserve   The portion of a resource that can be extracted profitably and legally with current technology.

Residual regolith   Regolith formed by the physical or chemical breakdown of the underlying materials.

Resisting forces   Those factors which act against the downslope movement of materials.

Resource   A geologic process or material that is useful to humans and is available for use.

Retrograde metamorphism   Metamorphism that occurs as temperatures and pressures decrease after the highest intensities attained during a metamorphic event.

Reverse fault   A dip-slip fault in which the footwall block has moved downwards relative to the hanging wall block.

Rhombohedral packing   The type of packing which is the tightest and has the least pore space.

Rhyolite   A fine-grained felsic igneous rock with the same composition as granite.

Rhythmic bedding   Stratification consisting of an alternation of two different types of sediment, implying a sequential alternation of two depositional conditions.

Riffles   Shallow barlike areas located at regular distances along straight stream channels.

Right-lateral fault   A strike-slip fault in which displacement of one block has been to the right as viewed from the other block.

Rill   A shallow, temporary channel formed in an area during early stages of stream system development.

Rilles   Linear, steep-walled depressions on the lunar maria and in the lunar highlands.

Rip currents   Currents moving seaward through the surf zone.

Ripple marks   Regularly spaced small ridges of sand resembling ripples of water and formed on the bedding surface of a sediment.

Rock   An aggregate of minerals or of many grains of a single mineral.

Rockburst   The violent discharge of rock into a tunnel or excavation site caused by the sudden release of large amounts of strain.

Rockfall   The fall of rock particles from the face of a cliff.

Rock flour   The very fine grained sediment resulting from abrasion at the base of a glacier.

Rock glacier   A lobe-shaped accumulation of boulder rubble formed by frost wedging in a periglacial climate.

Rockslide   Downslope movement of rock along planar surfaces.

Root wedging   A process of physical weathering in which the growth of plant roots into fractures in a rock dislodges pieces of the rock.

Roundness   The degree to which clastic sedimentary particles develop rounded surfaces.

Ruby   The red gem-quality variety of corundum.

S wave   A transverse seismic body wave that is propagated more slowly than P waves and cannot pass through a liquid.

Salinity   The concentration, in parts per thousand, of dissolved solids in seawater.

Saltation   The process by which particles are picked up and transported by wind or water, with intermittent contact with the bottom.

Salt dome   An upward projection of halite, formed by extreme ductile deformation, that arches and sometimes penetrates overlying sedimentary rocks.

Saltwater encroachment   The infiltration of seawater into coastal aquifers.

Sand   Sedimentary particles with diameters between $\frac{1}{16}$ and 2 mm.

Sandstone   Sedimentary rock made up of sand-sized particles.

Sapphire   A gem variety of corundum.

Satellites   Small bodies which orbit planets.

Saturation zone   That portion of the regolith and underlying rock in which all pore spaces are filled with water.

Schist   Strongly foliated metamorphic rock composed mainly of platy minerals.

Scoria   A vesicular (porous) dark-colored igneous rock with a composition similar to that of gabbro.

Sea arch   An archlike opening along a rocky coastline resulting from wave erosion cutting through opposite sides of a rocky headland.

Sea-floor spreading   In the plate tectonic model, the opening and enlargement of an ocean by continuous rifting at ocean ridges, accompanied by extrusion of basalt.

Seamounts   Submerged volcanoes that rise at least 1000 m above the sea floor.

Sea stack   A mass of rock isolated from others by wave erosion.

Secondary mineral   A new mineral formed by weathering processes.

Secondary porosity   Porosity that develops after rock formation.

Secondary recovery The pumping of water under pressure into reservoir rocks in order to free more oil.

Secondary sedimentary structure See **sedimentary structures.**

Secular variation Long-term variation in the strength of the earth's magnetic field.

Sediment Unconsolidated debris formed at the earth's surface. (See **clastic, biogenic,** and **chemical sediments.**)

Sedimentary environment An area with distinctive physical, chemical, and biologic conditions in which a unique type of sediment accumulates.

Sedimentary rocks Rocks formed at the earth's surface by deposition and lithification of particles eroded from previously existing rocks, biogenic materials, or dissolved ions.

Sedimentary structures A structure in a sedimentary rock formed either contemporaneously with deposition (a **primary sedimentary structure**) or subsequent to deposition (a **secondary sedimentary structure**).

Seismic methods The use of powerful sound sources to determine the structure of underlying sediments and rocks in the ocean basins or on land.

Seismic waves Pulses of energy released during an earthquake.

Seismogram The printed record of ground motion produced by a seismograph.

Seismograph An instrument that detects and records ground vibrations due to seismic waves.

Seismology A branch of geology that studies earthquakes and the passage of earthquake energy through the earth.

Semiarid area An area that receives between 25 and 50 cm of precipitation per year.

Series decay A sequence of nuclear reactions in which the parent element breaks down into a radioactive daughter which in turn undergoes nuclear reaction until a nonradioactive daughter is produced.

Settling velocity The speed at which a particle of a given size, density, and shape settles through still water or air.

Shadow zones Regions on the earth's surface that do not receive direct P-wave or S-wave transmission after an earthquake.

Shale Sedimentary rock composed of clay-sized particles.

Shards Small fragments of volcanic glass hurled into the air during volcanic eruptions.

Shear A type of stress produced by forces that do not oppose one another at right angles or are of unequal intensity in opposing directions.

Sheetflood Movement of water along broad expanses of the ground surface rather than in channels.

Shelf break The seaward edge of the continental shelf.

Shelf valleys Submarine channels cut into the continental shelf by stream erosion during a period when sea level was lower than it is today.

Shield volcano A volcano with very gentle slopes composed of lava.

Sial The uppermost layer of the continental crust, composed of rocks and minerals with the average composition of granite.

Silicate minerals Minerals that contain silicon and oxygen.

Siliceous ooze Biogenic sediment composed of the skeletons of organisms made of silica.

Silicified fossil Any fossil in which the original calcium carbonate of the shell has been replaced by silica.

Sill A tabular mass of intrusive igneous rock that parallels structures in its host rock.

Silt Sedimentary particles with a diameter between $\frac{1}{256}$ and $\frac{1}{16}$ mm.

Siltstone Sedimentary rock composed of silt-sized particles.

Sima The lower layer of continental crust (and the only layer of oceanic crust) composed of rocks with the average composition of basalt.

Simple melting See **congruent melting.**

Sinkhole A circular depression in the land surface caused by solution of the underlying rock and collapse of the land surface.

Sinuosity A measure of the degree of meandering of a stream.

Slate Fine-grained, foliated low-grade metamorphic rock.

Slickensides Polished and finely grooved surfaces caused by movement of pulverized material in fault planes. The grooves caused by relatively large grains indicate the sense of displacement (dip-slip, and so on).

Slide The mass wasting process in which rock or sediment moves downslope along a planar surface.

Slip face The steep, downwind side of a dune.

Slope failure The downslope movement of materials underlying a slope.

Slump A type of slide in which the sediments or rocks move downslope along a curved surface.

Snowfield An area in which winter snowfall exceeds summer melting so that snow covers the ground throughout the year.

Snow line In mountainous regions, the lowest altitude at which snow covers the ground throughout the year.

Soil Surficial material that has been weathered sufficiently to be able to support the growth of plants.

Soil horizons   Layers in the soil profile that are distinguished by color, organic content, mineralogy, or grain size as a result of weathering of the parent material.

Soil water zone   The thin belt of partially saturated pores within the root system.

Solar wind   Electrically charged particles that are generated by nuclear reactions in the sun and stream outward throughout the solar system.

Solid-solution series   A group of minerals that represent a mixture (by ionic substitution) of two or more end members, e.g., olivines, $(Mg,Fe)_2SiO_4$, a solid-solution series between the end members forsterite, Mg_2SiO_4, and fayalite, Fe_2SiO_4.

Solar system   A family of objects held in orbit around a star by gravitational attraction.

Solifluction   Slow downslope movement of water-saturated surface sediment; most common in areas where the underlying sediments are frozen.

Sorting   The degree of uniformity of size among sedimentary particles.   Particles of the same size are well sorted; if the particles cover a range of sizes, the sediment is poorly sorted.

Spheroidal weathering   The formation of rounded boulders as a result of chemical weathering during which the outermost layer of rock decomposes, crumbles, and falls off.

Spreading center   An ocean ridge segment which, according to the plate tectonic model, is the site of active rifting and sea-floor spreading.

Spit   An elongate and commonly curved bar of coarse material built by longshore drift from headlands or sandy islands.

Spring   A natural discharge of groundwater onto the land surface, or as a discrete flow into a body of surface water.

Stalactites   Projections extending downward from the roofs of caves, formed by evaporation of groundwater and precipitation of dissolved ions.

Stalagmites   Projections extending upward from the floors of caves, formed by evaporation of groundwater and precipitation of dissolved ions.

Steam tubes   Elongate cavities in volcanic rock that trace the path of gases that escaped from the lava.

Stock   A small, irregular, intrusive igneous body.

Stony-iron meteorite   A meteorite composed of an intergrowth of iron-nickel alloy and silicate minerals.

Storm surge   The elevation of the sea surface, resulting from the low pressure and high winds associated with a storm.

Strain   The amount a rock changes (in shape or volume) in response to deforming forces.

Stratification   A sequence of layers of sediment marked by differences in particle size, color, or other characteristics.

Stratified drift   The stratified deposits of glacial meltwater streams.

Stratovolcano   See **composite cone.**

Streak   The color of a mineral's powder.

Stream capture   The intersection and incorporation of one stream's drainage by another stream which has eroded headward through the drainage divide separating the two streams.

Stream hydrograph   A graph which shows the changes in the discharge of a stream with time.

Stream network   Large numbers of interconnected streams.

Stream terraces   Flat erosional remnants of valley fill sediment left behind when a stream begins downcutting.

Stress   The interaction between deforming forces and the internal cohesive forces of a rock; a measure of the intensity of deformation.

Strike   The compass orientation of a horizontal line drawn on any planar geologic structural feature (e.g., N25°E).   Strike is used with **dip** to describe the three-dimensional attitude of geologic structures.

Strike-slip fault   A fault in which displacement has been essentially horizontal, parallel to the strike of the fault plane.

Subduction   A process in which one lithosphere plate is thrust downward into the asthenosphere beneath a second plate.

Subduction zone   A region in which one lithosphere plate is thrust into the asthenosphere beneath another plate.

Sublimation   The direct transformation of a solid, such as snow (or ice), into vapor without intermediate melting.

Submarine canyons   Deep V-shaped valleys carved into the continental slope by underwater erosion.

Subsidence   Sinking of the ground surface owing to removal of large amounts of water or petroleum from the pores of underlying sediment or rocks.

Superposition   A principle that states that unless there has been tectonic overturning, the oldest rock in a sequence of sedimentary rocks will be at the bottom, the youngest at the top.

Surface currents   Circulation of ocean water caused by the prevailing wind system in a region.

Surface creep   The lateral movement of sand particles along a surface resulting from the impact of wind-transported particles.   Surface creep is the main mechanism which moves sand grains up the windward face of a dune.

Surface tension   A force acting parallel to the water

surface which enables fluids to remain as thin films within the pores of rocks and sediments.

Surface waves Seismic waves that transmit energy along the earth's surface. (See also **Love waves** and **Rayleigh waves.**)

Surf zone The portion of the nearshore zone where oversteepened waves topple over and crash onto the beach.

Surging glacier A glacier which moves with a high velocity (several kilometers a year rather than several meters) and whose rapid movement is short-lived (from a few months to a few years).

Suspended particles Those sedimentary particles which are carried within the body of a transporting agent (wind, water, or ice).

Swelling clay Mixtures of clay minerals that can expand to several times their normal volume by absorbing water.

Swells Regular, long-period, and long-wavelength waves which have moved away from the area in which they were generated.

Syncline A fold in which the limbs dip toward one another so that young rocks are found in the axial region of the fold.

Synfuels Synthetically produced energy equivalents of natural gas and petroleum.

Talus The accumulation of debris by rockfall at the base of a cliff.

Tar sands See **oil sands.**

Tension A type of stress produced when deforming forces pull a rock apart.

Tephra All solid particles formed from material ejected into the atmosphere during volcanic eruptions (synonym, **pyroclastic** debris).

Terminal moraine A deposit formed at the maximal extent of a valley or continental glacier.

Terrestrial planets The four inner planets (including Earth) that are relatively small but denser than the outer, larger Jovian planets. They are composed of silicate minerals and iron-nickel alloy (Mercury, Venus, Earth, and Mars).

Terrigenous sediment Sedimentary particles derived from landmasses.

Tertiary recovery The pumping of water and chemicals which reduce surface tension into reservoirs to increase oil recovery.

Texture A property of a rock encompassing its grain size, grain shape, and the manner in which the individual mineral grains (or crystals) are related to each other.

Thalweg A line connecting the deepest points along a stream channel.

Thermal metamorphism See **contact metamorphism.**

Thermal pollution Environmental problems resulting from the return of heated water to a body of surface water or groundwater.

Thermal springs Springs that bring heated water to the surface.

Thermoremanent magnetism Traces of ancient earth magnetism recorded by the minerals of a cooling igneous rock as they pass below their Curie points.

Thin section An extremely thin slice of rock mounted on a glass slide so that it can be studied with a microscope.

Threshold velocity The lowest velocity at which sand grains begin to move either in water or in air.

Thrust fault A gently dipping form of reverse fault with extensive horizontal displacement at high angles to the strike of the fault plane.

Tidal flats Lower portions of wetlands consisting of nearly horizontal and generally unvegetated expanses of fine-grained sediments which are covered completely at high tide and emerge at low tide.

Tidal inlets Openings in the shoreline through which estuarine and oceanic waters move back and forth during ebb and flood tides.

Tidal marsh Higher portions of wetlands covered by salt-tolerant vegetation and partially or completely covered during high tide.

Tidal wetlands Portions of the coastal zone which are covered partially or completely by water during part of the tidal cycle.

Tight fold A fold with a small angle between the two limbs.

Till Unstratified drift deposited by actively moving glacial ice.

Tillite An unstratified and poorly sorted rock of glacial origin.

Titus-Bode law A statement that the spacing of the planets is regular, with each planet's orbit being 75 percent larger than that of the next inner one.

Tombolo Bars of sediment which connect offshore islands with the mainland or with each other.

Top and bottom structures Sedimentary structures that can be used to determine the original tops and bottoms of beds.

Topographic desert An area that is arid because it is separated by mountain ranges from sources of moisture.

Topset beds The thin and horizontally layered sandy sediments covering the top surface of a delta.

Trace fossils Sedimentary structures resulting from the life activities of organisms. Some common trace fossils are tracks on the surface of a bed and burrows through the bed.

Tracks (within crystals)   Linear areas formed by the passage of high-energy particles resulting from the decay of radioactive isotopes such as uranium 238.

Track density (within crystals)   The number of tracks per unit area.

Traction   Transportation of particles by wind or water, where the particles move in constant contact with the bottom of the stream or the ground surface.

Transform fault   (*a*) A type of strike-slip fault that appears to offset ocean ridge segments, island arcs, or other transform faults.   Actual displacement is far more complex than in other strike-slip faults.   (*b*) A neutral plate boundary on which oceanic crust is neither created nor destroyed.

Transported regolith   Sediment deposited on bedrock by water, wind, ice, or mass wasting.

Transverse dune   A sand dune that forms at right angles to the dominant wind direction.

Transverse wave   A seismic wave (S wave) propagated by the straight-line, back-and-forth motion of particles in directions at right angles to the direction of wave propagation.

Travel-time curves   Charts plotting the relative travel times of the different seismic waves against distance from earthquake epicenters.   They are used to calculate the location of epicenters.

Travertine   A banded form of calcium carbonate characteristic of cavern floors.

Tree-ring dating   Dates determined from counting the number of annual rings in trees.

Trellis pattern   A stream network composed of long major streams and short tributaries that join the main stream at high angles; these patterns developed best in areas of folded rocks.

Trunk glacier   A type of mountain glacier formed by the merging of two smaller valley glaciers.

Tsunami   A seismic sea wave (often incorrectly called a tidal wave) generated when there is vertical displacement of the sea floor.

Tuff   A volcanic rock formed from aggregates of shards.

Turbidite   A graded bed deposited by a turbidity current.

Turbidity current   Submarine downslope movement of a dense water-sediment mixture.

Turbulent flow   A type of fluid flow in which there is both horizontal and vertical movement.

U-shaped valley   A valley with a U-shaped cross section resulting from erosion by a valley glacier.

Ultramafic rocks   Rocks composed almost entirely of ferromagnesian minerals, with little or no feldspar.

Ultramylonite   A glassy metamorphic rock formed by extreme grinding and pulverizing of grains during dynamic metamorphism.

Unconfined aquifer   An aquifer in which the water is at atmospheric pressure.

Unconformity   A substantial gap in the local rock record, indicating a lapse in time during which rocks either were not deposited or were removed later by erosion.

Uniformitarianism   A conceptual view of earth history that nolds that processes active on and in the earth today were active in the geologic past and thus can be used to interpret the features found in rocks.

Unstable slope   A slope along which there is a potential for mass movement.

Unstratified drift   A glacial deposit which lacks stratification and which generally is composed of poorly sorted and angular particles.

Upright fold   A fold with a vertical axial plane.

Valley glacier   A glacier that forms in mountainous areas and extends like a tongue down mountain valleys.

Valley train   A long narrow deposit of outwash deposited by a retreating valley glacier.

Van Allen Belt   The inner part of the magnetosphere.

Van der Waals' bonding   A weak electrostatic attraction between atoms caused by positioning of their orbiting electrons.

Varve   A type of rhythmic stratification common in the deposits of glacial meltwater lakes which consists of a yearly couplet of a coarser-grained (summer) bed and a finer-grained (winter) bed.

Ventifacts   Faceted rock fragments shaped and smoothed by sedimentary particles carried by the wind.

Vertical exaggeration   A deliberate increase in the vertical scale of a topographic section, relative to the horizontal scale, in order to show subdued features more clearly.   The vertical exaggeration (V.E.) can be calculated by dividing the horizontal scale by the vertical scale (both in the same units) and is expressed in multiples (2X, 4X, and so on).

Vertical tectonics   A theory that states that mountain-building events can be explained by dominantly vertical movements of deep mantle materials rather than the lateral movements involved in plate tectonics.

Viscosity   A measurement of the sluggishness with which a fluid moves.

Vitrophyre   An igneous rock containing both glass and minerals.

Volcanic rocks   Igneous rocks formed by solidification of magma at the earth's surface (synonym, **extrusive rocks**).

Volcanoclastic debris Sediment derived by erosion of a volcano.

Vulcanists People who believed that all rocks formed by solidification of magma.

Wacke Sedimentary rock composed of sand-sized particles in a finer-grained matrix.

Water masses Large volumes of ocean water with distinctive characteristics which move through the ocean basins.

Water table The surface that separates the groundwater aeration zone from the saturation zone. Below the water table, all pores are filled with water.

Wave base The downward limit of orbital water movement in a wave. Wave base occurs at a depth equal to about one-half the wavelength.

Wave-built terrace A gently sloping depositional terrace extending seaward from a coastal cliff.

Wave crest The highest part of a wave.

Wave-cut bench A gently sloping erosional platform extending toward the sea from the base of a coastal cliff.

Wave-cut notch An indentation cut at the base of a coastal cliff by wave erosion and abrasion from wave-transported debris.

Wave fronts The crest lines of moving waves.

Wave height The vertical distance between the wave crest and wave trough.

Wavelength The horizontal distance between a similar point on two waves, such as between two wave crests. Wavelength is represented by the Greek letter lambda (λ).

Wave normal A line at right angles to the wave crest which indicates the direction of wave movement.

Wave period The time required for two wave crests to pass a given point.

Wave refraction The bending of wave fronts as they enter shallow water and pass over submerged topography. Wave refraction concentrates wave energy on headlands, tending to smooth out the shoreline.

Wave trough The lowest part of a wave.

Weathering Physical and chemical changes that occur in sediments and rocks when they are exposed to the atmosphere and biosphere.

Wentworth scale A series of descriptive terms used to describe sizes of clastic sediment particles, e.g., pebble, sand, and silt.

Whole-rock date Radiometric age determined by the parent to daughter ratio in an entire rock rather than in a single mineral.

Wind shadows Zones of quiet air formed in front of and behind an object which obstructs the airflow.

Xenolith An inclusion of host rock in an igneous rock.

Index

Page numbers in *italic* indicate illustrations.